FRACTURE MECHANICS OF CONCRETE

FRACTURE MECHANICS OF CONCRETE: APPLICATIONS OF FRACTURE MECHANICS TO CONCRETE, ROCK, AND OTHER QUASI-BRITTLE MATERIALS

Surendra P. Shah

Walter P. Murphy Professor
Department of Civil Engineering
Northwestern University

Stuart E. Swartz

Professor and Head
Department of Civil Engineering
Kansas State University

Chengsheng Ouyang

Engineer
Office of Materials
Iowa Department of Transportation

A WILEY-INTERSCIENCE PUBLICATION

JOHN WILEY & SONS, INC.

New York / Chichester / Brisbane / Toronto / Singapore

This text is printed on acid-free paper.

Library of Congress Cataloging in Publication Data:

Shah, S. P. (Surendra P.)
Fracture mechanics of concrete : applications of fracture mechanics to concrete, rock and other quasi-brittle materials / Surendra P. Shah, Stuart E. Swartz, Chengsheng Ouyang.
p. cm.
Includes index.
ISBN 0-471-30311-9 (alk. paper)
1. Concrete—Fracture. 2. Rocks—Fracture. 3. Fracture mechanics. I. Swartz, Stuart E. II. Ouyang, Chengsheng. III. Title.
TA440.S465 1995
620.1'366—dc20 95-24676

Printed in the United States of America

10 9 8 7 6 5 4 3 2 1

CONTENTS

PREFACE

There has been literally an explosion of research activity and interest in the last few years of those in the academic and research community who see the advantages of fracture mechanics in describing material and structural behavior leading to failure. However, such developments have not taken place in a pedagogical sense either in academia or in practice, because suitable educational materials have not been available. This book seeks to fill this void and is intended to be a textbook for graduate students in civil engineering, engineering mechanics, and allied disciplines. It may be used as a stand-alone text for a 3-credit-hour course or as supplementary material for a course in advanced concrete theory. It is also a useful and comprehensive reference work on this subject for the practicing engineer. It offers, for the first time, a compendium of material distilled from over 20 years of research by the authors and others dealing with the applications of fracture mechanics concepts to cracking, damage, and fracture of plain and reinforced concrete structures and rock masses.

The book is divided into two main parts. The first part, comprising Chapters 1–8, is the exposition and development of the fundamentals of fracture mechanics of concrete and rock. The second part, Chapters 9–12, presents applications of this fundamental theory. These chapters are developed on an independent basis. After finishing the first part, readers can selectively review topics in the second part on an individual chapter basis.

The need for a fracture mechanics approach is demonstrated through consideration of experimental evidence of damage localization and size effects for a number of important structural design cases. Following this, a brief background of the equations for stress states due to mode I deformation and crack propagation in terms of linear elastic frcture mechanics (LEFM) is presented. The concepts of LEFM fracture toughness and critical energy release rates are introduced. A listing of commonly used formulas for stress intensity and examples of failures and dependency on size of certain material properties are presented.

The limitations of LEFM for quasi-brittle materials are demonstrated through a presentation of experimentally observed fracture mechanisms and models that are proposed to describe these. These result in concepts of nonlinear fracture mechanics (NLFM), which are described in detail in Chapters 3–5. Testing methods for obtaining the necessary mode I fracture properties to implement these models are presented in Chapter 6. These include methods proposed by RILEM (the International Union of Testing and Research Laboratories for Materials and Structures) for testing concrete.

This is followed by a discussion of the concept of *R*-curves and how these may be applied to describe cracking and fracture of quasi-brittle materials subjected to mode I deformation at the crack location. Fracture when other deformation modes are present is described in Chapter 8. The discussion is primarily for planar problems and includes LEFM and NLFM approaches for mode II and mixed-mode behavior. Applications to concrete and rock are presented that include a variety of proposed testing methods.

The next chapter describes applications of fracture mechanics to reinforced concrete structures. The resulting changes to current design practice and applicable building codes in such applications as shear, torsion, crack control, bar pullout, and minimum reinforcement are presented in detail with examples.

The remaining chapters deal with advanced topics that are presently at the forefront of research activities. These include applications at the aggregate/grain/paste level and correlations to macrobehavior; interactions between theories of damage mechanics and fracture mechanics; and applications to advanced cement-based composite materials. The final sections deal with time–rate–fatigue effects on crack propagation and fracture.

SURENDRA P. SHAH
STUART E. SWARTZ
CHENGSHENG OUYANG

LIST OF SYMBOLS

a	Crack length
a_c	Critical crack length
a_{cf}	Critical crack length for finite-size specimens
$a_{c\infty}$	Critical crack length for infinitely large specimens
a_d	Interfacial debonding crack length between a fiber and a matrix
a_{dc}	Critical interfacial debonding crack length
a_{d0}	Initial debonding length between a fiber and a matrix
a_{dc}^*	Interfacial debonding crack length of reference pullout specimen
a_0	Initial crack length
a_1	Long radius of an ellipse
a_2	Short radius of an ellipse
a_{11}, a_{12}, a_{22}, a_{33}	Coefficients of stress intensity factors for mixed-mode fracture
A	Gross sectional area of concrete
A^*	Net sectional area of concrete
A_B	Statistical coefficient in size effect model
A_{eff}	Effective concrete area surrounding one steel bar
A^{IF}	Coefficient for mixed-mode fracture
$A_{\min}$	Nondimensional stress intensity factor
A_{pl}	Quantity defined by ASTM E 1152
A_s	Area of reinforcing steel
A_0	A constant
A_1, A_2	Areas under load–LPD curves for specimens 1 and 2, respectively
b	Depth of beam, or width of tensile specimen
b_c	Thickness of concrete cover

b_0	Coefficient in crack band model
b_1	Effective depth of beam
B	Coefficient in crack band model
B^{IF}	Coefficient for mixed-mode fracture
B_s	Brittleness number
B_1, B_2, B_3, B_4	Coefficients
c	Crack extension
c_f	Crack extension for infinitely large specimen based on size effect model
c_F	Crack extension at $w = w_c$
c_a	Coefficient representing different types of structures
c_0, c_1, c_2	Coefficients
C	Compliance
C_B	Statistical coefficient in size effect model
C_F	Coefficient
C_i	Initial compliance for load–CMOD curve
C^n_{ijkl}	Elastic compliance tensor for nth strain increment
C^*_{ijkl}	Elastic compliance tensor
C_{int}	Compliance for pullout load and slip curve
C_{LL}	Compliance for load line displacement given by ASTM E 11-52
C_u	Unloading compliance at peak load for load–CMOD curve
C_0	Compliance of specimen with initial crack
C_1, C_2, C_3	Coefficients
CMD	Crack mouth displacement
CMOD	Crack mouth opening displacement
CMOD_e	Elastic part of crack mouth opening displacement
CMOD_ρ	Inelastic part of crack mouth opening displacement
CMOD_c	Critical crack mouth opening displacement
CMOD^e_c, CMOD^p_c	Elastic and inelastic components of CMOD_c
COD	Crack opening displacement
CTD	Crack tip displacement
CTD_c	Critical crack tip displacement
CTOD	Crack tip opening displacement
CTOD_e	Elastic part of crack tip opening displacement

CTOD_p	Inelastic part of crack tip opening displacement
CTOD_c	Critical crack tip opening displacement
CTOD_c^e, CMOD_c^p	Elastic and inelastic components of CTOD_c
CTSD	Crack tip slip displacement
d	Debonded length
d_{em}	Embedded length of anchor bolt
d_H	Distance between centers of two colinear slits
d_i, d_o	Inside and outside diameters of pipe, respectively
d_{io}	Pipe diameter defined by middle surface of wall
d_m	Mean distance between bridges in fracture process zone
d_n	Constants in R-curve, $n = 1, 2, 3, \ldots$
d_s	Distance between bottom surface of concrete and centroid of steel bar
d_t	Representative dimension of beam under torsion
d_0	A constant
d_1	Thickness of cementitious matrix
D	Characteristic size of structure (depth of beam, width of plate, diameter of chevron bend specimen, or diameter of short–rod specimen
D^{IF}	Coefficient for mixed-mode fracture
D_{min}	Minimum size of specimen
D_0	A constant
E	Modulus of elasticity of concrete
E_d	Modulus of elasticity under fatigue load
E_{dL}	Linear limit of E_d under fatigue load
E_{d0}	Reference value of E_d under fatigue load
E_n	Intrinsic modulus of elasticity for material around crack tip
E_f	Modulus of elasticity of fiber
E_s	Modulus of elasticity of steel
E_t	Strain softening modulus
E^*	Modulus of elasticity for undamaged material
f	Coefficient of friction
f_c	Compressive strength of concrete
f_{cc}	Maximum compressive stress at top of beam
f_{ca}	Allowable stress for concrete
f_{cs}	Allowable stress for steel
f_{ct}	Maximum tensile stress at bottom of beam

f'_c	Uniaxial compressive strength of concrete
f_{cr}	Nominal flexural strength of concrete pipe
f_f	Fracture strength of concrete beam for determining minimum reinforcement ratio
$f_{\mathrm{I}}, f_{\mathrm{II}}$	Shape functions for mixed-mode fracture
$f_N(n_F)$	Weibull probability density function for fatigue
f_r	Modulus of rupture of concrete
$f_r^{\max}$	Maximum stress in fatigue cycling
f'_r	Static strength of material
f_s	Tensile stress of steel in beam
$f_{st}(t)$	Time-decaying function
f_t	Tensile strength of concrete
f_{tf}	Tensile strength for finite-size specimen
f_y	Yielding stress of steel
$f_\sigma(t)$	Time-varying function
f_1, f_2	Geometric Factors in R-curve
F	Work done by applied force or fictitious force
F_c	Resultant of compressive stress in concrete beam
F_D	Safety factor
$F_N(n_F)$	Cumulative distribution function for fatigue
$F_{\max}$	Failure load, ISRM method
F_s	Total steel force in beam
$F_{s,\max}$	Maximum steel force at midspan of beam
F_v, F_{sp}	Vertical and horizontal (splitting) components of wedge force
F_1, F_{1r}, F_2, F_{2r}	Applied forces for mixed-mode fracture (Fig. 8-36)
$g(a/b)$	Geometric function for strain energy release rate
g_F	Geometric function for mixed-mode fracture
g_0	Coefficient for ISRM method
$g_1(a/b)$	Geometric function for stress intensity factor
$g_2(a/b)$	Geometric function for crack mouth opening displacement
$g_3(a/b)$	Geometric function for crack tip opening displacement
$g_4(a/b)$	Geometric function in effective crack model
$g'(\alpha_0)$	Value of first derivative of $g(a/b)$ with respect to a/b at $a = a_0$
g_{10}	$g_1(a_0/b)$

g'_{10}	$g'(\alpha_0)$
G_c	Critical strain energy release rate for linear elastic materials
G_d	Fracture energy per unit area
G_f	Critical strain energy release rate for infinitely large specimen based on size effect model
G_F	Critical strain energy release rate based on fictitious crack model
G_q	Mode I strain energy release rate for quasi-brittle materials
G_{I}	Strain energy release rate for mode I crack
G_{int}	Strain energy release rate for interfacial crack (debonding)
G_{intb}	Additional strain energy release rate due to bending of fiber for interfacial crack
$G_{\text{I}c}$	Critical strain energy release rate for mode I crack
G_{II}	Strain energy release rate for mode II crack
G_{III}	Strain energy release rate for mode III crack
$G^s_{\text{I}c}$	Critical strain energy release rate based on two-parameter fracture model
G_σ	Strain energy release rate due to closing pressure
h_c	Width of crack band
h_1	Distance between two midspan opposite forces for mixed-mode test
h_2, h_3	Distances defined in Fig. 8-21
H_0	A constant
I	Centroidal moment of inertia
I_{cr}	Cracked section moment of inertia for beam
I_c	Effective moment of inertia for beam
I_g	Moment of inertia of gross concrete section neglecting reinforcement
I_{ut}	Uncracked transformed moment of inertia for beam
$j(x)$	Variable coefficient for shearing
J	J-integral
J_{el}, J_{pl}	Elastic and plastic components of J-integral, respectively
k	Coefficient
k_c	Coefficient

k_d	Coefficient depending on maximum aggregate size
k_f	K_{min}/K_{max} for fatigue loading
k_F	$K_{II\,max}/K_{I\,max}$
k_{Fc}	K_{IIc}/K_{Ic}
k_n, k_s	Coefficients of normal and shear stress concentration
k_{int}	Initial slope between shear stress and slip
k_1, k_2, k_3, k_4, k_5, k_6	Coefficients
K_{CB}	Fracture toughness from chevron bend specimen
$K_{f,\,max}$	Maximum stress intensity factor during fatigue cracking
K_I	Mode I stress intensity factor
$K_I(t)$	Time-dependent mode I stress intensity factor
K_{II}	Mode II stress intensity factor
K_{III}	Mode III stress intensity factor
K_{Ic}	Critical stress intensity factor for mode I crack
K_f	Critical stress intenisty factor for infinitely large specimen, $(EG_f)^{1/2}$
K_{IIc}	Critical stress intensity factor for mode II crack
K_I^{ref}	Reference, time-independent stress intensity factor
K_{Ic}^e	Critical stress intensity factor based on effective crack models
K_{IA}	Stress intensity factor due to axial force
K_{Icd}	Dynamic, mode I critical stress intensity factor
K_{IF}	Stress intensity factor due to fictitious force
K_{IM}	Stress intensity factor due to bending moemnt
K_{IP}	Stress intensity factor due to applied load
K_I^{kink}	Mode I stress intensity factor for kink
K_{II}^{kink}	Mode II stress intensity factor for kink
K_I^p	Mode I stress intensity factor for concentrated load
K_I^q	Mode I stress intensity factor for distributed load
K_{Ic}^s	Critical stress intensity factor based on two-parameter fracture model
K_I^ρ	Mode I stress intensity factor due to applied load
K_I^σ	Mode I stress intensity factor due to yield stress
K_{ISRM}	Level I fracture toughness, ISRM method

K_{ISRM}^{c}	Level II fracture toughness, ISRM method
K_{I}^{*}, K_{II}^{*}	Dimensionless stress intensity factors
$K_{\mathrm{I}c}^{\mathrm{IF}}$	Mode I critical stress intensity factor of interface
$K_{\mathrm{II}c}^{\mathrm{IF}}$	Mode II critical stress intensity factor of interface
$K_{\mathrm{I}c}^{M}$	Mode I critical stress intensity factor of matrix
$K_{\mathrm{II}c}^{M}$	Mode II critical stress intensity factor of matrix
$K_{\mathrm{I\,max}}$, $K_{\mathrm{II\,max}}$	Maximum stress intensity factors for modes I and II, respectively
$K_{\max}$, $K_{\min}$	Maximum and minimum stress intensity factors under fatigue loading, respectively
K_m	Critical stress intensity factor of matrix
K_Q, P_Q	Stress intensity factor and load defined by ASTM E 399, respectively
K_R	Fracture toughness based on R-curve
K_{SR}	Fracture toughness from short-rod specimen
K_t	Stress concentration factor
K_σ	Stress intensity factor rate due to closing pressure
K_∞	Value of K_R at $w = w_c$
l	Half-length of colinear slit
l_e	Elastic component of load point displacement, ISRM method
l_h	Length of slit associated with circular hole
l_{th}	Material characteristic length based on fictitious crack model
l_0	Initial half-length of coline slit
l_1	Length of inclined crack in concrete
l_2	Length of branching crack in concrete
L	Length of specimen
L_b	Length of bend zone of fiber
L_f	Fiber embedded length
L_f^*	Fiber embedded length of reference specimen
L_1	Length of edge of aggregate
LPD	Load point displacement
m	A constant
m_b	Coefficient relating length of bend zone of fiber to slip
m_F	Coefficient in Paris's fatigue law
m_s	Ratio of principal stresses (σ_1/σ_2)
M	Moment

M_b	Moment for beam failure of pipe
M_c	Compliance of concrete with N cracks
M_f	Fracture moment of beam
M_n	Compliance of material around crack tip
M_{nf}	Nominal bending strength of beam
n	Number of tests, E_s/E
n_d	Parameter in Zhang et al.'s fatigue model
n_a	Coefficient relating maximum aggregate size to crack band width
n_F	Variable for Weibull probability density function
N	Number of cracks or number of fatigue cycles
N_f	Fatigue life
N_{I}, N_{II}	Coefficients for modes I and II stress intensity factors for split-tension cylinder, respectively
N_L	Number of fatigue cycles for linear limit of E_d
N_p	brittleness number defined by Carpinteri
N_1	Coefficient for F_s
N_2	Coefficient depending on r_s
p	Applied force
p_f	Closing pressure from fiber pullout and slip curve
p_F	Coefficient, ISRM method
P	Applied force
P_a	Applied load
P_c^*	Critical pullout load of reference specimen
P_m, P_s	Applied forces resisted by concrete and longitudinal steel, respectively
P_c	Critical (maximum) load
$P_{\max}$, $P_{\min}$	Maximum and minimum load in fatigue loading
P_M	Load defined by ASTM E 1152
P_x, P_y	Components of inclined fiber pullout force
P_w	Load due to self-weight
P_θ	Inclined fiber pullout force
q	Applied distributed force
$q_{\mathrm{I}}^{\mathrm{IF}}$	Compressive load due to mode I failure
$q_{\mathrm{II}}^{\mathrm{IF}}$	Compressive load due to mode II failure
q_0	Scale factor for applied distributed force
q_1, q_2	Coefficients
Q	Material length based on two-parameter fracture model

r	Distance to crack tip or radius of microvoid
r_a	Radius of model aggregate
r_b	Radius of split-tension specimen
r_c	Radius of a penny-shape crack or curvature of beam
r_d	Radius of semicircular fracture specimen
r_f	Fiber radius
r_h	Radius of circular hole representing material defect
r_e, r_i, r_1, r_2, r_3	Radius parameters for direct load core specimen of rock
r_{p0}, r_p	Sizes of plastic zone at crack tip
r_s	Radius of reinforcing steel bar
r_{sb}	Strength ratio
r_σ	Ratio of principal stresses (σ_{22}/σ_{11})
R	Fracture resistance curve
$R_{\mathrm{I}cf}$	Fracture toughness for finite-size specimen in R-curve
R_{int}	Fracture resistance curve for interface debonding
$R_{\mathrm{I}c}^s$	Fracture toughness for infinitely large specimen in R-curve
s	Sample standard deviation
s_c	Crack spacing
s_m	Coefficient, ISRM method
s_v	Shear span
s_x, s_Y, $s_{X\|Y}$	Statistical coefficients for size effect model
s_1, s_2	Creep constants
S	Span
S_d	Ratio of fatigue modulus of elasticity to its reference value, E_d/E_{d0}
S_{dL}	Linear limit of S_d
S_F	Stiffness, ISRM method
S_{F0}	Initial tangent slope of load cycling, ISRM method
S_G	Strain energy density factor
S_{Gc}	Critical strain energy density factor
S_p	Stress level for fatigue
t	Thickness
t_c	Clear cover of concrete

$\mathbf{T}$	Tension vector in J-integral, $=(T_x, T_y)$
T	Tensile force resultant for shearing
T_p	Torsion
T_s	Total sliding force on inclined crack
T_u	Ultimate torque
T_0	Loading capacity of beam under pure torsion
T_1	Prepeak tensile stress resultant of concrete
T_2	Postpeak tensile stress resultant of concrete
u, v	Displacement components in x and y directions, respectively
$\mathbf{u}$	Displacement vector in J-integral, $=(u_x, u_y)$
u_{c0}	Slip of concrete after interfacial debonding between steel and concrete
u_{int}	Slip between fiber and matrix
u_F	Parameter in cumulative fatigue function
u_{s0}	Elongation of steel after interfacial debonding between steel and concrete
U, U^*	Strain energy and complementary strain energy, respectively
U_a	Change in strain energy due to cracking
U_d, U_d^*	Densities of strain energy and complementary strain energy, respectively
U_N	Density of strain energy for solid with N cracks
U_s	Change in elastic surface energy due to formation of new crack surfaces
U_0	Strain energy of uncracked body
U_x	Function of unloading/reloading elastic compliance given by ASTM E 1152
v_c	Nominal shear strength of concrete
v_s	Total shear stress
v_u	Ultimate shear stress
V	Volume or shear force
V_a	Volume fraction of coarse aggregate in concrete
V_c	Shear strength provided by concrete
V_n	Nominal shear strength of beam
V_1, V_2	Components of shear force
w	Crack separation displacement or crack width
w_a	Average crack opening (separation) displacement

w_c	Crack separation displacement corresponding to $\sigma(w) = 0$
w_{eu}, w_L, w_{inc}	Crack opening variables in Hordijk and Reinhardt fatigue curve
w_{max}	Maximum crack width of beam
w_t	Crack separation displacement at initial crack tip
W	Energy for crack formation
W_t	Total fracture energy of beam for work-of-fracture method
W_0, W_1, W_2	Parts of total fracture energy W_t
x, y, z	Components in cartesian coordinate
X, Y	Components of body force
y_1	Distance from top of beam to neutral axis
Y_{IM}	Geometric function for K_{IM}
Y_{IA}	Geometric function for K_{IA}
z, $\bar{z}$	Complex number and its conjugate
Z	Complex stress function
Z_i, Z_r	Imaginary and real parts of complex stress function
Z_I	Complex function

Greek Letters

α	Ratio of critical crack length to initial crack length for infinitely large specimen
α_c	Ratio of critical crack length to beam depth
α_d	Parameter in fatigue model of Zhang et al.
α_F	Shape parameter for cumulative fatigue function
α_D	Coefficient, ISRM method
α_p	Angle for mixed-mode Iosipescu specimen
α_0	Ratio of initial crack length to beam depth
α_1, α_2	Parameters defining loading path for mixed-mode fracture
β	Inclined angle
β_b	Brittleness number based on size effect model
β_c	Coefficient, $=(b-\xi_0 y_1)/(b_1-\xi_0 y_1)$
β_n	Constants in R-curve, $n = 1, 2, 3, \ldots$
β_0	Constant in R-curve or ratio of initial crack length to critical crack length
γ_b	Debonding surface energy

γ_s	Fracture surface energy
$\gamma_1, \gamma_2, \gamma_3, \gamma_4$	Coefficients
$\gamma_{xy}, \gamma_{xz}, \gamma_{yz}$	Shear strain components in cartesian coordinate system
δ	Load line displacement
δ_c	Load point displacement at peak load
δ_f	Critical crack tip opening displacement for infinitely large specimen
δ_F	Load point displacement, ISRM method
δ_F^r	Residual displacement, ISRM method
δ_{ij}	Kronecker delta
δ_{pl}	Plastic part of load line displacement
δ_0	Load point displacement due to self-weight of beam
δ_1, δ_2	Normal and shear displacements for mixed-mode fracture, respectively
δa	Crack length correction
Δa	Crack extension
ΔK	Increment of stress intensity factor for fatigue loading
ΔU_a	Strain energy change due to creation of crack under constant load
$\Delta\delta_F$	Increment of load point displacement, ISRM method
ϵ^a	Effective strain
$\epsilon^{a\,\text{th}}$	Threshold effective strain
ϵ_c	Tensile strain of concrete
ϵ_{c0}	Tensile strain of concrete at cracked section
ϵ_{c1}	Tensile strain of concrete at uncracked section
ϵ^e	Equivalent strain
ϵ_{ij}	Strain tensor
ϵ_n	Compressive strain of concrete at top of beam
ϵ_s	Tensile strain of steel
ϵ_{s0}	Tensile strain of reinforcing steel at cracked section
ϵ_{s1}	Tensile strain of reinforcing steel at uncracked section
ϵ_t	Fatigue strain
$\epsilon_x, \epsilon_y, \epsilon_z$	Normal strain components in cartesian coordinate system

ϵ_θ	Circumferential strain
$\epsilon_{\theta c}$	Critical value of circumferential strain
ϵ_0	Strain of concrete corresponding to maximum compressive stress
ϵ_1	Compressive strain of concrete including contribution from crack width
$\epsilon_{11}^{\text{th}}$, $\epsilon_{22}^{\text{th}}$	Principal threshold strains of concrete subjected to uniaxial tension
η	Coefficient
θ	Angle
θ_a	Angle of aggregate face for propagating interface crack
θ_d	Direction for maximum value of K_{II}
θ_i	Inclined angle of fiber with respect to pullout force
θ_k	Direction of inclined crack
θ_0	Direction for minimum strain energy density factor
θ_1, θ_2	Angles
λ, ν	Lamé constants
λ^*, ν^*	Lamé constants for undamaged material
λ_F	Load line compliance, ISRM method
λ_h	Relative crack length
λ_p	Length of plastic zone
λ_0	Constant in size effect model for torsion
λ_1	Relative size of microvoid and microcrack
λ_{10}	Initial value of λ_1 (at peak load)
λ_{12}, λ_{21}	Constants
Λ	A constant
κ	Angle of kink
κ_p	Rotational factor for CTOD_p
μ	Poisson's ratio
μ^*	Poisson's ratio for undamaged material
μ_n	Intrinsic Poisson's ratio for material around crack tip
ξ	Coordinate along crack length, or a coefficient
$\xi_0 y_1$	Distance from compressive surface to neutral axis of cracked beam cross section
ξ_1, ξ_2	Coefficients

Π	Total potential energy
ρ	Reinforcement ratio
ρ_b	Balanced reinforcement ratio
$\rho_{\min}$	Minimum reinforcement ratio
ρ_p	Length of fracture process zone
σ	Nominal normal stress
σ^*	Net normal stress
$\sigma(w)$	Closing pressure in fracture process zone
σ_c, σ_f, σ_{Nc}	Critical nominal stress (nominal strength)
σ_{eu}, σ_L	Stress variables in Hordijk and Reinhardt fatigue curve
σ_{cr}, σ_{sr}	Tensile stresses in concrete and steel bars immediately after first crack forms, respectively
σ_{ij}	Stress tensor
σ_{ij}^*	Net stress tensor
$\sigma_{\max}$	Maximum stress
σ_{mc}	Critical tensile stress of matrix
σ_{Nr}, σ_{Nb}	Nominal failure stresses for ring failure and beam failure of concrete pipe, respectively
σ_r, σ_θ, $\tau_{r\theta}$	Stress components in two-dimensional polar coordinate system
σ_s	Stress in reinforcing bar
σ_x, σ_y, σ_z	Stress components in cartesian coordinate system
σ_w	Net bending stress
σ_{yf}	Yield stress of fiber
σ_{ys}	Yield stress of steel
σ_0	$(\sigma_r + \sigma_\theta)/3$
σ_1, σ_2	Principal stresses
σ_{11}, σ_{22}	Principal stresses
σ_β, τ_β	Normal and shear stress applied on inclined crack
τ	Shear stress
τ_F	Frictional stress between fiber and matrix
τ_{xy}, τ_{xz}, τ_{yz}	Shear stress components in cartesian coordinate system
τ_u	Ultimate shear stress resisted by concrete
τ_s	Frictional stress between concrete and steel bar (or fiber)
τ_0	Octahedral shear stress
$\phi(a, x)$	Green's function

ϕ_N	Strain energy of member with N cracks
ψ	Airy's stress function
ψ_a	Function in R-curve
φ_F	Fatigue damage function, current stiffness parameter
φ_{F0}	Original reference state for fatigue
$\varphi_1(z)$, $\varphi_2(z)$, $\varphi_3(z)$	Analytic functions
ω	Crack density
ω_A, ω_C, ω_E, ω_G, ω_X, $\omega_{X\|Y}$	Statistical coefficients for size effect model
ω_{int}	Parameter to quantify stiffness of weak interface layer
ω_p	Accumulated damage at peak load, $=2l_0/d_H$
ω_1	Damage variable of a plate with microvoid of radius r and microcrack of length of l, $=2(l+r)/d_H$
ω_{10}	Initial value of ω_1 (at peakload)
Ω	Scalar damage variable
Ω_F	Fatigue damage variable
Ω_0	Initial scalar damage variable
φ_c	Strain energy of concrete
φ_d	Total debonding energy
φ_s	Total sliding energy

1

FAILURE OF STRUCTURES

1.1 STRUCTURAL FAILURE BASED ON MATERIAL PERFORMANCE

The mechanical behavior of structures is greatly influenced by the materials used. Based on their tensile stress–deformation response, most engineering materials can be categorized into brittle, ductile, or quasi-brittle, as shown in Fig. 1-1. The stress suddenly drops to zero when a brittle material fractures (Fig. 1-1*a*), whereas stress stays a constant when a ductile material yields, shown in Fig. 1-1*b*. On the other hand, a quasi-brittle material is characterized by a gradually decreasing stress after the peak stress. The failure of structures greatly depends on the properties of the materials used in construction. This may be understood conceptually by considering an infinitely wide plate with an elliptical hole subjected to a far-field tensile stress, as in Fig. 1-2, where the material defect is modeled by the hole. The presence of the hole in the plate alters the stress distribution such that the maximum stress along the edge of the hole, $\sigma_{\max}$, is greater than the nominal far-field stress σ_N. From an elastic stress analysis, the relationship between $\sigma_{\max}$ and σ_N is obtained as

$$\sigma_{\max} = \left(1 + \frac{2a_1}{a_2}\right)\sigma_N = K_t\sigma_N \tag{1.1}$$

where a_1 and a_2 are the long and short radii of the ellipse, respectively, and K_t is the stress concentration factor. If the plate is made of a perfectly brittle material, it fails catastrophically whenever the value of $\sigma_{\max}$ reaches the material tensile strength f_t (Fig. 1-2*a*). On the other hand, if the plate is made of a ductile material, the applied tensile load can continuously increase after $\sigma_{\max}$

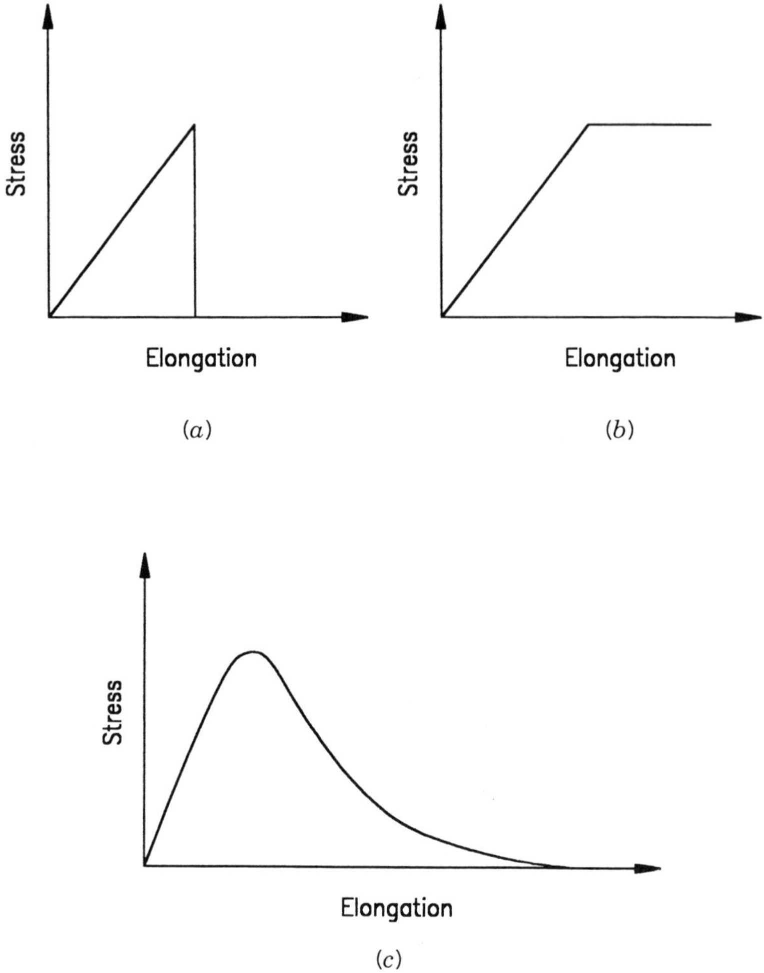

Figure 1-1 Different types of material response, uniaxial stress, and deformation: (*a*) elastic–brittle material, (*b*) elastic–plastic material, and (*c*) elastic–quasi-brittle material.

$= f_t$ due to plastic stress redistribution. The plate fails when the normal stress in entire section *A–A* reaches f_t (Fig. 1-2*b*). For the plate made of a quasi-brittle material, an inelastic zone develops at the location of the maximum stress when the plate fails. This inelastic zone is often referred to as the fracture process zone. The normal tensile stress decreases toward the tip of the hole within the fracture process zone. The development of the fracture process zone usually results in a ''softening'' behavior in the load–deformation response.

The above example indicates that failure characterization of structures generally relates not only to structural geometry but also to materials used. When a ductile material is used, a structure fails only when the nominal stress on the

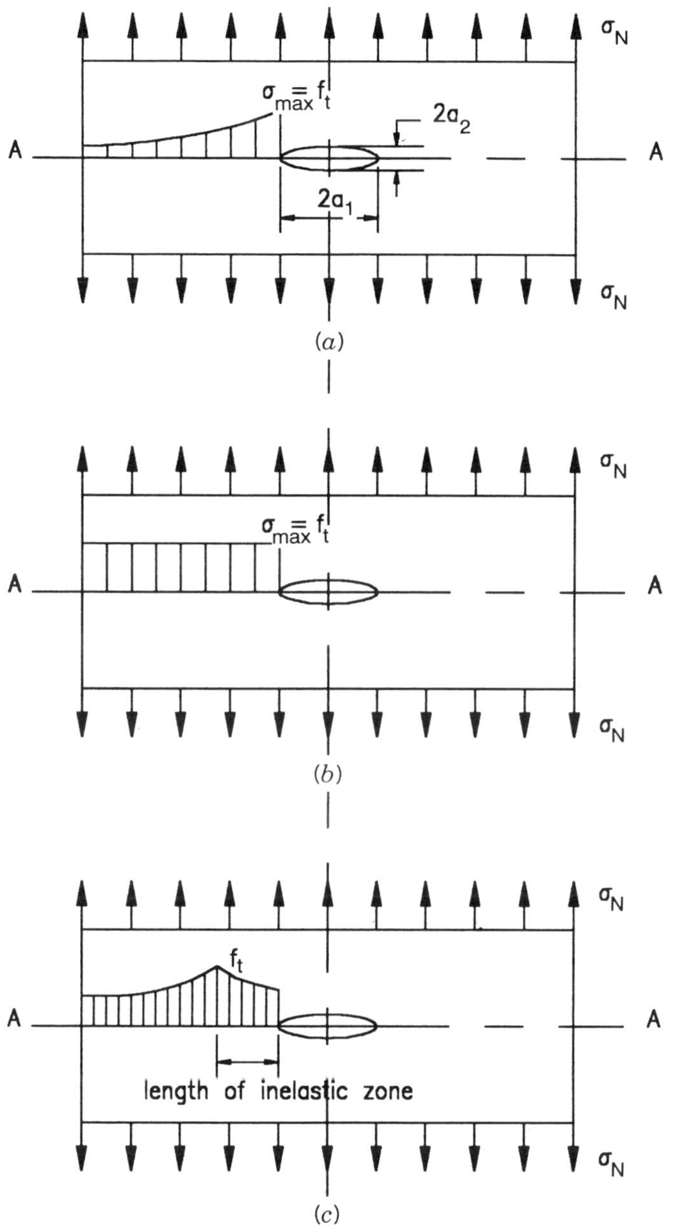

Figure 1-2 Structural failures based on different materials: (*a*) brittle failure, (*b*) plastic failure, and (*c*) quasi-brittle failure.

entire critical cross section reaches the material yield strength. Therefore a strength-based failure criterion in terms of the nominal stress can be simply used to describe failure of the structure. When a perfectly brittle material is used, the structure in this simple example can be assumed to fail whenever the maximum stress is equal to the tensile strength of the material. Since the maximum stress depends not only on this material property but also on the structural geometry and boundary conditions (values of a_1 and a_2 in the above example), the strength failure criterion in terms of the nominal stress is not suitable for a structure made with a brittle material. It will be shown later that for this case the use of the strength-based failure criterion is not always appropriate and can result in unconservative designs for large-size structures. The failure process for a brittle material can be described by the *elastic* energy dissipated in the structure. Since only the elastic energy is involved, a single energy criterion is enough to describe the failure of structures made of brittle materials. When a quasi-brittle material is used, failure of the structure is additionally characterized by developing a fracture process zone at the critical section. Similar to the brittle material, failure of the structure should be described in terms of energy rather than strength. Since not only the elastic energy but also the *inelastic* energy in the fracture process zone should be considered, two (or more) energy criteria are usually needed to completely describe failure of this type of structure. The energy criterion (or criteria) for failure of structures can be established using principles of linear elastic (or nonlinear) fracture mechanics.

1.2 CONCEPTS OF LINEAR ELASTIC FRACTURE MECHANICS

Fracture mechanics is the study of the response and failure of structures as a consequence of crack initiation and propagation. The classical uses of fracture mechanics as listed by Broek[1] are to answer the following questions:

(a) What is the residual strength as a function of crack size?
(b) What size of crack can be tolerated at the expected service loads; that is, what is the critical crack size?
(c) How long does it take for a crack to grow from a certain initial size to the critical size?
(d) What size of preexisting flaw can be permitted at the moment the structure starts its service life?
(e) How often should the structure be inspected for cracks?

Rolfe and Barsom[2] state: "Fracture mechanics is a method of characterizing the fracture behavior in structural parameters that can be used directly by the engineer, namely stress and flaw size" and "the science of fracture mechanics can be used to describe quantitatively the trade-offs among these three factors (stress, materials toughness and flaw size) so that the designer can determine the relative importance of each of them during design rather than during failure analysis.''

These ideas were developed initially for very brittle materials such as glass and some metals where flaws—deliberate or accidental—in the material initiated the growth of a single crack. Such materials are classified as elastic–brittle and display uniaxial, tension stress–strain behavior very similar to that shown in Fig. 1-1*a*. These materials also exhibit *notch sensitivity*. This is defined by Shah and McGarry[3] and Mindess[4] as: ''A material may be considered to be notch sensitive if the presence of a notch causes a change in the net section strength (σ_{net}) of the material (calculated on the basis of the reduced cross-section but neglecting the stress concentrating effect of the notch).'' The effect of this notch sensitivity is to decrease the net section strength for the elastic–brittle material. For such a material, the net section strength would vary with notch length, as shown in Fig. 1-3, whereas it is usually unaltered for ductile materials.

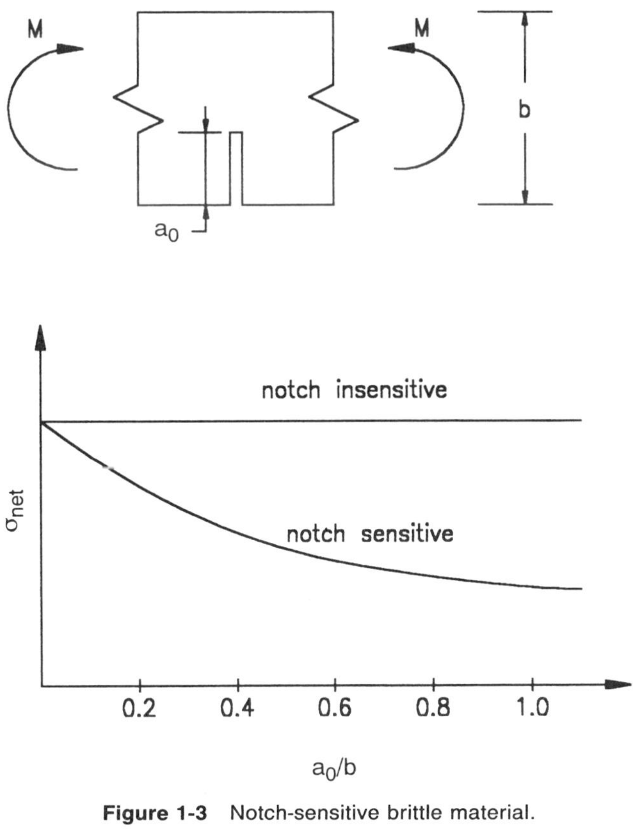

Figure 1-3 Notch-sensitive brittle material.

Even normally ductile, elastic–plastic materials such as mild steel could also exhibit rapid crack growth from a notch in a manner similar to that of brittle materials. This could be due to a combination of stress raisers (sharp stress concentrations at abrupt changes in geometry with low, but still ambient, temperatures and load-induced stresses) or, especially, to fatigue crack growth. In many cases, a single crack (or crack front) would grow to a critical length and then propagate in an unstable manner to complete failure.

Most materials contain some defects. Propagation of these initial defects results in failure of a structure. Conventional strength-based design approaches consider the influence of these defects by introducing a stress concentration factor K_t as in the example given by Eq. (1-1). It is seen that the value of K_t approaches infinity for a very narrow ellipse or sharp crack. This indicates that the conventional analysis based on the stress concentration factor is not valid for a structure with a sharp crack. The fact that $\sigma_{max} \to \infty$ as $a_2/a_1 \to 0$ leads to a basic idea in linear elastic fracture mechanics (LEFM): If one looks close enough at the crack tip, the stresses have the same distribution for all cracks and all applied loads. This is shown in Fig. 1-4 and leads to[1]

$$\sigma(r) = C_F r^{-1/2} \tag{1.2}$$

where (as derived in Chapter 2)

$$C_F = \frac{K_I}{\sqrt{2\pi}} \tag{1.3}$$

As a result, Eq. (1.2) becomes

$$\sigma(r) = \frac{K_I}{\sqrt{2\pi r}} \tag{1.4}$$

The coefficient K_I is called the opening-mode stress intensity factor. It depends on crack length a, structural geometry, and load. Formulas for K_I may be found in many references, for example, Broek[1] and Rolfe and Barsom.[2] For the infinitely wide, cracked plate with crack length of $2a$ (Fig. 1-5) the value of K_I is given by

$$K_I = \sigma\sqrt{\pi a} \tag{1.5}$$

The value of K_I can be related to elastic strain energy released during crack propagation by

$$G_I = \frac{K_I^2}{E'} \tag{1.6}$$

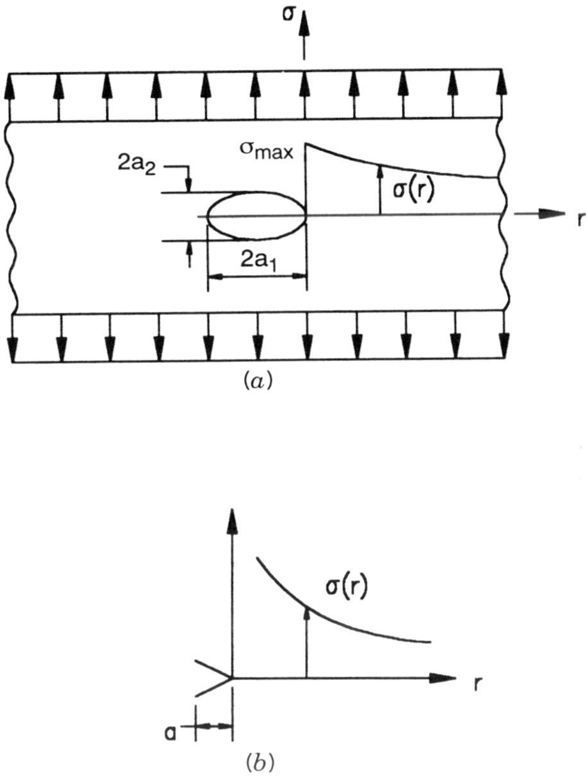

Figure 1-4 Distribution of internal stress in region of flaw: (*a*) elliptical flaw and (*b*) sharp crack.

where $E' = E$ for plane stress and $E' = E/(1 - \mu^2)$ for plane strain and μ is Poisson's ratio. The term G_{I} is called the opening-mode strain energy release rate and it represents the amount of energy required for propagation of a unit area of the crack surface.

If one were to load this plate (Fig. 1-5) to failure stress σ_f and if $\sigma = \sigma_f$ is the value of the nominal far-field stress at failure, then the value of K_{I} associated with σ_f and a could be determined. It is postulated that this is a material fracture property called the *fracture toughness* $K_{\mathrm{I}c}$. If this toughness property is available for the matrerial, then a *failure state stress* or strength can be expressed.[1]

$$\sigma_f = \frac{K_{\mathrm{I}c}}{\sqrt{\pi a}} \tag{1.7}$$

This is plotted against the values of a in Fig. 1-5. A design approach can now be developed relating strength to applied loads, geometry, and crack length.[2]

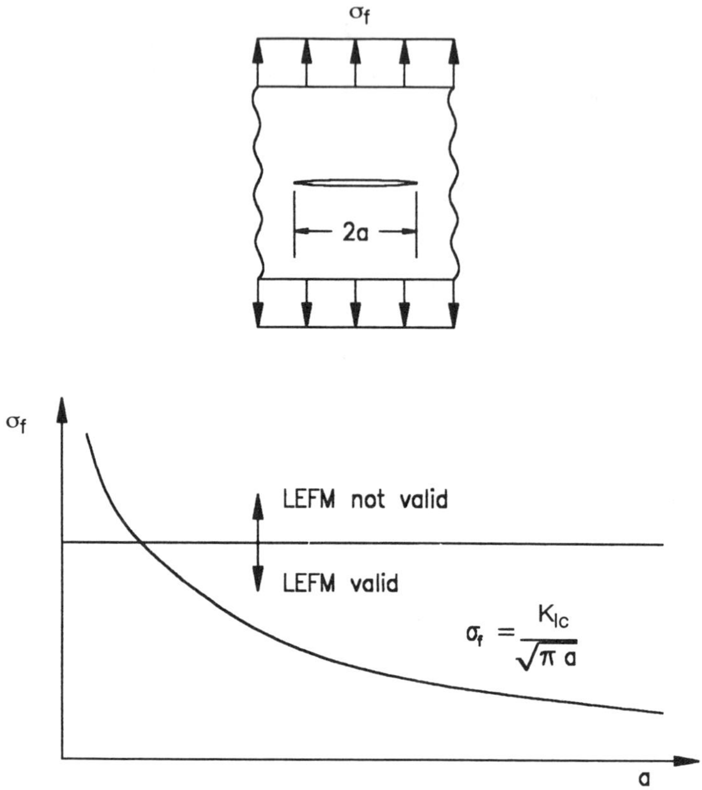

Figure 1-5 Failure stress related to crack size for infinitely wide plate subjected to tension.

Notice that this concept is identical to the idea of structural stability (the Euler buckling curve) and in fact is derivable strictly from a stability viewpoint.[5] It is also seen in Fig. 1-5 that when the obtained value of σ_f exceeds the tensile strength of the material, f_t (or when $a < a_c$, the critical crack length), for small crack lengths, LEFM cannot be valid. On the other hand, it is seen in Fig. 1-4 that the value of σ approaches infinity when r tends to zero at a crack tip based on LEFM. Since no material can sustain infinitely large stress, a certain size of the inelastic zone (the fracture process zone) always exists at a crack tip for a real material. If this inelastic zone is small compared to the fracture dimension in the structure, LEFM can be approximately used. Otherwise energy dissipated in the fracture process zone should be considered.

For materials obeying the failure criterion given by Eq. (1.7), when the flaw size is larger than the critical crack length, the material is notch sensitive. That means that the failure criteria based on net section and the nominal tensile strength of the material will result in an unconservative prediction.

1.3 FRACTURE MECHANICS OF CONCRETE AND ROCK

Since the fracture process zone in quasi-brittle materials such as concrete and rock is not small, the validity of LEFM for these materials is limited to large structures. This localized damage zone consumes part of the energy provided by the applied load and induces nonlinearity into the response.

Conceptually it would seem that such materials are ideally suited to be analyzed through LEFM methods because of their inherent brittleness and multiplicity of flaws: microcracks or macrocracks. This very multiplicity of cracks actually mitigates against the use of LEFM for small structures such as laboratory-size specimens and is the cause of the strain-softening behavior characterized by prepeak and postpeak response of the stress–displacement curve shown in Fig. 1-1*c*. This is due to a combination of distributed cracking leading to localization of cracking into large fractures that include toughening mechanisms of crack bridging and interlock within the fracture process zone. This phenomenon, termed *strain localization*, which we now understand and know how to model, greatly hindered early researchers working in this field who tried to apply the LEFM principles directly to concrete. These included Kaplan[6] and Kesler et al.[7] The evolution of this early work into the present level of understanding is described in reports of the Internatioal Union of Testing and Research Laboratories for Materials and Structures (RILEM)[8] and the American Concrete Institute (ACI) Committee 446, Fracture Mechanics.[9] An explanation of fracture mechanisms in the fracture process zone for a mode I crack may be helpful in understanding how recent research has overcome this problem.

Consider the region near a crack of a beam in bending, as depicted in Fig. 1-6. The load–deformation behavior of such a beam is shown in Fig. 1-7 along with idealized sketches of the growing crack. Part of the confusion associated with attempting to applying LEFM concepts to this situation dealt with a misuse in terminology, that is, the crack length a and notch length a_0 were used interchangeably. As demonstrated by Swartz et al.[10] for concrete and Schmidt[11] for rock, *a notch cannot be used to simulate a crack because of the crack closure phenomenon*. We now know that this effect is related to *crack bridging* and *aggregate interlock* as well as the development of a *process zone*, as shown in Fig. 1-6. The micromechanics and various mechanisms of fracture within the process zone are detailed in Chapter 4. The concepts of crack arrest, coalescence, and need to consider slow crack growth in some nonlinear way were presented by Shah and McGarry[3] as early as 1971. Thus, as shown in Fig. 1-6, the crack length a may be considered as the region that is *traction free*. Preceding this is the process zone l_p, which includes bridging, interlock, and coalescing microcracks. Preceding this is material in which microcracks may exist but they have not yet started to coalesce. Clearly, for a crack to grow from a notch, the process of microcrack coalescence must develop followed by localization and discrete, but not necessarily continuous, cracks. Thus, it

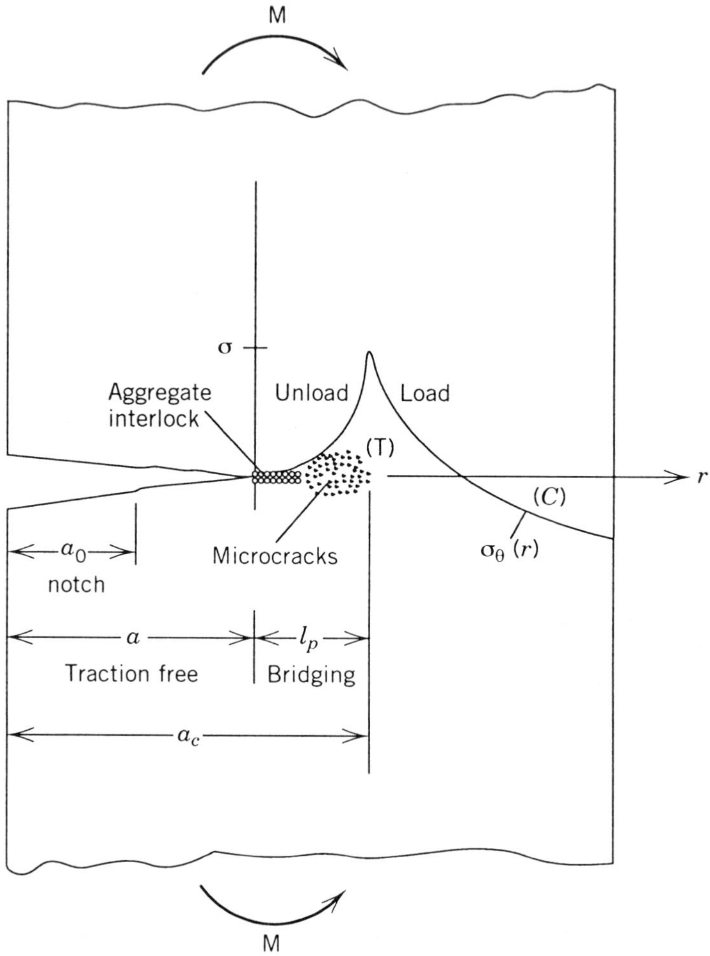

Figure 1-6 Growth of mode I crack in concrete or rock.

is impossible to correlate crack length with stress intensity factor and load to obtain fracture toughness in the LEFM sense if a notch length a_0 is used, unless the notch is very long compared to l_p. In fact, it usually is not sufficient to use the open crack length but rather an *equivalent effective crack length* must be used, as described in Chapter 5. Since the value of the equivalent effective crack length, instead of a_0, is unknown now, the single fracture parameter K_{Ic}, is also not sufficient to determine the failure stress σ_f as shown in Eq. (1.7). At least an additional parameter is needed, and this will be discussed in Chapter 5.

From a testing standpoint, the cracking process can be controlled and monitored using the *crack mouth opening displacement* (CMOD) shown in Fig.

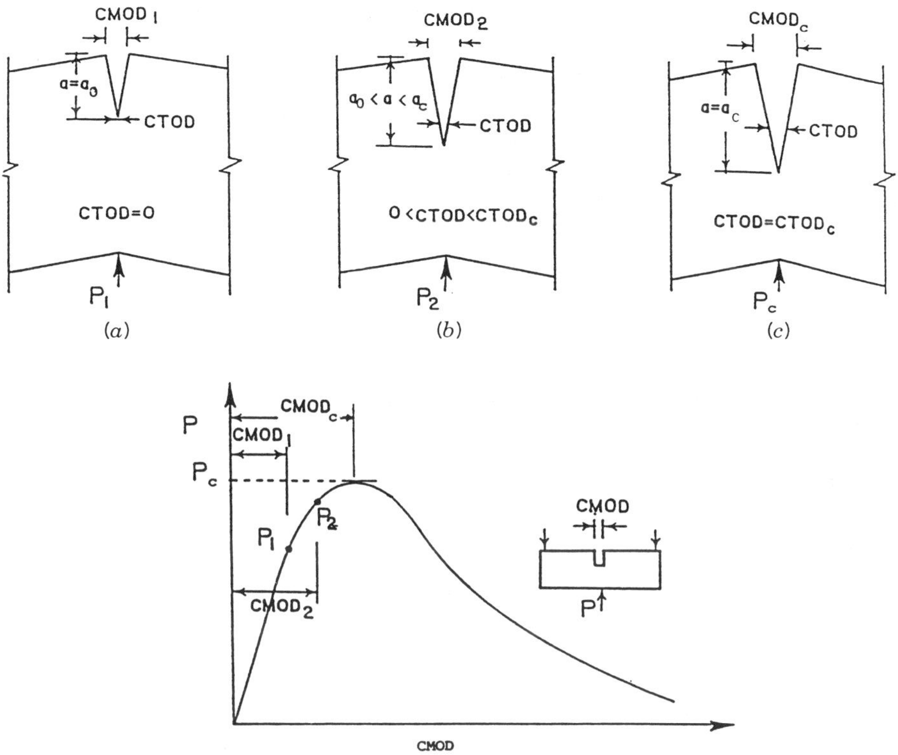

Figure 1-7 Stages of mode I crack growth: (*a*) initiation of stable crack growth, (*b*) process zone partially developed, and (*c*) start of unstable crack growth at peak load.

1-7. This displacement can be related theoretically to notch length a_0 and effective critical crack length a_c. At the same time, another parameter, the *crack tip opening displacement* (CTOD), can be defined as the displacement at tip of the notch a_0.

The stages of opening (mode I) crack growth for a notched specimen shown in Fig. 1-7 are related to the load–displacement curve shown. For convenience the CMOD is used although other displacements could be plotted, for example, the load point–displacement. Initially, and up to some load level P_1, the crack does not grow from the notch and the CTOD is zero. For all practical purposes the response in this range is linearly elastic. Load level P_2 is at an intermediate stage where the process zone is starting to develop and some separation occurs that is reflected by CTOD > 0. The structural response becomes nonlinear. Following this the peak load is reached and the value of P starts to decrease. The process zone may or may not be fully developed at the peak load depending upon the geometry and size of the structure. It is noted that $a = a_c$ and CTOD $=$ CTOD$_c$ at the peak load. Here CTOD$_c$ is the critical CTOD and may be considered as another material fracture parameter. After the peak load, unstable crack growth occurs if the testing system has no way to control it. Here unstable crack growth means that the crack automatically propagates even though the load remains constant or decreases. After the peak load, a softening load–displacement curve is observed. The fracture toughness K_{Ic}, which is a fracture parameter used to describe fracture response at the peak load, may be determined from an appropriate LEFM formula using P_c and a_c. Certainly, ideas of fracture mechanics can be used for structural designs. This will be discussed in the following section.

1.4 STRUCTURAL DESIGN IN TERMS OF FAILURE PROCESS

A structure is built to serve a certain function. If it does not in some way fulfill its intended purpose, then it has "failed." In most civil engineering structures failures may be classified as *serviceability* or *strength*. In the case of serviceability the designer, owner, and user are generally concerned with such things as deflections, appearance, and environmental impacts, which must be satisfied for the desired economic life of the structure. Overriding these concerns usually is the question of safety. This consideration is associated with the strength of the structure. The term strength implies some sort of limit—usually related to level of internal stresses—beyond which additional loads will cause material distress. This distress might be the initiation and continuation of slippage of crystal planes, as in ductile materials such as mild steel, or it might be the actual separation of the material continuum. In the former case, the slipping of crystal planes does not actually cause the structure to fail, but rather deflections of a large scale are introduced that violate the serviceability state. Nevertheless, this case is still called the strength state or the strength limit. Similarly, in concrete design the strength state may or may not actually correspond to collapse.

1.4.1 Serviceability State

Historically, this is the typical design state used from the early origins of structural engineering even to the present day. It is characterized by the following limits: maximum stress $\leq$ allowable stress; maximum deflection $\leq$ allowable deflection; maximum crack size $\leq$ allowable crack size. Traditionally, the material properties governing this state were modulus of elasticity E, Poisson's ratio μ, and allowable stress (which was taken to be a failure stress divided by a factor of safety). The maximum stresses, deflections, and crack sizes are obtained from critical combinations of actual, applied loads called the service loads in conjunction with the structure's geometry and support (boundary) conditions. As will be seen later, the allowable crack size depends on such factors as type of load (static or dynamic or repeated), material resistance to crack growth (which is not the same as material strength), and structural geometry (including type and location of reinforcing present).

For reinforced concrete the service state design consists of calculations of stress, deflection and cracking. The current method given by ACI 318[12] for calculating these under service loading may be illustrated by considering the cracked beam in bending shown in Fig. 1-8*a*. Considering the materials (concrete and steel) to be linearly elastic up to some stress level and the Euler–Bernoulli hypothesis to be valid, the stresses in the steel and concrete in Fig. 1-8*b* are[13]

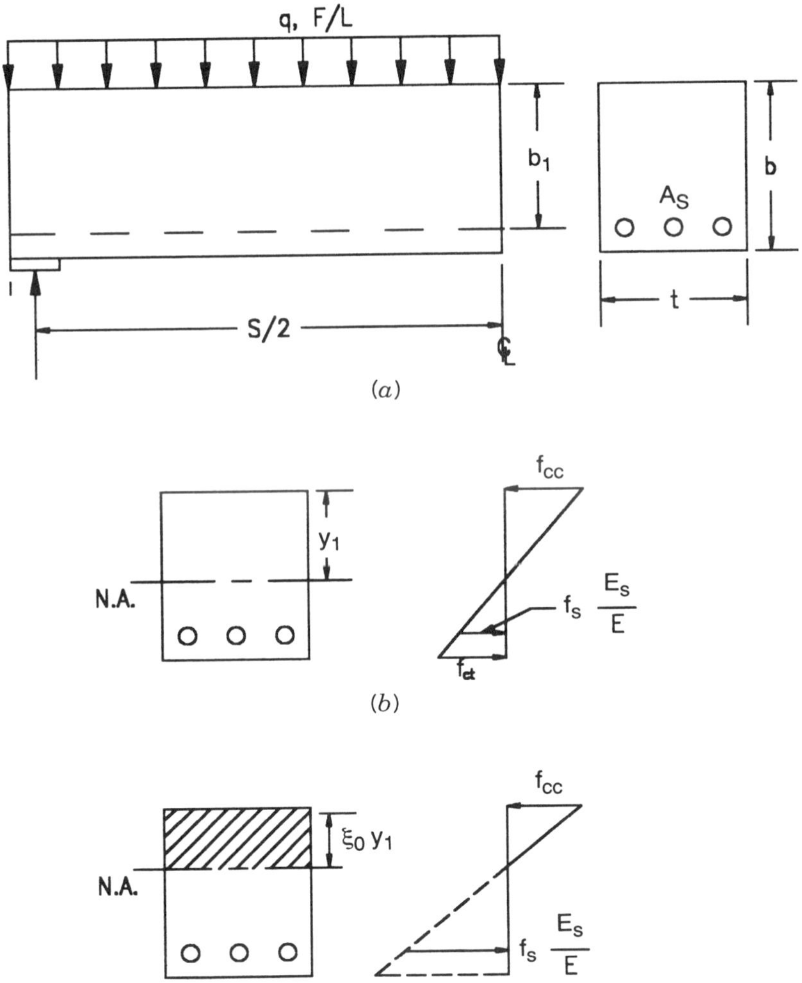

Figure 1-8 Reinforced concrete beam, service level load cases: (*a*) beam subjected to uniform load; (*b*) service level, uncracked location, distribution of stresses; and (*c*) service level, cracked location, distribution of stresses based on commonly used simplified assumptions.

$$\begin{aligned} \text{Steel:} \qquad & f_s = n\,\frac{M(b_1 - y_1)}{I_{ut}} \\ \text{Concrete:} \qquad & f_{cc} = \frac{My_1}{I_{ut}} \\ & f_{ct} = \frac{M(b_1 - y_1)}{I_{ut}} \quad \text{if } f_{ct} < f_r \end{aligned} \tag{1.8}$$

where M is the applied moment due to load q, n is the ratio of the steel Young's modulus E_s to the concrete Young's modulus E, b_1 is the effective depth of the beam, f_s is the tensile stress of the steel bar, f_{cc} is the concrete compressive stress at the top of the beam, f_{ct} is the concrete tensile stress at the bottom of the beam, and f_r is the modulus of rupture (MOR). Thus Eq. (1.8) is applicable to beam locations that are not cracked. In Eq. (1.8), y_1 is the distance from the top of the beam to the neutral axis and I_{ut} is the uncracked transformed moment of inertia. For the rectangular section shown in Fig. 1-8*b* this is

$$I_{ut} = \tfrac{1}{12}b^3t + bt\,(\tfrac{1}{2}b - y_1)^2 + (n - 1)A_s(b_1 - y_1)^2 \tag{1.9}$$

where A_s is the area of the reinforcing steel and t is the beam thickness. It is understood that Eq. (1.8) is valid if $f_s \le f_{sa}$ and $f_{cc} \le f_{ca}$, the "allowable" values of steel and concrete stresses, respectively.

But what happens if the concrete cracks? At the location of the crack, the cross section is normally assumed to be as shown in Fig. 1-8*c* and the stresses are

$$\begin{aligned} \text{Steel:} \qquad & f_s = n\,\frac{M(b_1 - \xi_0 y_1)}{I_{cr}} \\ \text{Concrete:} \qquad & f_{cc} = \frac{M\xi_0 y_1}{I_{cr}} \\ & f_{ct} = 0 \end{aligned} \tag{1.10}$$

where $\xi_0 y_1$ is the distance from the compressive surface to the neutral axis of the cracked beam cross section. It is seen that with this commonly used assumption *all tension in the concrete is neglected, or the crack is considered to be "open" up to the level of the neutral axis*. The cracked section moment of inertia is

$$I_{cr} = \tfrac{1}{3}t(\xi_0 y_1)^3 + nA_s(b_1 - \xi_0 y_1)^2 \tag{1.11}$$

Again, the maximum stresses are less than allowable values: $f_s \leq f_{sa}$ and $f_{cc} \leq f_{ca}$.

The uncracked and cracked service load cases are still used in design (see ACI 318[12], Chapters 9, 18, and 19 and Appendix A, where the cracked case is called the "alternate design method").

What are some consequences of using the cracked section analysis based upon Eqs. (1.10) and (1.11)? Consider the following example: a simply supported beam of span $S = 12$ ft (3.66 m), $b = 25$ in. (635 mm), $t = 10$ in. (254 mm), $b_1 = 23$ in. (584 mm), $f_{ca} = 1.8$ ksi (12.4 MPa), $f_{sa} = 24$ ksi (165 MPa), and $n = 8$. Three No. 8 steel bars with $A_s = 2.35$ in.^{2} (1516 mm^2) are used, and the applied load $q = 5$ kips/ft (73 kN/m). Further assume a tensile strength $f_t = 0.475$ ksi (3.27 MPa). It can readily be determined that the cross-sectional properties are as follows: for the uncracked section, $y_1 = 13.2$ in. (335 mm), $I_{ut} = 14710$ in.^{4} (6.12×10^9 mm^4); for the cracked section, $\xi_0 y_1 = 7.6$ in. (193 mm), $I_{cr} = 5910$ in.^{4} (2.46×10^9 mm^4).

At midspan, the steel and concrete stresses based on an uncracked assumption [Eq. (1.8)] are $f_s = 5.76$ ksi (39.7 MPa), $f_{cc} = 0.969$ ksi (6.68 MPa), and $f_{ct} = 0.866$ ksi (5.97 MPa). Since f_{ct} exceeds the tensile strength, a crack is assumed at midspan and the stresses are recomputed using Eq. (1.10) as $f_s = 22.5$ ksi (155 MPa) and $f_{cc} = 1.389$ ksi (9.57 MPa). It is seen that these stresses are less than the allowable values.

Let us now assume that cracks exist at 1-ft intervals along the span provided that f_{ct} from Eq. (1.8) exceeds the tensile strength. The consequences of this (reasonable) assumption are shown in Fig. 1-9. At the locations of cracks it is possible to calculate stresses without violating the maximum tensile strength requirement. In between the cracks, this is not possible; see $x = 3.5, 4.5, 5.5$ ft. This type of cracking behavior has been observed many times[13] in experiments. Thus the ordinary assumption of the cracked section is not sufficiently accurate. This may be the source of the problem in modeling deflections and bond and crack width in concrete beams.

In order to calculate deflections according to ACI 318[12], the value of I_{cr} obtained may be substituted into the following empirical equation to calculate an *effective moment of inertia* I_e:

$$I_e = \left(\frac{M_{cr}}{M_{\max}}\right)^3 I_g + \left[1 - \left(\frac{M_{cr}}{M_{\max}}\right)^3\right] I_{cr} \tag{1.12}$$

where $M_{cr} = 2f_r I_g/b$ is the cracking moment of the section, $M_{\max}$ is the maximum moment resulting from the applied load, and I_g is the moment of inertia of the gross concrete section neglecting reinforcement. The value of I_e is used to determine the deflection of the beam.

On the other hand, the value of f_s may be substituted into the following empirical equation to evaluate the maximum crack width of the beam $w_{\max}$[14].

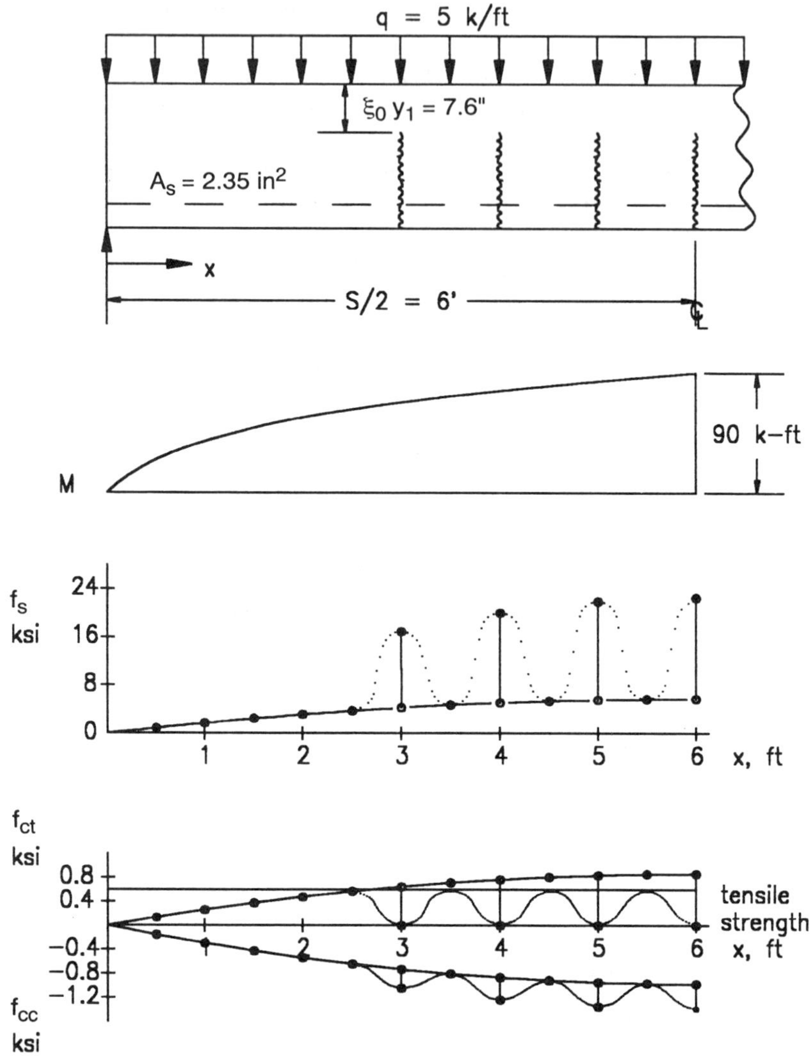

Figure 1-9 Calculated distribution of steel and concrete stresses; usual cracked section assumption.

$$w_{\max} = 0.076\beta_c f_s (d_c A)^{1/3} \tag{1.13}$$

where d_c is the thickness of concrete cover, $\beta_c = (b - \xi_0 y_1)/(b_1 - \xi_0 y_1)$, and A is the effective concrete area surrounding one steel bar. The value of A is equal to the total effective tension area of concrete surrounding reinforcement and having the same centroid divided by the number of bars.

The above design procedures for serviceability of reinforced concrete structures are semianalytical and semiempirical. Some problems associated with these design methods for serviceability of reinforced concrete structures are discussed in the following.

As mentioned earlier, the first problem with this commonly used procedure is that all tension in the concrete is neglected. The influence of the tensile resistance of concrete is certainly significant at the service state where the crack opening is relatively small. The tensile resistance of concrete can be taken into account by introducing a softening stress–separation curve based on fracture mechanics. This will be presented in Chapter 5.

The initiation of the crack is described by a strength criterion $f_{ct} = f_r$. Experimental results indicate that the value of the MOR, f_r, depends on the size of the beam. This is illustrated in Fig. 1-10 for two concrete mixes with compressive strengths of 13.7 and 25.4 MPa, respectively.[15] It is seen that the beam with the greater depth has the smaller value of f_r. As a result, the use of the value of f_r measured from a small-size beam for a large structure can lead to an unconservative design. In fact, since concrete is a heterogeneous material, cracking of concrete is a very complex phenomenon. A sizable fracture process zone is formed before microcracks coalesce into a major crack. Mechanisms in the fracture process zone include microcrack shielding and aggregate bridging and crack branching. All these mechanisms can be reasonably quantified by energy dissipated during the fracture process. Therefore, energy criteria, instead of the strength criterion, should be used to describe cracking of concrete. Energy criteria for propagation of a crack will be discussed in Chapters 3 and 7.

Multiple cracking is usually observed in an adequately reinforced concrete structure under service loading. Debonding and slipping at the interface be-

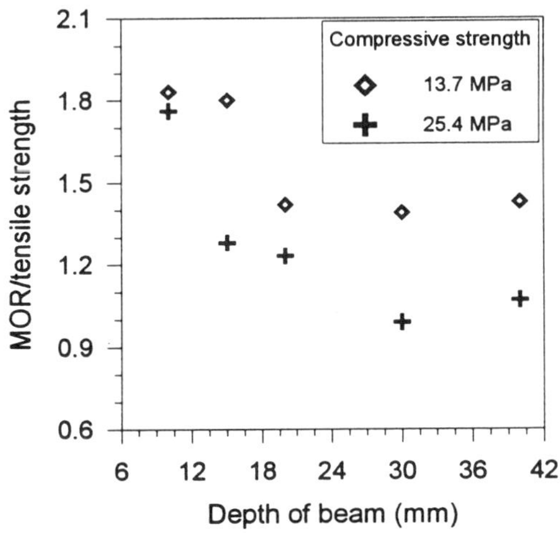

Figure 1-10 Modulus of a rupture for different size of beams.[15]

tween the steel and concrete are usually accompanied by multiple transverse cracks in concrete structures. Both debonding and slipping may greatly influence the development of the transverse cracks. Only energy criteria can simultaneously describe debonding and slipping at the interface and transversely cracking in concrete. This energy approach will be discussed in Chapter 9.

It is obvious that cracking behavior is influenced by structural geometry. The current design method uses Eq. (1.13) to calculate the crack width for both tensile members and beams (for tensile members $\beta_c = 1.316$). Reinforcement is located in the area near the bottom for a beam, whereas it is more or less uniformly distributed in the entire section for a tensile member. In Eq. (1.13) this geometric effect on cracking of reinforced concrete members is taken into account by assigning different values for the empirical parameter β_c and by only considering the area surrounding steel bars for calculating the value of A for the beams. This is obviously not a rational approach. Furthermore, a recent study has shown the cracking behavior of reinforced concrete structures to be structural size dependent.[16] Values of the crack width at a steel strain of 0.001 are plotted against specimen widths for a series of reinforced concrete tensile members in Fig. 1-11, which is predicted using an energy approach based on fracture mechanics. The reinforcement ratio is 0.015 and the diameter of the steel bars is 16 mm for all members. It is seen in the figure that the crack width decreases with increasing size of the members. Both the geometric and size effects on cracking behavior of reinforced concrete members, which may be responsible for the great deviation of experimental values of crack widths previously reported, can be uniquely considered by using a fracture mechanics approach.

In addition, fibers have been currently incorporated into concrete to reduce

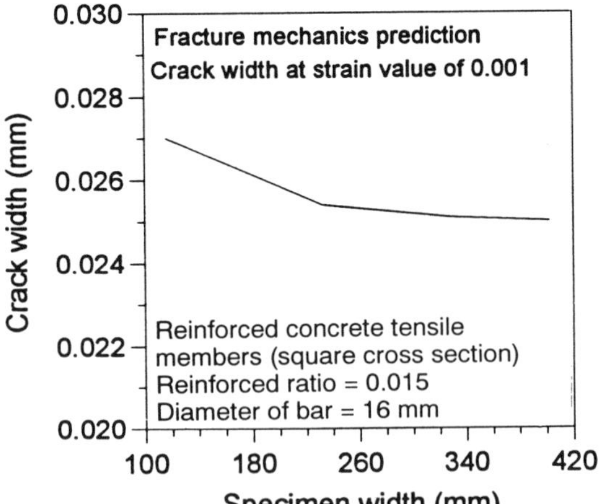

Figure 1-11 Crack width predicted by fracture mechanics for different size of reinforced concrete tensile members.

the crack width at the service state. Fiber amounts up to approximately 2% by volume of concrete are usually used for most applications. Typical stress–displacement curves for a cementitious mix with 0, 1, and 1.5% of steel fiber by volume, respectively, are shown in Fig. 1-12.[17] It is seen that with such a relatively low fiber volume fraction, the fiber only influences the stress–displacement behavior after peak stress, whereas the peak stresses for three volume fractions of the fiber are almost the same. As a result, one can conclude that in this case fibers only influence the toughness of the material, which is defined as the area under the stress–displacement curve. Obviously, the strength-based approach cannot predict the increase of toughness after the peak stress. Postpeak response of concrete should be considered in order to characterize the property of fiber-reinforced concrete. Applications of fracture mechanics to fiber-reinforced concrete will be given in Chapter 12.

1.4.2 Strength State

The loading capacity of concrete structures depends on not only structural geometry and loading conditions but also material performance. The strength design methods of ACI 318[12] are primarily based on ultimate limit analysis. Generally, to develop a design method based on the ultimate limit analysis for a structural geometry, the related failure mechanism should first be experimentally determined. The design formulas are then derived using both equilibrium and simplified compatibility equations based on the observed failure mechanisms. Both factored loads and design strengths are used for safety. Even though the current strength-based design methods are capable of predicting

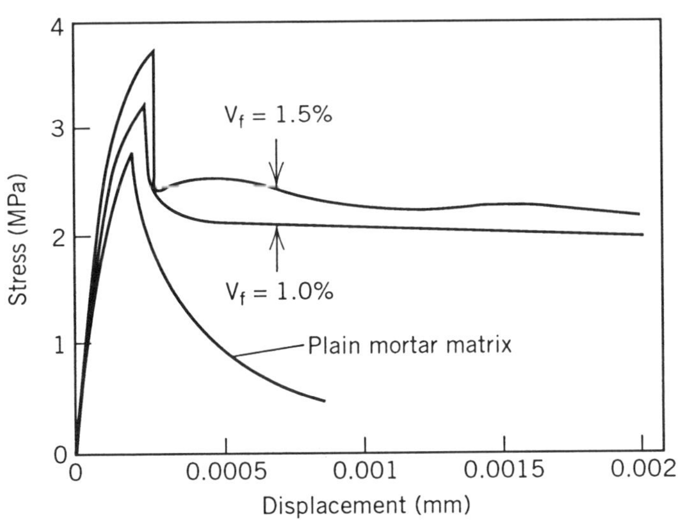

Figure 1-12 Tensile response of fiber-reinforced mortar.[17]

loading capacity of different structures by introducing some empirical constants, fracture processes and failure mechanisms of concrete and reinforced concrete structures are not completely and explicitly taken into account by these approaches. Generally, failure of reinforced concrete structures may be categorized into "ductile" and quasi-brittle. For example, the bending failure of an adequately reinforced concrete beam is generally regarded as ductile, whereas both shear failure of a beam without stirrups and punching-shear failure of a slab are quasi-brittle. The former shows an elastic–plastic load–deformation curve, and the latter exhibit a softening type of load–deformation curve. Problems associated with the current strength-based design methods for these two types of failure are examined below.

The current design method for bending failure is possibly the best developed theory for reinforced concrete structures. It is generally regarded that under-reinforced concrete beams undergo ductile failure. A plastic zone is developed at the location of the maximum moment when a beam fails, as shown in Fig. 1-13*a*. The reinforcing steel yields and the compressive concrete reaches its ultimate capacity before the beam fails. The beam nominal strength M_{nf} is found from Fig. 1-13*c* based on the analysis of Hognestad et al.[18]:

$$M_{nf} = F_c(b_1 - k_2 y_1) = A_s f_y(b_1 - k_2 y_1) \tag{1.14}$$

where F_c is the resultant of compressive stress in the concrete compressive zone and

$$y_1 = \frac{A_s f_y}{k_1 k_3 f'_c b} \tag{1.15}$$

The coefficients k_1, k_2, and k_3 were obtained experimentally,[18, 19] and f'_c is the uniaxial compressive strength.

The balanced reinforcement ratio, at which the reinforcing steel and the compressive concrete simultaneously reach their ultimate strength capacities, should be determined in order to prevent compressive failure of the beam. From the strain distribution given in Fig. 1-13*c*, one can obtain

$$y_1 = \frac{\epsilon_n}{\epsilon_s + \epsilon_n} b_1 \tag{1.16}$$

where ϵ_n is the compressive strain of concrete at the top of the beam and ϵ_s is the tensile strain of the steel bar in the beam. By equating Eqs. (1.14) and (1.15) and letting $\epsilon_s = f_y/E_s$, the balanced reinforcement ratio ρ_b is obtained,

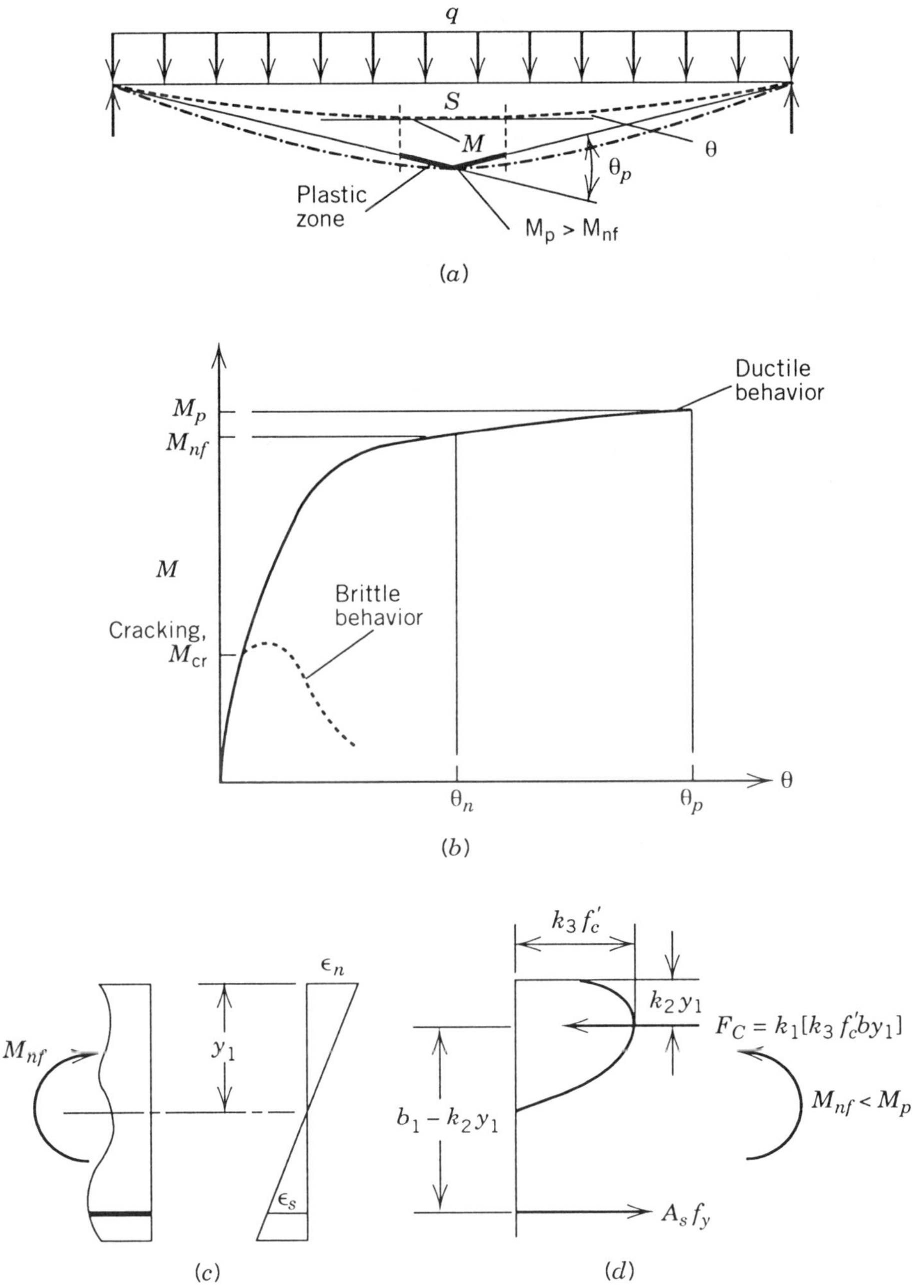

Figure 1-13 Beam bending strength: (*a*) deflections, (*b*) moment–rotation relationship, (*c*) strain distribution, and (*d*) stress distribution.

$$\rho_b = \frac{k_3 f'_c}{f_y} \frac{\epsilon_n}{f_y/E_s + \epsilon_n} \tag{1.17}$$

The value of $\epsilon_n = 0.003$ is used in ACI 318.[12]

Similar to the service state design, the tensile resistance of concrete should be considered and energy criteria should be used to describe the failure process of concrete. In addition, the following problems are obviously associated with the above design method for reinforced concrete beams.

In Fig. 1-13*c*, the softening compressive stress–strain curve is used as a material property. Coefficients k_1, k_2, and k_3 have been used in the strength-based design method to account for the shape of the compressive stress–strain diagram. However, experimental results by van Mier[20] have indicated that the softening part of the compressive stress–strain curves depends on the length of the specimens, as shown in Fig. 1-14*a*. The longer the compressive specimen, the sharper the decrease in the softening stress–strain curve. Alternatively, if the stress is plotted against deformation for the postpeak behavior using the same test results, one obtains basically the same postpeak stress–displacement curves shown in Fig. 1-14*b*. Therefore, the stress–deformation curve, instead of the stress–strain curve, should be used to describe the softening behavior of concrete. As a result, values of coefficients k_1, k_2, and k_3 depend on not only the f'_c value[18] but also the size of the beam. Clearly, these parameters are related to the compressive stress–deformation response of the concrete. This response actually is due to tensile failures related to microcrack growth and coalescence. This phenomenon can be modeled through the use of methods of fracture mechanics as seen in Chapter 10.

If the reinforcement ratio is very low, the beam may fail immediately upon formation of the first crack. Therefore, a minimum reinforcement ratio $\rho_{\min}$ must be determined to ensure against this type of failure. In ACI 318[12] this is specified as $\rho_{\min} = 200/f_y$, where f_y is expressed in psi units. The minimum reinforcement ratio $\rho_{\min}$ is determined by equating the cracking moment, computed from the MOR of the concrete, to the strength of the cracked reinforced section. The value of the MOR has been found to depend on the beam size tested.[21] Therefore, the value of $\rho_{\min}$ also depends on the beam size. This has been experimentally observed and theoretically predicted using fracture mechanics methods, as shown in Fig. 1-15.[22] This will be discussed in detail in Chapter 9.

It is often assumed by the ultimate-limit analysis that a plastic hinge is formed at the section with the maximum moment (see Fig. 1-13*a*). The failure mechanism determined by locations of the plastic hinges is used to calculate the loading capacity of the structure. Therefore, the moment–rotation relationship (Fig. 1-13*b*) is critical for determining the load-carrying capacity of the structure. It has been found that such a moment–rotation relationship depends on the loading configuration of the beams. The same reinforced concrete beams subjected to a uniform-load, three-point bending or four-point bending exhibit

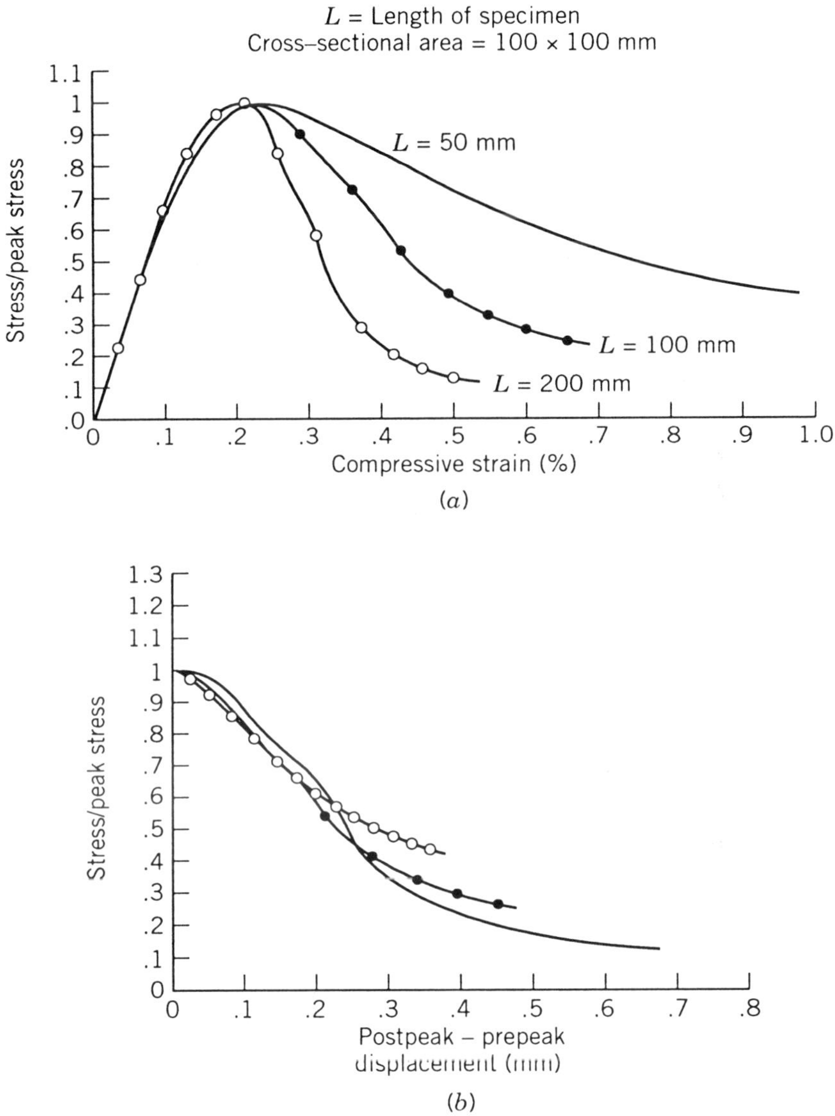

Figure 1-14 Compressive response of concrete prisms: (*a*) stress–strain relationships and (*b*) postpeak stress–displacement curves.[20]

different moment–rotation curves. Although the currently used strength-based method may introduce some empirical parameters to account for this geometry effect, the fracture mechanics approach can solve this problem by considering strain localization at the maximum-moment section.

Many reinforced concrete structures exhibit softening-type load–deflection

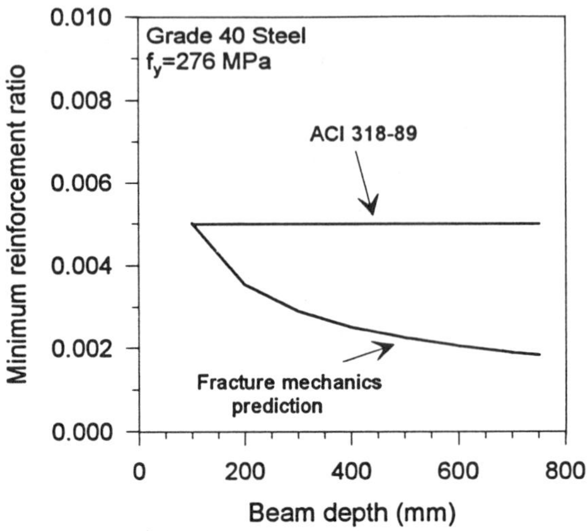

Figure 1-15 Minimum reinforcement ratio for different beam sizes.[22]

curves when they fail. This type of failure is termed quasi-brittle failure. The most serious problem for using the strength-based criterion for quasi-brittle failure of structure is possibly the fact that it cannot correctly predict the effect of *the size of the structure on unit failure strength (size effect)*.[5] This effect may be defined through a comparison of geometrically similar structures of different sizes and is conveniently characterized in terms of a nominal stress σ_N at maximum load. This size effect on the nominal strength may be illustrated by a series of concrete beams reinforced for bending but not for shear, as shown in Fig. 1-16. The nominal shear strength given by the ACI Building Code[12] is

$$V_n = V_c = 2b_1 t\sqrt{f_c'} \quad \text{or} \quad \tau_u = \frac{V_c}{b_1 t} = 2\sqrt{f_c'} \tag{1.18}$$

where V_c is the shear strength resisted by concrete and τ_u is the ultimate shear stress resisted by concrete. The independence of unit shear strength with respect to beam size was investigated through these tests. The effective depth b_1 varies from 100 to 3000 mm with a constant span–depth ratio of 12, as indicated in Fig. 1-16.[23] The results of the tests are shown in Fig. 1-17, where the unit strength $\tau_u = V_n/(b_1 t)$ is plotted versus effective depth b_1. The open symbols are experimental results for a maximum aggregate size of 10 mm and the closed symbols are for a maximum aggregate size of 25 mm. The ACI formula Eq. (1.18) is shown as a horizontal, dashed line. It is seen that there exists a strong *size effect* on the results and when beam sizes are large enough, that is, $b_1 \geq 1000$ mm (39.4 in.), the ACI formula is unconservative. It will be seen

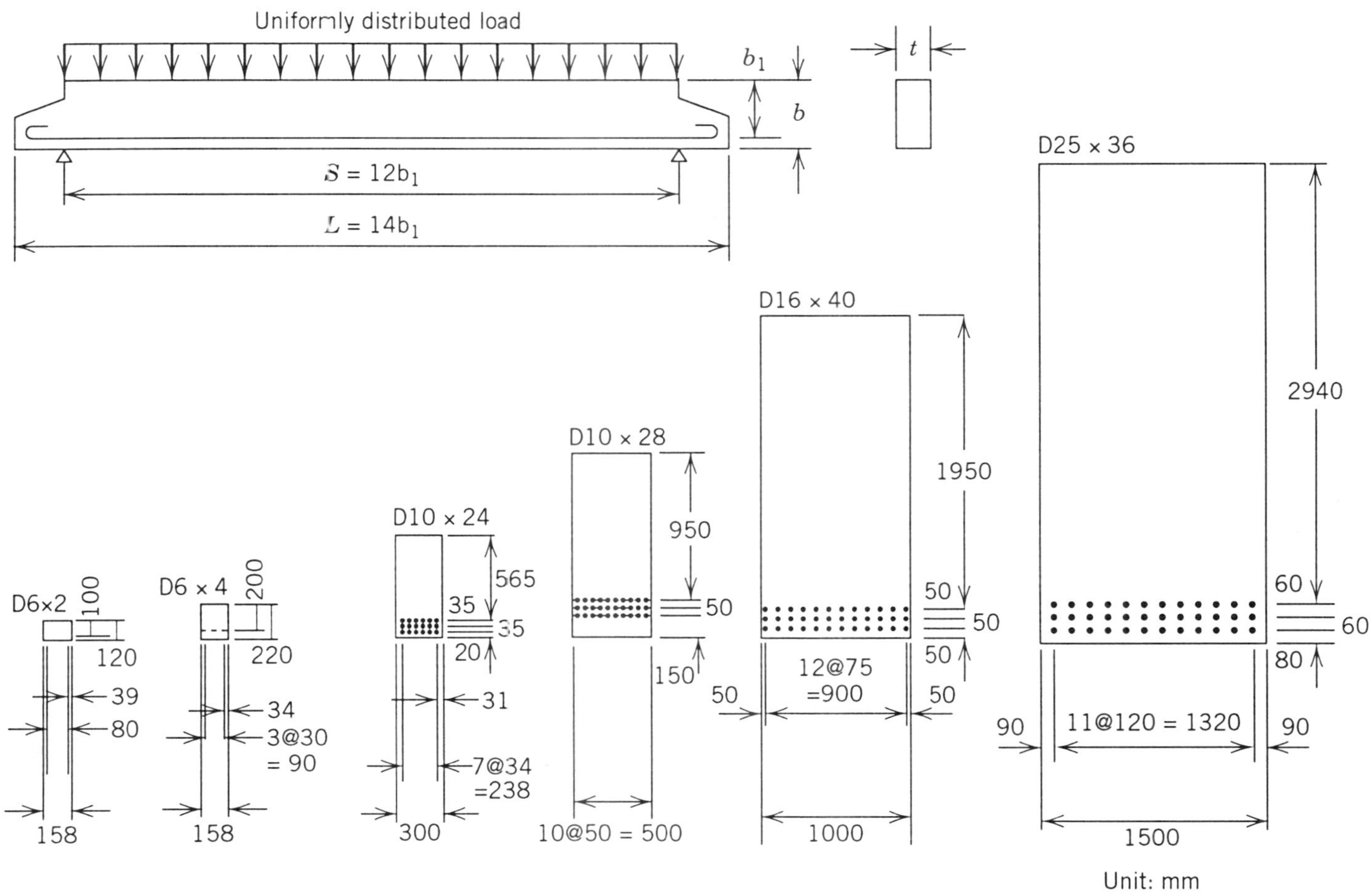

Figure 1-16 Size effect shear tests of reinforced concrete beams.[23]

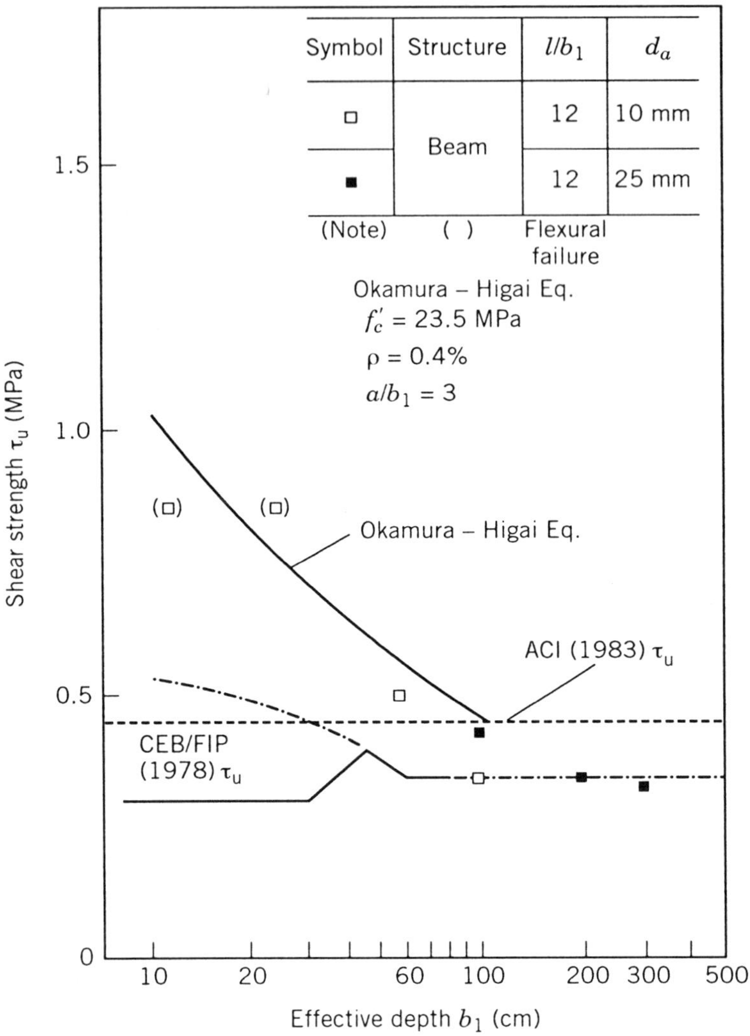

Figure 1-17 Unit shearing stresses.[23]

in many chapters of the book that this size effect on nominal strength can be described using fracture mechanics.

Since no plastic failure mechanisms are developed, the application of the ultimate-limit analysis for the structures exhibiting quasi-brittle failure is conceptually wrong. Failure of these structures primarily results from cracking of concrete, which leads to softening-type load–deformation behavior. This implies that the failure process cannot develop a single-degree-of-freedom mechanism but must consist of propagation of the failure zone (the process zone) throughout the structure. As an example, consider a punching shear failure in a slab as shown in Fig. 1-18. In part (*a*) is shown the geometry of the failure surface. A plastic material response is shown in part (*b*), where one sees that eventually the entire failure surface reaches the same plastic stress throughout. In parts (*c*) and (*d*) are shown the types of stress distributions corresponding to the new crack positions 1, 2, and 3, shown in part (*e*) for a quasi-brittle (softening) material. Notice that by the time the crack has extended to position 3 a substantial portion of the crack plane is stress free. This effect depends greatly on the relative size of the process zone versus the size of the structure. For a small structure, the stress distribution will still be as in Fig. 1-18*b* even for a quasi-brittle material. But for a large structure of this material it will become as shown in part (*d*). Obviously this difference will have a great effect on the overall load-carrying capacity of the structure.

In addition, in finite element (FE) calculations where cracking behavior is modeled, including the strain-softening response, it turns out that use of a strength criterion for failure yields results that are not objective with respect to the size of the FE mesh used.[24,25] This is a particularly serious problem as this method is used extensively to model unusual or special structures, for example, reinforced concrete shells, nuclear containment vessels, and dams. Fortunately, the energy criteria of fracture mechanics provide the control nec-

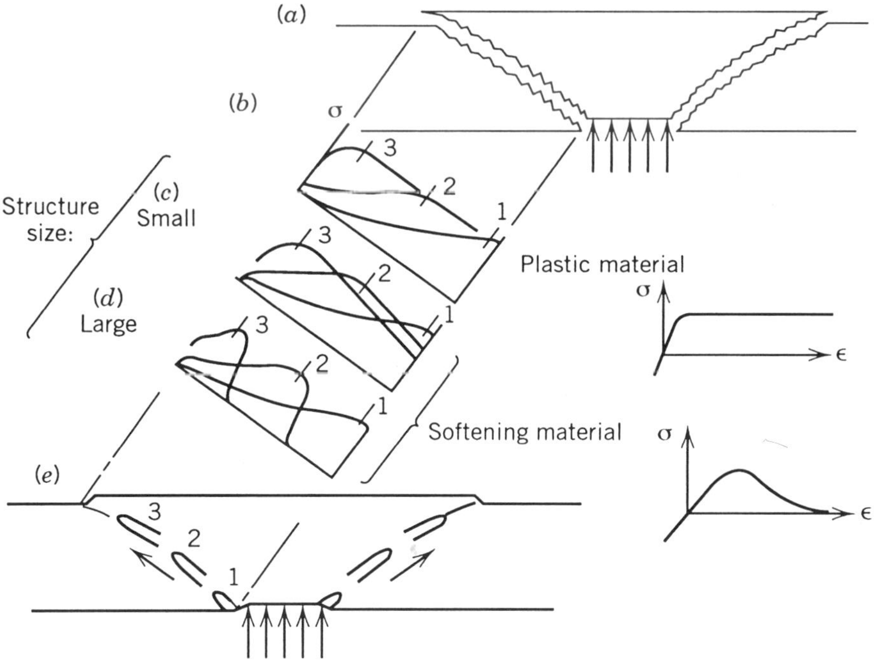

Figure 1-18 Plastic material compared to quasi-brittle material; response to punching failure.[26]

essary to this numerical approach to assure stable and convergent solutions. So, from a numerical modeling standpoint the need for a fracture mechanics approach is mandatory for applications to concrete and rock structures. Fracture mechanics approaches based on FE analysis will be presented in Chapter 5.

Many of these problems related to the service state and strength state designs may be addressed by using energy criteria based on fracture mechanics. Experimental observations have revealed that failure of a structure usually undergoes several stages, including the formation of distributed microcracks (in micrometer scale), coalescence of these microcracks into major cracks (damage localization, in millimeter to centimeter scale), and failure of the structure (in centimeter to meter scale). Fracture mechanics is capable of consistently describing both the service state and strength state through these stages. Many studies on applications of fracture mechanics to concrete and rock have been conducted during the past several decades. It is emphasized that applications of fracture mechanics to concrete and rock started with the general principles of LEFM that were first applied to elastic–brittle materials. From the time of the work of Kaplan[6] in 1961 up until the late 1980s many contradictory results were obtained, K_{I_c} varied with crack length and specimen size, the variation trends were different for various investigators, and confusion existed in the terminology, especially with respect to crack length versus notch length. Many of these problems were due to lack of sophisticated testing equipment and techniques to investigate fracture processes in a controlled manner at the macrolevel and, especially, at the microlevel. Nevertheless, LEFM is still valid for failure of large structures such as dams provided the experimental data are available for such things as fracture toughness, fracture energy, and other fracture parameters. However, these parameters must be obtained on small structures tested in the laboratory. These problems have been solved through extensive studies of fracture processes and use of nonlinear fracture mechanics (NLFM).

Describing the effects of the formation and growth of localized cracks on the response of structures in the serviceability state and in the failure state in a consistent manner is a major contribution of fracture mechanics to concrete and rock. These problems are quite analogous to similar problems in steel structures, such as stable crack growth in fatigue leading to unstable crack propagation and failure. However, an even greater contribution of the fracture mechanics approach lies in the area of *size effects on unit strength*. As presented by ACI 446 in a forthcoming report,[26] consideration of methods of fracture mechanics is expected to lead to changes to the following sections of the Building Code.[12]: 7.12, temperature and shrinkage reinforcement; 9.5.2.3, modulus of rupture; 10.5, minimum reinforcement; 11.3, 11.4, 11.5, beam shear; 11.6, torsion; 11.1.2, punching shear; and 15.8, anchor bolt pullout. In addition to these sections, fracture mechanics modeling of the following problems may very well lead to Building Code changes: deflections, crack width and spacing, development length of reinforcement, cover requirements on bond and cracking, and moment–ductility requirements related to moment redistribution in concrete beams and frames.

1.5 STRUCTURE OF BOOK

This book provides not only fundamental principles of linear and nonlinear fracture mechanics but also state-of-the-art knowledge on applications of fracture mechanics to concrete and rock. The book consists of 12 chapters, and contents of different chapters are briefly presented in the following.

Chapter 1 discusses failure of structures in terms of structural geometry and material performance. After introducing concepts of LEFM, applications of fracture mechanics to concrete are presented. After indicating various problems associated with the currently used design methods, it is emphasized why fracture mechanics is needed for structural design.

Principles of LEFM are reviewed in Chapter 2. Elastic stress and displacement fields at a crack tip are discussed using the stress function approach. Stress intensity factors for commonly used structural geometries are given. The influence of the plastic zone at a crack tip is estimated. The energy principle for crack propagation in linearly elastic materials is discussed. The American Society for Testing and Materials (ASTM) method for measuring the critical stress intensity factor of metallic materials is presented. The design philosophy based on fracture mechanics is introduced.

Chapter 3 mentions principles of NLFM. General energy principles for crack propagation in nonlinear materials are discussed. The J-integral, R-curve, and CTOD are presented. The ASTM methods for determining J–R curves and CTOD for metallic materials are briefly described.

Chapter 4 deals with the structure and fracture process of concrete. The material structure of concrete is described. Various stages of crack propagation in concrete as well as the concept of strain localization are discussed based on experimental observations. After examining different toughening mechanisms in the fracture process zone, experimental methods for measuring the fracture process zone are presented.

Chapter 5 discusses principles of NLFM for mode I quasi-brittle fracture. After a general description of quasi-brittle fracture is introduced, the fictitious crack approach and the equivalent-elastic crack approach are described. The fictitious crack approach discussed includes the Hillerborg et al. fictitious crack model and the crack band model, whereas the equivalent-elastic crack approach discussed includes the two-parameter fracture model, the size effect model, and other effective crack models. Critical comments are made for different models. The FE method of analysis for fracture of concrete is also presented.

Test methods for determining mode I material fracture parameters are presented in Chapter 6. For concrete, these include three RILEM methods and methods by Karihaloo and Nallathambi, Refai and Swartz using three-point bend beams, the Tang, et al. peak load method for the two-parameter model, Li et al. using the J-integral, and Brühwiler et al. using wedge splitting specimens. For rock, these include the International Society for Rock Mechanics (ISRM) methods, the Chong et al. method using semicircular core specimen, and the Luong method using a direct load core.

Chapter 7 discusses the R-curve approach for fracture of quasi-brittle materials in detail. After a general description, R-curves based on bridging mechanisms are discussed. Two R-curves, proposed by Ouyang et al. and Bazant et al., respectively, using the concept of an equivalent elastic crack are also presented.

Fracture mechanics approaches for other deformation modes are described in Chapter 8. After reviewing the stress field at the tip of a planar mixed-mode crack, different mixed-mode failure criteria are presented. Various test methods for evaluating the mixed-mode fracture toughness for concrete and rock are discussed.

Chapter 9 discusses applications of fracture mechanics to various concrete and reinforced concrete structures. These include plain concrete specimens, reinforced concrete tensile members, reinforced concrete beams under bending, minimum reinforcement ratio of concrete members, pullout of steel bar from concrete block, shear of reinforced concrete beams, torsion for reinforced concrete beams, combined loadings and interaction diagrams, anchor bolts, concrete pipes, concrete dams, and failure of dissimilar interfaces and joints.

Chapter 10 presents fracture mechanics approaches to simulate compressive failure of concrete structures. Three approaches, models for failure resulting from interfacial bond cracks, the multilevel model of Zaitsev and Wittmann, and the mixed-mode fracture model, are presented in this chapter.

The relationships between fracture mechanics and damage mechanics are discussed in Chapter 11. Both phenomenological and micromechanical damage models are presented. Particularly, a model proposed by Karihaloo et al. for both prepeak and postpeak responses of concrete under tension is presented in detail. The influence of microcracks on fracture toughness is discussed based on the Karihaloo model.

Chapter 12 discusses applications of fracture mechanics to toughening of fiber-reinforced concrete, fatigue fracture of concrete, and the effect of creep on concrete fracture.

1.6 SUMMARY

This introductory chapter has presented the basic ideas of fracture mechanics—both LEFM and NLFM—and some applications to concrete and reinforced structures. It is seen that the commonly used methods of analysis—the cracked section elastic or plastic approach and the strength criterion—both have limitations. These are because not only have simplifying assumptions been used in these conventional methods but also the actual failure process of such structures cannot be reasonably described by these methods. The methods of fracture mechanics are capable of overcoming these problems. As stated by Charles H. Scholer, a former Head of Engineering Mechanics at Kansas State University and President of the American Concrete Institute, "No material fails for lack of compressive strength." As will be seen in other chapters of this text, the approaches of fracture mechanics can describe failure response of structures under tension, bending, shear, torsion, and compression by considering explicitly the fracture process of materials.

REFERENCES

1. Broek, D., *Elementary Engineering Fracture Mechanics*, 4th rev. ed., Martinus Nijhoff, Dordrecht, 1986.
2. Rolfe, S. T., and Barsom, J. M., *Fracture and Fatigue Control in Structures: Applications of Fracture Mechanics*, Prentice-Hall, Englewood Cliffs, NJ, 1977, pp. 7, 12.
3. Shah, S. P., and McGarry, F. J., "Griffith Fracture Criterion and Concrete," *Journal of the Engineering Mechanics Division, ASCE*, Vol. 97, 1971, pp. 1663–1676.
4. Mindess, S., "Fracture Toughness Testing of Cement and Concrete," in *Fracture Mechanics of Concrete*, edited by Carpinteri, A., and Ingraffea, A. R., Martinus Nijhoff, The Hague, 1984, pp. 67–110.
5. Bazant, Z. P., and Cedolin, L., *Stability of Structures*, Oxford University Press, New York, 1991.
6. Kaplan, M. F., "Crack Propagation and Fracture of Concrete," *ACI Journal*, Vol. 58, No. 5, 1961, pp. 591–610.
7. Kesler, C. E., Naus, D. J., and Lott, J. L., "Fracture Mechanics: Its Applicability to Concrete," in *Proceedings of the International Conference on Mechanical Behavior of Materials*, Vol. IV, Kyoto, Japan, Society of Materials Science, 1972, pp. 113–124.
8. Shah, S. P., and Carpinteri, A. (eds.), *Fracture Mechanics Test Methods for Concrete*, RILEM Report 5, Chapman & Hall, London, 1991.
9. ACI Committee 446, "Fracture Mechanics of Concrete: Concepts, Models and Determination of Materials Properties," ACI 446.1R-91, American Concrete Institute, Detroit, MI, 1991.
10. Swartz, S. E., Hu, K. K., and Jones, G. L., "Compliance Monitoring of Crack Growth in Concrete," *Journal of the Engineering Mechanics Division, ASCE*, Vol. 104, 1978, pp. 789–800.
11. Schmidt, R. A., "Fracture Toughness Testing of Limestone," *Experimental Mechanics*, Vol. 16, No. 5, 1976, pp. 161–167.
12. ACI Committee 318, "Building Code Requirements for Reinforced Concrete [ACI 318-89] and Commentary [ACI 318R-89]," American Concrete Institute, Detroit, MI, 1989.
13. Nilson, A. H., and Winter, G., *Design of Concrete Structures*, 11th ed., McGraw-Hill, New York, 1991.

14. Gergely, P., and Lutz, L. A., "Maximum Crack Width in Reinforced Concrete Flexural Members," in *Causes, Mechanisms and Control of Cracking in Concrete*, ACI publication SP-20, American Concrete Institute, Detroit, MI, 1968, pp. 1–17.
15. Avram, C., Facaoaru, R. E., Filimon, I., Mirsu, O., and Tertea, I., *Concrete Strength and Strains*, Elsevier, New York, 1981.
16. Ouyang, C., and Shah, S. P., "A Fracture Energy Approach for Predicting Cracking of Reinforced Concrete Tensile Members," *ACI Structural Journal*, Vol. 91, No. 1, 1994, pp. 69–78.
17. Gopalaratnam, V. S., and Shah, S. P., "Tensile Failure of Steel-Fiber-Reinforced Mortar, *Journal of Engineering Mechanics Division, ASCE*, Vol. 113, No. 5, 1987, pp. 636–652.
18. Hognestad, E., Hanson, N. W., and McHenry, D., "Concrete Stress Distribution in Ultimate Strength Design," *ACI Journal*, Vol. 27, No. 4, 1955, pp. 455–479.
19. Mattock, A. H., Kriz, L. B., and Hognestad, E., "Rectangular Concrete Stress Distribution in Ultimate Strength Design," *ACI Journal*, Vol. 32, No. 8, 1961, pp. 875–928.
20. van Mier, J. G. M., "Multiaxial Strain-Softening of Concrete, Part I: Fracture," *Materials and Structures*, Vol. 19, No. 111, 1986, pp. 179–190.
21. Jenq, Y.-S., and Shah, S. P., "Features of Mechanics of Quasi-Brittle Crack Propagation in Concrete," *International Journal of Fracture*, Vol. 51, 1991, pp. 103–120.
22. Bosco, C., Carpinteri, A., and Debernardi, P. G., "Minimum Reinforcement in High-Strength Concrete," *Journal of Structural Engineering, ASCE*, Vol. 116, 1990, pp. 427–437.
23. Shioya, T., Iguro, M., Nojiri, Y., Akiyama, H., and Okada, T., "Shear Strength of Large Reinforced Concrete Beams," in *Fracture Mechanics: Application to Concrete*, edited by Li, V. C., and Bazant, Z. P., ACI SP-118, American Concrete Institute, Detroit, MI, 1989, pp. 259–279.
24. Bazant, Z. P., and Cedolin, L., "Fracture Mechanics of Reinforced Concrete," *Journal of Engineering Mechanics Division, ASCE*, Vol. 106, No. EM6, 1980, pp. 1287–1306.
25. Rots, J. G., Nauta, P., Kusters, G. M. A., and Blaauwendraad, J., "Smeared Crack Approach and Fracture Localization in Concrete," *Heron*, Vol. 30, No. 1, 1985, pp. 48.
26. ACI Committee 446, "Fracture Mechanics: Applications to Concrete Structures and Implications with Regard to the Code," American Concrete Institute, Detroit, MI, 1995, in review.

2

PRINCIPLES OF LINEAR ELASTIC FRACTURE MECHANICS

2.1 AIRY STRESS FUNCTIONS FOR PLANE PROBLEMS IN ELASTICITY

Stress components at any point (x, y, z) in a stressed body in a cartesian coordinate system are defined as σ_x, σ_y, σ_z, τ_{xy}, τ_{xz}, τ_{yz}, as shown in Fig. 2-1. The corresponding strain components are ϵ_x, ϵ_y, ϵ_z, γ_{xy}, γ_{xz}, γ_{yz}. In a condition of plane stress $\sigma_z = \tau_{xz} = \tau_{yz} = 0$, whereas in a condition of plane strain $\epsilon_z = 0$.

For planar problems the equilibrium equations are

$$\frac{\partial \sigma_x}{\partial x} + \frac{\partial \tau_{xy}}{\partial y} + X = 0 \qquad \frac{\partial \tau_{xy}}{\partial x} + \frac{\partial \sigma_y}{\partial y} + Y = 0 \tag{2.1}$$

where X and Y are the body forces in the x and y directions, respectively.

The strain–displacement equations are

$$\epsilon_x = \frac{\partial u}{\partial x} \qquad \epsilon_y = \frac{\partial v}{\partial y}$$
$$\gamma_{xy} = \frac{\partial u}{\partial y} + \frac{\partial v}{\partial x} \tag{2.2}$$

where u and v are the displacement components in the x and y directions, respectively.

The stress–strain relations for an elastic, isotropic material under the plane stress condition are

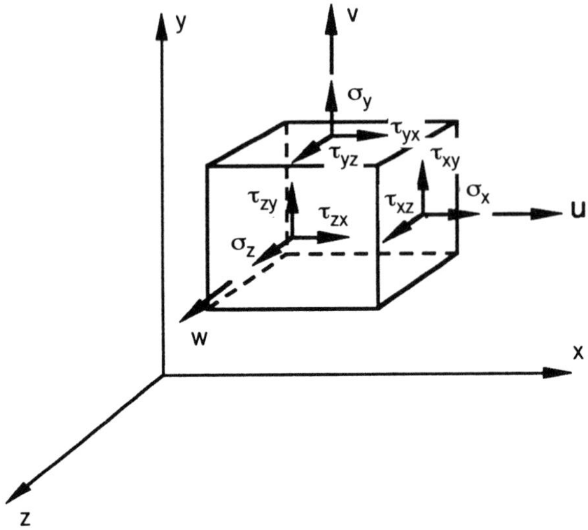

Figure 2-1 Stress and displacement components in a cartesian coordinate system.

$$\epsilon_x = \frac{1}{E}(\sigma_x - \mu\sigma_y) \qquad \epsilon_y = \frac{1}{E}(\sigma_y - \mu\sigma_x)$$

$$\gamma_{xy} = \frac{2(1 + \mu)}{E}\tau_{xy} \tag{2.3}$$

where E and μ are the modulus of elasticity and Poisson's ratio, respectively. The stress–strain relations for the plain strain condition can be obtained by replacing E with $E/(1 - \mu^2)$, and μ with $\mu/(1 - \mu)$ in Eq. (2.3), respectively.

Equations (2.1)–(2.3) are the governing equations describing plane problems in elasticity. Stress and strain distributions in an elastic body subjected to actions can be solved from these equations associated with given boundary conditions.

It is difficult to solve directly the governing equations for a given problem. However, these equations can be simplified if the body forces X and Y are constant within an elastic body. The equilibrium equations (2.1) are automatically satisfied by introducing

$$\sigma_x = \frac{\partial^2\Psi}{\partial y^2} - Xx \qquad \sigma_y = \frac{\partial^2\Psi}{\partial x^2} - Yy$$

$$\tau_{xy} = -\frac{\partial^2\Psi}{\partial x\,\partial y} \tag{2.4}$$

where Ψ is called the Airy stress function. Substituting Eqs. (2.2) and (2.4) into Eq. (2.3) and differentiating twice result in the compatibility equation

$$\frac{\partial^4 \Psi}{\partial x^4} + 2\frac{\partial^4 \Psi}{\partial x^2\, \partial y^2} + \frac{\partial^4 \Psi}{\partial y^4} = 0 \tag{2.5}$$

Introduction of the Airy stress function greatly simplifies solutions of planar problems. As a result, solving a planar problem becomes one of finding an Airy stress function satisfying Eq. (2.5) and the given boundary conditions.

2.2 COMPLEX STRESS FUNCTION

The complex stress function approach presented in this section is based on Muskhelishvili.[1] Westergaard[2] has proposed a similar approach.

A complex function $Z(z)$ is defined by

$$Z(z) = Z_r(x, y) + iZ_i(x, y) \tag{2.6}$$

where $z = x + iy$, $Z_r(x, y)$ is the real part of $Z(z)$, and $Z_i(x, y)$ is the imaginary part of $Z(z)$. For $Z(z)$ to be an analytic function, the following Cauchy–Riemann conditions should be satisfied:

$$\frac{\partial Z_r(x, y)}{\partial x} = \frac{\partial Z_i(x, y)}{\partial y} \qquad \frac{\partial Z_i(x, y)}{\partial x} = -\frac{\partial Z_r(x, y)}{\partial y} \tag{2.7}$$

By introducing the transformation of $z = x + iy$, Eq. (2.5) becomes

$$\frac{\partial^4 \Psi}{\partial z^2\, \partial \bar{z}^2} = 0 \tag{2.8}$$

A quantity with an overbar represents its conjugate from hereafter (e.g., $z = x + iy$ and $\bar{z} = x - iy$). Equation (2.8) is the compatibility condition based on the complex stress function. Equation (2.8) is automatically satisfied if the following expression is used:

$$\Psi = \tfrac{1}{2}[\bar{z}\varphi_1(z) + z\overline{\varphi_1(z)} + \varphi_2(z) + \overline{\varphi_2(z)}] \tag{2.9}$$

where $\varphi_1(z)$ and $\varphi_2(z)$ are the analytic functions satisfying Cauchy–Riemann conditions [Eq. (2.7)].

Substituting Eq. (2.9) into Eq. (2.4) and neglecting body forces result in

$$\sigma_y + \sigma_x = 2[\varphi_1'(z) + \overline{\varphi_1'(z)}]$$
$$\sigma_y - \sigma_x + 2i\tau_{xy} = 2[\bar{z}\varphi_1''(z) + \varphi_3'(z)] \tag{2.10}$$

where a function with a prime represents its derivative with respect to z and $\varphi_3(z) = \partial\varphi_2(z)/\partial z$ is used. The stress components can be obtained by separating the real and imaginary parts of Eq. (2.10).

Displacements for the plane stress condition can be expressed as

$$\frac{E}{1 + \mu}(u + iv) = \frac{3 - \mu}{1 + \mu}\varphi_1(z) - \overline{z\varphi_1'(z)} - \overline{\varphi_3(z)} \tag{2.11}$$

The displacement components can be obtained by separating the real and imaginary parts of Eq. (2.11). For the condition of plane strain, E and μ should be replaced by $E/(1 - \mu^2)$ and $\mu/(1 - \mu)$ in Eq. (2.11), respectively.

2.3 ELASTIC STRESS AND DISPLACEMENT FIELDS AT CRACK TIP

Consider a mode I crack in an infinite plate under tension as shown in Fig. 2-2. The following expressions of $\varphi_1(z)$ and $\varphi_3(z)$ can be used for this problem:

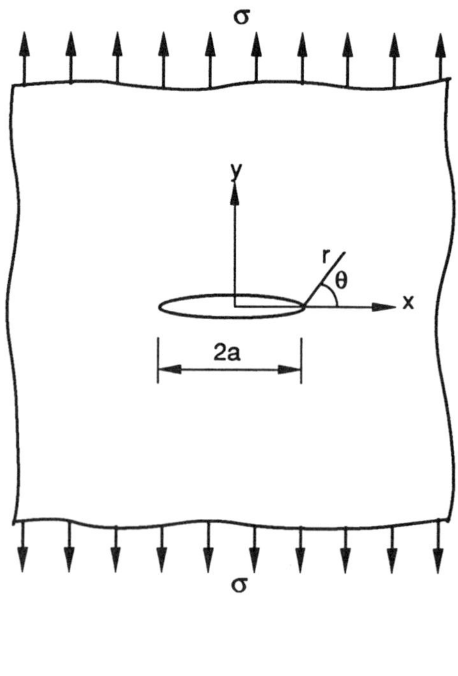

Figure 2-2 Infinite plate with crack subjected to tension.

$$\varphi_1 = \frac{\sigma}{4}\left(2\sqrt{z^2 - a^2} - z\right) \qquad \varphi_3 = \frac{\sigma}{2}\left(z - \frac{a^2}{\sqrt{z^2 - a^2}}\right) \tag{2.12}$$

It can be proven that $\varphi_1(z)$ and $\varphi_3(z)$ selected above satisfy Eq. (2.7) and the boundary conditions. The stress components can be derived by substituting Eq. (2.12) into Eq. (2.10),

$$\begin{aligned} \sigma_y + \sigma_x &= \sigma\left[2\,\mathrm{Re}\left(\frac{z}{\sqrt{z^2 - a^2}}\right) - 1\right] \\ \sigma_y - \sigma_x + 2i\tau_{xy} &= \sigma\left[\frac{2ia^2y}{(z^2 - a^2)^{3/2}} + 1\right] \end{aligned} \tag{2.13}$$

where Re $Z_1(z)$ means the real part of $Z_1(z)$ and $Z_1(z)$ can be any complex function.

Since only the stress field at the crack tip is of interest, one may use the polar coordinates instead of the cartesian coordinates, as shown in Fig. 2-2, for convenience. Transformations between the two coordinate systems are

$$x = a + r\cos\theta \qquad y = r\sin\theta \qquad z = a + r\cos\theta + ir\sin\theta \tag{2.14}$$

Substituting Eq. (2.14) into Eq. (2.13) and neglecting higher order terms of r result in

$$\begin{aligned} \sigma_y + \sigma_x &= \sigma\sqrt{\frac{2a}{r}}\cos\frac{\theta}{2} \\ \sigma_y - \sigma_x + 2i\tau_{xy} &= \sigma\sqrt{\frac{2a}{r}}\sin\frac{\theta}{2}\cos\frac{\theta}{2}\left(\sin\frac{3\theta}{2} + i\cos\frac{3\theta}{2}\right) \end{aligned} \tag{2.15}$$

By separating the real and imaginary parts from Eq. (2.15), the stress components for the infinite cracked plate under tension are obtained:

$$\begin{aligned} \sigma_x &= \frac{K_{\mathrm{I}}}{\sqrt{2\pi r}}\cos\frac{\theta}{2}\left(1 - \sin\frac{\theta}{2}\sin\frac{3\theta}{2}\right) \\ \sigma_y &= \frac{K_{\mathrm{I}}}{\sqrt{2\pi r}}\cos\frac{\theta}{2}\left(1 + \sin\frac{\theta}{2}\sin\frac{3\theta}{2}\right) \\ \tau_{xy} &= \frac{K_{\mathrm{I}}}{\sqrt{2\pi r}}\sin\frac{\theta}{2}\cos\frac{\theta}{2}\cos\frac{3\theta}{2} \end{aligned} \tag{2.16}$$

The stress intensity factor K_{I} for a mode I crack may be defined as

$$K_{\mathrm{I}} = \lim_{r \to 0} \sqrt{2\pi r}\sigma_y(r, \theta = 0) \tag{2.17}$$

For the infinite cracked plate this is

$$K_{\mathrm{I}} = \sigma\sqrt{\pi a} \tag{2.18}$$

The displacement components at the crack tip for the plane stress condition are similarly obtained as

$$u = 2(1 + \mu)\frac{K_{\mathrm{I}}}{E}\sqrt{\frac{r}{2\pi}}\cos\frac{\theta}{2}\left(1 - 2\mu + \sin^2\frac{\theta}{2}\right)$$

$$v = 2(1 + \mu)\frac{K_{\mathrm{I}}}{E}\sqrt{\frac{r}{2\pi}}\sin\frac{\theta}{2}\left(2 - 2\mu - \cos^2\frac{\theta}{2}\right) \tag{2.19}$$

Similarly, for a mode II crack in an infinite cracked plate subjected to in-plane shearing, as shown in Fig. 2-3, the stress components at the crack tip are

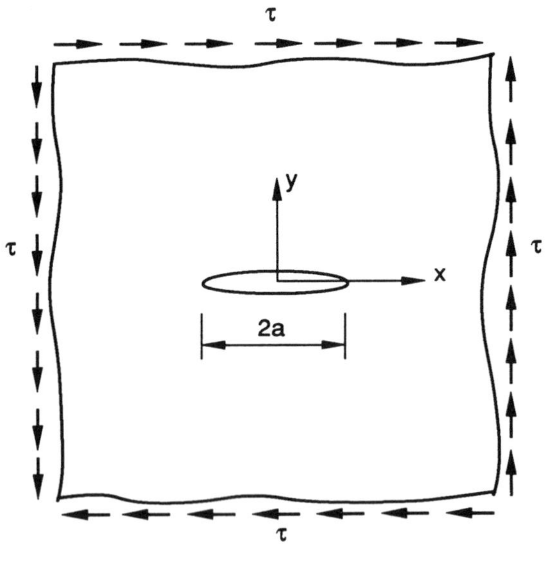

Figure 2-3 Infinite plate with crack subjected to shear.

$$\sigma_x = -\frac{K_{\rm II}}{\sqrt{2\pi r}} \sin\frac{\theta}{2}\left(2 + \cos\frac{\theta}{2}\cos\frac{3\theta}{2}\right)$$

$$\sigma_y = \frac{K_{\rm II}}{\sqrt{2\pi r}} \sin\frac{\theta}{2}\cos\frac{\theta}{2}\cos\frac{3\theta}{2}$$

$$\tau_{xy} = \frac{K_{\rm II}}{\sqrt{2\pi r}}\left(1 - \sin\frac{\theta}{2}\sin\frac{3\theta}{2}\right) \tag{2.20}$$

where the stress intensity factor $K_{\rm II}$ for a mode II crack is given by

$$K_{\rm II} = \tau\sqrt{\pi a} \tag{2.21}$$

The values of $K_{\rm I}$ and $K_{\rm II}$ account for singularity of the stress field at the crack tip and are functions of load, specimen geometry, boundary condition, and crack size. Stress intensity factors can serve as a fracture criterion. A crack propagates whenever $K_{\rm I}$ (or $K_{\rm II}$) is equal to a threshold value, that is,

$$K_{\rm I} = K_{{\rm I}c} \qquad K_{\rm II} = K_{{\rm II}c} \tag{2.22}$$

where $K_{{\rm I}c}$ and $K_{{\rm II}c}$ are the critical stress intensity factors for mode I and mode II cracks, respectively. The values of $K_{{\rm I}c}$ and $K_{{\rm II}c}$ are regarded as material fracture properties based on LEFM.

Stress intensity factors can also be used to characterize structures. Based on how stress intensity factors change with crack lengths, structures can be classified as positive and negative geometries.[3] The stress intensity factor increases monotonically with the increasing crack length for the positive-geometry structures. On the other hand, the stress intensity factor initially decreases and then increases with the crack propagation for the negative-geometry structures. Therefore, based on linear elastic solutions, any propagation of the crack means a catastrophic failure for a positive-geometry structure. However, for a negative-geometry structure, the crack can steadily propagate to a certain threshold value due to the initial decrease of the stress intensity factor.

Expressions of stress intensity factors for different structures and boundary conditions are necessary before LEFM can be applied to actual structures.

2.4 STRESS INTENSITY FACTORS AND CRACK OPENING DISPLACEMENTS FOR SOME USEFUL GEOMETRIES

In the previous sections, it has been shown how to obtain stress intensity factors for a mode I crack in an infinite plate subjected to tension using the stress function approach. Generally, stress intensity factors and crack opening displacements can be determined from stress functions, boundary collocation techniques, or finite element analysis. Stress intensity factors are a function of applied load, boundary conditions, crack length, and structural geometry. Extensive stress intensity factors and crack opening displacements for different geometries and loading conditions have been presented in handbooks.[4,5] Unless indicated, the following expressions are selected from reference.[4] Stress intensity factors and crack opening displacements for some commonly used geometries and loading conditions are listed below. All displacements given in this section are for the condition of plane stress. By replacing E with $E/(1 - \mu^2)$, corresponding displacements for the condition of plane strain will be obtained.

2.4.1 Uniaxial Tensile Plate with Center Crack

The stress intensity factor for a uniaxial tensile plate with a center crack (unit thickness) as shown in Fig. 2-4 is

$$K_{\mathrm{I}} = \sigma\sqrt{\pi a}\, g_1\left(\frac{a}{b}\right) \qquad g_1\left(\frac{a}{b}\right) = \sqrt{\sec\left(\frac{\pi a}{2b}\right)} \tag{2.23}$$

where $g_1(a/b)$ is a geometry factor that approaches unity when a/b tends to zero.

The crack opening displacement at the center (CODC) is[5]

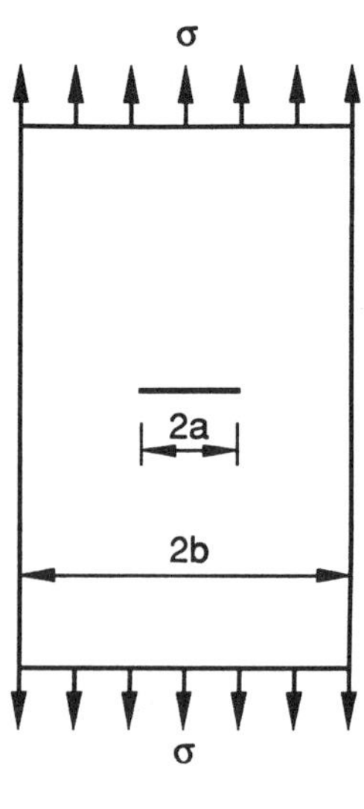

Figure 2-4 Center-cracked plate subjected to tension.

$$\text{CODC} = \frac{4\sigma a}{E} g_2\left(\frac{a}{b}\right)$$

$$g_2\left(\frac{a}{b}\right) = 1 + 0.065\frac{a}{b} - 0.241\left(\frac{a}{b}\right)^2 + 3.76\left(\frac{a}{b}\right)^3 - 6.63\left(\frac{a}{b}\right)^4 + 4.93\left(\frac{a}{b}\right)^5 \quad (2.24)$$

The crack opening displacement along the crack surface (COD) is

$$\text{COD} = \text{CODC}\sqrt{1 - \left(\frac{x}{a}\right)^2} \quad (2.25)$$

2.4.2 Uniaxial Tensile Plate with Double-Edge Crack

The stress intensity factor for a uniaxial tensile plate with double-edge crack (unit thickness) as shown in Fig. 2-5 is

$$K_I = \sigma\sqrt{\pi a}\, g_1\left(\frac{a}{b}\right)$$

$$g_1\left(\frac{a}{b}\right) = \left[1 + 0.122\cos^2\left(\frac{\pi a}{2b}\right)\right]\sqrt{\frac{2b}{\pi a}\tan\left(\frac{\pi a}{2b}\right)} \quad (2.26)$$

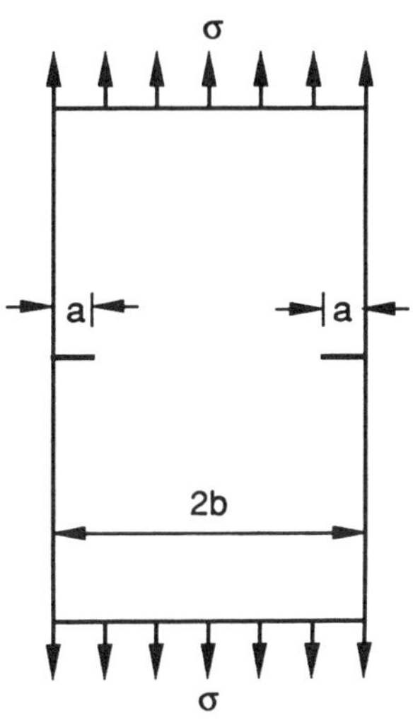

Figure 2-5 Double-edge cracked plate subjected to tension.

The crack mouth opening displacement is[5]

$$\text{CMOD} = \frac{4\sigma a}{E} g_2\left(\frac{a}{b}\right)$$

$$g_2\left(\frac{a}{b}\right) = 1.458 - 0.308\frac{a}{b} + 0.985\left(\frac{a}{b}\right)^2 - 1.869\left(\frac{a}{b}\right)^3 + 2.009\left(\frac{a}{b}\right)^4 \tag{2.27}$$

The crack opening displacement is[3]

$$\text{COD}(x) = \text{CMOD}\, g_3\left(\frac{a}{b}, \frac{x}{a}\right)$$

$$g_3\left(\frac{a}{b}, \frac{x}{a}\right) = \left[1 + \left(0.407\frac{a}{b} - 0.959\right)\frac{x}{a} + \left(0.450\frac{a}{b} - 0.528\right)\left(\frac{x}{a}\right)^2 - \left(0.857\frac{a}{b} - 0.487\right)\left(\frac{x}{a}\right)^3\right]^{1/2} \tag{2.28}$$

2.4.3 Uniaxial Tensile Plate with Single-Edge Crack

The stress intensity factor for a uniaxial tensile plate with a single-edge crack (unit thickness) as shown in Fig. 2-6 is

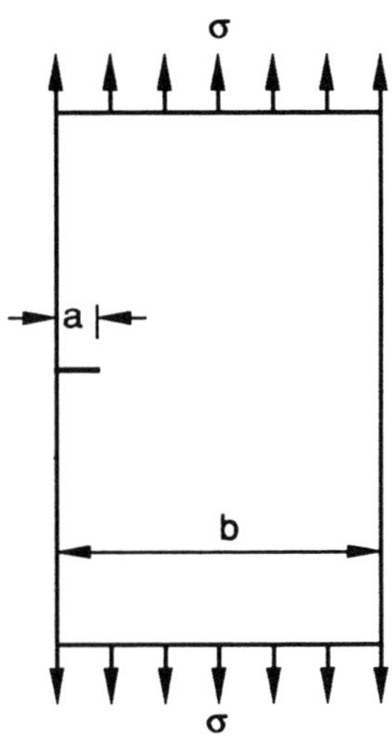

Figure 2-6 Single-edge cracked plate subjected to tension.

$$K_I = \sigma\sqrt{\pi a}\, g_1\left(\frac{a}{b}\right)$$

$$g_1\left(\frac{a}{b}\right) = 1.12 - 0.231\,\frac{a}{b} + 10.55\left(\frac{a}{b}\right)^2 - 21.72\left(\frac{a}{b}\right)^3 + 30.39\left(\frac{a}{b}\right)^4 \tag{2.29}$$

The crack mouth opening displacement is[5]

$$\text{CMOD} = \frac{4\sigma a}{E}\, g_2\left(\frac{a}{b}\right)$$

$$g_2\left(\frac{a}{b}\right) = 1.46 - 0.70\,\frac{a}{b} + 25.93\left(\frac{a}{b}\right)^2 - 143.0\left(\frac{a}{b}\right)^3 + 538.6\left(\frac{a}{b}\right)^4 - 907.5\left(\frac{a}{b}\right)^5 + 633.7\left(\frac{a}{b}\right)^6 \tag{2.30}$$

2.4.4 Three-Point Bend Beams

A three-point bend beam is shown in Fig. 2-7. The stress intensity factor for the beam is

$$K_I = \sigma\sqrt{\pi a}\, g_1\left(\frac{a}{b}\right) \qquad \sigma = \frac{3PS}{2b^2 t} \tag{2.31}$$

where the geometry factor $g_1(a/b)$ depends on the ratio of span to depth of the beam and is given below[6]:

$$g_1\left(\frac{a}{b}\right) = \frac{1.0 - 2.5a/b + 4.49(a/b)^2 - 3.98(a/b)^3 + 1.33(a/b)^4}{(1 - a/b)^{3/2}} \quad \text{for } S = 2.5b \tag{2.32}$$

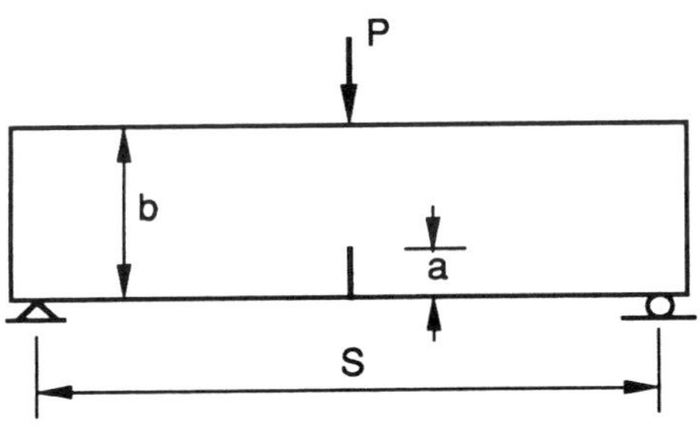

Figure 2-7 Three-point bend beam.

$$g_1\left(\frac{a}{b}\right) = \frac{1.99 - (a/b)(1 - a/b)[2.15 - 3.93\,a/b + 2.70(a/b)^2]}{\sqrt{\pi}(1 + 2a/b)(1 - a/b)^{3/2}} \quad \text{for } S = 4b \tag{2.33}$$

Expressions for $g_1(a/b)$ for other span–depth ratios are given by Go et al.[7] The crack mouth opening displacement is

$$\text{CMOD} = \frac{4\sigma a}{E}\, g_2\left(\frac{a}{b}\right) \tag{2.34}$$

where $g_2(a/b)$ depends on the ratio of span to depth of the beam and is given below:

$$g_2(a/b) = \frac{1.73 - 8.56a/b + 31.2(a/b)^2 - 46.3(a/b)^3 + 25.1(a/b)^4}{(1 - a/b)^{3/2}} \quad \text{for } S = 2.5b \tag{2.35}$$

$$g_2(a/b) = 0.76 - 2.28a/b + 3.87(a/b)^2 - 2.04(a/b)^3 + \frac{0.66}{(1 - a/b)^2} \quad \text{for } S = 4b \tag{2.36}$$

The crack opening displacement for the beam with $S = 4b$ is

$$\text{COD}(x) = \text{CMOD}\; g_3\left(\frac{a}{b}, \frac{x}{a}\right)$$

$$g_3\left(\frac{a}{b}, \frac{x}{a}\right) = \left\{\left(1 - \frac{x}{a}\right)^2 + \left(1.081 - 1.149\frac{a}{b}\right)\left[\frac{x}{a} - \left(\frac{x}{a}\right)^2\right]\right\}^{1/2} \tag{2.37}$$

2.4.5 Center-Crack Plate Subjected to Wedge Forces

An infinite-size plate with a center crack (unit thickness) subjected to wedge forces is shown in Fig. 2-8. The stress intensity factor for the specimen is

$$K_{\text{I},A} = \frac{p}{\sqrt{\pi a}}\sqrt{\frac{a + x}{a - x}} \qquad K_{\text{I},B} = \frac{p}{\sqrt{\pi a}}\sqrt{\frac{a - x}{a + x}} \tag{2.38}$$

where p is the wedge force pair per unit thickness acting on the crack surface.

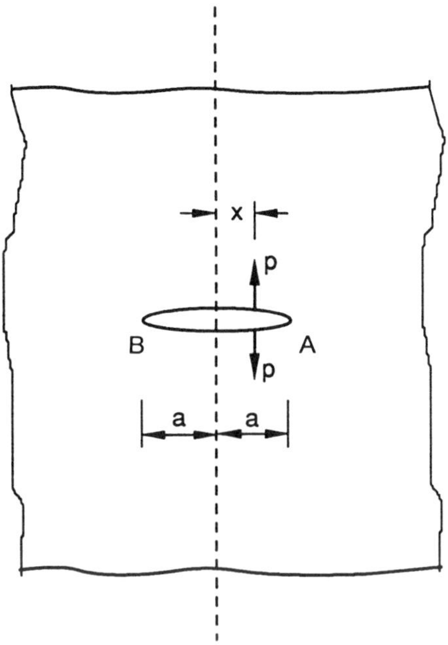

Figure 2-8 Infinite plate with center crack subjected to wedge forces.

2.4.6 Single-Edge Crack Plate Subjected to Wedge Forces

A finite-size plate with a single-edge crack subjected to wedge forces is shown in Fig. 2-9. The stress intensity factor for the specimen is

$$K_{\mathrm{I}} = \frac{2p}{\sqrt{\pi a}} g_1\left(\frac{a}{b}, \frac{x}{a}\right)$$

$$g_1\left(\frac{a}{b}, \frac{x}{a}\right) = \frac{3.52(1 - x/a)}{(1 - a/b)^{3/2}} - \frac{4.35 - 5.28x/a}{(1 - a/b)^{1/2}} + \left[\frac{1.30 - 0.30(x/a)^{3/2}}{\sqrt{1 - (x/a)^2}} + 0.83 - 1.76\frac{x}{a}\right]\left[1 - \left(1 - \frac{x}{a}\right)\frac{a}{b}\right] \tag{2.39}$$

Equations (2.38) and (2.39) can be used as a Green's function to obtain the stress intensity factor of a specimen subjected to distributed surface forces, which will be demonstrated in the following section.

2.5 SUPERPOSITION OF STRESS INTENSITY FACTORS

Since the stress field equations and the stress intensity factor are a linear functions of the applied load [e.g., Eqs. (2.16) and (2.18)], superposition can be

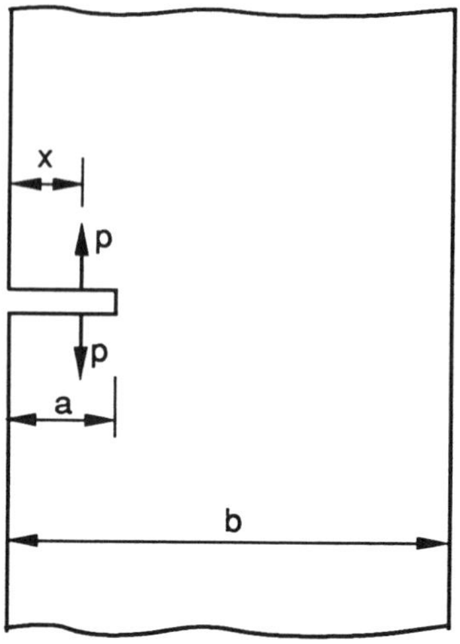

Figure 2-9 Single-edge cracked plate subjected to wedge forces.

used to obtain the stress intensity factor for a specimen subjected to multiple loads associated with the same fracture deformation mode.

A beam with a center-edge crack subjected to a concentrated load at the midspan and linearly distributed compressive load, $q = q_0 x/a$, on the crack surface is shown in Fig. 2-10, where q_0 is a scale factor. Based on superposition, the stress intensity factor for the beam under two kinds of loads can be evaluated from the beam subjected to the concentrated load and the distributed load, individually, as shown in Fig. 2-10. This results in

$$K_{\mathrm{I}} = K_{\mathrm{I}}^{P} + K_{\mathrm{I}}^{q} \tag{2.40}$$

where K_{I}^{P} is the stress intensity factor for the concentrated load and K_{I}^{q} is the stress intensity factor for the distributed load. The value of K_{I}^{P} can be directly obtained from Eqs. (2.31) and (2.32) or (2.33). The value of K_{I}^{q} can be obtained by using Eq. (2.39) as the Green's function. By setting $q = dq = (q_0 x/a)\,dx$, the increment of the stress intensity factor, dK_{I}^{q}, due to dq can be obtained from Eq. (2.39),

$$dK_{\mathrm{I}}^{q} = \frac{2}{\sqrt{\pi a}} \frac{q_0 x\, dx}{a} g_1\left(\frac{a}{b}, \frac{x}{a}\right) \tag{2.41}$$

Integrating dq over the crack length a leads to

$$K_{\mathrm{I}}^{q} = \int_0^a dK_{\mathrm{I}}^{q} = \int_0^a \frac{2q_0 x}{a\sqrt{\pi a}} g_1\left(\frac{a}{b}, \frac{x}{a}\right) dx \tag{2.42}$$

Equation (2.42) can be evaluated using an analytical or numerical method.

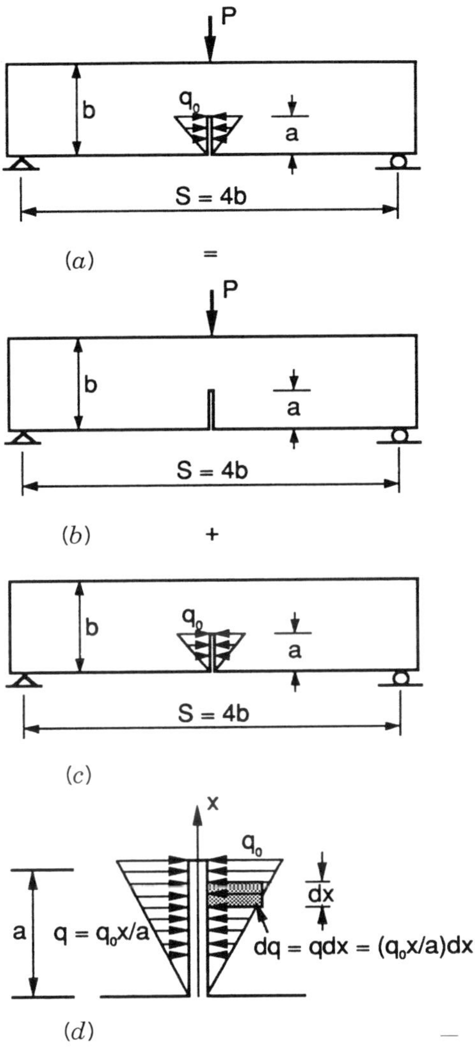

Figure 2-10 Illustration of superposition principle.

Similarly, the superposition can also be used to obtain the crack opening displacement of a specimen subjected to multiple loads.

2.6 PLASTIC ZONE AT CRACK TIP

According to the elastic stress field given by Eq. (2.16), the stress approaches infinity when r tends to zero. This phenomenon is termed as the stress singularity at the tip of an elastic crack. Since infinite stress cannot develop in real materials, a certain range of plastic zone must exist at the crack tip for yielding type of materials.

A rough estimate of the plastic zone can be made based on Fig. 2-11a, which shows the distribution of the stress σ_y in the plane $\theta = 0$ based on Eq. (2.16). The value of σ_y is greater than the material yield stress σ_{ys} for the region $r < r_{p0}$. The distance r_{p0} indicates the size of the crack tip plastic zone, and its value can be estimated from the condition of $\sigma_y = \sigma_{ys}$ at $r = r_{p0}$. Combining this condition and Eq. (2.16) with $\theta = 0$ leads to

$$r_{p0} = \frac{K_{\mathrm{I}}^2}{2\pi\sigma_{ys}^2} = \frac{a}{2}\left(\frac{\sigma}{\sigma_{ys}}\right)^2 \tag{2.43}$$

It is quite clear that the actual size of the plastic zone is larger than the value of r_{p0} obtained, because the stress represented by the shaded area in Fig. 2-11*a* still needs to be carried.

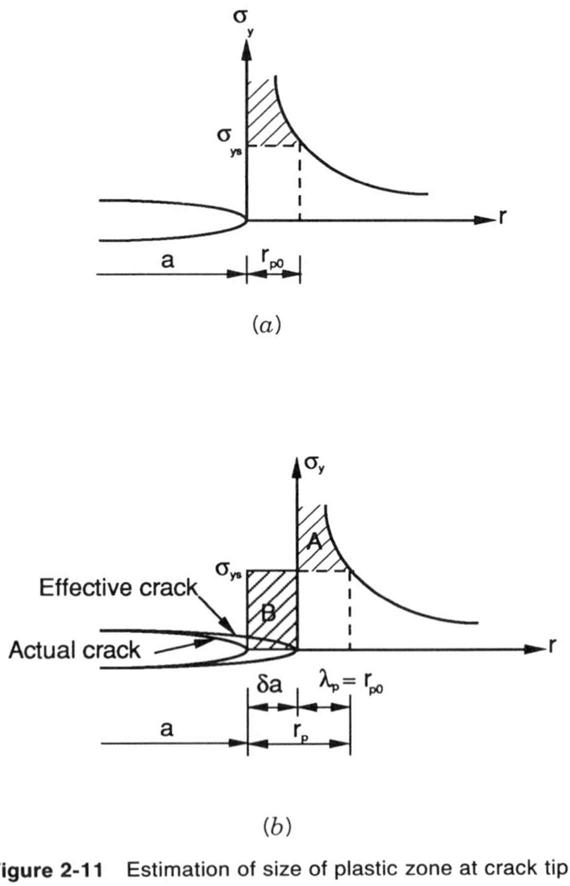

Figure 2-11 Estimation of size of plastic zone at crack tip.

Irwin[8] proposed that the presence of the crack tip plastic zone reduces the stiffness of the structure, and this is equivalent to that of a structure containing a longer crack. Therefore, the effective crack length $a_e = a + \delta a$, where δa is the crack length correction, rather than the actual crack length, should be used to estimate the size of the crack tip plastic zone r_p, as shown in Fig. 2-11*b*. The distribution of stress σ_y at the tip of the effective crack is given in Fig. 2-11*b*, and the stress over the length of δa and at the tip of the effective crack is limited to σ_{ys}. Again using the condition $\sigma_y = \sigma_{ys}$ at $r = \lambda_p$ from the tip of the effective crack, it can be proven that the value of λ_p is approximately equal to r_{p0} for the case that δa is much smaller than a. Consequently, the value of δa should be large enough to carry the stress released by cutting the area A; that is, the area A should be equal to the area B. This condition results in

$$\int_0^{\lambda_p} \sigma \sqrt{\frac{a + \delta a}{2r}}\, dr - \sigma_{ys}\lambda_p = \sigma_{ys}\, \delta a \tag{2.44}$$

Using $\lambda_p = r_{p0}$ and neglecting δa compared to a, Eq. (2.44) becomes

$$r_p = \delta a + r_{p0} = a\left(\frac{\sigma}{\sigma_{ys}}\right)^2 = \frac{1}{\pi}\left(\frac{K_{\text{I}}}{\sigma_{ys}}\right)^2 \tag{2.45}$$

It is seen that the size of the plastic zone, r_p, based on the above second-order estimation is twice as large as the first-order estimation, r_{p0}.

Dugdale[9] used a different approach to estimate the size of the crack tip plastic zone. He also considered an effective crack that is longer than the actual crack, as shown in Fig. 2-12. Since the fictitious crack extensions ρ_p at both crack edges are not really cracked, the parts of the material can still sustain the yield stress. As a result, the parts ρ_p carry the yield stresses that tend to close the crack. The magnitude of ρ_p is chosen in such a way that total stress intensity due to the applied load σ and the yield stress σ_{ys} should be equal to zero. Based on superposition this condition leads to

$$K_{\text{I}} = K_{\text{I}}^{\sigma} + K_{\text{I}}^{\rho} = 0 \tag{2.46}$$

where K_{I}^{σ} and K_{I}^{ρ} are the stress intensity factors due to the applied load and the yield stress, respectively. For an infinite plate with a center crack as shown in Fig. 2-2,

$$K_{\text{I}}^{\sigma} = \sigma\sqrt{\pi(a + \rho_p)} \tag{2.47}$$

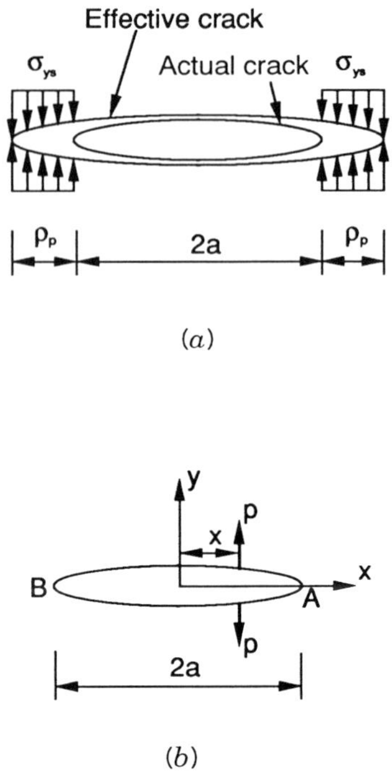

Figure 2-12 Dugdale approach for size of plastic zone.

It is noted that the effective crack length rather than the actual crack length is used here. The term K_{I}^{ρ} can be calculated based on Eq. (2.38) using superposition, which results in

$$
\begin{aligned}
K_{\mathrm{I}}^{\rho} &= \frac{-\sigma_{ys}}{\sqrt{\pi(a+\rho)}} \int_{a}^{a+\rho_p} \left(\sqrt{\frac{a+x}{a-x}} + \sqrt{\frac{a-x}{a+x}} \right) dx \\
&= -2\sigma_{ys} \sqrt{\frac{a+\rho_p}{\pi}} \arccos\left(\frac{a}{a+\rho_p} \right)
\end{aligned}
\tag{2.48}
$$

It is noted in Eq. (2.48) that the yield stress on both edges has a contribution to K_{I}^{ρ}, and the negative sign is due to the fact that the yield stress tends to close the crack. Substituting Eqs. (2.47) and (2.48) into Eq. (2.46) results in

$$
\frac{a}{a+\rho_p} = \cos\left(\frac{\pi\sigma}{2\sigma_{ys}} \right) \tag{2.49}
$$

Neglecting the higher order terms in the series expansion of the cosine, ρ_p is obtained as

$$\rho_p = \frac{\pi^2\sigma^2 a}{8\sigma_{ys}^2} = \frac{\pi}{8}\left(\frac{K_I}{\sigma_{ys}}\right)^2 \tag{2.50}$$

This result can be compared with Eq. (2.45) previously obtained.

The aforementioned estimations of the size of the plastic zone are only for the case of plane stress. The size of the plastic zone in the case of plane strain is smaller than that for the plane stress due to the presence of a plane-strain constraint. Irwin[10] suggested that the effect of the plane-strain constraint may be considered by the increase in the plastic yield stress of the material. The yield stress is estimated to increase by a factor of $\sqrt{3}$. The estimations of the size of the crack tip plastic zone are also based on the assumption that the size of the plastic zone is small compared to the actual crack, which may be true for some materials and geometries. The above estimations cannot be applied to quasi-brittle materials such as concrete, rock, and ceramics because these materials do not exhibit a yielding limit for increasing load. As a result, an inelastic zone that consists of many crack arrest mechanisms such as microcracks and particle bridging, instead of the yield zone, exists at the crack tip in these materials. The size of the inelastic zone in quasi-brittle materials is usually not small compared to the initial crack.

2.7 GRIFFITH FRACTURE THEORY

Fracture of a structure can also be described using energy principles. The first analysis of fracture behavior of structures with a sharp crack was probably developed by Griffith[11] based on an energy consideration. Consider an infinite plate of unit thickness that contains a crack length of $2a$ subjected to uniform tensile stress σ, as shown in Fig. 2-13. The energy change due to the introduction of a crack is

$$U - U_0 = -U_a + U_s \tag{2.51}$$

where U and U_0 are the elastic energies of the cracked and uncracked plate, respectively; U_a is the change in the strain energy due to introducing the crack in the plate, which is equal to $\pi\sigma^2 a^2/E$ based on Inglis[12]; U_s is the change in the elastic surface energy due to formation of the crack surfaces, which is equal to $4a\gamma_s$; E is the modulus of elasticity of the material; and γ_s is the elastic surface energy due to formation of a unit crack surface. Therefore, Eq. (2.51) can be rewritten as

$$U - U_0 = -\frac{\pi\sigma^2 a^2}{E} + 4a\gamma_s \tag{2.52}$$

The energy equilibrium condition for crack extension is obtained by mini-

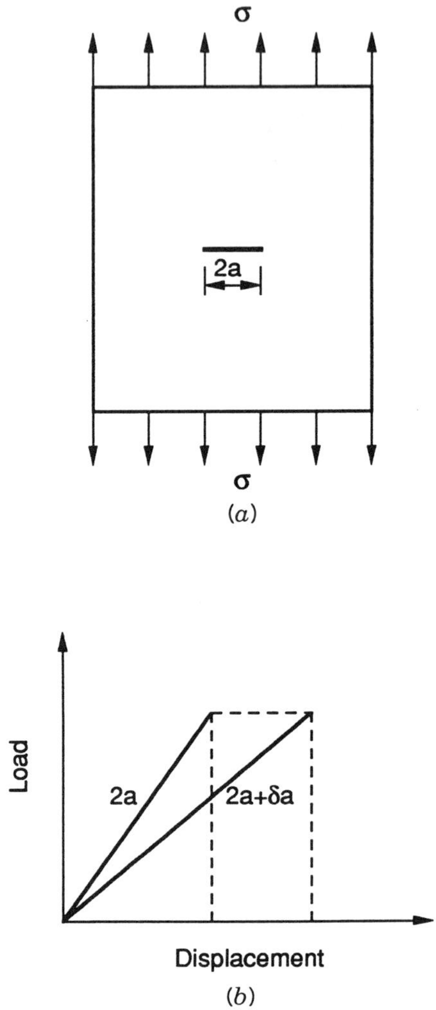

Figure 2-13 Cracked plate subjected to tension: (*a*) plate with crack and (*b*) load–displacement curve.

mizing the energy change with respect to change in the crack length δa (Fig. 2.13). This is done by setting the first derivative of Eq. (2.52) with respect to a equal to zero. This results in

$$\frac{\pi a \sigma^2}{E} = 2\gamma_s \tag{2.53}$$

Equation (2.53) is referred to as the Griffith fracture criterion, which can only be used for ideally brittle materials.

2.8 STRAIN ENERGY RELEASE RATE FOR CRACK PROPAGATION

The energy release rate concept proposed by Griffith can be generalized by introducing the strain energy release rate for crack propagation. To demonstrate the equilibrium states of a structure with a crack, one may consider a plate with a crack subjected to a load P as shown in Fig. 2-14. For simplicity, body forces are neglected and only a single concentrated load is considered below. The total potential energy Π in the structure can be expressed as

$$\Pi = U - F + W \tag{2.54}$$

where U is the strain energy of the structure, F is the work done by the applied force, and W is the energy for crack formation. The necessary condition for the structure in equilibrium states is that the first-order derivative of Π is equal to zero during an infinitesimal crack extension da, that is,

$$\frac{\partial}{\partial a}(U - F + W) = 0 \tag{2.55}$$

or

$$\frac{\partial}{\partial a}(F - U) = \frac{\partial W}{\partial a} \tag{2.56}$$

It is noted that since a change of the equilibrium states results from a change

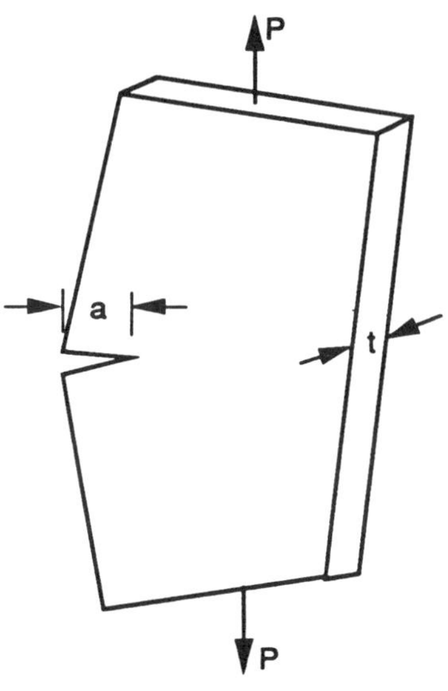

Figure 2-14 Cracked plate subjected to tension.

of the crack length, Π is taken as a derivative with respect to the crack length a. Two types of boundary conditions may be considered for Eq. (2.56). One is that the load point displacement is fixed during crack extension, and the other is that the load keeps constant during crack extension. The work done by the applied force F can be written as a product of the applied load P and the load point displacement u, that is, $F = Pu$. It is noted that $F = Pu = 0$ for the condition of fixed load point displacement. As a result, Eq. (2.56) for the condition of the fixed load point displacement becomes

$$-\left[\frac{\partial U(a, u)}{\partial a}\right]_u = \frac{\partial W}{\partial a} \tag{2.57}$$

where $U(a, u)$ is a function of the crack length and the displacement, and the subscript u indicates that the derivative is to be calculated at constant u in Eq. (2.57). On the other hand, for the condition of the constant applied load Eq. (2.56) is given as

$$\left\{\frac{\partial}{\partial a}[Pu(P) - U(a, P)]\right\}_P = \left[\frac{\partial U^*(a, P)}{\partial a}\right]_P = \frac{\partial W}{\partial a} \tag{2.58}$$

where $u(P)$ indicates that u is a function of P, $U^*(a, P) = Pu(P) - U(a, P)$ is the complementary strain energy and is a function of the crack length and the applied load. The subscript P indicates that the derivative is to be calculated at constant P in Eq. (2.58). By introducing G as the strain energy release rate for the propagation of a unit length of crack in a structure with unit thickness and $G_c = 1/t\, \partial W/\partial a$ as the critical strain energy release rate of materials, where t is the thickness of the specimen, Eqs. (2.56)–(2.58) can be rewritten in a uniform form as

$$G = G_c \tag{2.59}$$

where

$$G = -\frac{1}{t}\left[\frac{\partial U(a, u)}{\partial a}\right]_u = \frac{1}{t}\left[\frac{\partial U^*(a, P)}{\partial a}\right]_P \tag{2.60}$$

It is noted that a negative sign is necessary when the strain energy is expressed as the function of the crack length and the displacement (or strain). For linearly elastic materials, the strain energy U is equal to the complementary strain energy U^*. As a result, Eq. (2.60) reduces to

$$G = -\frac{1}{t}\left[\frac{\partial U(a, u)}{\partial a}\right]_u = \frac{1}{t}\left[\frac{\partial U(a, P)}{\partial a}\right]_P \tag{2.61}$$

Equation (2.59) presents a necessary condition for equilibrium states of a structure during crack propagation. Since any propagation of an initial crack means catastrophic failure of a structure made of linearly elastic materials, Eq. (2.59) can be used as a failure criterion for such materials. Generally, the strain energy release rate G is a function of applied load, structural geometry, and boundary conditions, whereas the critical strain energy release rate G_c is a material fracture constant for linearly elastic materials. Since G provides energy for crack extension, it is also termed as the crack driving force.

In order to use energy principles to characterize crack propagation, the energy release rate G should be calculated. As given by Eq. (2.61), G can be evaluated based on the strain energy U. When the strain energy is used, it should be expressed as a function of the crack length and the strain (or displacement). On the other hand, when the complementary strain energy is used, it should be expressed as a function of the crack length and the stress (or applied load). The strain energy can be obtained from

$$U = \int_V U_d(a, \epsilon)\, dV = \int_V U_d(a, \sigma)\, dV \tag{2.62}$$

where V is the volume of the specimen and $U_d(a, \epsilon)$ and $U_d(a, \sigma)$ are the densities of the strain energy. As shown in Fig. 2-15, $U_d(a, \epsilon)$ and $U_d(a, \sigma)$ can be calculated from a given stress–strain curve for a linearly elastic material,

$$\begin{aligned} U_d(a, \epsilon) &= \int \sigma(a, \epsilon)\, d\epsilon \\ U_d^*(a, \sigma) &= U_d(a, \sigma) = \int \epsilon(a, \sigma)\, d\sigma \end{aligned} \tag{2.63}$$

Although G can be calculated from the condition of either the fixed load or the fixed displacement as shown in Eq. (2.61), values of G obtained from both

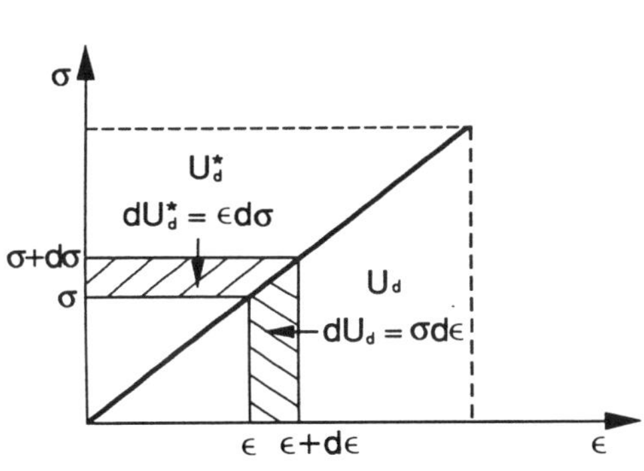

Figure 2-15 Densities for strain energy and complementary strain energy.

conditions are identical. This is illustrated in Fig. 2-16. For the condition of the fixed displacement, the energy release due to the crack extension *da* is represented by the area *OABO* (Fig. 2-16*a*). For the condition of the fixed load, this energy release is represented by the area *OACBO* (Fig. 2-16*b*). Since the energy difference (the small area *ACBA* in Fig. 2-16*b*) is negligible for the crack extension *da*, both conditions arrive at the same result.

The strain energy release rate *G* can be calculated based on compliance of a structure for linearly elastic materials. As shown in Fig. 2-17, the strain energy release rate *G* can be also expressed as the function of structural compliance,

$$U(a, P) = \tfrac{1}{2}Pu = \tfrac{1}{2}CP^2 \tag{2.64}$$

where $C = u/P$ is the structural compliance and is a function of the crack length. Substituting Eq. (2.64) into Eq. (2.61) results in

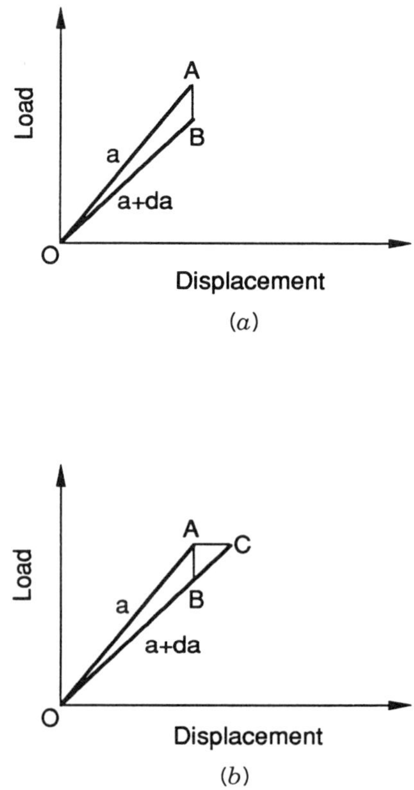

Figure 2-16 Energy evaluations based on (*a*) constant displacement and (*b*) constant load.

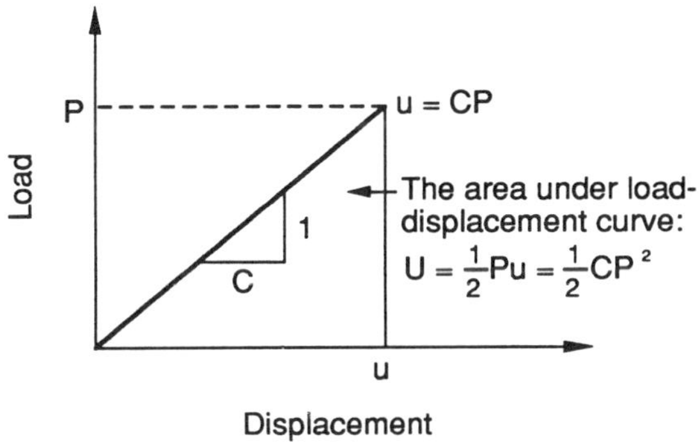

Figure 2-17 Strain energy release rate based on compliance.

$$G = \frac{1}{t}\left[\frac{\partial U(a, P)}{\partial a}\right]_P = \frac{P^2}{2t}\frac{\partial C}{\partial a} \tag{2.65}$$

It can be proven that Eq. (2.65) is valid regardless of whether or not the applied load is constant.[13]

2.9 RELATIONSHIP BETWEEN STRESS INTENSITY FACTOR AND STRAIN ENERGY RELEASE RATE

Since both the strain energy release rate G and the stress intensity factor K are the crack driving forces to open a crack, a relationship between G and K must exist. To derive such a relationship, let us consider an infinite plate with a crack of size $2a$ under tension as shown in Fig. 2-18. The terms G and K are denoted as G_I and K_I, respectively, because the problem discussed here is a mode I fracture. Since the energy released from the crack extension, Δa, is equal to the energy required to close, Δa, one can calculate the G_I value from the energy required to close the crack, Δa, under a closing pressure $\sigma_y = K_\text{I}/(2\pi r)^{1/2}$. As a result, it follows

$$G_\text{I} = \frac{2}{\Delta a}\int_0^{\Delta a} \frac{1}{2}\sigma_y v \, dr = \frac{1}{\Delta a}\int_0^{\Delta a} \frac{K_\text{I} v}{\sqrt{2\pi r}} \, dr \tag{2.66}$$

where v is the displacement in the y direction. The factor 2 is due to the fact that the crack has two surfaces, and the $\frac{1}{2}$ means that the stress increases from zero in proportion with v.

The value of $2v$ is equal to the COD. For the infinite plate subjected to tension, the value of v is given as

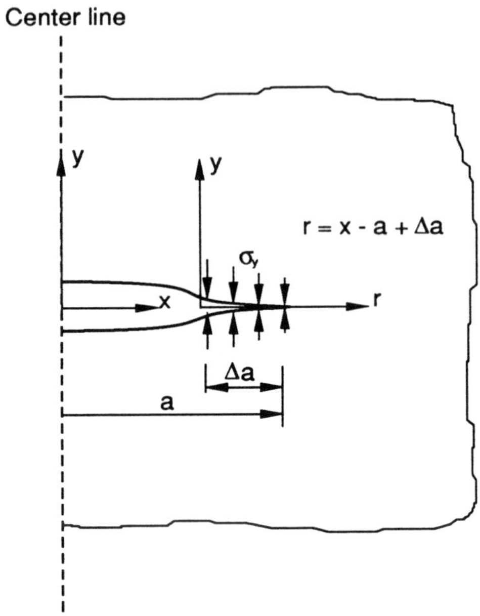

Figure 2-18 Closure of crack tip.

$$v = \frac{\text{COD}}{2} = \frac{2\sigma}{E}\sqrt{a^2 - x^2} = \frac{2K_\text{I}}{E}\sqrt{\frac{a^2 - x^2}{\pi a}} \tag{2.67}$$

Substituting $r = x - a + \Delta a$ into Eq. (2.67) and neglecting second-order terms result in

$$v = \frac{2K_\text{I}}{E\sqrt{\pi}}\sqrt{2\,\Delta a - 2r + \frac{2r\,\Delta a}{a} - \frac{r^2}{a}} \simeq \frac{2K_\text{I}}{E\sqrt{\pi}}\sqrt{2(\Delta a - r)} \tag{2.68}$$

By using Eq. (2.68), Eq. (2.66) becomes

$$G_\text{I} = \frac{2K_\text{I}^2}{E\pi\,\Delta a}\int_0^{\Delta a}\sqrt{\frac{\Delta a - r}{r}}\,dr \tag{2.69}$$

Integrating Eq. (2.69) and letting Δa approach zero lead to

$$G_\text{I} = \frac{K_\text{I}^2}{E} \tag{2.70}$$

It can be proven that Eq. (2.70) is valid for all geometries of specimens under plane stress. For plane strain, E should be replaced by $(1 - \mu^2)/E$, which leads to

$$G_I = (1 - \mu^2)\frac{K_I^2}{E} \tag{2.71}$$

Similarly, it can be shown that for a mode II crack,

$$G_{II} = \begin{cases} \dfrac{K_{II}^2}{E} & \text{for plane stress} \\ (1 - \mu^2)\dfrac{K_{II}^2}{E} & \text{for plane strain} \end{cases} \tag{2.72}$$

and for mode III,

$$G_{III} = \begin{cases} \dfrac{1}{1 - \mu}\dfrac{K_{III}^2}{E} & \text{for plane stress} \\ (1 + \mu)\dfrac{K_{III}^2}{E} & \text{for plane strain} \end{cases} \tag{2.73}$$

If superposition is used for obtaining the total strain energy in combined crack modes, then for plane stress,

$$G = G_I + G_{II} + G_{III} = \frac{1}{E}\left(K_I^2 + K_{II}^2 + \frac{1}{1 - \mu}K_{III}^2\right) \tag{2.74}$$

and for plane strain,

$$G = G_I + G_{II} + G_{III} = \frac{1 - \mu^2}{E}\left(K_I^2 + K_{II}^2 + \frac{1}{1 - \mu}K_{III}^2\right) \tag{2.75}$$

The strain energy release rate G discussed may be valid for linearly elastic materials. These G expressions may still be used when a small inelastic zone is present at a crack tip.[14] Influence of the inelastic zone on the energy release rate will be discussed in the next chapter.

2.10 MEASUREMENT OF K_{Ic} FOR METALLIC MATERIALS

The critical stress intensity factor K_{Ic} may be determined following the concept of Eq. (2.22). For a perfectly brittle material, the critical stress intensity factor is a material constant. However, a certain size of the inelastic zone always exists at the tip of an initial crack for real materials due to plasticity, grain bridging, etc., and the K_{Ic} value is influenced by the size of the inelastic zone. Therefore, to measure the K_{Ic} value for a real material, one needs to design a test procedure that minimizes the influence of the inelastic zone on the measured value of K_{Ic}. The size of the inelastic zone associated with crack propagation in quasi-brittle materials such as concrete and rock is usually larger than that in metallic materials. Introduction of measurement of K_{Ic} for metallic materials, which is briefly presented here, may lead to a better understanding of the measurement of fracture toughness for quasi-brittle materials, which will be discussed in Chapter 6.

It has been reported that the K_{Ic} value for metals decreases with increasing specimen thickness and reaches a rather constant minimum value when the plane-strain condition is approached, as illustrated in Fig. 2-19 for metallic materials.[13] This is due to the fact that the size of the crack tip plastic zone in the plane-strain condition is small compared to that in the plane-stress condition, as discussed in the previous section, and its effect on the K_{Ic} value can be neglected. Therefore, K_{Ic} is termed as the plane-strain critical stress intensity factor or plane-strain fracture toughness and is regarded as a material fracture property. In this chapter K_{Ic} is simply referred to as the critical stress intensity factor or fracture toughness. After considerable study and experimental verification, the ASTM committee on fracture testing, Committee E-24, has issued "Standard Test Method for Plane Strain Fracture Toughness of Metallic Materials" (ASTM Designation E399-83),[14] which is briefly outlined below.

The ASTM standard E399 allows one to use different geometry specimens, such as bend specimens and compact tension specimens, to measure the K_{Ic} value. However, only the bend specimens (three-point bend beams) with a straight-through initial notch, as shown in Fig. 2-20, are mentioned in this

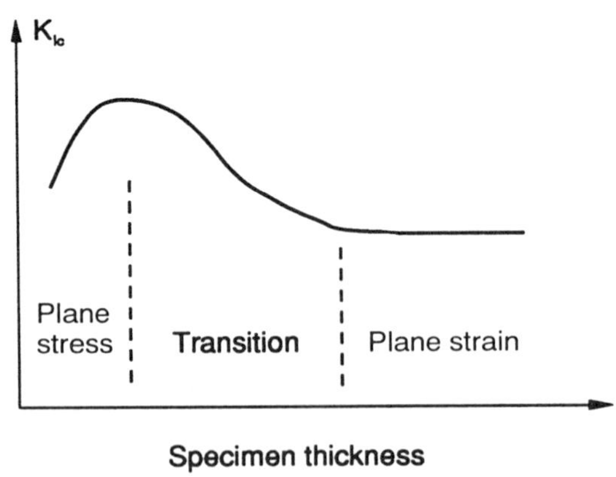

Figure 2-19 Critical stress intensity factor for different conditions.

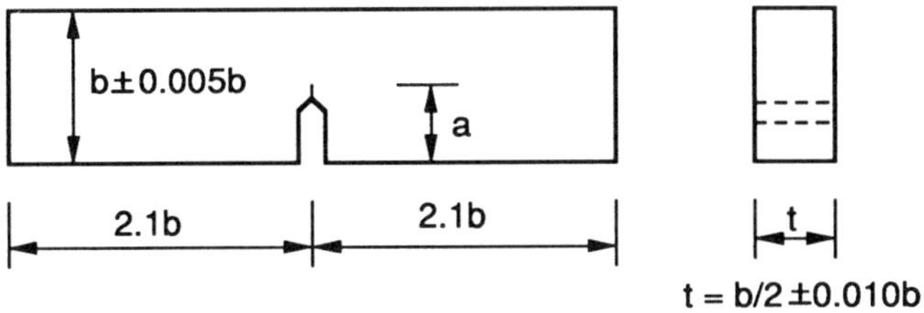

(*a*)

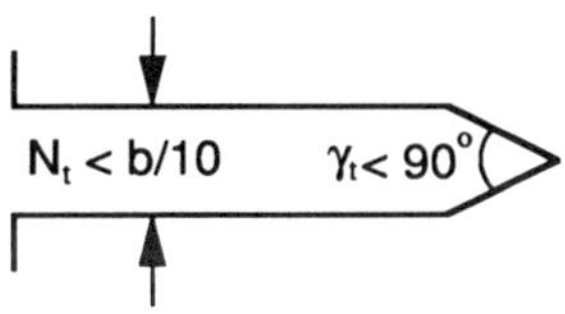

(*b*)

Figure 2-20 Three-point bend beam by ASTM E399: (*a*) beam geometry and (*b*) machine notch geometry.

section. To minimize the influence of the crack tip plastic zone, it is required that

$$\begin{aligned} \text{Crack length } a &= \geq 2.5 \left(\frac{K_{\text{Ic}}}{\sigma_{ys}}\right)^2 \\ \text{Specimen thickness } t &= \geq 2.5 \left(\frac{K_{\text{Ic}}}{\sigma_{ys}}\right)^2 \\ \text{Specimen depth } b &= \geq 5.0 \left(\frac{K_{\text{Ic}}}{\sigma_{ys}}\right)^2 \end{aligned} \tag{2.76}$$

To simulate an ideal plane crack that is assumed in the K_{I} calculation, fatigue cracking beyond the initial machined notch is necessary. Therefore, the crack length is the sum of the initial notch and the fatigue crack. The fatigue crack should extend at least *0.05b* beyond the initial machined notch. The maximum stress intensity factor during fatigue cracking, $K_{f,\max}$ must not exceed 60% of K_Q, which is a conditional calculation of K_{Ic} and will be explained later. The value of $K_{f,\max}/E$ should not exceed *0.002* in.$^{1/2}$ (*0.0032* m$^{1/2}$).

The crack length a is normally equal to the thickness t and is between 0.45 and 0.55 times the width b. The ratio b/t is normally equal to 2.

The fracture test can be performed after the fatigue cracking, and load–displacement curves are recorded during the fracture test. The specimen is loaded at a rate such that the rate of increase of stress intensity is within the range of 30–150 ksi·in.$^{1/2}$/min (0.55–2.75 MPa·m$^{1/2}$/min). From the measured load–displacement curve, the line whose slope is 5% less than the original slope is drawn, and the load P_5 is obtained as shown in Fig. 2-21. The load P_Q is then defined for three principal types of load–displacement curves measured as follows: If the load at every point on the record, which precedes P_5, is less than P_5, then $P_Q = P_5$ (Fig. 2-21, type *1*); if, however, there is a maximum load preceding P_5 that is larger than P_5, then $P_Q = P_{max}$ (Fig. 2-21, types *2* and *3*). The ratio of P_{max}/P_Q usually does not exceed 1.10. For this case, the stress intensity factor K_Q can be calculated using the obtained P_Q:

$$K_Q = \frac{3P_Q S\sqrt{\pi a}}{b^2 t}\, g_1\left(\frac{a}{b}\right)$$

$$g_1\left(\frac{a}{b}\right) = \frac{1.99 - (a/b)(1 - a/b)[2.15 - 3.93a/b + 2.70(a/b)^2]}{\sqrt{\pi}(1 + 2a/b)(1 - a/b)^{3/2}} \tag{2.77}$$

It is noted that the value of a is the sum of the initial machine notch and the fatigue crack.

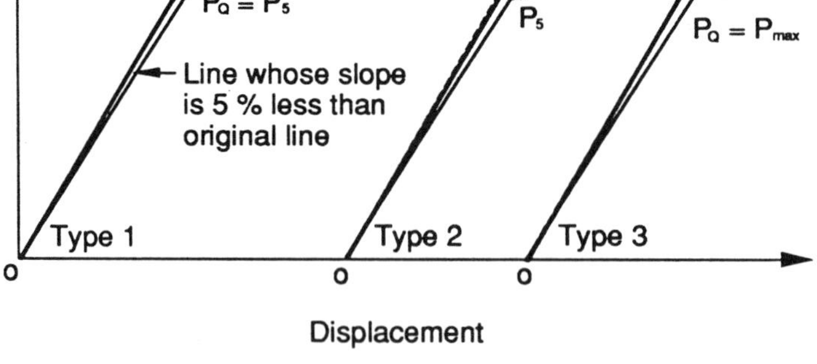

Figure 2-21 Types of load–displacement curves for determining K_{Ic}.

If the obtained value of K_Q meets Eq. (2.76) and other subsequent requirements aforementioned, then $K_Q = K_{Ic}$.

If the ratio P_{max}/P_Q does exceed 1.10, proceed to calculate the strength ratio r_{sb} as follows:

$$r_{sb} = \frac{6P_{max}b}{t(b - a)^2\sigma_{ys}} \tag{2.78}$$

where σ_{ys} is the 0.2% offset yield strength of the material in tension. If $r_{sb} < 1$, the specimen has failed before yielding, which means that the behavior is linear elastic, and the K_Q value should be equal to K_{Ic}. For $r_{sb} > 1.5$ or above, the behavior is elastic–plastic.

2.11 DESIGN BASED ON LINEAR ELASTIC FRACTURE MECHANICS

As mentioned earlier, propagation of a crack is governed by Eq. (2.22) (e.g., $K_I = K_{Ic}$). For brittle materials, propagation of the initial crack may mean catastrophic failure of the structure. Therefore, Eq. (2.22) can also be used as a design criterion to select materials or structural dimensions. This is similar to the stress criterion used in a conventional design. Recalling the left side of Eq. (2.22), K (K_I *or* K_{II}) is a function of load, specimen geometry, boundary condition, and crack size and can generally be expressed as

$$K = \sigma\sqrt{a}g_1 \tag{2.79}$$

where σ is the applied load, a is the initial flaw of materials, and g_1 is the function depending on specimen geometry and size. For the right side of Eq. (2.22), K_c (K_{Ic}, *or* K_{IIc}) is a material fracture property. After introducing a safety factor F_D, Eq. (2.22) can be rearranged in the form of an inequality as

$$F_D\sigma_c\sqrt{a}g_1 \leq K_c \tag{2.80}$$

where σ_c is the load capacity of the structure. Equation (2.80) is convenient for design. For a given material with a known initial flaw a, the load capacity of the structure, σ_c, can be determined from Eq. (2.80). For a prescribed load capacity, the material characterized by the values of K_c and a can be selected.

The relationship among the load capacity σ_c, the flaw size a, and the fracture toughness K_c governed by Eq. (2.80) is illustrated in Fig. 2-22, where the safety factor is equal to unity. The load capacity decreases with increasing flaw size but increases with increasing fracture toughness. Flaws are always present in materials, and their sizes are usually determined by the material manufacturing process. The a value may be selected based on the maximum flaw size

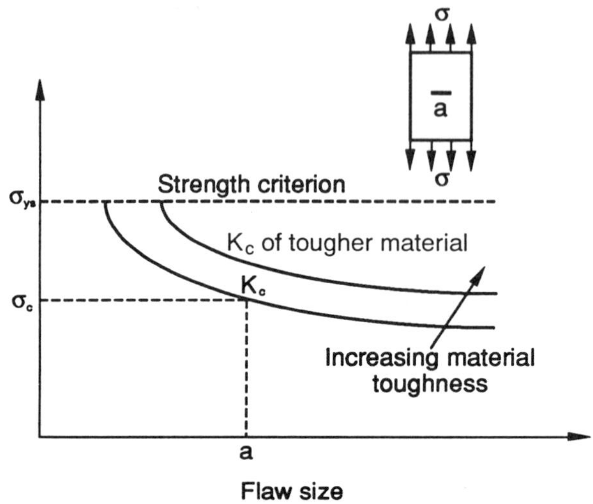

Figure 2-22 Illustration of relationship among load capacity, flaw size, and fracture toughness.

of materials. One can determine the maximum flaw size that can be tolerated during a certain period of service using Eq. (2.80). A conventional strength criterion for design is also shown in Fig. 2-22. Since the stress criterion does not account for the stress singularity at the crack tip, it usually results in an unsafe design when the flaw size is larger than a certain value. It is noted that the value of the material fracture toughness K_c also depends on temperature, loading rate, and load history.[15]

2.12 SUMMARY

Principles of LEFM have been presented in this chapter. The elastic stress field around a crack tip has been derived using a complex stress function approach. The elastic stress component in the direction perpendicular to the crack was found to approach infinity at the crack tip. This crack tip stress singularity can be characterized by a stress intensity factor, which can be derived using elasticity for different structures with cracks. Stress intensity factors for some useful geometries have been given. Plastic zones in real materials resulting from the stress singularity at the crack tip were discussed. Analysis of crack propagation based on equilibrium of energies was discussed. The concept of the strain energy release rate was introduced, and the relationship between the stress intensity factor and the strain energy release rate was derived. Measurement of the critical stress intensity factor for metals as recommended by ASTM was briefly presented. Fracture criteria in the design approach were discussed.

REFERENCES

1. Muskhelishvili, N. I., *Some Basic Problems of the Mathematical Theory of Elasticity*, Noordhoff, Holland, 1953 [English translation].
2. Westergaard, H. M., "Bearing Pressures and Cracks," *Journal of Applied Mechanics*, Vol. 61, 1939, pp. A49–53.
3. Jenq, Y. S., and Shah, S. P., "Two Parameter Fracture Model for Concrete," *Journal of Engineering Mechanics*, Vol. 111, No. 10, 1985, pp. 1227–1241.
4. Tada, H., Paris, P. C., and Irwin, G. R., *The Stress Analysis of Cracks Handbook*, 2nd ed., Paris Productions, St. Louis, MO, 1985.
5. Murakami, Y., et al., *Stress Intensity Factors Handbook*, Pergamon, New York, 1987.
6. Gettu, R., Bazant, Z. P., and Karr, M. E., "Fracture Properties and Brittleness of High-Strength Concrete," *ACI Materials Journal*, Vol. 87, No. 6, 1990, pp. 608–618.
7. Go, C. G., Hu, K. K., and Swartz, S. E., "Stress Intensity Factors for Single Edge Notch Beams," *Journal of Engineering Mechanics, ASCE*, Vol. 110, No. 4, 1984, pp. 629–634.
8. Irwin, G. R., "Plastic Zone Near a Crack and Fracture Toughness," in *Proceedings of the Seventh Sagamore Ordnance Materials Conference*, Syracuse University, 1960, p. IV-63.
9. Dugdale, D. S., "Yielding of Steel Sheets Containing Slits," *Journal of Mechanics, Physics and Solids*, Vol. 8, 1960, pp. 100–108.
10. Irwin, G. R., "Linear Fracture Mechanics, Fracture Transition, and Fracture Control," *Engineering Fracture Mechanics*, Vol. 1, No. 2, 1968, pp. 241–257.
11. Griffith, A. A., "The Phenomena of Rupture and Flow in Solids," *Philosophical Transactions of the Royal Society of London*, Series A 221, 1920, pp. 163–198.
12. Inglis, C. E., "Stresses in a Plate due to the Presence of Cracks and Sharp Corners, *Transaction Institute, Naval Architects*, Vol. 55, 1913, pp. 219–241.
13. Broek, D., *Elementary Engineering Fracture Mechanics*, 4th rev. ed., Martinus Nijhoff, Dordrecht, 1986.
14. "Standard Method of Test for Plane-Strain Fracture Toughness of Metallic Materials," ASTM Designation E 399-83, in *ASTM Annual Standards*, American Society for Testing and Materials, Philadelphia, PA, 1990, pp. 488–512.
15. Barson, J. M., and Rolfe, S. T., *Fracture and Fatigue Control in Structures*, 2nd ed., Prentice-Hall, Englewood Cliffs, NJ, 1987.

3

PRINCIPLES OF NONLINEAR FRACTURE MECHANICS

3.1 ENERGY PRINCIPLES FOR CRACK PROPAGATION IN NONLINEAR MATERIALS

The performance of a material can be evaluated by its stress–strain response. For a linear elastic material, elastic response is terminated and stress suddenly drops to zero when it reaches fracture strength of the material, as shown in Fig. 3-1*a*. For an elastic–plastic material, shown in Fig. 3-1*b*, stress approaches a constant after a certain value of strain. However, stress in a quasi-brittle material gradually decreases after the maximum stress. A softening-type stress–strain curve as shown in Fig. 3-1*c* is usually observed for quasi-brittle materials such as concrete and ceramics.

The strain energy release rate has been introduced to describe fracture of a linear elastic material in the previous chapter. The similar energy principles can be used to describe crack propagation in nonlinear materials, which can be nonlinear elastic, elastic–plastic, or quasi-brittle. In a linear elastic material, any propagation of a crack means catastrophic failure of the material. However, in a nonlinear material, a crack may steadily propagate until it reaches a critical length. The use of energy principles to describe the stable crack propagation and the critical fracture for nonlinear materials will be discussed in this section.

As in Chapter 2, let us consider a plate with a crack subjected to a load P. The necessary condition for the plate in equilibrium states during crack propagation is that the first-order derivative of the potential energy Π with respect to the crack length a is equal to zero:

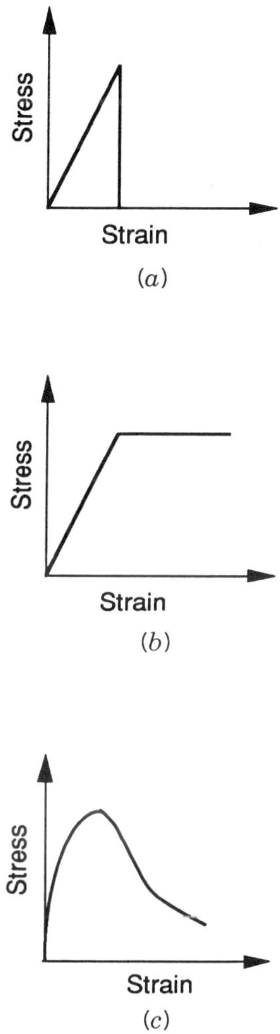

Figure 3-1 Stress–strain curves for different materials: (*a*) linear elastic (brittle) material, (*b*) elastic–plastic material, and (*c*) quasi-brittle material.

$$\begin{aligned} \frac{\partial \Pi}{\partial a} &= \frac{\partial}{\partial a}(U - F + W) = 0 \\ \frac{\partial}{\partial a}(F - U) &= \frac{\partial W}{\partial a} \end{aligned} \tag{3.1}$$

where U is the strain energy of the structure, F is the work done by the applied force, and W is the energy required for crack formation. By introducing two types of boundary conditions, the fixed-load-point displacement or the fixed-load, during crack extension, Eq. (3.1) becomes

$$G_q = R \tag{3.2}$$

and

$$G_q = -\frac{1}{t}\left[\frac{\partial U(a,\, u)}{\partial a}\right]_u = \frac{1}{t}\left[\frac{\partial U^*(a,\, P)}{\partial a}\right]_P \tag{3.3}$$

where G_q is the energy release rate for propagation of unit length of crack in a structure with unit thickness, $R = 1/t\ \partial W/\partial a$ is fracture resistance of the material, and t is the thickness of the structure. In Eq. (3.3) the subscripts u and P indicate that the derivatives are to be calculated at a constant displacement or a constant load, respectively. It is noted that although Eq. (3.3) is similar to the expression for G given in Chapter 2, G_q is calculated from the strain energy or the complementary strain energy for nonlinear materials rather than from linear elastic materials. Equation (3.2) presents a necessary condition for equilibrium states of the structure during crack propagation. Generally, the energy release rate G_q is a function of applied load, structure geometry, and boundary conditions, whereas the fracture resistance R is a function of material fracture property and structural geometry. It is noted that different from linear elastic materials where $R = G_c$ is a material constant, R is a rising function of the crack length for nonlinear materials.

In nonlinear materials, propagation of a crack may be stable, stationary, or unstable because of the presence of crack arrest mechanisms in the inelastic zone around the crack tip. The stable crack growth means that the crack propagates only when the applied load or displacement increases, whereas the unstable crack growth means that the crack may propagate even though the applied load decreases or keeps constant. The stationary crack propagation is a critical state between the stable and unstable crack growths. These three different equilibrium states are illustrated in Fig. 3-2 by a ball on different surfaces, where the potential energy Π is assumed to be a function of position x. Mathematically, the second-order derivative of Π should be further considered to obtain sufficient conditions identifying different equilibrium states for crack propagation. Noting $\partial\Pi/\partial a = R - G_q$, these sufficient conditions can be expressed as

$$\frac{\partial^2\Pi}{\partial a^2} = \begin{cases} \dfrac{\partial R}{\partial a} - \dfrac{\partial G_q}{\partial a} > 0 & \text{for stable crack growth} \\[2ex] \dfrac{\partial R}{\partial a} - \dfrac{\partial G_q}{\partial a} = 0 & \text{for stationary crack growth} \\[2ex] \dfrac{\partial R}{\partial a} - \dfrac{\partial G_q}{\partial a} < 0 & \text{for unstable crack growth} \end{cases} \tag{3.4}$$

In order to use the energy principles to characterize crack propagation, the energy release rate G_q should be calculated. As given by Eq. (3.3), G_q can be evaluated based on the strain energy U or the complementary strain energy U^*.

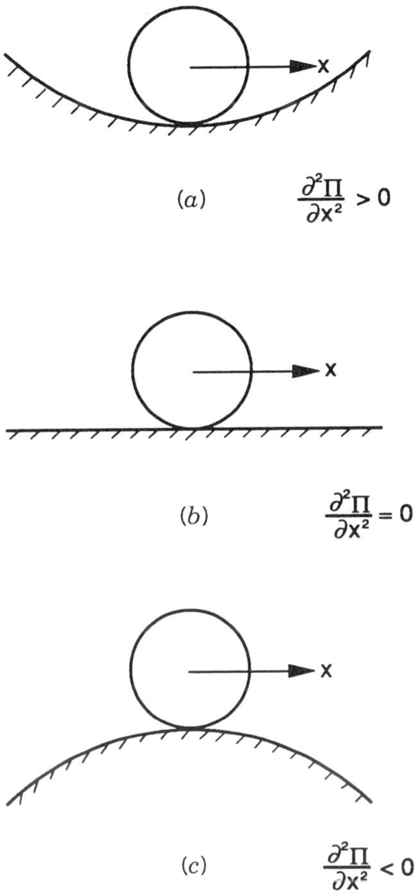

Figure 3-2 Illustration of different equilibrium states: (*a*) stable equilibrium state, (*b*) stationary equilibrium state, and (*c*) unstable equilibrium state.

It is noted that the value of U is not equal to the value of U^* for a nonlinear material. When the strain energy is used, it should be expressed as the function of the crack length and the strain (or displacement). On the other hand, when the complementary strain energy is used, it should be expressed as the function of the crack length and the stress (or applied load). The strain energy and the complementary strain energy can be obtained from

$$U(a, \epsilon) = \int_V U_d(a, \epsilon)\, dV \qquad U^*(a, \sigma) = \int_V U_d^*(a, \sigma)\, dV \tag{3.5}$$

where V is the volume of the specimen, $U_d(a, \epsilon)$ and $U_d^*(a, \sigma)$ are densities of the strain energy and the complementary strain energy, respectively. Both $U_d(a, \epsilon)$ and $U_d^*(a, \sigma)$ can be calculated from a given stress–strain curve for a material, as shown in Fig. 3-3:

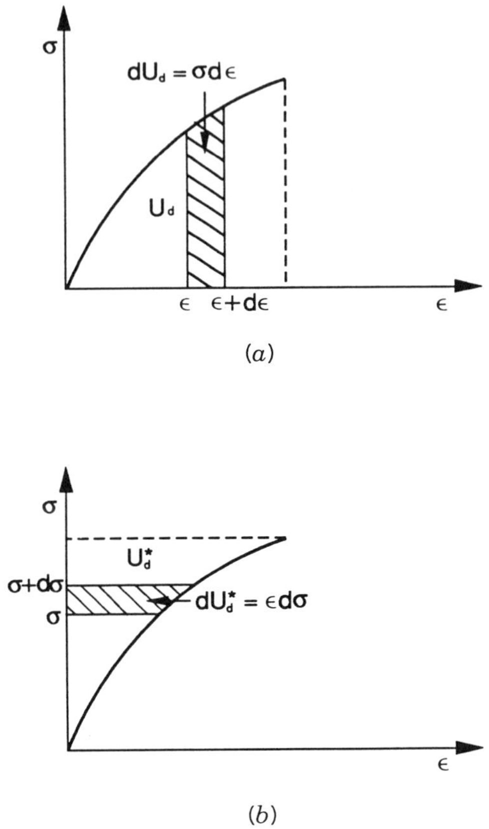

Figure 3-3 Densities of (*a*) strain energy and (*b*) complementary strain energy.

$$U_d(a, \epsilon) = \int \sigma(\epsilon)\, d\epsilon \qquad U_d^*(a, \sigma) = \int \epsilon(\sigma)\, d\sigma \tag{3.6}$$

Although G_q can be calculated from the condition of either the fixed load or the fixed displacement, as shown in Eq. (3.3), values of G_q obtained from both conditions are identical. This is similar to G given in Chapter 2.

It is noted that since no material properties are involved, the energy principles discussed above can be used for any nonlinear material including nonlinear elastic, elastic–plastic, and quasi-brittle materials.

3.2 *J*-INTEGRAL FOR NONLINEAR ELASTIC MATERIALS

For nonlinear elastic materials, the energy release rate G_q is equal to the *J*-integral proposed by Rice,[1] which is defined as

$$J = \int_\Gamma \left[U_d(a, \epsilon)\, dy - \mathbf{T} \frac{\partial \mathbf{u}}{\partial x}\, ds \right] \tag{3.7}$$

where Γ is any closed contour followed counterclockwise surrounding the crack tip in a stressed body, as shown in Fig. 3-4; U_d is the density of strain energy given by Eq. (3.6); $\mathbf{T} = (T_x, T_y)$ is the tension vector perpendicular to Γ in the outside direction; $\mathbf{u} = (u_x, u_y)$ is the displacement vector at ds; and ds is an arc element along Γ. It can be shown that the J-integral defined by Eq. (3.7) is independent of selection of Γ if stress–strain behavior of materials can be described by the plastic deformation theory, which is equivalent to nonlinear elastic and requires that the unloading curve should be the same for loading, as shown in Fig. 3-5*a*. For elastic–plastic and quasi-brittle materials, the value of the J-integral is not path independent and is not equal to the energy release rate G_q. It is noted that $J = G$ for a linear elastic material.

The J-integral can be used as a fracture criterion due to its path independence, which is valid only for materials exhibiting elastic stress–strain behavior. When a crack extends, some unloading may take place in the inelastic zones in the wake and front of the newly formed crack tip. For most real materials the unloading is linearly elastic and is governed by the plastic flow theory, as shown in Fig. 3-5*b*. The presence of the inelastic zone in real materials often violates the requirement for the path independence in the J-integral. Therefore, the J-integral can only partially account for the influence of the crack tip inelastic zone in real materials and be used when this inelastic zone is relative small with respect to the size of the region within the contour Γ in Eq. (3.7). Otherwise, the energy required for propagation of unit area of a crack cannot be calculated based on a selected local contour Γ, and G_q, which was obtained based on the equilibrium of energies dissipated in the whole structure, should be used.

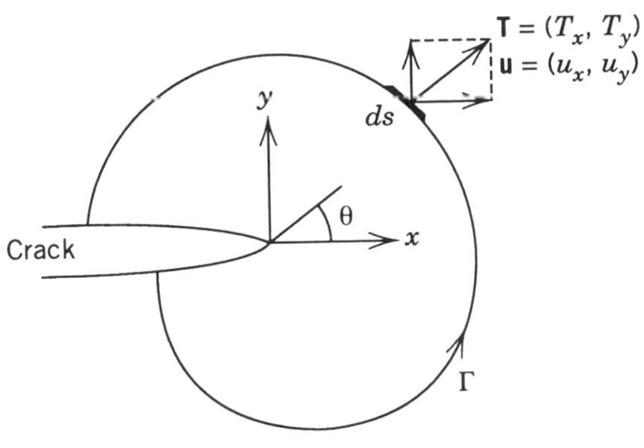

Figure 3-4 Contour around crack tip for J-integral.

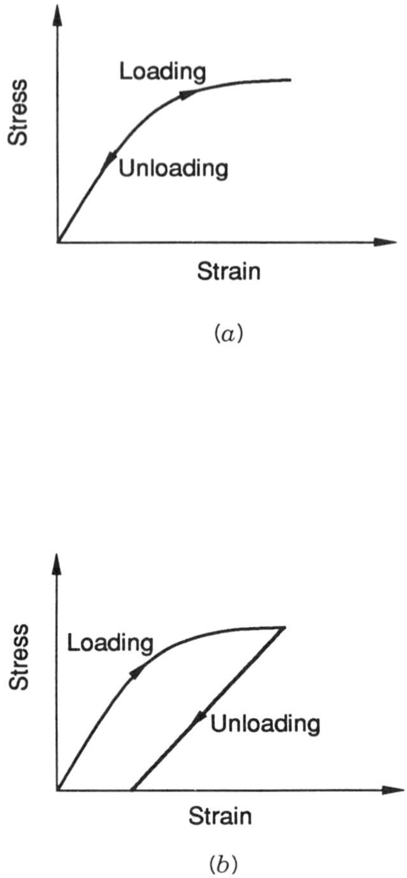

Figure 3-5 Loading and unloading curves for different elastic–plastic materials: (*a*) nonlinear–elastic or elastic–plastic governed by deformation theory and (*b*) elastic–plastic governed by flow theory.

3.3 FRACTURE RESISTANCE (*R*-CURVE)

Based on Eqs. (3.2) and (3.4), the fracture resistance R can be used to characterize the energy required to create a unit crack area in a structure. Therefore, R is a crack resistance force and a function of R is usually termed an R-curve. Generally, R-curves are a function of structure geometry and material fracture property. However, for materials with a very small inelastic crack tip zone, R-curves may be regarded as only material dependent. R-curves are a useful tool to account for stable crack propagation resulting from the presence of crack arrest mechanisms in the inelastic zone around a crack tip.

To further understand the implications of an R-curve, consider a structure with an existing initial crack of length a_0 and unit thickness. Application of load to the structure produces the strain energy U. The energy release rate $G_q = \partial U/\partial a$ is a function of the crack length a and the applied load P for the given structure. On the other hand, the crack propagation at the crack tip needs to consume some energy W, and the fracture resistance $R = \partial W/\partial a$. Both G_q and R are expressed as functions of crack extension, as illustrated in Fig. 3-6. First consider linear elastic materials. Since no inelastic zone forms at the crack tip, an ideally linear elastic material is perfectly brittle. As a result, R is a material constant regardless of specimen geometry. As P increases from zero to P_1, G_q increases along the vertical axis until point 0, and the crack initiates at load P_1, where $G_q = R$ as given in Eq. (3.2). Since any increase in crack length makes $G_q > R$, the propagation of the crack in the brittle materials is unstable and leads to catastrophic failure of the structure. However, due to the presence of crack arrest mechanisms in the inelastic zone around the crack tip, the R-curve for inelastic materials rises monotonically, as shown in Fig. 3-6. Again, with the increase of P the crack begins to extend at the point 0 for inelastic materials. Assume that a small crack extension Δa_1 occurs at load P_1. Then, R increases from 0 to 2, and G_q increases from 0 to 1 along line G_{q1}. No further fracture can occur at the same load since $G_q < R$. Let P continuously increase such that G_q increases from 1 to 2 where again some crack extension occurs, and so on. Here, $G_q = R$ and $\partial G_q/\partial a < \partial R/\partial a$ at points 0 and 2 indicate that stable crack propagation occurs at these points; that is, the crack propagates only when P increases. It can be seen that there is stable crack propagation until point 3, where $G_q = R$ and $\partial G_q/\partial a = \partial R/\partial a$ are satisfied. The point 3 is referred to as the critical point, and Δa_3 is termed the critical crack extension. After the critical point it is always $\partial G_q/\partial a > \partial R/\partial a$, but $G_q = R$ may be achieved by decreasing P. As a result, the crack may further propagate even though the load decreases. Since the load decreases with increasing displacement after the critical point, the softening type of load–displacement relationship is obtained. As discussed above, an R-curve can be used to account for stable crack growth prior to the critical fracture and is expressed as a function of crack extension (or total crack length) and material fracture property for a given structure. Therefore, an R-curve can first be phys-

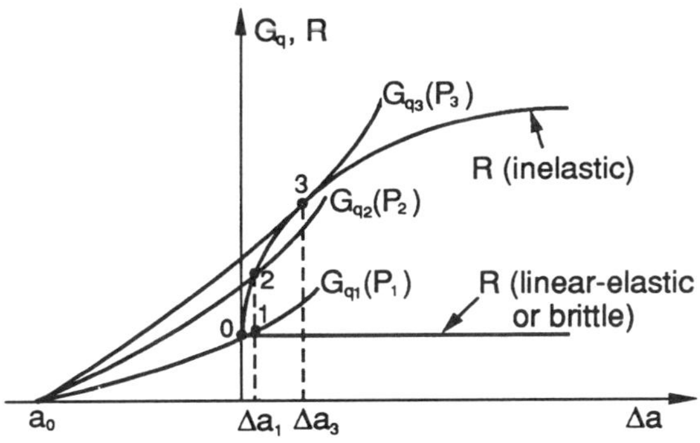

Figure 3-6 Illustration of relationship between G_q-curves and R-curves.

ically interpreted as the fracture resistance to stable crack growth after initiation from an initial crack for a given structure.

It is seen from the above discussion that conditions for a critical crack propagation are

$$(G_q)_c = (R)_c \qquad \left(\frac{\partial G_q}{\partial a}\right)_c = \left(\frac{\partial R}{\partial a}\right)_c \tag{3.8}$$

where the subscript c represents the critical point. It has been suggested that R-curves may be invariant for a certain category of structures. For example, Krafft et al.[2] had proposed that for a given geometry a series of structures with different initial crack lengths may follow the same R-curve. In this sense Eq. (3.8) implies that an R-curve is tangent to G_q-curves for a series of structures at their critical points. Therefore, an R-curve can also be mathematically defined as an envelope of critical energy release rates of G_q for a certain category of structures. Since G_q is specimen geometry dependent, R-curves generally also depend on structure geometry based on Eq. (3.8) except for ideally brittle materials.

Two kinds of interpretations for R-curves have been discussed above. An R-curve is physically interpreted as fracture toughness for stable crack growth in terms of the crack extension for a given structure. On the other hand, an R-curve is mathematically defined as an envelope of G_q-curves for a certain category of structures because it is tangent to these G_q-curves at their critical points, as indicated in Eq. (3.8). If one can derive an R-curve based on Eq. (3.8) and substitute the R-curve into the condition $G_q = R$ to describe the stable crack growth, the two kinds of interpretations of R-curves are equivalent. Generally, the crack extension in a structure is influenced by the material fracture property, applied load, initial crack length, and a boundary condition that includes the geometry and size of structures. It is noted that R-curves should be independent of the applied load and a single R-curve can only be used for a series of structures with the same material fracture property and the same geometry. As a result, R-curves are a function of the initial crack length a_0 and the characteristic size of structures, b. Here the characteristic size of structures, b, may be interpreted as the width of a tensile specimen or the depth of a notched beam, as illustrated in Fig. 3-7. Since different values of a_0 and b result in different crack extensions, it is possible to mathematically categorize R-curves as different types of envelopes of G_q-curves by changing parameters b and a_0 for a given geometry. As a result, R-curves may be categorized into three types of envelopes of G_q-curves for a given structure geometry and material fracture property[3]:

(i) *Fix b and Change a_0.* an R-curve is defined as an envelope of G_q-curves for a series of structures with the same size but different initial crack lengths, as shown in Fig. 3-8.

(ii) *Fix a_0 and Change b.* an R-curve is defined as an envelope of G_q-

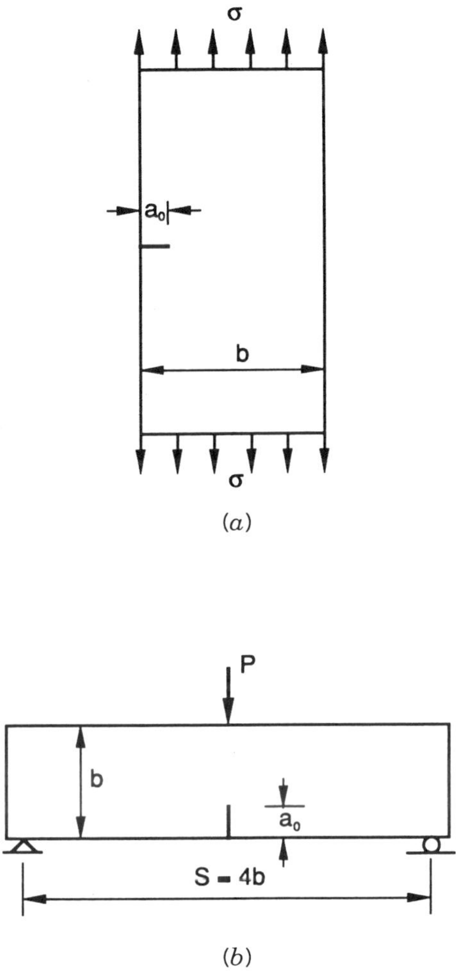

Figure 3-7 Definition of characteristic size *b* for some geometries: (*a*) single-edge cracked tensile specimen and (*b*) three-point bend beam.

curves for a series of structures with different sizes but the same initial crack length, as shown in Fig. 3-9.

(iii) *Proportionally Change both a_0 and b (Geometrically Similar).* an *R*-curve is defined as an envelope of G_q-curves for a series of geometrically similar structures with increasing sizes (that is, both b and a_0 are varied but the ratio of these two quantities is kept the same) as shown in Fig. 3-10.

The first category of *R*-curves was initially suggested by Krafft et al.[2] and has been widely used to experimentally measure the fracture toughness of ceramics.[4,5] The *R*-curve is fitted using data points obtained from testing of specimens with the same size but different initial crack lengths. The different initial

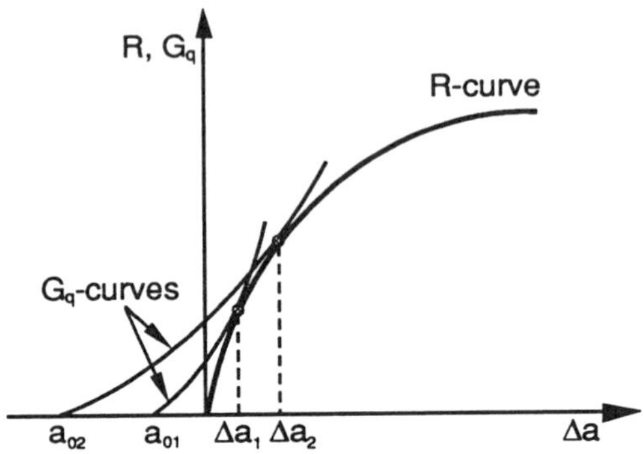

Figure 3-8 First category of *R*-curves as envelope of G_q-curves for series of structures with same size but different initial crack lengths.

crack lengths are used to obtain different critical crack extensions. Since size of structures, b, is fixed in this case, ligament sizes of specimens, $b - a_0$, decrease with increasing initial crack lengths a_0. The values of $b - a_0$ may become comparable with size of the crack tip inelastic zone when the a_0 values are large enough. This may result in size-dependent *R*-values from testing. In order to obtain an *R*-curve that is independent of structure size, it is required that the inelastic zone around a crack should be small compared to the ligament size of the tested specimen. Therefore, the first category of the *R*-curve may only be used when the crack tip inelastic zone is small compared to the structure size.

The second category of *R*-curves may be represented by the *R*-curve devel-

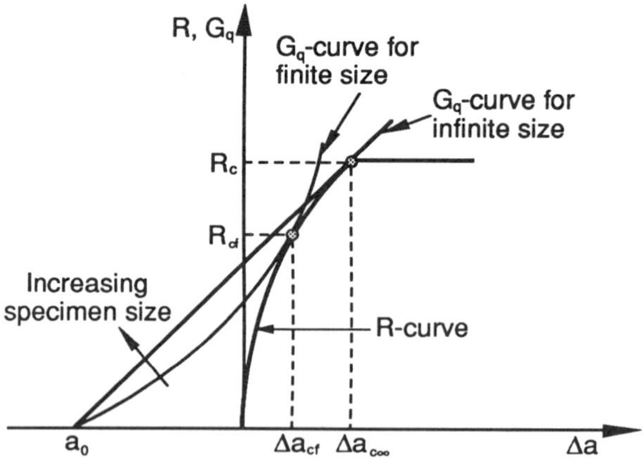

Figure 3-9 Second category of *R*-curves as envelope of G_q-curves for series of structures with same initial crack length but different sizes.

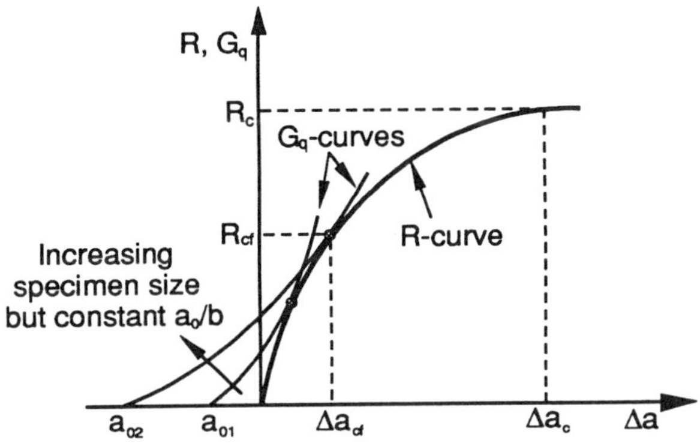

Figure 3-10 Third category of R-curves as envelope of G_q-curves for series of geometrically similar structures (change specimen size but keep ratio of a_0/b constant).

oped by Ouyang and Shah,[3] whereas the R-curve proposed by Bazant and Kazemi[6] is classified as the third category. These two categories of R-curves are interpreted as envelopes of G_q-curves for a series of structures with increasing sizes and are constructed in semianalytical and semiexperimental ways. Mathematical functions of R-curves containing some constants are first derived based on Eq. (3.8), which is regarded as the basic definition of R-curves. These unknown constants are then determined based on values of R_c and Δa_c (or $\Delta a_{c\infty}$), as indicated in Figs. 3-9 and 3-10. These $\Delta a_{c\infty}$ are obtained from the critical conditions of infinite-size specimens for a given geometry. Since the crack tip inelastic zone is always small compared to the ligament size when specimen sizes approach infinity, the second and third categories of R-curves can be used regardless of size of the crack tip inelastic zone. Detailed derivations of the second and third categories of R-curves will be presented in Chapter 7.

It is noted that since R-curves may be defined as envelopes of G_q-curves for different categories of structures, the R-curve is not unique for a given structure. In other words, different categories of R-curves can be used to describe fracture behavior for a given structure. Different categories of R-curves will be different even for the same structure. However, different categories of R-curves should predict the same critical fracture properties for the given structure.

3.4 CRACK TIP OPENING DISPLACEMENT

Since the J-integral can only be applied to fracture of a structure with a small crack tip inelastic zone for real materials that are often elastic–plastic or quasi-brittle, one needs to seek another approach to describe fracture with a large scale of crack tip inelastic zone. Since the presence of the crack tip inelastic strain can be taken into account by extending the originally physical crack by r_p, some opening displacement occurs at the position of original crack tip, as shown in Fig. 3-11. This CTOD may be used as a measurement of the magnitude of the crack tip inelastic strain. Wells[7] proposed that the CTOD can be used to describe fracture toughness of inelastic materials. The CTOD criterion is particularly suitable for some metals that have such high toughness that the values of K_{Ic} cannot be measured.

To better understand the CTOD fracture criterion, let us consider an infinite plate with a crack of size $2a$ subjected to tension. The COD for the structure is shown in Fig. 3-11 and is expressed as

$$\text{COD} = \frac{4\sigma}{E}\sqrt{a^2 - x^2} \tag{3.9}$$

By introducing the crack length correction r_p, which accounts for the effect of the crack tip inelastic zone, Eq. (3.9) becomes

$$\text{COD} = \frac{4\sigma}{E}\sqrt{(a + r_p)^2 - x^2} \tag{3.10}$$

where $a + r_p$ is the effective crack size. Since CTOD is the value of COD at the position of $x = a$, it follows that

$$\text{CTOD} = \frac{4\sigma}{E}\sqrt{2ar_p} \tag{3.11}$$

where the second-order term of r_p is neglected. Substituting $r_p = K_{\mathrm{I}}^2/(\pi\sigma_{ys}^2)$ into Eq. (3.11) and using $K_{\mathrm{I}} = \sigma\sqrt{2a}$ results in

$$\text{CTOD} = \frac{4\sqrt{2}}{\pi\sigma_{ys}}\frac{K_{\mathrm{I}}^2}{E} = \frac{4\sqrt{2}}{\pi\sigma_{ys}}G_{\mathrm{I}} \tag{3.12}$$

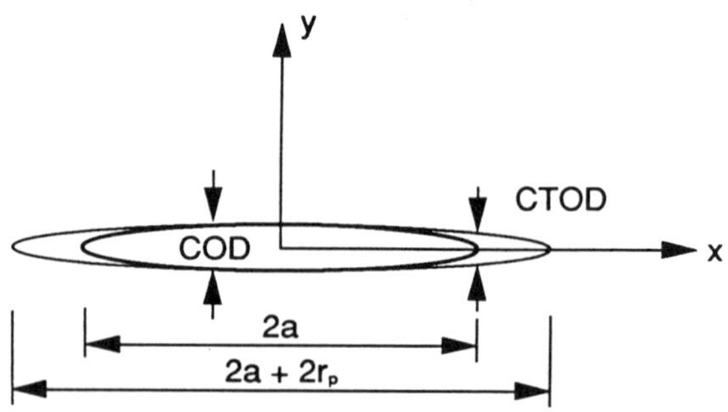

Figure 3-11 Crack opening displacement and CTOD.

Equation (3.12) can be rewritten as

$$\mathrm{CTOD} = \frac{K_{\mathrm{I}}^2}{m\sigma_{ys}E} = \frac{G_{\mathrm{I}}}{m\sigma_{ys}} \tag{3.13}$$

where m is a constant and its value has been found to lie between 1 and 2 based on different estimations of the crack tip inelastic zone.[8] Equation (3.13) is for the plane-stress condition, and E should be replaced by $E/(1 - \mu^2)$ for the plane strain condition. Equation (3.13) indicates that for linear elastic materials the fracture criterion based on CTOD is equivalent to that based on the strain energy release rate or the stress intensity factor.

Since CTOD is related to the stress intensity factor by Eq. (3.13), the use of CTOD for linear elastic materials is straightforward. However, the application of CTOD for fracture of inelastic materials is somewhat intuitive. With the presence of the crack tip inelastic zone, CTOD consists of an elastic component, CTOD_e, and an inelastic component, CTOD_p. The values of CTOD_e and CTOD_p can be calculated from the corresponding measurement of CMOD for a given crack profile. The elastic and inelastic components of CMOD, CMOD_e and CMOD_p, respectively, can be obtained from a loading and unloading procedure demonstrated in Fig. 3-12*a*. The CTOD_e value can be calculated using Eq. (3.13). On the other hand, for ductile materials inelasticity usually results from plasticity. In this case the CTOD_p value may be evaluated by using the plastic hinge model, as shown in Fig. 3-12*b*. This leads to

$$\mathrm{CTOD}_p = \frac{\kappa_p(b - a_0)\mathrm{CMOD}_p}{\kappa_p(b - a_0) + a_0} \tag{3.14}$$

where κ_p is a rotational factor defined in Fig. 3-12*b*, a_0 is an initial crack length, and b is a depth of a specimen. It is noted that a linear crack profile for an inelastic component is assumed here. Therefore, the total CTOD value can be obtained[9]:

$$\begin{aligned} \mathrm{CTOD} &= \mathrm{CTOD}_e + \mathrm{CTOD}_p \\ &= \frac{K_{\mathrm{I}}^2}{2\sigma_{ys}E} + \frac{\kappa_p(b - a_0)\mathrm{CMOD}_p}{\kappa_p(b - a_0) + a_0} \end{aligned} \tag{3.15}$$

where m is taken to equal to 2 in Eq. (3.13). Equation (3.15) has been adapted in the standard methods for CTOD testing such as BS 5762[10] and ASTM E1290.[11] It is noted that Eqs. (3.14) and (3.15) may only be valid for ductile fracture in elastic–plastic materials, where plasticity is a dominant mechanism in the crack tip inelastic zone. Equations (3.14) and (3.15) cannot be directly applied to quasi-brittle materials such as concrete and rock, because different crack arrest mechanisms exist in the inelastic fracture zone in such materials.

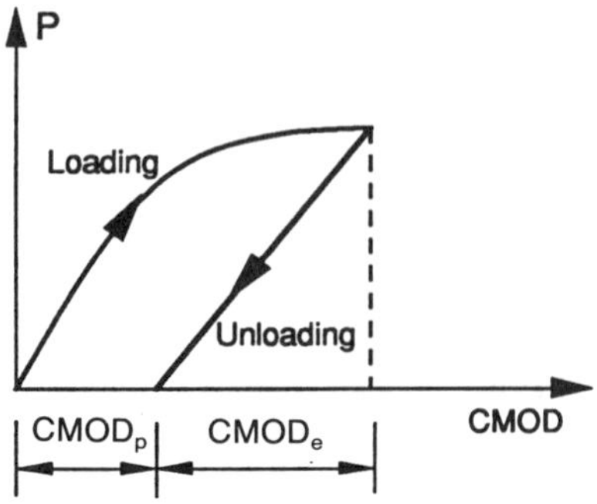

(a)

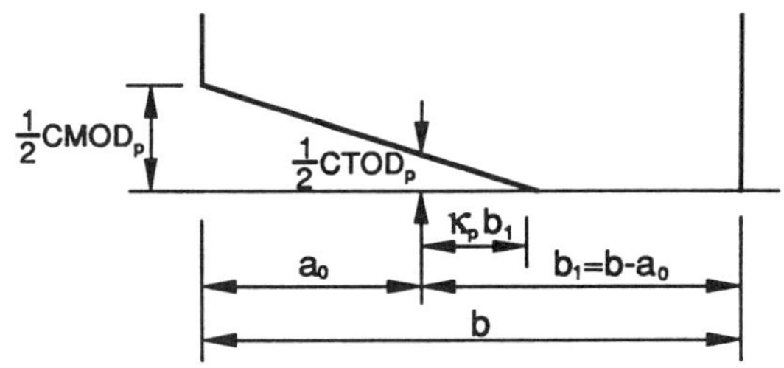

(b)

Figure 3-12 Definition and determination of elastic–plastic components of CTOD: (*a*) elastic–plastic components of CMOD and (*b*) relationship between $CMOD_p$ and $CTOD_p$.

Many studies have shown that the CTOD values are overestimated by using Eq. (3.15) for ductile fracture. Kolednik[12, 13] indicated that the relative error resulting from using Eq. (3.15) is particularly large for specimens with small values of $CTOD_p$. The inadequacy of Eq. (3.15) with respect to the crack size measurement was analytically demonstrated by Kumar.[14] Bhattacharya and Kumar[15] proposed that the total CTOD may be divided into three parts, namely elastic, plastic, and crack growth. It is also found that the $CTOD_c$ values obtained from Eq. (3.15) sometimes depend on the geometry and size of the tested specimens. It is noted that Eq. (3.14) is based on the plastic hinge model. However, under certain conditions, plasticity may spread to the other specimen surface. In such cases the plastic hinge model is not valid. This may partly explain the effect of specimen geometry on the $CTOD_c$ value. To establish CTOD as a powerful tool for describing stable crack growth resulting from fracture with large-scale inelasticity, further studies are required to develop the

CTOD approach that gives a unique toughness fracture parameter for a material.

3.5 MEASUREMENT OF *J–R* CURVES AND CTOD FOR METALLIC MATERIALS

Two approaches, the J-integral and the CTOD, have been presented to characterize nonlinear fracture. For a better understanding of the stable crack propagation in a real material, measurements of the J-integral and CTOD relations for metallic materials will be briefly presented in this section. Based on considerable studies and experiments, the ASTM committee on fracture testing, Committee E-24, has issued two-standard testing methods for measuring the J-integral and CTOD for metallic materials. These include "Standard Test Method for Determining *J–R* Curves" (ASTM designation E1152-87),[16] and "Standard Test Method for Crack-Tip Opening Displacement (CTOD) Fracture Toughness Measurement" (ASTM designation E1290-89).[11] The ASTM Standards E1152 and E1290 allow the use of specimens with different geometries, such as bend specimens and compact tension specimens. However, only the method using a single bend specimen (three-point bend beams) is presented here.

The three-point bend beam and the profile of the initial crack used are shown in Fig. 3-13. The ratio of span to width of the beam, S/b, is equal to 4. The thickness t is recommended to be equal to $0.5b$ for measuring the *J–R* curve and to b for measuring CTOD. The initial crack length a_0 is suggested to range from $0.5b$ to $0.55b$.

The beam should be precracked in fatigue. The length of the fatigue crack extension from the initial machined notch should not be less than $0.05a_0$ and not less than 1.3 mm (0.05 in.). For the final 50% of the fatigue precracking extension or 1.3 mm (0.05 in.), whichever is less, the maximum precracking load should not exceed the value of P_M, which is given by

$$P_M = \frac{t(b - a_0)^2\sigma_y}{2S} \tag{3.16}$$

where σ_y is the average of the 0.2% offset tensile yield strength σ_{ys} and the ultimate tensile strength σ_{ts}. The ratio of minimum precracking load to maximum precracking load should not exceed 0.1.

The fracture test can be performed after the fatigue cracking, and the load and load line displacement curve and load and CMOD curve should be recorded during the test. The former will be used to calculate the J-value, whereas the latter will be used to determine the crack extension. The load and load line displacement measurement is not necessary if only CTOD–R curves are of interest. The specimen is loaded at a rate such that the time taken to reach the

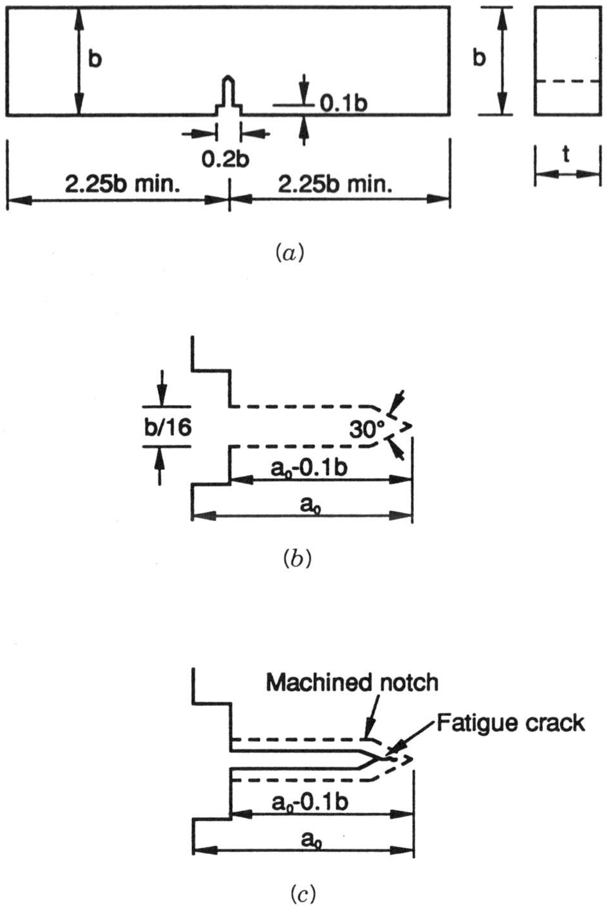

Figure 3-13 Geometry of three-point bend beam for testing: (*a*) dimension of a three-point bend beam, (*b*) envelope of notch, and (*c*) accepted notch.[16]

load P_M lies between 0.1 and 10.0 min. To obtain the J-values and the corresponding crack extensions, a series of unloading/reloading should be performed during testing. The unloading/reloading compliances obtained can be used to calculate the crack extensions at different loading stages. The maximum range of unload/reload for the crack extension measurement should not exceed the smaller of $0.5P_M$ or 50% of the current load. The loading history from the fatigue precracking to the fracture testing is demonstrated in Fig. 3-14.

After testing, values of the J-integral can be calculated from the obtained load and load line displacement curve, whereas values of the crack extensions can be evaluated from the load and CMOD curve using the elastic compliance method. The value of the J-integral is

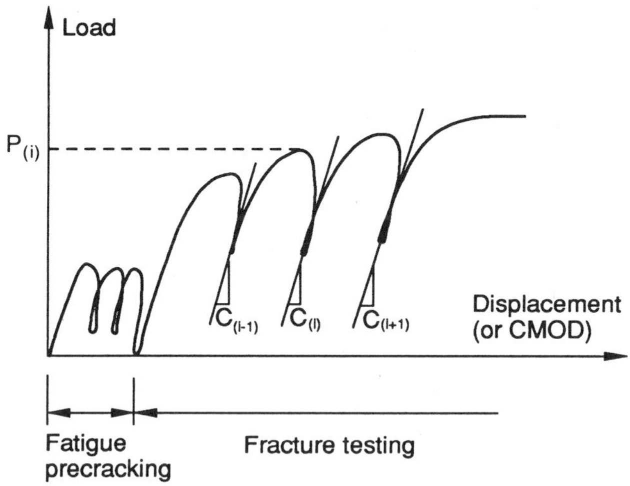

Figure 3-14 Illustration of testing procedures.

$$J = J_{el} + J_{pl} \tag{3.17}$$

where J_{el} and J_{pl} are the elastic and the plastic components of J, respectively. When the values of load $P_{(i)}$, load line displacement $\delta_{(i)}$, and crack length $a_{(i)}$ are given, the corresponding $J_{(i)}$ value can be calculated from

$$J_{(i)} = J_{el(i)} + J_{pl(i)} = \frac{(1 - \mu^2)K_{(i)}^2}{E} + J_{pl(i)} \tag{3.18}$$

where the plane-strain condition is considered and the $K_{(i)}$ value can be calculated from

$$K_{(i)} = \frac{3P_{(i)}S\sqrt{\pi a_{(i)}}}{b^2 t}\, g_1\left(\frac{a_{(i)}}{b}\right)$$

$$g_1\left(\frac{a_{(i)}}{b}\right) = \frac{1.99 - (a_{(i)}/b)(1 - a_{(i)}/b)[2.15 - 3.93\, a_{(i)}/b + 2.70\,(a_{(i)}/b)^2]}{\sqrt{\pi}\,(1 + 2a_{(i)}/b)(1 - a_{(i)}/b)^{3/2}} \tag{3.19}$$

The $J_{pl(i)}$ value is calculated from[17]

$$J_{pl(i)} = \left[J_{pl(i-1)} + \frac{\eta_{(i)}(A_{pl(i)} - A_{pl(i-1)})}{(b - a_{(i)})t}\right]\left[1 - \frac{\gamma_{(i)}(a_{(i)} - a_{(i-1)})}{b - a_{(i)}}\right] \tag{3.20}$$

where $\eta_{(i)} = 2.0$, $\gamma_{(i)} = 1.0$, and the quantity $A_{pl(i)} - A_{pl(i-1)}$ is the increment of plastic area under the load and displacement curve between lines of constant displacement at points $i - 1$ and i. The quantity $A_{pl(i)}$ can be calculated from

$$A_{pl(i)} = A_{pl(i-1)} + \tfrac{1}{2}(P_{(i)} + P_{(i-1)})(\delta_{pl(i)} - \delta_{pl(i-1)}) \tag{3.21}$$

where $\delta_{pl(i)} = \delta_{(i)} - P_{(i)}C_{LL(i)}$, and its value can be determined from knowledge of $a_{(i)}$ using

$$C_{LL(i)} = \frac{1}{Et}\left(\frac{S}{b - a_{(i)}}\right)^2 \left[1.193 - 1.980\,\frac{a_{(i)}}{b} + 4.478\left(\frac{a_{(i)}}{b}\right)^2 - 4.443\left(\frac{a_{(i)}}{b}\right)^3 + 1.739\left(\frac{a_{(i)}}{b}\right)^4\right] \tag{3.22}$$

Using the measured load and CMOD curve, the crack length $a_{(i)}$ can be evaluated as

$$\frac{a_{(i)}}{b} = 0.999748 - 3.9504U_x + 2.9821U_x^2 - 3.21408U_x^3 + 51.51564U_x^4 - 113.031U_x^5 \tag{3.23}$$

where

$$U_x = \frac{1}{(4tbEC_{(i)}/S)^{1/2} + 1} \tag{3.24}$$

and $C_{(i)}$ is the unloading/reloading elastic compliance from the load and CMOD curve.

The upper bound of the J-values for the J–R curve is the smaller of $J_{max} = (b - a)\sigma_y/20$ or $J_{max} = t\sigma_y/20$. The maximum crack extension is given by $\Delta a_{max} = 0.1(b - a_0)$. After calculating J-values for different crack extensions for a given specimen, the J–R curve is established by fitting the data points bounded by the coordinate axes and the J_{max} and Δa_{max} limits using the following power law relationship:

$$J = C_1(\Delta a)^{C_2} \tag{3.25}$$

where C_1 and C_2 are constants determined using a least-squares method.

Only the load and CMOD curve is needed to obtain CTOD. After the crack length $a_{(i)}$ is calculated from Eq. (3.23), the corresponding $\text{CTOD}_{(i)}$ values can be determined as

$$\mathrm{CTOD}_{(i)} = \frac{K_{\mathrm{I}(i)}^2}{2\sigma_{ys}E} + \frac{\kappa_p(b - a_{(i)})\mathrm{CMOD}_{p(i)}}{\kappa_p(b - a_{(i)}) + a_{(i)}} \tag{3.26}$$

where the plastic rotation factor $\kappa_p = 0.44$ and the plane-strain condition is specified. A series of values of $a_{(i)}$ and $\mathrm{CTOD}_{(i)}$ obtained can be fitted by the following suggested relationship:

$$\mathrm{CTOD} = C_1(C_2 + \Delta a)^{C_3} \tag{3.27}$$

where C_1, C_2, and C_3 are constants determined using the least-squares method. The CTOD–R curve obtained here may depend on specimen geometry and dimension.

The measurements of the J–R curve and the CTOD–R curve based on a single specimen have been briefly presented. The J–R curves and CTOD–R curves obtained here represent the fracture resistance of materials to slow stable crack growth after initiation from an initial crack. Since every unloading/reloading procedure during testing is equivalent to testing a new specimen with a different initial crack length, the J–R curve obtained using the above method may be classified as the first category of the R-curve. The obtained J–R curve may be size independent due to the limitation of $\Delta a_{\max} \leq 0.1(b - a_0)$. According to ASTM E1152, the J–R curve obtained from a three-point bend beam is conservative compared to those obtained with tensile specimens. The difference between J–R curves obtained using three-point bend beams and compact tension specimens has also been reported by Gudas and Davis.[18] These results further suggest that R-curves are specimen geometry dependent.

3.6 SUMMARY

Principles of NLFM have been briefly reviewed in this chapter. The concepts of stable, stationary, and unstable crack propagation have been discussed in terms of energy principles. Based on the energy principles, these three types of crack propagation can be described using the energy release rate G_q and the fracture resistance R. The energy release rate G_q is equal to G for linear elastic materials and to J for nonlinear elastic materials. The fracture resistance R is a rising curve for nonlinear materials due to the presence of crack arrest mechanisms in the inelastic zone around a crack tip and depends on material fracture properties and specimen geometry. Since J can only be used for materials exhibiting a small plastic zone at a crack tip, fracture toughness in terms of the CTOD was introduced. The ASTM standard testing methods for measurements of J–R and CTOD–R curves for metallic materials have also been presented.

REFERENCES

1. Rice, J. R., "A Path Independent Integral and the Approximate Analysis of Strain Concentration by Notches and Cracks," *Journal of Applied Mechanics, ASME*, Vol. 35, No. 6, 1968, pp. 379–386.
2. Krafft, J. M., Sullivan, A. M., and Boyle, R. W., "Effect of Dimensions of Fast Fracture Instability of Notched Sheets," *Proceedings of Cranfield Crack Propagation Symposium*, Vol. 1, 1961, pp. 8–28.
3. Ouyang, C., and Shah, S. P., "Geometry Dependent *R*-Curve for Quasi-Brittle Materials," *Journal of the American Ceramic Society*, Vol. 74, No. 11, 1991, pp. 2831–2836.
4. Homeny, J., and Vaughn, W. L., "*R*-Curve Behavior in a Silicon Carbide Whisker/Alumina Matrix Composite," *Journal of the American Ceramic Society*, Vol. 73, No. 7, 1990, pp. 2060–2062.
5. Anderson, R. M., and Braun, L. M., "Technique for the *R*-Curve Determination of *Y-TZP* Using Indentation-Produced Flaws," *Journal of the American Ceramic Society*, Vol. 73, No. 10, 1990, pp. 3059–3062.
6. Bazant, Z. P., and Kazemi, M. T., "Determination of Fracture Energy, Process Zone Length and Brittleness Number from Size Effect, with Application to Rock and Concrete," *International Journal of Fracture*, Vol. 44, 1990, pp. 111–131.
7. Wells, A. A., "Unstable Crack Propagation in Metals: Cleavage and Fast Fracture," *Proceedings of Cranfield Crack Propagation Symposium*, Vol. 1, 1961, pp. 210–230.
8. Rice, J. R., "Mathematical Analysis in the Mechanics of Fracture," in *Fracture: An Advanced Treatise*, Vol. II, edited by Liebowits, H., Academic, New York, 1968, pp. 191–311.
9. Dawes, M. G., "Elastic–Plastic Fracture Toughness Based on the COD and J-Contour Integral Concepts," in *Elastic–Plastic Fracture*, ASTM STP 668, American Society for Testing and Materials, Philadelphia, PA, 1979, pp. 307–333.
10. British Standards Institution (BSI), "Method for Crack-Tip Opening Displacement (CTOD) Testing," BS 5762, BSI, 1979.
11. "Standard Method of Test for Crack-Tip Opening Displacement (CTOD) Fracture Toughness Measurement," ASTM Designation E1290-89, in *ASTM Annual Standards*, American Society for Testing and Materials, Philadelphia, PA, 1990, pp. 911–925.
12. Kolednik, O., "On the Calculation of COD from the Clip-Gauge Displacement in CT and Bend Specimens," *Engineering Fracture Mechanics*, Vol. 29, No. 2, 1988, pp. 173–188.
13. Kolednik, O., "An Improved Procedure for Calculating COD in Bend and CT Specimens," *Engineering Fracture Mechanics*, Vol. 33, No. 5, 1989, pp. 813–826.
14. Kumar, A. N., "Effect of Physical Crack Growth on CTOD Measurement," *International Journal of Fracture*, Vol. 35, 1987, pp. R3–R8.
15. Bhattacharya, S., and Kumar, A. N., "A New Approach for CTOD Evaluation

in Slow Crack Growth Situation,'' *Engineering Fracture Mechanics*, Vol. 40, No. 6, 1991, pp. 1089–1103.

16. ''Standard Method of Test for Determining *J–R* Curves,'' ASTM Designation E1152-87, in *ASTM Annual Standards*, American Society for Testing and Materials, Philadelphia, PA, 1990, pp. 812–822.
17. Ernst, H. A., Paris, P. C., and Landes, J. D., ''Estimations on *J*-Integral and Tearing Modulus *T* from a Single Specimen's Test Record,'' in *Fracture Mechanics*, STP 743, edited by Roberts, R., American Society for Testing and Materials, Philadelphia, PA, 1981, pp. 476–502.
18. Gudas, J. P., and Davis, D. A., ''Evaluation of the Tentative *J–R* Curve Testing Procedure by Round Robin Tests of *HY* 130 Steel,'' *Journal of Testing and Evaluation*, Vol. 10, No. 6, 1982, pp. 252–262.

4

STRUCTURE AND FRACTURE PROCESS OF CONCRETE

4.1 CONSTITUENTS AND MICROSTRUCTURE OF CONCRETE

Constituents and the micro-, mezo-, and macrostructures of cementitious materials are briefly presented in the section. Based on different constituents, cementitious materials can be classified as paste, mortar, or concrete. Paste is a mixture of cement and water. Mortar is a mixture of fine aggregate (sand), cement, and water. Concrete is a composite consisting of cement, fine aggregate, coarse aggregate, and water. Different admixtures may be used to improve properties of any of these cementitious materials. The cement hydrates when it is mixed with water and forms a hard matrix after a certain period of curing. When fluid, the matrix fills the space among aggregates and bonds them together to form hardened concrete.

Hydration products of the cement primarily include calcium silicate hydrate (C–S–H), calcium hydroxide (CH), and calcium sulfoaluminates (ettringite).[1] The C–S–H is a purely crystalline material and is a principal hydration component that makes up about one-half to two-thirds of the volume of the hydrated paste. The C–S–H forms extremely small particles of size less than 1 μm in any dimension. The CH is a crystalline material and occupies about 20–25% of the paste volume. The CH crystals have a distinctive hexagonal prism morphology and typical crystal dimensions in the cement paste range from 0.01 to 0.1 mm. Ettringite usually crystallizes as long, slender prismatic needles, typically 10×0.5 μm, in the cement paste. Ettringite makes up perhaps 10–15% by volume of the cement paste. Porosity is another major component of the microstructure of the hardened cement paste. Porosities in the cement paste are usually classified as gel porosity and capillary porosity. The gel porosities, ranging approximately from 0.5 to 10 nm in size, can be regarded as part of C–S–H, whereas the capillary porosities, varying from 0.01 to 10 μm in size, are remnants of water-filled spaces that exist between the partially hydrated cement grains. The volume of porosity decreases during hydration.

Aggregates are mixed with the cement paste matrix to form concrete. Physically, concrete is a two-phase material consisting of the matrix and the aggregate. Both the matrix and the aggregate are bonded together at the interface. Since bleeding or segregation usually occurs for fresh concrete, some cracks form at the interface between the matrix and the aggregate when the concrete is hardened. On the other hand, since the matrix and aggregate are so different in modulus of elasticity, thermal coefficient, and response to change of moisture content, the interfacial transition zone between the matrix and the aggregate often has more voids and is weaker compared to the bulk cement matrix.[2,3] This weak interfacial zone may be improved by using admixtures and by improving processing techniques for concrete. Since properties of concrete are greatly influenced by this weak interfacial transition zone, concrete may be regarded as a three-phase material including the matrix, the aggregate, and the weak transition zone.

Hardening of fresh concrete is accompanied by a loss of moisture in the cement paste, which causes shrinkage of the concrete. Shrinkage only happens within the cement paste. Shrinkage strain is usually not uniform over a concrete structure because the loss of moisture is different for the surface area and for the interior part of the concrete. On the other hand, shrinkage in the paste is usually restrained by aggregates and boundary conditions of the structure during casting. Both the nonuniform distribution of the shrinkage strain and the restraints from the aggregates and the boundary condition can develop a certain value of tensile stress in some parts of the concrete, which may cause shrinkage cracks before loading.

Properties of concrete are influenced by its chemical constituents and micro-, mezo-, and macrostructures, which are characterized by amounts and distributions of internal pores and cracks. However, only the influence of internal cracks on the mechanical properties of cement-based materials is discussed here. Sizes of porosities that affect the fracture processes and properties depend on various internal structures of materials. A hierarchy of fracture processes in cement-based materials is presented in Fig. 4-1. Since the microstructure of cement paste is in a scale of nanometers, as shown in Fig. 4-1*a*, fracture processes in cement pastes are influenced by particles and voids in the same scale. Internal structures of mortar are shown in Fig. 4-1*b*. Use of sands or fine aggregates results in voids in mortars up to a scale of a micrometer. As a result, fracture processes in mortars may primarily involve initiation and propagation of internal voids on the scale of a micrometer. Interfacial cracks and a weak zone at a scale of a millimeter are major defects for concrete with coarse aggregates, as shown in Fig. 4-1*c*. Therefore, fracture processes in concrete may primarily depend on stability of these interfacial cracks. Although concrete is an anisotropic and heterogeneous material consisting of ma-

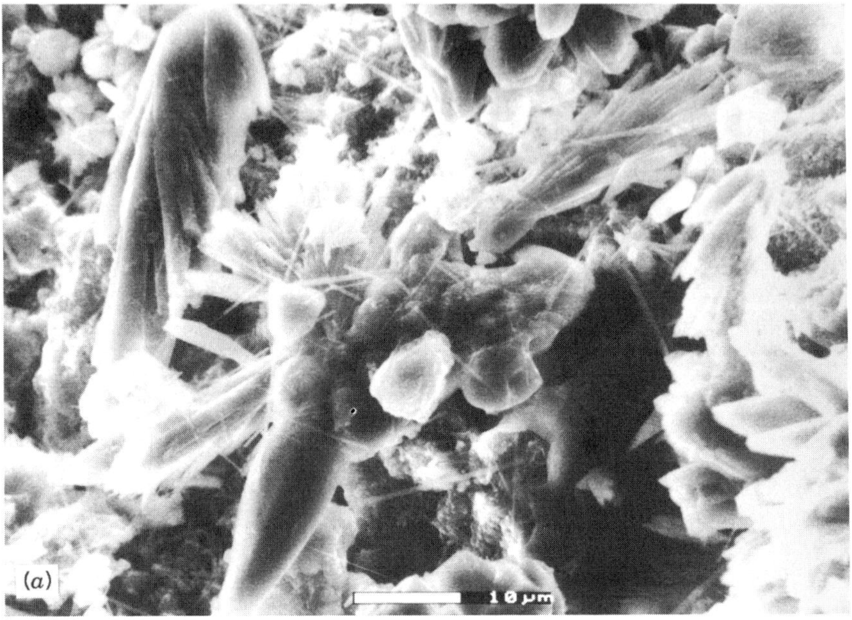

Figure 4-1 Hierarchy of fracture process in cement-based materials: (*a*) cement paste, (*b*) mortar, and (*c*) concrete.

trix and aggregates, it is generally accepted that concrete may be approximately treated as a uniform material before cracking if the minimum dimension of the concrete structure is greater than five times the maximum size of aggregates used.

4.2 FRACTURE BEHAVIOR AND STRAIN LOCALIZATION OF CONCRETE

Concrete is a tension-weak material, and its tensile strength approximately ranges from 8 to 15% of its compressive strength. As discussed earlier, many internal flaws and cracks exist in concrete prior to loading. The mechanical behavior of concrete subjected to different loading conditions is governed by the initiation and propagation of these internal cracks and flaws during loading. For a concrete structure subjected to tension, the cracks propagate in a direction perpendicular to the applied tensile load. On the other hand, for a concrete structure subjected to purely uniaxial compression, the cracks propagate primarily in the same direction as the applied compressive load. Since different mechanical responses of concrete structures can be explained by fracture processes at different loading conditions, one needs to understand when the internal cracks initiate and how they propagate with increasing load.

A concrete plate with a tested area of 126 × 126 mm (5 × 5 in.) subjected

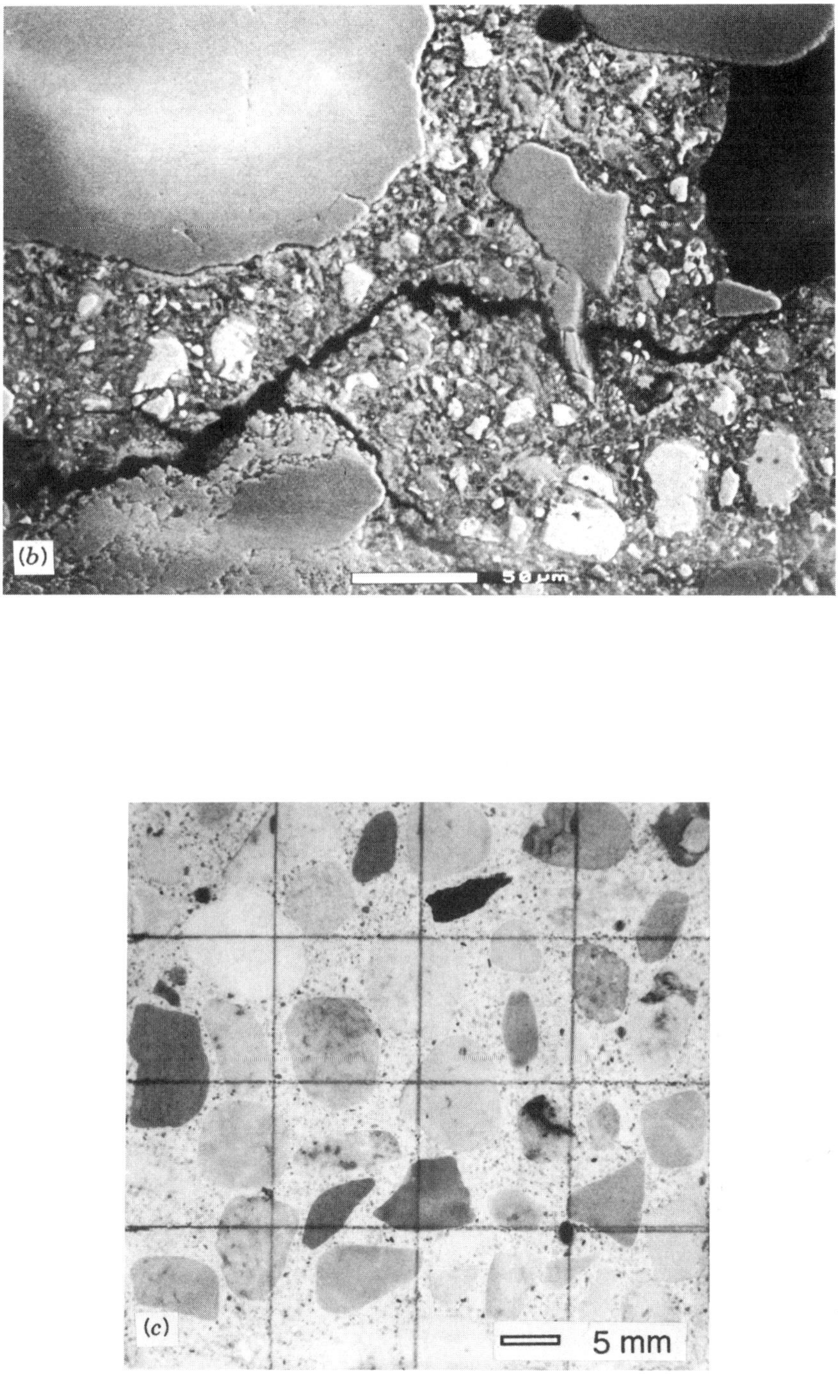

Figure 4-1 (*Continued*)

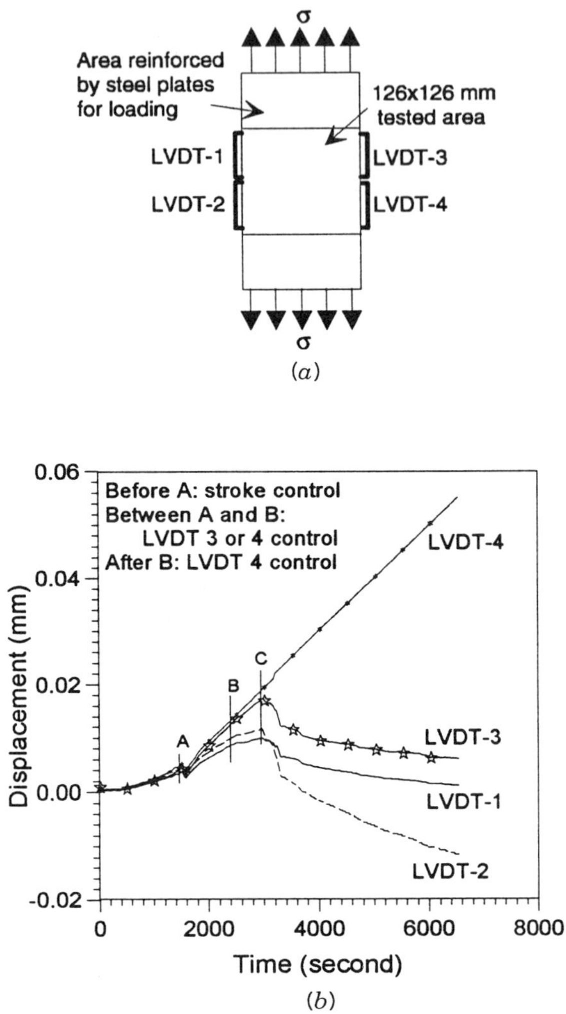

Figure 4-2 Concrete plate subjected to tension: (*a*) specimen geometry and arrangement of LVDTs, (*b*) relationships between displacements and tested time.[4]

to uniaxial tension is shown in Fig. 4-2*a*.[4] Displacements of the plate under tension were measured using four linear variable differential transducers (LVDTs). The plate was loaded in a closed-loop testing machine that allows one to switch feedback control signals during testing. Initiation and propagation of internal cracks were located using an acoustic emission detecting system that will be presented in detail later. The displacements measured from different LVDTs were recorded as functions of the testing time, as shown in Fig. 4-2*b*. The corresponding stress–displacement relationships measured from

LVDT-3 and LVDT-4 are shown in Fig. 4-3. On the other hand, microcracks located in the tested area (see Fig. 4-2*a*) for different loading stages detected by the acoustic emission system are indicated in Fig. 4-4. It is seen that prior to point *A* the specimen behaves elastically and the displacements measured from four LVDTs are essentially the same and linearly increase with the loading (Fig. 4-2*b*). The acoustic emission measurement indicates that initiation of internal cracks is negligible before point *A*. Initiation of the internal cracks was detected at point *A*. As a result of the initiation of the internal cracks, the displacements from different LVDTs start to exhibit some differences, and the readings from LVDT-3 and LVDT-4 are higher than those from LVDT-1 and LVDT-2, as indicated in Fig. 4-2*b*. As the load increased, more and more internal cracks were detected. However, the internal cracks were isolated and more or less uniformly distributed up to point *B*, as shown in Fig. 4-4*a*. After point *B*, the displacement from LVDT-4 increases at a faster rate than that from LVDT-3 (Fig. 4-2*b*). Correspondingly, a narrow band of internal cracks was detected within the measurement range of LVDT-4 by the acoustic emission system (Fig. 4-4*b*). Appearance of this narrow band of internal cracks indicates that damage starts to localize, which is equivalent to a continuous major crack starting to form at the location of the band of microcracks. The length of the band of internal cracks increased after point *B*. This indicates that the major crack was propagating. The peak load was reached at point *C* (Fig. 4-3). After point *C*, the displacement from LVDT-4 continuously increases (Fig. 4-2b). On the other hand, displacements from the rest of the three

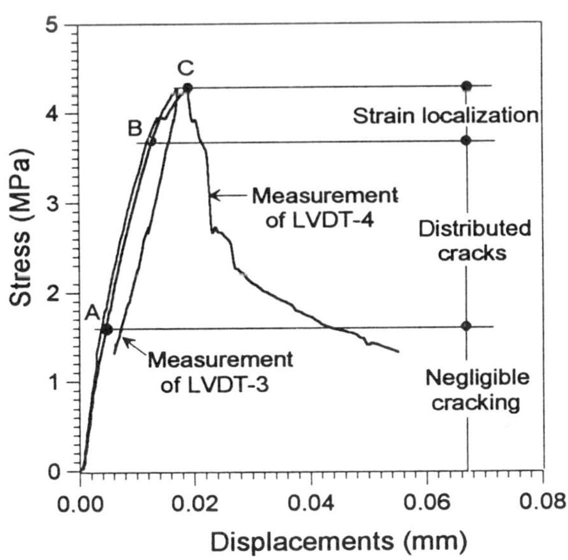

Figure 4-3 Stress–displacement relationships and different cracking stages.[4]

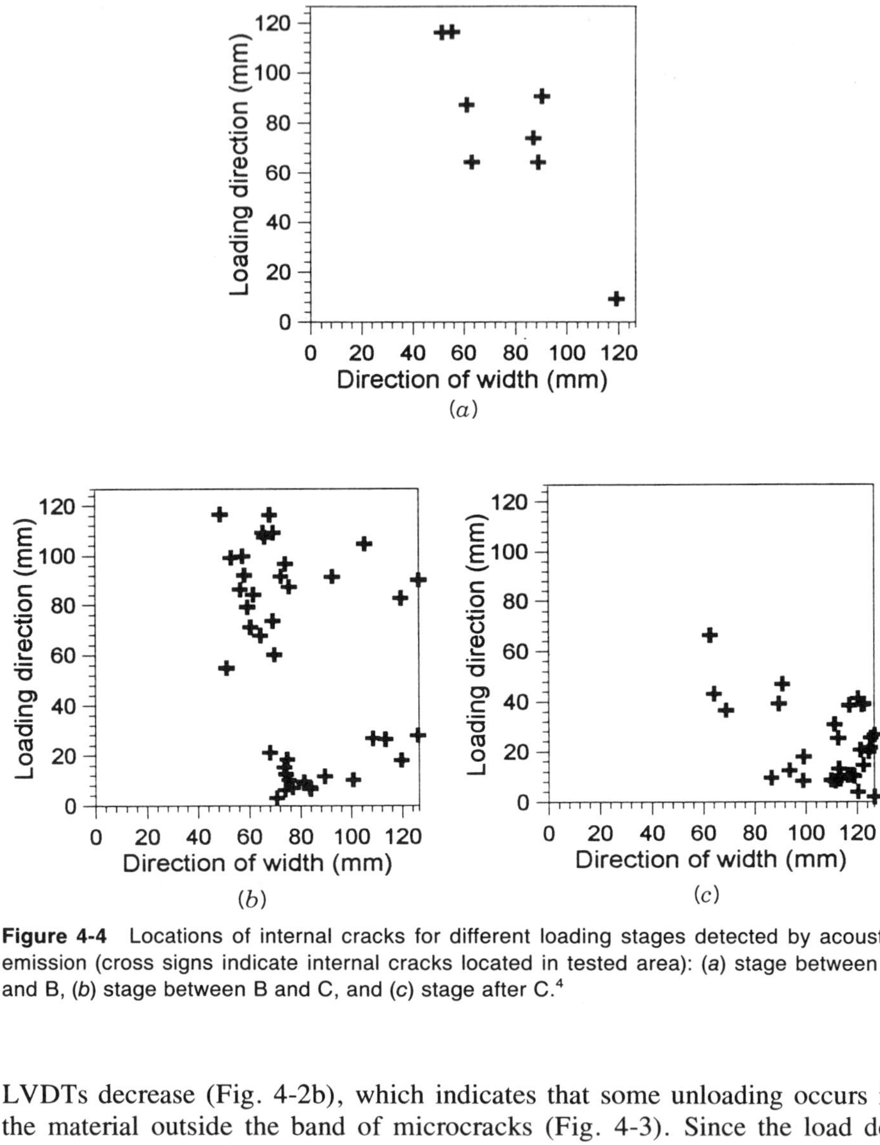

Figure 4-4 Locations of internal cracks for different loading stages detected by acoustic emission (cross signs indicate internal cracks located in tested area): (*a*) stage between A and B, (*b*) stage between B and C, and (*c*) stage after C.[4]

LVDTs decrease (Fig. 4-2b), which indicates that some unloading occurs in the material outside the band of microcracks (Fig. 4-3). Since the load decreases after point *C*, a softening-type stress–elongation curve was obtained.

The above experimental observations indicate that a stress–displacement relationship for a concrete subjected to uniaxial tension can be divided into four stages based on initiation and propagation of internal cracks, as shown in Fig. 4-3. The first stage is before point *A*, which is about 30% of the peak load. Initiation of internal cracks is negligible during the first stage. The second stage is from point *A* to point *B*, which is about 80% of the peak load. The internal cracks initiate and propagate during this stage. But these cracks are isolated and randomly distributed over the specimen volume during the second stage. The third stage is between point B and point C. The internal cracks start to localize into a major crack that propagates with increasing load. This phenomenon is referred to as damage localization or strain localization. The crack propagation is stable until the peak load, where the stable crack growth means that the crack propagates only when the load increases. The crack length at the peak load is referred to as the critical crack length. The fourth stage is after the peak load. The major crack continuously propagates even though the load decreases. The tensile strain of the material within the localized damage band increases, whereas unloading may occur for the material outside the damage band.

A similar trend for concrete under uniaxial compression has been observed.[5] It was reported that distribution of the compressive strain over the specimen is more or less uniform up to approximately 80% of peak load. This indicates that initiation and propagation of internal cracks until 80% of the peak load are approximately random and uniform. Examination of the distribution of internal cracks in the tested specimens confirms that strain localization appears after 80% of the peak load.

Based on these discussions, it can be concluded that the fracture behavior of concrete is characterized by the phenomenon of strain localization. The fact that the tensile strength of concrete is only approximately 8–15% of its compressive strength may be due to the fact that the strain localization appears in higher stress levels in compression compared to tension. Since the localized damage band can be physically simulated by a crack, the presence of the strain localization allows us to use fracture mechanics to describe failure of concrete.

4.3 FRACTURE PROCESS ZONE AND TOUGHENING MECHANISMS

Linear elastic fracture mechanics allows the stress to approach infinity at a crack tip. Since infinite stress cannot develop in real materials, a certain range of inelastic zone must exist at the crack tip. For metallic materials, this inelastic zone is a yielding zone. Since concrete is a heterogeneous material consisting of different phases, the inelastic zone around a crack tip is termed as fracture process zone and dominated by complicated mechanisms. Mindess and Diamond[6] have observed the crack tip in concrete using a scanning electron microscope. They reported that the crack surface in concrete is tortuous and the crack process in concrete is very complicated.

Many mechanisms that are responsible for fracture process in concrete have been reported. Some of these toughening mechanisms are indicated in Fig. 4-5. During fracture, the high-stress state near the crack tip causes microcracking at flaws. Such flaws result from water-filled pores, air voids acquired during casting, and shrinkage cracks due to the curing process. This phenomenon, known as microcrack shielding, is shown in Fig. 4-5*a* and consumes a part of external energy caused by the applied load.[7] Krstulovic-Opara[8] measured mi-

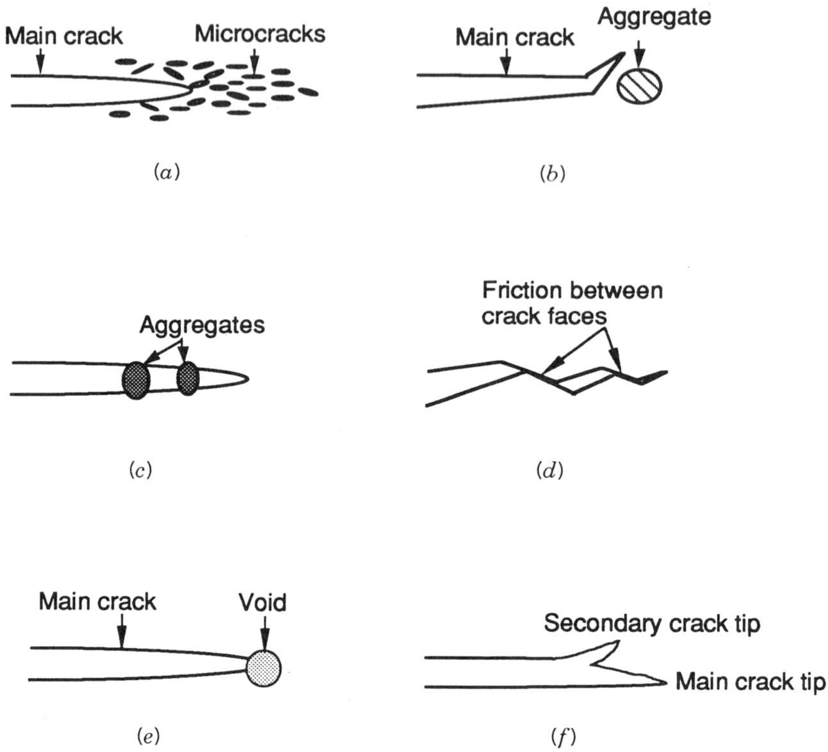

Figure 4-5 Some toughening mechanisms in fracture process zone: (*a*) crack shielding, (*b*) crack deflection, (*c*) aggregate bridging, (*d*) crack surface roughness-induced closure, (*e*) crack tip blunted by void, and (*f*) crack branching.

crocracks in the fracture process zone in compact tension mortar specimens using high-magnification scanning electron microscopy (SEM). He reported that microcracks have random orientation with respect to the main crack plane, and density of microcracks decrease with increasing distance from the face of the main crack. Shum and Hutchinson[9] also discussed toughening resulting from microcracks. Crack deflection occurs when the path of least resistance is around a relatively strong particle or along a weak interface (Fig. 4-5*b*). This mechanism has been studied in detail by Faber and Evans.[10] One of the important toughening processes in concrete is grain bridging[11] as shown in Fig. 4-5*c*. Bridging occurs when the crack has advanced beyond an aggregate that continues to transmit stresses across the crack until it ruptures or is pulled out. Also, during grain pullout or the opening of a tortuous crack, there may be some contact (or interlock) between the cracked faces (Fig. 4-5*d*). This causes energy dissipation through friction and some bridging across the crack. It has been reported that the crack tip is sometimes terminated by internal voids, which produces a blunt tip. This is called crack tip blunting by voids and is shown in Fig. 4-5*e*. Additional energy is required to propagate the crack with a new blunt tip. Another mechanism in the fracture process zone is crack branching, as indicated in Fig. 4-5*f*. The crack may propagate into several branches due to heterogeneity of concrete. More energy should be consumed to form new crack branches.

Microcracking, crack deflection, aggregate bridging, crack-face friction, crack tip blunting by voids, and crack branching are examples of inelastic toughening mechanisms that appear around a crack when it propagates. The presence of these mechanisms generally deter the use of principles of LEFM for concrete. Usually influences of all these mechanisms are lumped together and taken into account by a conceptual fracture process zone. Therefore, to apply fracture mechanics to concrete, one needs to understand and determine this fracture process zone, especially its size and how it changes with propagation of a crack. In the following sections, some experimental measurements on the fracture process zone in concrete are presented.

4.4 EXPERIMENTAL DETERMINATION OF FRACTURE PROCESS ZONE

Many experimental techniques have been used to determine the fracture process zone.[12, 13] No attempt is made to describe all these techniques. However, some experimental results for the fracture process zone in concrete obtained using three different techniques—laser holographic interferometry, acoustic emission, and dye penetration—are briefly presented here.

4.4.1 Laser Holographic Interferometry

Laser holographic interferometry in combination with digital image analysis is a very sensitive method for measuring surface crack propagation.[14] Crack openings as small as 0.25 μm can be detected as interruptions in the fringe patterns. One set of the specimens used as well as the experimental setup for the laser holographic interferometry developed at Northwestern University are shown in Fig. 4-6. Holograms are obtained from interference of two coherence beams: the object beam and the reference beam. Holograms are made at several load steps during testing. Two holograms from two consecutive load steps can be combined to make a sandwich interferogram whose fringe pattern describes displacement occurring between two load steps. Each interferogram provides one component of the displacement vector. By arranging the test setup to provide illumination from three directions, it is possible to create three different interferograms to obtain a complete two-dimensional displacement field, a process made possible by computerized digital image analysis.

Strain fields obtained for loads of 5.9, 6.5, and 7.1 kN are shown in Fig. 4-7. The strain fields show a growing zone of very high strain (>100 microstrain)

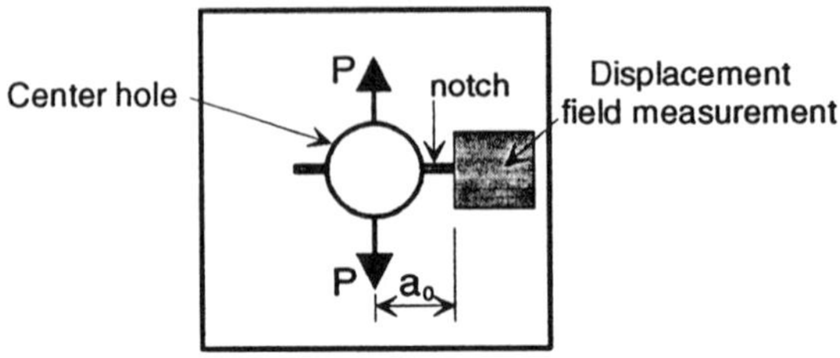

(a)

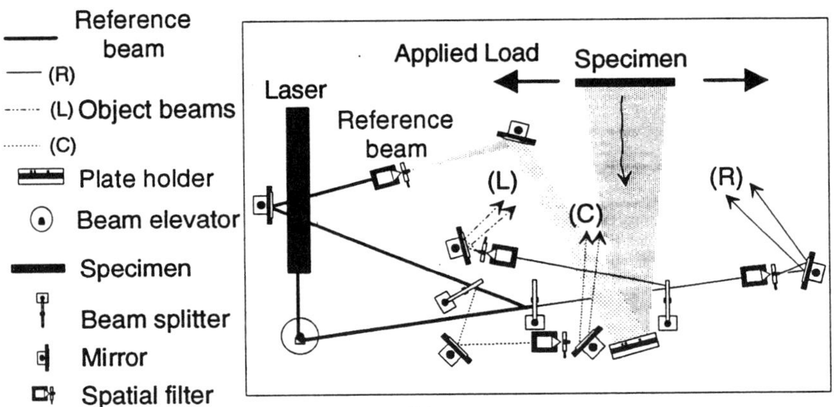

(b)

Figure 4-6 Specimen and experimental set-up for laser holographic interferometry (*a*) a center cracked plate under tension, and (*b*) experimental set-up for laser holographic interferometry.[14]

at the crack tip. It is noted that the line of zero strain is behind the crack tip, indicating that tension is being transmitted across the crack. It is clear that the crack faces are not totally traction free. According to the linear elastic solution, stresses and strains approach infinity at a crack tip and suddenly drop to zero behind the crack tip. This is impossible in a real material. A fracture process zone may be interpreted as the difference between experimentally measured and linear elastic strain fields. The difference between the linear elastic solution and measured values is shown in Fig. 4-8. Positive areas (zone *A*) indicate regions where the linear elastic solution exceeds the measured value by 60 microstrain or more. These areas are near the crack tip and show the difference between the infinite linear elastic solution and the finite strains in the fracture process zone of concrete. Note that this zone is basically a constant size re-

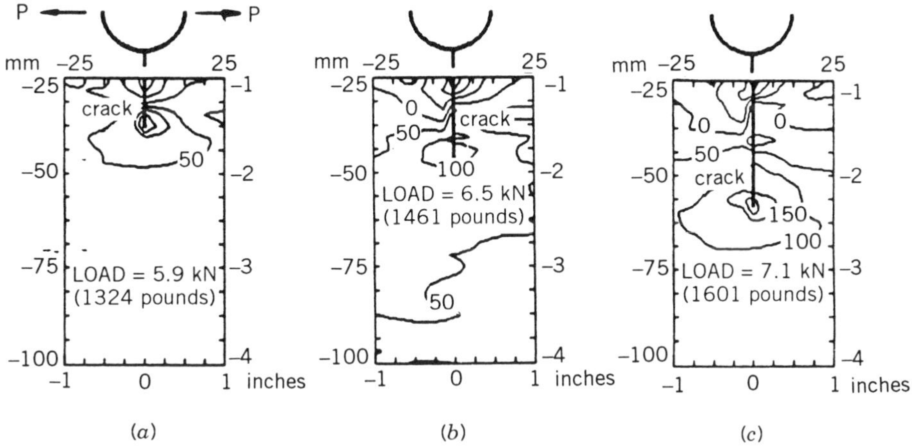

Figure 4-7 Strain fields measured by laser holographic interferometry: (*a*) crack length = 1.375 in. (34.9 mm) and load = 1324 lb (5890 N); (*b*) crack length = 1.875 in. (47.6 mm) and load = 1461 lb (6500 N); and (*c*) crack length = 2.250 in. (57.2 mm) and load = 1601 lb (7120 N).[14]

gardless of the length of the crack. Zone *B* located at the wake of the propagating crack tip is where the holographic measurements exceed the linear elastic solution by more than 60 microstrain. This zone is substantial and increases with crack extension. From such observations it can be concluded that most of the toughening in concrete occurs in the wake of the propagating crack tip. This fact may suggest that the increasing size of the wake process zone rather

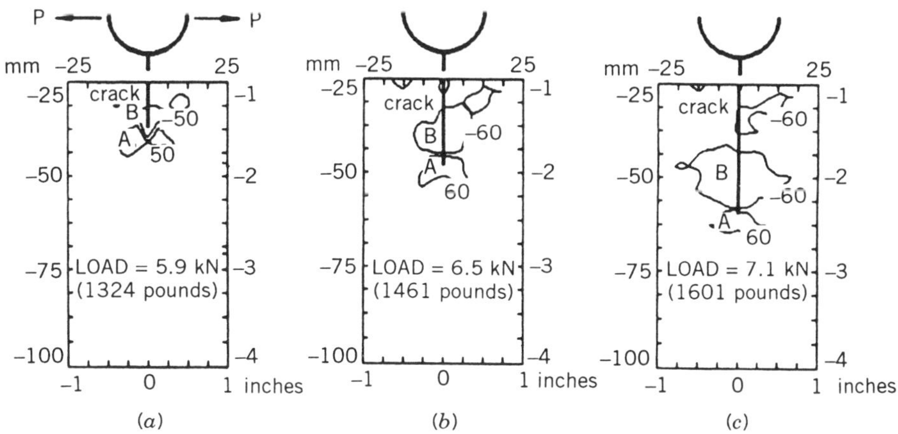

Figure 4-8 Difference between holographic and LEFM strain fields: (*a*) crack length = 1.375 in. (34.9 mm) and load = 1324 lb (5890 N); (*b*) crack length = 1.875 in. (47.6 mm) and load = 1461 lb (6500 N); and (*c*) crack length = 2.250 in. (57.2 mm) and load = 1601 lb (7120 N).[14]

than the constant size of the crack tip process zone should be primarily responsible for the growing R-curve behavior in concrete. Van Mier[15] has obtained similar results by using vacuum impregnation techniques. This result was further confirmed by Krstulovic-Opara[8] using a scanning electronic microscope. The same observations were also reported for other quasi-brittle materials.[16,17]

4.4.2 Acoustic Emission

An acoustic emission is a localized, rapid release of strain energy in a stressed material. This energy release causes stress waves to propagate through the specimen. These waves can be detected on the specimen surface and analyzed to deduce the magnitude and the nature of damage present in materials.[18,19] A typical experimental setup as developed at Northwestern University for an acoustic emission measurement is shown in Fig. 4-9. The acoustic emission setup usually consists of acoustic emission sensors, a data acquisition system, and a control unit such as a personal computer. When a specimen is loaded, acoustic emission signals resulting from microcracking can be detected by acoustic sensors mounted on the specimen at different locations. The signals are picked up by the data acquisition system, which is controlled by the computer, and stored in the computer. Useful insights about microcracking can be drawn from the acoustic measurement. From arrival times of the first P-wave of the acoustic emission signals, which were detected by the sensors mounted at different positions in a single event, the locations of microcracking can be

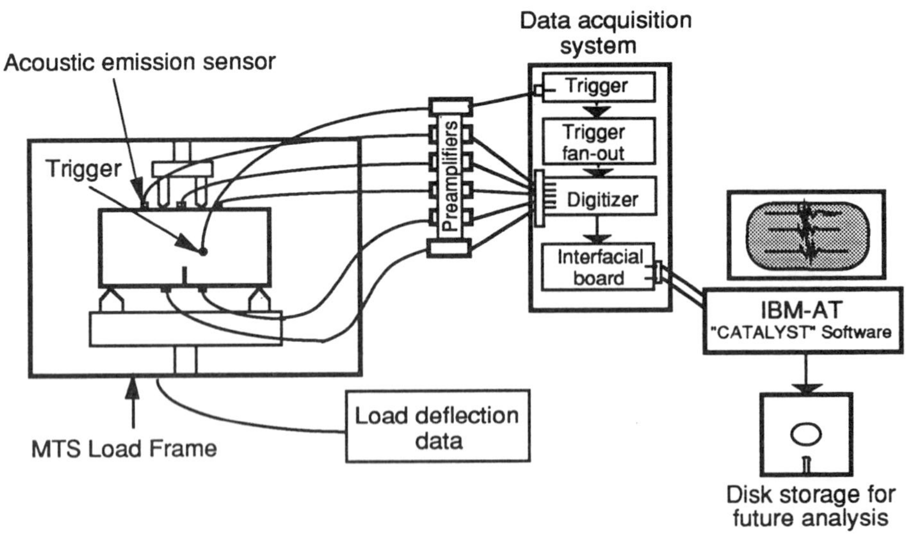

Figure 4-9 Experimental set-up for acoustic emission.[19]

determined. The signals may be further characterized to infer orientation, movement, and mode of microcracks.

The locations of acoustic emission events from a center-notched concrete beam subjected to four-point bending are indicated in Fig. 4-10*a*. Most of the acoustic emission events were located in the front of the notch. The fact that the acoustic emission events are located at both sides of the plane containing the notch indicates that the fracture process zone may have a substantial width. This observation coincides with zone *B* observed from laser holography (see Fig. 4-8). The locations of acoustic events along the cross section of the beam are also indicated in Fig. 4-10*b*.[20] The saddle shape of the crack front shown in the figure is similar to that observed by dye penetration testing discussed later (Fig. 4-13).

4.4.3 Dye Penetration

This technique was first developed by Swartz and Go[21] and modified by Swartz and Refai.[22,23] The technique has been used on three-point bend beams and

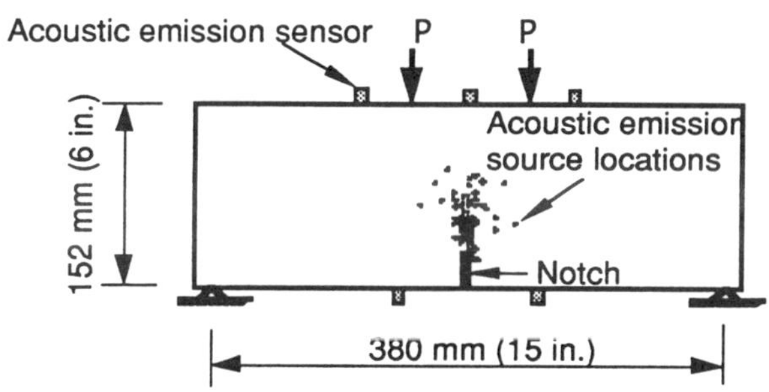

(*a*)

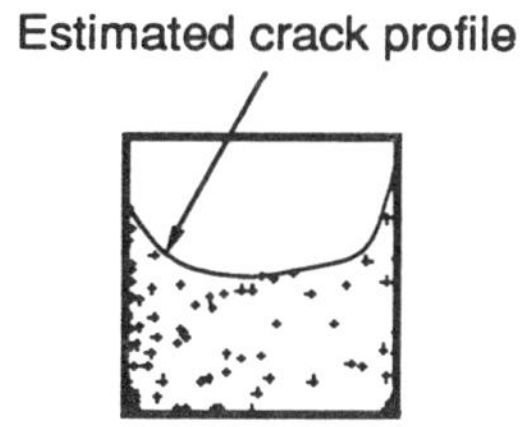

(*b*)

Figure 4-10 Microcracks in concrete beams detected by acoustic emission techniques: (*a*) acoustic emission source locations, and (*b*) distribution of acoustic emission events along width of a beam.[20]

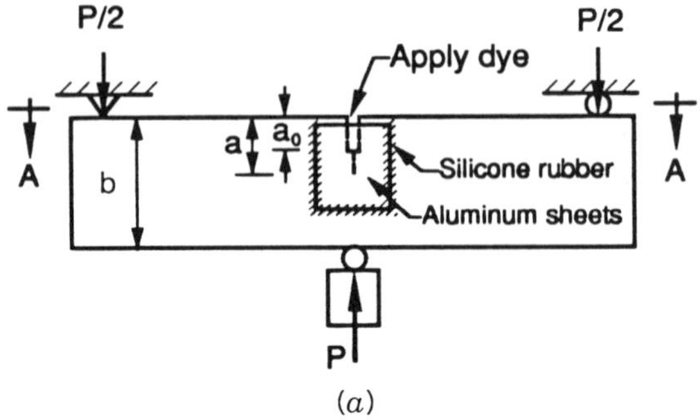

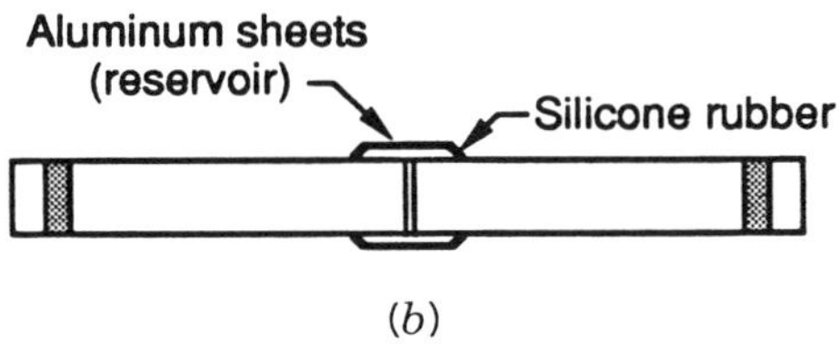

Figure 4-11 Test specimen and dye reservoir: (*a*) elevation view, and (*b*) section A-A view.[22]

utilizes a vivid, blue dye. This has a viscosity of $1.5 \times 10^{-2} \rho^0$ with molecular dimensions of 2–3 × 18 μm. As shown in Fig. 4-11, a container (reservoir) is bonded to the beam on both sides of the area expected to be damaged. This is done after a starter notch is cut but prior to loading the beam to induce cracking.

The beam is placed in the testing machine and loaded in the inverted position shown in Fig. 4-12. Cracking in a controlled manner is induced and then

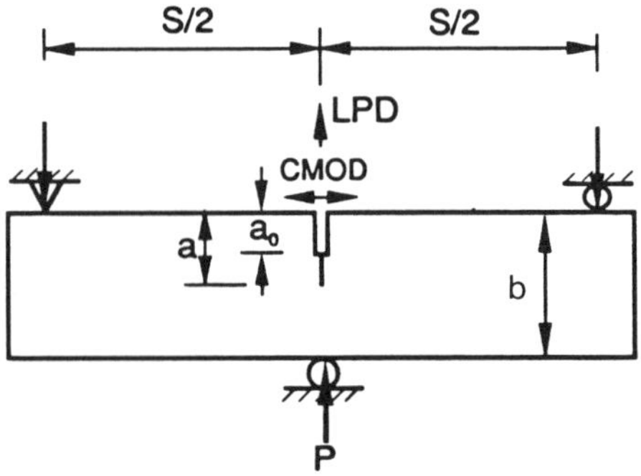

Figure 4-12 Testing configuration.[22]

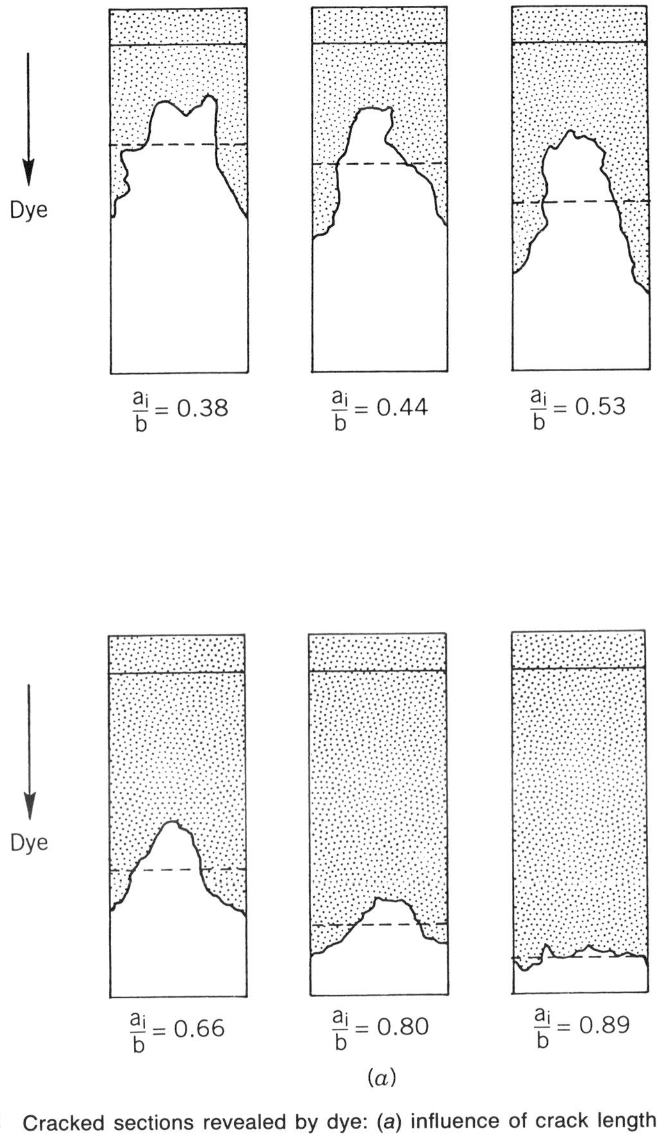

Figure 4-13 Cracked sections revealed by dye: (*a*) influence of crack length on shape of fracture surface[23] and (*b*) influence of beam width on shape of fracture surface.[24]

dye is inserted, filling the reservoir. A small load is applied to open the crack and allow the dye to penetrate. Following this, the dye is removed, the plates of the reservoir are removed, and the beam is dried. This then is a precracked specimen to be loaded to failure, still in the inverted position. Although some dye may penetrate into the uncracked zone, variation of the fracture process zone along the thickness direction can be qualitatively revealed by the dye

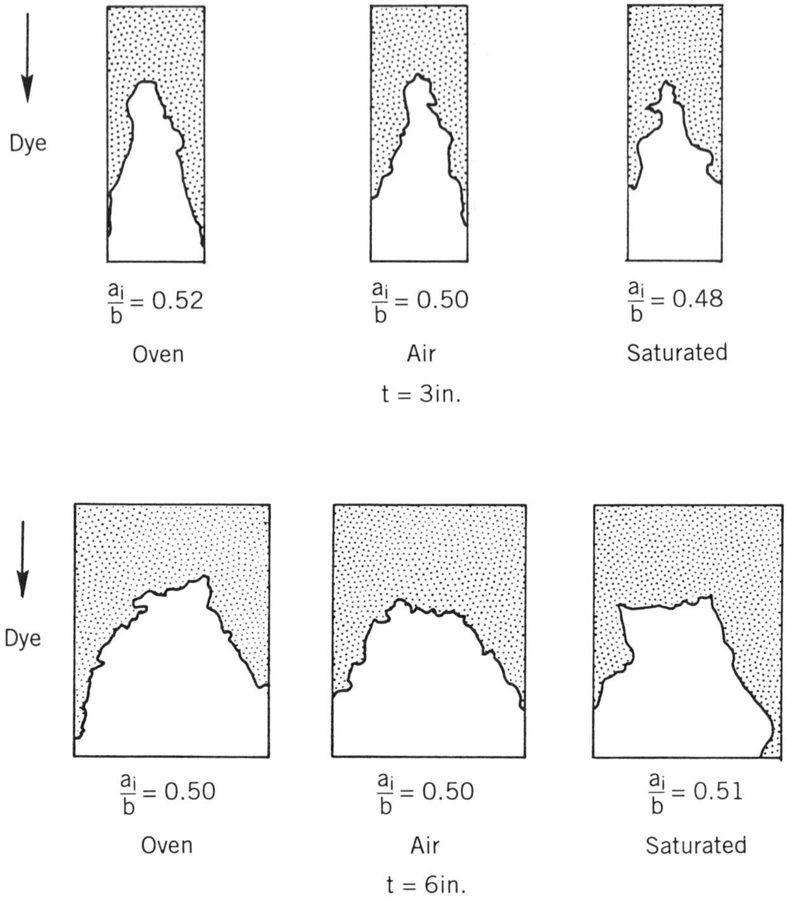

(*b*)

Figure 4-13 (*Continued*)

penetration technique. After failure, typical, dyed beam surfaces are shown in Fig. 4-13. A group of crack fronts is shown in Fig. 4-13*a* to illustrate the changing patterns observed as the crack progresses from the tension side (top) to the compression side of the beam. In general, the crack length on the beam surface is greater than that in the interior of the beam. It is shown by Kan and Swartz[24] that this also varies with beam width, as presented in Fig. 4-13*b*. Similar results for micro-concrete beams using a dye technique were obtained by Bascoul et al.[25] A similar result has also been observed using a three-dimensional acoustic emission technique, as shown in Fig. 4-10*b*.

4.5 INFLUENCE OF FRACTURE PROCESS ZONE ON FRACTURE BEHAVIOR OF CONCRETE

Toughening mechanisms and size of the fracture process zone in concrete have been discussed in the previous sections. Fracture behavior of concrete is greatly influenced by the fracture process zone. Since the fracture process zone consumes a substantial amount of the energy supplied by the applied load, a crack can propagate steadily prior to the peak load. Therefore, the presence of the fracture process zone results in stable crack growth before the peak load. The fracture process zone is also responsible for the quasi-brittle fracture response of concrete after the peak load. Since some parts of crack surfaces may still be in contact after cracking, catastrophic failure of concrete structures immediately after the peak load is prevented. As a result, stress gradually decreases after the peak load, and a softening type of concrete stress–strain relationship is obtained.

Experimental results have indicated that the strength of concrete usually decreases with increasing size of structures and then remains constant, as shown in Fig. 4-14,[26] where the nominal strength of a three-point bend beam is equal to its failure load (peak load) divided by its cross-sectional area. This size effect on concrete strength may be primarily explained by the fracture process zone. When a concrete structure is loaded, the strain energy produced by the applied load is converted to the energy consumed to create a new fracture surface and the energy absorbed in the fracture process zone. For large-size structures that have long ligament lengths on the cracked sections, the latter is

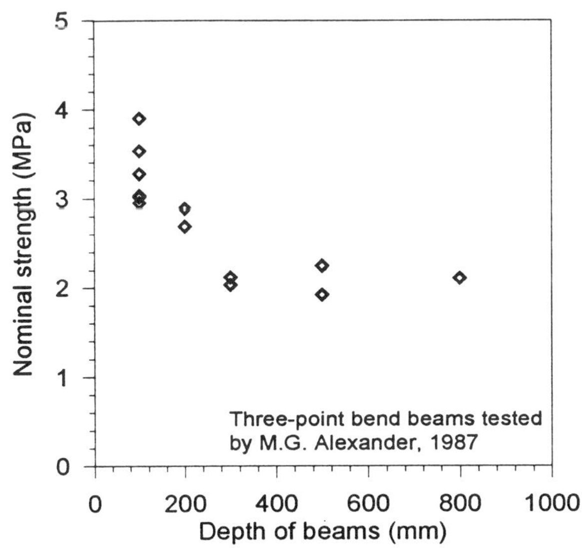

Figure 4-14 Strength of concrete beams with different sizes.

negligible compared to the former, whereas for small-size structures the latter can be of the same magnitude as the former. Therefore, the larger the structure's size, the lower the nominal strength. However, the concrete strength approaches a constant when sizes of concrete structures become very large.

The presence of the fracture process zone hinders the application of LEFM to cement-based materials.[27] Higgins and Bailey[28] tested three-point bend beams with different depths but the same material and calculated the critical stress intensity factor using LEFM. However, they found values of the critical stress intensity factor in this manner to be size dependent, as shown in Fig. 4-15. This is due to the fact that LEFM does not take the stable crack growth associated with the fracture process zone into account. In other words, in LEFM the initial crack length, rather than the sum of the initial crack length and the crack extension, is used to determine the critical stress intensity factor. Therefore, principles of NLFM presented in Chapter 3 should be used for modeling concrete fracture.

Although both the fracture process zone in concrete and the plastic zone in metals result from the stress singularity at a crack tip, these inelastic zones are dominated by different mechanisms. Generally, the fracture process zone in concrete is much more complicated and larger than the plastic zone in metals. Yielding is a primary mechanism for the plastic zone in metals, whereas toughening mechanisms for the fracture process zone in concrete include microcracking, crack deflection, bridging, crack-face friction, crack tip blunting by voids, and crack branching. As discussed in Chapter 2, the size of the plastic

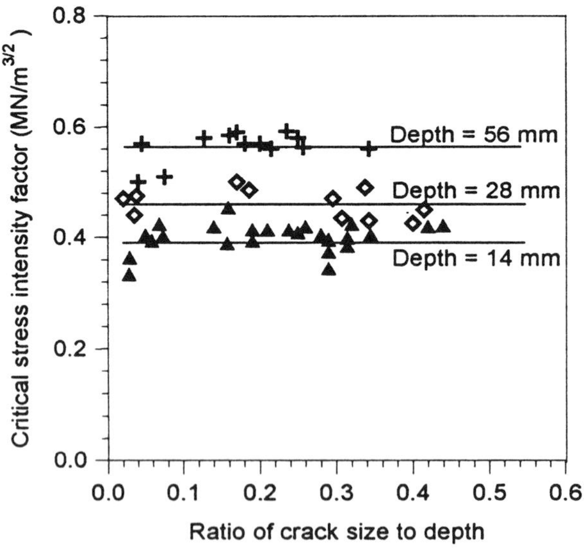

Figure 4-15 Critical stress intensity factor obtained from tests of different sizes of beams using LEFM.[28]

zone decreases with increasing specimen thickness due to the plane-strain restraint. However, it has been reported that measured values of fracture toughness for concrete almost do not vary with change of thickness of tested specimens. Also the shape of the crack front in the thickness direction for metallic specimen is such that the crack is deeper in the middle of the specimen. Such a crack profile in the thickness direction is opposite to that observed for the concrete specimen, as shown in Figs. 4-10 and 4-13. This result indicates that sizes of the fracture process zone in concrete may not greatly change with the specimen thickness. Therefore, some experimental techniques and analytical approaches that are valid for fracture of metals may not be directly used for fracture of concrete. A theoretical description and simulation of the fracture behavior of concrete structures will be discussed in detail in the next chapter.

4.6 SUMMARY

The constituents and microstructure of concrete have been briefly discussed. Many internal cracks exist in concrete prior to loading. The fracture behavior of concrete depends on when and how these internal cracks initiate and propagate under different loading conditions. The fracture behavior of concrete is characterized by the phenomenon of strain localization. The presence of strain localization allows one to apply fracture mechanics methods to describe failure of concrete. The inelastic zone around a propagating crack tip in concrete is called the fracture process zone. Toughening mechanisms in the fracture process zone include microcracking, crack deflection, bridging, crack-face friction, crack tip blunting by voids, and crack branching. Some experimental results describing the fracture process zone have been presented. The influence of the fracture process zone on the mechanical behavior of concrete structures has been discussed.

REFERENCES

1. Mindess, S., and Young, J. F., *Concrete*, Prentice-Hall, Englewood Cliffs, NJ, 1981.
2. Mehta, P. K., and Monteiro, P. J. M., *Concrete: Structure, Properties, and Materials*, 2nd ed., Prentice-Hall, Englewood Cliffs, NJ, 1993.
3. Mitsui, K., Li, Z., Lange, D. A., and Shah, S. P., "Relationship between Microstructure and Mechanical Properties of the Paste–Aggregate Interface," *ACI Materials Journal*, Vol. 91, No. 1, Jan.–Feb. 1994, pp. 30–39.
4. Li, Z., Kulkarni, S. M., and Shah, S. P., "New Test Method for Obtaining Softening Response of Unnotched Concrete Specimen under Uniaxial Tension," *Experimental Mechanics*, Vol. 33, No. 3, Sept. 1993, pp. 181–188.
5. Shah, S. P., and Sankar, R., "Internal Cracking and Strain-Softening Response

of Concrete under Uniaxial Compression,'' *ACI Materials Journal*, Vol. 84, No. 3, 1987, pp. 200–212.

6. Mindess, S., and Diamond, S., ''The Cracking and Fracture of Mortar,'' *Materials and Structures*, Vol. 15, No. 86, 1982, pp. 107–113.
7. Kachanov, M., ''Interaction of a Crack with Some Microcrack Systems,'' in *Fracture Toughness and Fracture Energy of Concrete*, edited by Wittmann, F. H., Elsevier Science, Amsterdam, 1986, pp. 3–10.
8. Krstulovic-Opara, N., ''Fracture Process Zone Presence and Behavior in Mortar Specimens,'' *ACI Materials Journal*, Vol. 90, No. 6, 1993, pp. 618–626.
9. Shum, K. M., and Hutchinson, J. W., ''On Toughening by Microcracks,'' *Mechanics of Materials*, Vol. 9, 1990, pp. 83–91.
10. Faber, K. T., and Evans, A. G., ''Crack Deflection Processes: I. Theory,'' *Acta Metallurgy*, Vol. 31, No. 4, 1983, pp. 565–576.
11. van Mier, J. G. M., ''Mode I Fracture of Concrete: Discontinuous Crack Growth and Crack Interface Grain Bridging,'' *Cement and Concrete Research*, Vol. 21, No. 1, 1991, pp. 1–15.
12. Mindess, S., ''The Fracture Process Zone in Concrete,'' in *Toughening Mechanisms in Quasi-Brittle Materials*, edited by Shah, S. P., Kluwer Academic, Dordrecht, The Netherlands, 1991, pp. 271–286.
13. Shah, S. P., and Ouyang, C., ''Measurement and Modeling of Fracture Processes in Concrete,'' in *Materials Science of Concrete*, Vol. III, edited by Scalny, J., American Ceramic Society, Westerville, OH, 1992, pp. 243–270.
14. Castro-Montero, A., Shah, S. P., and Miller, R. A., ''Strain Field Measurement in Fracture Process Zone,'' *Journal of Engineering Mechanics, ASCE*, Vol. 116, No. 11, 1990, pp. 2463–2484.
15. van Mier, J. G. M., ''Mode I Fracture of Concrete Discontinuous Crack Growth and Crack Interface Grain Bridging,'' *Cement and Concrete Research*, Vol. 21, No. 1, 1991, pp. 1–15.
16. Sakai, M., Yoshimura, J.-I., Goto, Y., and Inagaki, M., ''*R*-curve Behavior of a Polycrystalline Graphite: Microcracking and Grain Bridging in the Wake Region,'' *Journal of the American Ceramic Society*, Vol. 71, No. 8, 1988, pp. 609–616.
17. Homeny, J., and Vaughn, W. L., ''*R*-Curve Behavior in a Silican Carbide Whisker/Alumina Matrix Composite,'' *Journal of the American Ceramic Society*, Vol. 72, No. 7, 1990, pp. 2060–2062.
18. Maji, A. K., Ouyang, C., and Shah, S. P., ''Fracture Mechanism of Quasi-Brittle Materials Based on Acoustic Emission,'' *Journal of Materials Research*, Vol. 5, No. 1, 1990, pp. 206–217.
19. Ouyang, C., Landis, E., and Shah, S. P., ''Damage Assessment in Concrete Using Quantitative Acoustic Emission,'' *Journal of Engineering Mechanics*, ASCE, Vol. 117, No. 11, 1991, pp. 2681–2698.
20. Landis, E., and Shah, S. P., ''Recovery of Microcrack Parameter in Mortar Using Quantitative Acoustic Emission,'' *Journal of Nondestructive Evaluation*, Vol. 12, No. 4, 1993, pp. 219–232.
21. Swartz, S. E., and Go, C. G., ''Validity of Compliance Calibration to Cracked

Concrete Beams in Bending,'' *Experimental Mechanics*, Vol. 24, No. 2, 1984, pp. 129–134.

22. Swartz, S. E., and Refai, T. M., ''Influence of Size Effects on Opening Mode Fracture Parameters for Precracked Concrete Beams in Bending,'' in *Fracture of Concrete and Rock*, edited by Shah, S. P., and Swartz, S. E., Springer-Verlag, New York, 1989, pp. 242–254.
23. Swartz, S. E., and Refai, T., ''Cracked Surface Revealed by Dye and Its Utility in Determining Fracture Parameters,'' in *Fracture Toughness and Fracture Energy: Test Methods for Concrete and Rock*, edited by Mihashi, H., et al., Balkema, Brookfield, VT, 1989, pp. 509–520.
24. Kan, Y.-C., and Swartz, S. E., ''Influence of Curing Conditions and Width on the Fracture of Concrete Beams,'' in *Proceedings of the 1989 Spring Conference on Experimental Mechanics*, Cambridge, MA, Society for Experimental Mechanics, Bethel, CT, May 28–June 1, 1989, pp. 196–201.
25. Bascoul, A., Kharchi, F., and Maso, J. C., ''Concerning the Measurement of the Fracture Energy of a Micro-Concrete According to the Crack Growth in a Three-Point Bending Test on Notched Beams,'' in *Fracture of Concrete and Rock*, edited by Shah, S. P., and Swartz, S. E., Springer-Verlag, New York, 1989, pp. 396–408.
26. Karihaloo, B. L., and Nallathambi, P., ''Notched Beam Test: Mode I Fracture Toughness,'' in RILEM Report 5, *Fracture Mechanics Test Methods for Concrete*, edited by Shah, S. P., and Carpinteri, A., Chapman & Hall, London, 1991, pp. 1–86.
27. Shah, S. P., and McGarry, F. J., ''Griffith Fracture Criterion and Concrete,'' *Journal of Engineering Mechanics Division, ASCE*, Vol. 97, No. EM6, 1971, pp. 1663–1676.
28. Higgins, D. D., and Bailey, J. E., ''Fracture Measurements on Cement Paste,'' *Journal of Materials Science*, Vol. 11, No. 11, 1976, pp. 1995–2003.

5

NONLINEAR FRACTURE MECHANICS FOR MODE I QUASI-BRITTLE FRACTURE

5.1 GENERAL DESCRIPTION OF QUASI-BRITTLE FRACTURE

For an ideally brittle material, the stress–strain curve is linearly elastic up to the maximum stress, at which point an initial flaw catastrophically propagates, leading to failure. A typical tensile stress–elongation curve for an ideally brittle material for which linear elastic fracture mechanics is valid is shown in Fig. 5-1*a*. For a quasi-brittle material like concrete, a substantial nonlinearity exists before the maximum stress. The mechanisms of deformation beyond the proportional limit f_y (see Fig. 5-1*b*) are not clearly understood. A commonly accepted view, as summarized for Fig. 5-1*b*, has been discussed in detail in the previous chapter. Initially, randomly distributed microcracks are formed. At some point before the peak stress, microcracks begin to localize into a macrocrack that critically propagates at the peak stress. Strain softening is observed under steady-state propagation of this crack. Under a closed-loop displacement-controlled testing condition, it is observed that the displacement during the postpeak stage consists of opening of the major crack accompanied by unloading of the rest of the specimen.

To model this behavior using discrete crack fracture mechanics, it is assumed that an initial crack begins to propagate at the proportional limit f_y and continues to propagate in a stable manner until the peak stress. When the crack extends in concrete, new crack surfaces are formed along the path of the initial crack tip. The newly formed crack surfaces may be in contact and are tortuous in nature. This leads to toughening mechanisms in the fracture process zone such as aggregate bridging, which have been discussed in detail in the previous chapter. Since the newly formed crack surfaces may be in contact, they may continue to sustain some normal tensile stress that is characterized by a ma-

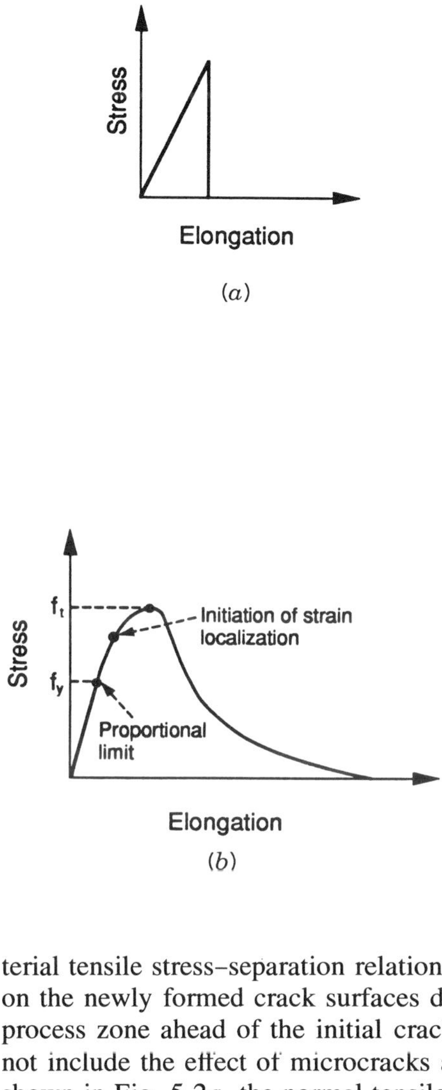

Figure 5-1 Tensile stress–elongation curves for (*a*) linear elastic material and (*b*) quasi-brittle material.

terial tensile stress–separation relationship. Distribution of the tensile stresses on the newly formed crack surfaces depends on the definition of the fracture process zone ahead of the initial crack tip. If the fracture process zone does not include the effect of microcracks ahead of the newly formed crack tip as shown in Fig. 5-2*a*, the normal tensile stress gradually increases from the initial (open) crack tip and reaches the *tensile strength* of the material, f_t, at the end of the fracture process zone. It is noted that the tensile strength f_t proposed here is different than the conventional concrete tensile strength f_t' obtained from a regular tensile test. The former is regarded as a material fracture parameter, whereas the latter depends on the material as well as the size and geometry of the tested specimen and the testing procedure. If the fracture process zone is defined such that it includes effects of microcracks ahead of the newly formed crack tip as shown in Fig. 5-2*b*, the normal tensile stress gradually increases from the initial crack tip and reaches its maximum value (the tensile strength of the material) and drops to the proportional limit f_y at the end

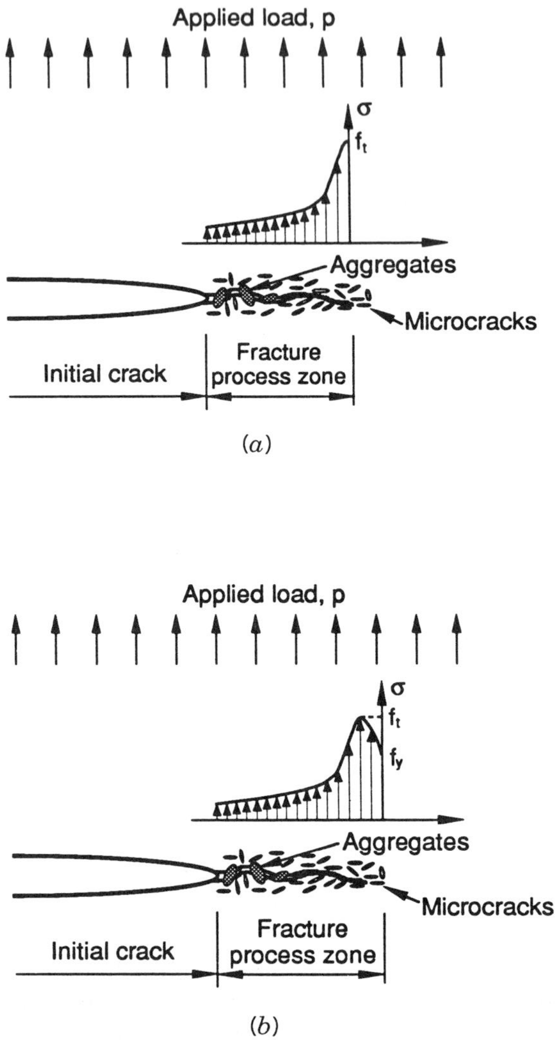

Figure 5-2 Sketch of concrete crack and fracture process zone: (*a*) not including effect of crack-tip microcracks, and (*b*) including effect of crack-tip microcracks.

of the fracture process zone, as shown in Fig. 5-2*b*, where the proportional limit f_y corresponds to initiation of microcracks in the material.

The fracture behavior of concrete is greatly influenced by this fracture process zone. Since the work of Kaplan[1], many attempts have been made to apply fracture mechanics to concrete. As realized by Shah and McGarry,[2] LEFM cannot be directly applied to concrete due to the presence of the sizable fracture process zone. Other difficulties for using fracture mechanics to describe failure of concrete include (i) the crack path in concrete is tortuous and (ii) it is difficult to determine the crack tip in concrete due to particles bridging and variation of the fracture process zone along the thickness direction. An accurate description of concrete fracture should include the tortuous crack path, the three-dimensional aspects of the crack profile, and the inelastic material response within the fracture process zone. To apply fracture mechanics concepts to the failure of concrete, most currently available models attempt to simulate mode I concrete fracture with an effective line crack. The variation of the fracture process zone along the structure thickness or width is usually neglected. The inelastic fracture response due to the presence of the fracture process zone may then be taken into account by a cohesive pressure acting on the crack faces.

An effective quasi-brittle crack is shown in Fig. 5-3*a*, where an initial crack and the associated fracture process zone are presented by a crack with length a. The toughening mechanisms in the fracture process zone are modeled by a cohesive pressure acting on the crack surfaces.[3] The cohesive pressure $\sigma(w)$ is a monotonic decreasing function of crack separation displacement w. The value of $\sigma(w)$ is equal to material tensile strength f_t for $w = 0$ at the crack "tip" (the end of the fracture process zone). This implies that microcracks ahead of the crack tip are not included in the fracture process zone. This will be reasonable if the size of the crack tip fracture process zone is small compared to the size of the crack wake process zone. This cohesive pressure tends to close the crack and is a function of the crack opening displacement w. It is noted that the crack extension is usually not equal to the length of the fracture process zone because of the variation of the actual fracture process zone along the thickness or the width direction, as described in Chapter 4. When a concrete structure with a quasi-brittle crack is subjected to loading, the applied load results in an energy release rate G_q at the tip of the effective quasi-brittle crack, where the subscript q stands for quasi-brittle materials. The energy release rate G_q may be divided into two portions: (i) the energy rate consumed during material fracturing in creating two surfaces, G_{Ic}, which is equivalent to the material surface energy, and (ii) the energy rate to overcome the cohesive pressure $\sigma(w)$ in separating the surfaces, G_σ, where the subscript σ indicates this part of energy overcomes the cohesive pressure $\sigma(w)$ to open the crack. As a result, the energy release rate for a mode I quasi-brittle crack, G_q, may be expressed as[3]

$$G_q = G_{Ic} + G_\sigma \tag{5.1}$$

The value of G_{Ic}, may be evaluated based on LEFM and is called the critical energy release rate. Since G_σ is equal to the work done by the cohesive pressure over a unit length of the crack for a structure with a unit thickness, its value can be calculated as[4]

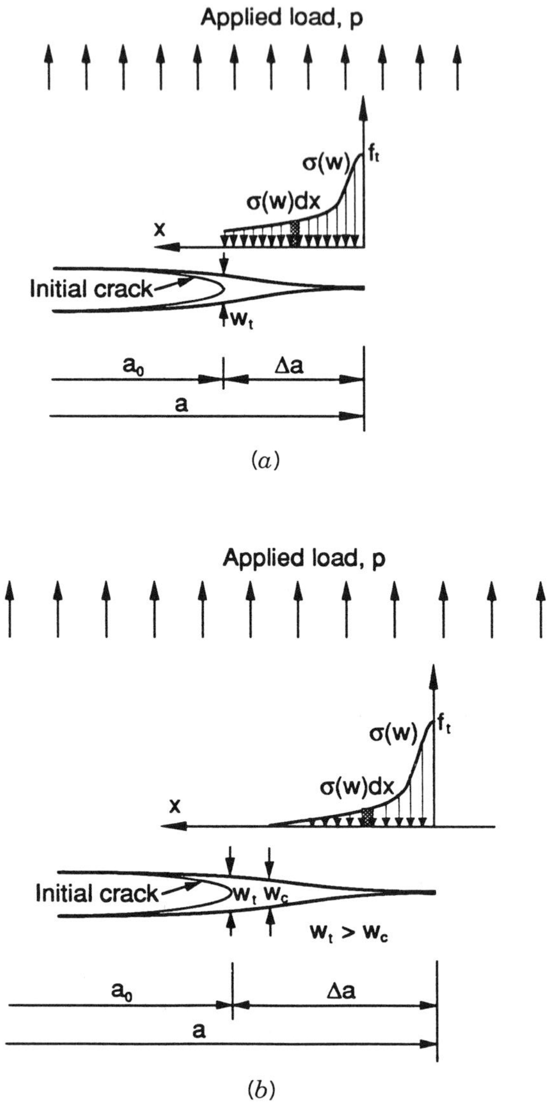

Figure 5-3 Modeling of quasi-brittle crack: (*a*) a cohesive crack with crack surfaces in contact, and (*b*) a cohesive crack with partially separated crack surfaces.

$$
\begin{aligned}
G_\sigma &= \frac{1}{\Delta a} \int_0^{\Delta a} \int_0^{w} \sigma(w)\, dx\, dw = \frac{1}{\Delta a} \int_0^{\Delta a} dx \int_0^{w} \sigma(w)\, dw \\
&= \int_0^{w_t} \sigma(w)\, dw
\end{aligned}
\tag{5.2}
$$

where $\sigma(w)$ is the normal cohesive pressure and w_t is the crack separation displacement at the initial crack tip, as shown in Fig. 5-3*a*. Since dx is taken out of the integral with respect to w, Eq. (5.2) implies that the profile (shape) of crack opening displacement w does not vary significantly with change of crack length. It is noted that Eq. (5.2) is valid only if the material at the initial crack tip still keeps in contact. If the crack separation is large enough so that some part of the newly formed crack surfaces is separated (i.e., $w_t > w_c$, as shown in Fig. 5-3*b*), the upper integral limit w_t in Eq. (5.2) should be replaced by w_c, where w_c is the crack separation corresponding to $\sigma(w) = 0$ and is termed the critical crack separation displacement. At the peak load w_t is likely to be much less than w_c. Determination and influence of $\sigma(w)$ on the fracture behavior of concrete will be discussed later. Substituting Eq. (5.2) into Eq. (5.1) leads to

$$G_q = G_{\mathrm{Ic}} + \int_0^{w_t} \sigma(w)\, dw \tag{5.3}$$

It is noted that when $w_t > w_c$ as shown in Fig. 5-3*b*, the upper integral limit w_t in Eq. (5.3) should be replaced by w_c. Since Eq. (5.3) is based on the effective quasi-brittle crack subjected to two sets of forces, that is, the applied load and the cohesive force on the fracture process zone, as shown in Fig. 5-3, the crack separation displacement w in Eq. (5.3) should include contributions from the two sets of forces. In other words, Eq. (5.3) is valid only when the total value of w under the two sets of forces is used. Equation (5.3) is a general energy equilibrium condition for propagating a mode I quasi-brittle crack. Equation (5.3) indicates that for quasi-brittle fracturing, the energy release rate due to the applied load G_q is balanced by two fracture energy dissipation mechanisms. The Griffith–Irwin energy dissipation mechanism is represented by the fracture energy release rate G_{Ic}, whereas the Dugdale–Barenblatt energy dissipation mechanism is represented by the material traction term G_σ. Jenq and Shah[3] and Kobayashi et al.[5] have used Eq. (5.3) to describe fracture of concrete. Similar fracture models including two energy dissipation mechanisms have also been proposed by Cook et al.[6] and Cox and Marshall.[7] It is noted that if the second term in Eq. (5.3) is constant because the fracture process zone is constant, then G_q is constant and also satisfies the classical definition of energy release rate.

Although propagation of a quasi-brittle crack can be described by Eq. (5.3), one may approximately use only a single fraction energy dissipation mechanism, either the Griffith–Irwin mechanism by assuming $\sigma(w) = 0$ or the Dugdale–Barenblatt mechanism by assuming $G_{\mathrm{Ic}} = 0$. Based on different energy dissipation mechanisms used, nonlinear fracture mechanics models for quasi-brittle materials may be classified as a fictitious crack approach[8,9] and an equivalent-elastic crack approach (or an effective-elastic crack approach).[10,11] Fracture mechanics models using only the Dugdale–Barenblatt energy dissipation mechanism are usually referred to as the fictitious crack approach, whereas fracture mechanics models using only the Griffith–Irwin energy dissipation mechanism are usually referred to as the effective-elastic crack approach or equivalent-elastic crack approach. Both the fictitious crack approach and the effective-elastic crack approach will be discussed below.

5.2 FICTITIOUS CRACK APPROACH

5.2.1 Energy Dissipation for Fictitious Crack

The fictitious crack approach assumes that energy to create the new surfaces is small compared to that required to separate them, and the energy rate term $G_{\mathrm{I}c}$ vanishes in the fictitious crack approach. This is illustrated in Fig. 5-4, where the newly formed crack surfaces and the corresponding fracture process zone are simply simulated by a cohesive zone located in the front of the initial crack tip. As a result, the energy dissipation for crack propagation can be completely characterized by the cohesive stress–separation relationship $\sigma(w)$. Since all energy produced by the applied load is completely balanced by the cohesive pressure in this case, Eq. (5.3) is reduced to (with $G_{\mathrm{I}c} = 0$)

$$G_q = \int_0^{w_t} \sigma(w)\, dw \tag{5.4}$$

Equation (5.4) is valid for structures with a unit constant thickness. It is noted that the upper integral limit w_t in Eq. (5.4) should be replaced by w_c for $w_t > w_c$, which sometimes occurs after the peak load. The fictitious crack is assumed to initiate and propagate when the principal tensile stress reaches the tensile strength of material, f_t, here.

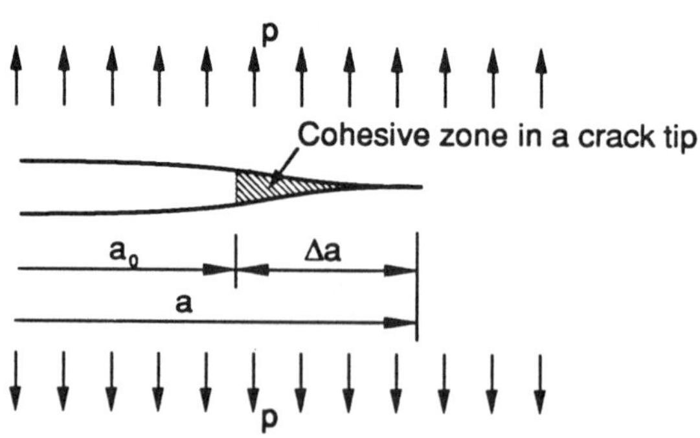

Figure 5-4 Mode I crack for fictitious crack approach.

5.2.2 Fictitious Crack Model by Hillerborg

Hillerborg et al.[8] first proposed a fictitious crack model for fracture of concrete. A typical stress–elongation curve for a concrete plate subjected to uniaxial tension is shown in Fig. 5-5*a*. The elongation of the plate is measured by two displacement gages, gage *A* and gage *B*, respectively. Gage *A* is assumed to measure the elongation including the cracked section, whereas gage *B* is as-

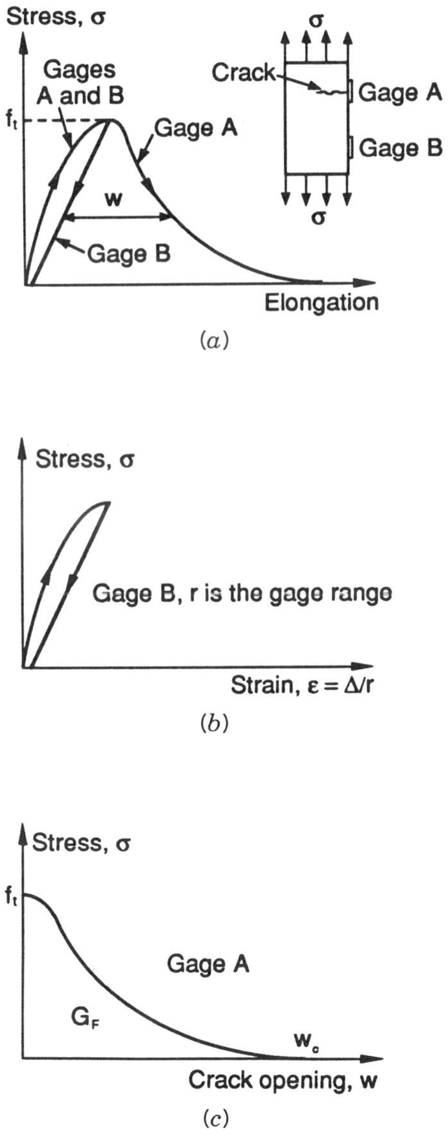

Figure 5-5 Principle for fictitious crack model by Hillerborg et al.: (*a*) a complete tensile stress-elongation curve (*b*) stress-strain curve for uncracked section, and (*c*) stress-elongation curve for cracked section.[8]

sumed to measure the elongation outside the cracked section. It is assumed that strain localization appears only after the maximum load is reached. As a result, the two gages measure the same elongation up to the maximum load. After the maximum load, the elongation at the location of gage *A* continuously increases, whereas unloading occurs at the location of gage *B* due to the strain localization, as shown in Fig. 5-5*a*. It was proposed that the prepeak tensile response of concrete can be described by a stress–strain curve as shown in Fig. 5-5*b*, whereas the postpeak fracture behavior or softening can be characterized by a stress–elongation curve as shown in Fig. 5-5*c*. The area under the entire softening stress-elongation curve, $\sigma(w)$, is denoted as G_F, which is given by

$$G_F = \int_0^{w_c} \sigma(w)\, dw \tag{5.5}$$

where w_c is the critical crack separation displacement when the softening stress is equal to zero. The material fracture toughness G_F represents the energy absorbed per unit area of crack and is regarded as a material fracture parameter.

In the fictitious crack model, the softening stress–separation curve $\sigma(w)$ is assumed to be a material property that is independent of structural geometry and size. It is noted that the softening curve $\sigma(w)$ can be completely determined if the material tensile strength f_t, the fracture toughness G_F, and the shape of the $\sigma(w)$ curve are known. Therefore, the fictitious crack model requires one to know three material parameters: the values of f_t and G_F and the shape of the $\sigma(w)$ curve, as shown in Fig. 5-5. When the shape of the $\sigma(w)$ curve is given, the material fracture property is determined by the values of f_t and G_F. In this case, it is also suitable to combine f_t and G_F into a characteristic length,[8]

$$l_{\text{ch}} = \frac{EG_F}{f_t^2} \tag{5.6}$$

where E is the modulus of elasticity for the material. The characteristic length l_{ch} is a purely material property and is proportional to the length of the fracture process zone based on the fictitious crack model. The value of l_{ch} for concrete approximately ranges from 100 mm (4 in.) to 400 mm (16 in.).[12] The length of the fracture process zone at complete separation of the initial crack tip ($w_t = w_c$) in concrete is of the order of $0.3l_{\text{ch}}$ to $0.5l_{\text{ch}}$ according to this model.

The fictitious crack model may be combined with a finite element analysis to predict fracture behavior of concrete structures. This is illustrated in Fig. 5-6 for a notched beam under three-point bending. Only half of the beam needs to be considered due to symmetry, as shown in Fig. 5-6*b*. One may assume the load $P = P_A$ produces $\sigma_{x1} = f_t$, where σ_{x1} is the normal tensile stress at point 1, as shown in Fig. 5-6*b*. This indicates that the initial crack starts to propagate at the load level of P_A. As a result, a cohesive force F_1 should be introduced at point 1, where the F_1 value is determined based on the $\sigma(w)$

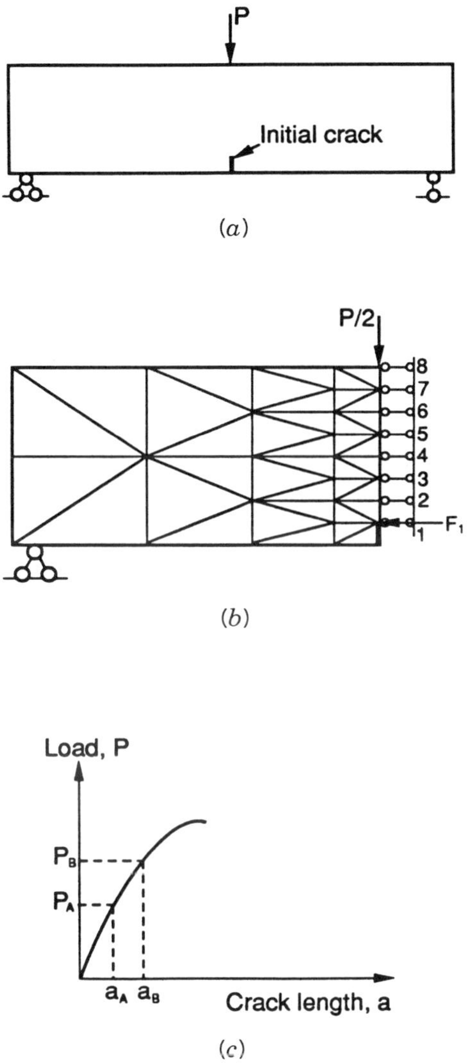

Figure 5-6 Illustration of using fictitious crack model in finite element analysis: (*a*) three-point bend beam (*b*) finite element mesh of the beam, and (*c*) load-crack length curve

relationship and tends to close the crack. Then, the new analysis can be performed for the beam under both F_1 and the increasing load P. It is noted that since the cohesive stress $\sigma(w)$ is a function of the crack opening displacement, the F_1 value may change with increasing P as the result of more crack opening displacement. After the load P_B, which leads to $\sigma_{x2} = f_t$ at point 2, is determined, another cohesive force, F_2, should be introduced at point 2. Proceeding with similar calculations by changing the load and the cohesive force, the fracture behavior of the beam can be predicted. It is noted that the crack propagation is determined by whether or not the principal stress exceeds the tensile strength of the material, instead of an energy criterion, in the above calculation. The finite element method is described in more detail in Section 5.5.

A comparison of load–displacement curves predicted by the fictitious crack model and experimentally measured for a three-point bend beam is shown in Fig. 5-7.[12] The span, depth and thickness of the beam are 400, 100, and 100 mm, respectively. A bilinear stress–separation curve recommended by the CEB-FIP Model Code was used, which will be described in Section 5.2.4. The maximum aggregate size used was 16 mm. Values of E = 35.2 GPa, G_F = 84.4 N/mm, and f_t = 3.25 MPa were used for the theoretical prediction. It is seen that by appropriately selecting fracture parameters, the fictitious crack model is capable of predicting fracture behavior of a structure.

It is noted that the cracked faces are assumed to be in contact during the stable crack propagation in the fictitious crack model. Hillerborg[13] has shown that at the critical crack propagation, which corresponds to the maximum load, the energy release rate due to the applied force G_q is equal to the value of G_F when specimen sizes approach infinity. Values of the energy release rate at the maximum load for finite-size structures are less than the value of G_F.

5.2.3 Crack Band Model by Bazant and Oh

Bazant and Oh[9, 14] modeled the fracture process zone by a band of uniformly and continuously distributed microcracks with a fixed width of h_c, as shown in Fig. 5-8*a*. Stable crack propagation is then simulated by progressively microcracking within this band, which is described by a simple stress–strain relationship, as shown in Fig. 5-8*b*. The crack opening displacement is equal to

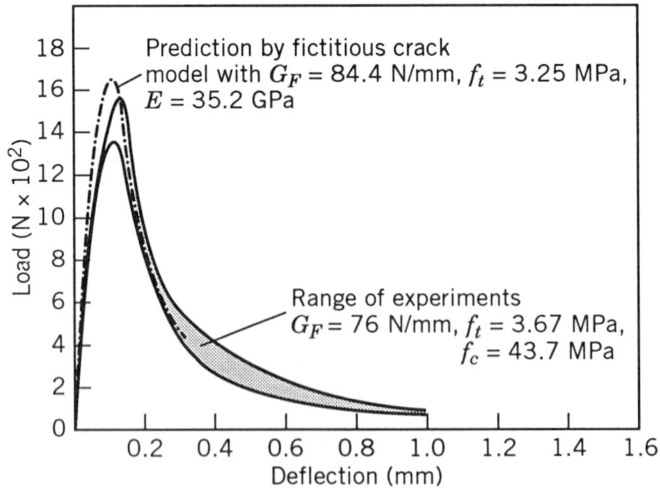

Figure 5-7 Comparison of predicted and measured load–displacement curves for three-point bend beam.[12]

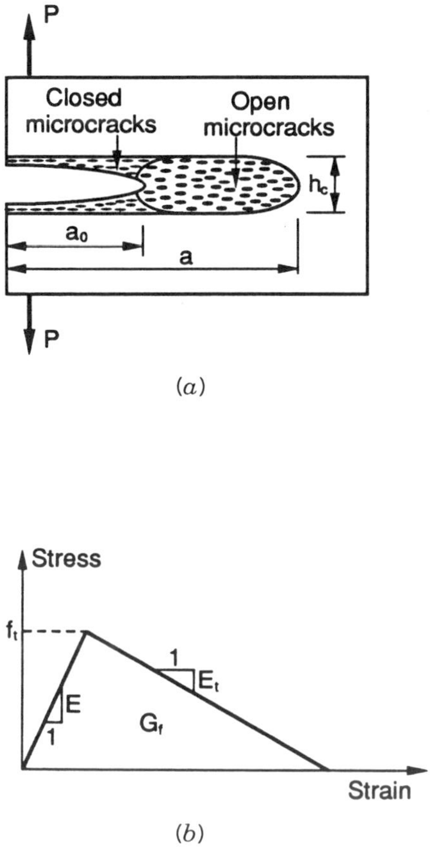

Figure 5-8 Crack band model for fracture of concrete and rock: (*a*) a microcrack band fracture and (*b*) stress-strain curve for the microcrack band.

the product of fracture strain and the width of the crack band. Based on Eq. (5.4), the energy consumed due to the crack advance per unit area of the crack band, G_f, is the product of the area under the stress–strain curve given by Fig. 5-8*b* and the width of the crack band, h_c. This leads to

$$G_f = h_c\left(1 + \frac{E}{E_t}\right)\frac{f_t'^2}{2E} \tag{5.7}$$

where E is the modulus of elasticity, E_t is the strain-softening modulus (E_t is positive here), and f_t is the tensile strength of material. The values of E and E_t are defined as positive. It is seen that, in addition to E, three material fracture parameters, h_c, f_t, and E_t, are required in the crack band model. An approximate function $h_c = n_a d_a$ has been proposed to calculate the h_c value, where d_a is the maximum aggregate size in concrete or maximum grain size in rock and n_a is an empirical constant. It was suggested that the value of n_a equals 3 for concrete and 5 for rock.

To demonstrate how to apply the crack band model to fracture of quasi-brittle materials, consider a rectangular plate of unit thickness with a center crack subjected to tension, as shown in Fig. 5-9. Both ends of the plate are considered to be fixed. The width of the plate is $2b$, and the length of the crack band is $2a$. Before cracking, the strain energy density in the plate is equal to $\sigma_{Nc}^2/(2E)$, where σ_{Nc} is the fracture stress of the plate. The formation of the crack band causes relief of the strain energy and stress in some area surrounding the crack band. This energy-relief area may be approximated by area 125346 in Fig. 5-9, where the slopes of lines 16, 64, 35, and 52 are assumed to equal unity. Thus, the total energy release due to the formation of the crack band is

$$U = 2a^2 \frac{\sigma_{Nc}^2}{2E} + 2h_c a \frac{\sigma_{Nc}^2}{2E} = \frac{a^2 \sigma_{Nc}^2}{E} + \frac{n_a d_a a \sigma_{Nc}^2}{E} \tag{5.8}$$

Since the top and bottom boundaries of the plate are fixed, the work done by the applied load at the boundaries is zero. As a result, the energy release rate of the plate due to the formation of a unit length of the crack band, G_f, is equal to $\partial U/\partial a$,

$$G_f = \frac{\partial U}{\partial a} = \frac{2a\sigma_{Nc}^2}{E} + \frac{n_a d_a \sigma_{Nc}^2}{E} \tag{5.9}$$

Substituting Eq. (5.7) into Eq. (5.9) leads to

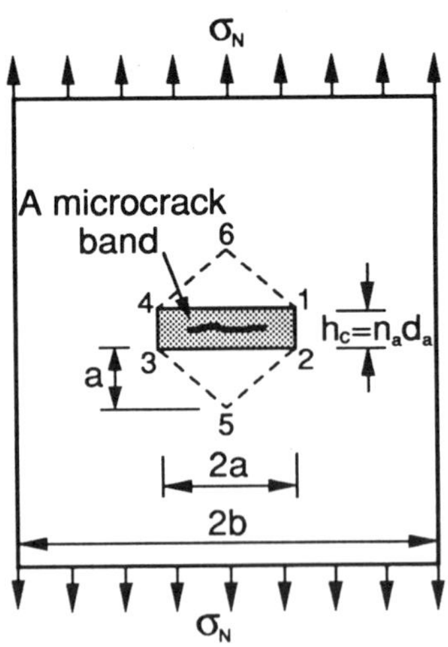

Figure 5-9 Crack band model for tensile plate with center crack.

$$\sigma_{Nc} = \frac{Bf_t}{\sqrt{1 + b/b_0}} \tag{5.10}$$

where

$$B = \sqrt{1 + \frac{E}{E_t}} \qquad b_0 = \frac{n_a d_a}{2} \frac{b}{a} \tag{5.11}$$

When the maximum size of aggregates is a constant (i.e., d_a = const), the value of b_0 is constant for a series of geometrically similar structures that have a constant value of b/a. Equation (5.10) indicates the relationship between the failure stress and size of structures for a given material and predicts a decrease of the fracture stress with increasing structure size. Since a is the crack length when the plate fails, Eq. (5.10) implies that for geometrically similar structures, the value of the critical crack length a, rather than the initial crack length a_0, should be proportional to the plate width b.

The crack band model can also be combined with finite element analysis to predict fracture response of different concrete structures.[9]

5.2.4 Determination and Influence of $\sigma(w)$ Relationship

The cohesive crack model requires a unique $\sigma(w)$ curve to quantify the value of energy dissipation. The choice of the $\sigma(w)$ function influences the prediction of the structural response significantly,[15, 16] and the local fracture behavior, for example the crack opening displacement, is particularly sensitive to the shape of $\sigma(w)$. Many different shapes of $\sigma(w)$ curves, including linear, bilinear, trilinear, exponential, and power functions, have been previously used. Some of the proposed $\sigma(w)$ curves are listed in Table 5-1 and are discussed below.

Hillborg et al.[8] and Cedolin et al.[17] proposed a simple linear $\sigma(w)$ curve as

$$\sigma(w) = f_t\left(1 - \frac{w}{w_c}\right) \tag{5.12}$$

The proposed linear curve depends on the parameters f_t and w_c. Hillerborg et al.[8] simply used w_c values of 0.01–0.02 mm (0.000394–0.000787 in.). On the other hand, Cedolin et al.[17] obtained values of f_t and w_c based on their test results and numerical modeling. They tested two mixes of concrete with a water-to-cement ratio of 0.5. The first mix had a maximum aggregate size of 12 mm (0.47 in.) and E = 34.2 GPa (4960 ksi) and f'_c = 41.6 MPa (6040 psi), where f'_c is the compressive strength of the concrete cylinder specimen. For the first concrete mix, Cedolin et al. estimated f_t = 4 MPa (581 psi) and w_c = 0.035 mm (0.00138 in.) using a least-square data-fitting procedure based on solution of an inverse problem and assumed shape of the stress–strain curve.

Table 5-1 Different Types of Closing Pressure for Fracture Process Zone

Type	Expression	Shape
Linear curve	$\sigma = f_t(1 - w/w_c)$	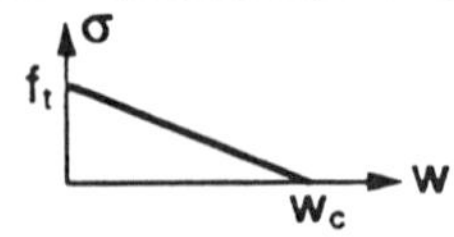
Bilinear curve	$\sigma = \begin{cases} f_t - (f_t - \sigma_1)w/w_1 & \text{for } w \le w_1 \\ \sigma_1 - \sigma_1(w - w_1)/(w_c - w_1) & \text{for } w_1 > w \end{cases}$	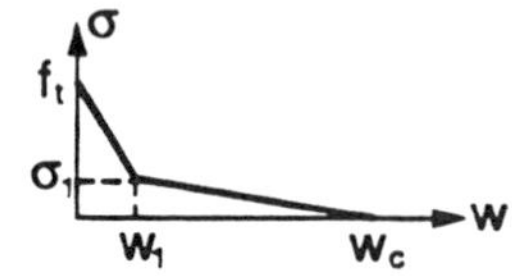
Trilinear curve	$\sigma = \begin{cases} f_t & \text{for } w \le w_1 \\ f_t - 0.7f_t(w - w_1)/(w_2 - w_1) & \text{for } w_1 < w \le w_2 \\ 0.3f_t(w_c - w)/(w_c - w_2) & \text{for } w_2 < w \le w_c \end{cases}$	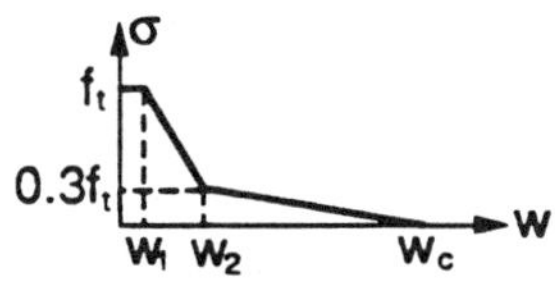
Exponential curve	$\sigma = f_t \exp(Aw^B)$ where $A, B =$ const	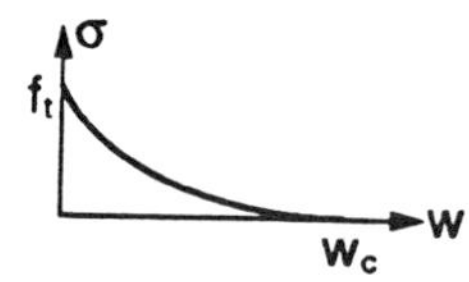
Power curve	$\sigma = 0.4f_t(1 - w/w_c)^{1.5}$	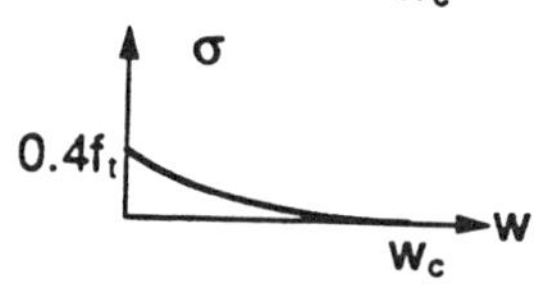
Bilinear curve with $w_1 = 0$	$\sigma = kf_t(1 - w/w_c)$	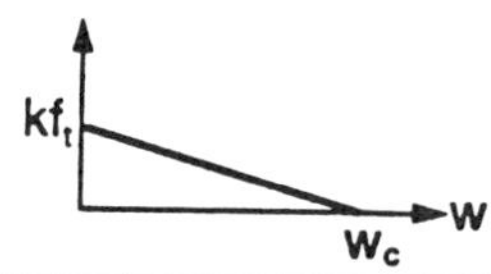

The second mix of concrete had a maximum aggregate size of 8 mm and $E =$ 30.0 GPa (4350 ksi) and $f'_c = 41.6$ MPa (6040 psi). For this second concrete mix, the values of $f_t = 3.6$ MPa (522 psi) and $w_c = 0.037$ mm (0.00146 in.) were estimated.

Roelfstra and Wittmann[15] proposed a bilinear curve for $\sigma(w)$ as

$$\sigma(w) = \begin{cases} f_t - (f_t - \sigma_1)\dfrac{w}{w_1} & \text{for } w \le w_1 \\[2ex] \sigma_1 - \sigma_1 \dfrac{w - w_1}{w_c - w_1} & \text{for } w > w_1 \end{cases} \tag{5.13}$$

where σ_1 and w_1 are defined in the corresponding figure given in Table 5-1. Roelfstra and Wittmann tested three-point bend concrete beams to determine the values of f_t, w_c, σ_1, and w_1. The concrete used had a water-to-cement ratio of 0.58 and maximum aggregate sizes of 3, 8, and 16 mm (0.118, 0.315, and 0.630 in.), respectively. For the concrete mix with a maximum aggregate size of 3 mm (0.118 in.), the values of $f_t = 4.7$ MPa (682 psi), $w_c = 0.077$ mm (0.0030 in.), $\sigma_1 = 0.57$ MPa (83 psi), and $w_1 = 0.017$ mm (0.00067 in.) were obtained. For the concrete mix with a maximum aggregate size of 8 mm (0.315 in.), the values of $f_t = 4.0$ MPa (581 psi), $w_c = 0.100$ mm (0.00394 in.), $\sigma_1 = 0.64$ MPa (93 psi), and $w_1 = 0.022$ mm (0.00087 in.) were obtained. For the concrete mix with a maximum aggregate size of 16 mm (0.630 in.), the values of $f_t = 4.4$ MPa (639 psi), $w_c = 0.123$ mm (0.00484 in.), $\sigma_1 = 0.82$ MPa (119 psi), and $w_1 = 0.020$ mm (0.00079 in.) were obtained.

The CEB-FIP Model Code[12, 18] also recommended a bilinear curve for $\sigma(w)$ as given in Eq. (5.13). However, the value of w_c depends on maximum aggregate size d_a. The values of $w_c = 0.12$ mm (0.00472 in.) for $d_a = 8$ mm (0.315 in.), $w_c = 0.15$ mm (0.00591 in.) for $d_a = 16$ mm (0.630 in.), and $w_c = 0.25$ mm (0.00984 in.) for $d_a = 32$ mm (1.260 in.) were given. The value of σ_1 was assigned to be equal to $0.15f_t$ and the value of w_1 in units of millimeters is given by

$$w_1 = \frac{G_F - 22w_c(G_F/k_d)^{0.95}}{150(G_F/k_d)^{0.95}} \tag{5.14}$$

where the coefficient k_d also depends on the maximum aggregate size d_a. The values of $k_d = 4$ for $d_a = 8$ mm (0.315 in.), $k_d = 6$ for $d_a = 16$ mm (0.630 in.), and $k_d = 10$ for $d_d = 32$ mm (1.260 in.) were given. For concrete with other maximum aggregate sizes, the corresponding values of w_c and k_d may be obtained by using a linear interpolation algorithm based on the above given values. In the absence of test data, the CEB-FIP Model Code also specifies empirical relations for the fracture parameters of the fictitious crack model in terms of the mean compressive strength f_c:

$$f_t = 0.3(f_c + 8)^{2/3} \qquad G_F = k_d f_c^{0.7} \qquad E = 10^4 f_c^{1/3} \tag{5.15}$$

in which f_c, f_t, and E are in megapascals and G_F is in newtons per meter. The commentary to the code cautions that these values are approximate and may be unconservative for large structures. Nevertheless, these relations are useful when other data are lacking.

Kobayashi et al.[19, 20] used a trilinear curve for $\sigma(w)$, which is given by

$$\sigma(w) = \begin{cases} f_t & \text{for } w \leq w_1 \\ f_t - 0.7f_t \dfrac{w - w_1}{w_2 - w_1} & \text{for } w_1 < w \leq w_2 \\ 0.3f_t \dfrac{w_c - w}{w_c - w_2} & \text{for } w_2 < w \leq w_c \end{cases} \tag{5.16}$$

where w_1 and w_2 are illustrated in the corresponding figure given in Table 5-1. The values of w_1 and w_2 depend on the compressive strength f'_c. It has been reported that for 23.9 MPa (3470 psi) $< f'_c <$ 34.5 MPa (5010 psi), $w_1 = 0.37f'_c - 2.54$, $w_2 = 5w_1$, and $w_c = w_2 + 228.6$, whereas for 34.5 MPa (5010 psi) $< f'_c <$ 62.5 MPa (9070 psi), $w_1 = 19 - 0.26f'_c$, $w_2 = (18.2f'_c - 1.25)w_1$, and $w_c = w_2 + 228.6$. Here all crack opening displacements are in micrometers.

Gopalaratnam and Shah[21] proposed an exponential curve for $\sigma(w)$ as

$$\sigma(w) = f_t \exp(Aw^B) \tag{5.17}$$

where A and B are constants depending on mixes of concrete. Based on the experimental data obtained, they reported $A = -0.06163$ and $B = 1.01$ for concrete with f'_c values of 33–47 MPa, and the crack opening displacement is in micrometers here. Note that no value of w_c is found using this approach. The value of w corresponding to $\sigma(w) = 0$ depends on the constants A and B selected in Eq. (5.17). A similar exponential relationship for $\sigma(w)$ was also suggested by Cedolin et al.[17]

Du et al.[22] suggested a power function for $\sigma(w)$,

$$\sigma(w) = 0.4f_t\left(1 - \frac{w}{w_c}\right)^{1.5} \tag{5.18}$$

It is noted that the $\sigma(w)$ function suggested by Du et al. has a sudden drop after the initial opening of the crack.

Figueiras and Owen[23] used a linear softening curve with a sudden drop after initial opening of the crack. The expression of their softening curve is similar to Eq. (5.13) but with $\sigma_1 = kf_t$ and $w_1 = 0$.

Since the assessment of the fracture behavior of a concrete structure is influenced by using different $\sigma(w)$ functions, reasonable and accurate determinations of the $\sigma(w)$ curve and the corresponding parameters become crucial for the cohesive crack approach. Experimental determination of $\sigma(w)$ directly from tension tests has been suggested,[21] but this is difficult and the results may vary with specimen size and shape. Li et al.[24] have proposed a J-integral-based method for obtaining the entire $\sigma(w)$ curve. Miller et al.[25] have computed the $\sigma(w)$ curve from parabolic crack profiles observed in fracture tests through laser holography interferometry. Hawkins et al.[26,27] have applied Moiré interferometry to deduce crack profiles that were used to determine the stress-opening relation in the process zone. The concept that the shape of the softening curve is a material property is currently not accepted by many researchers in the field.

5.2.5 Some Comments on Fictitious Crack Approach

It is noted that by assuming $G_{Ic} = 0$ some computational efficiency may be accomplished. However, the single Dudgale–Barenblatt energy dissipation mechanism may achieve global energy balance only by selecting suitable model parameters. The fictitious crack approach may physically make sense when the crack tip fracture process zone has the same fracture mechanisms as the crack wake process zone.[7] This is true in Dudgale's original model of yielding of metals, where the fracture process zone advances when hitherto elastic metal yields and stresses in the whole process zone are governed by the same yielding process. Since different fracture mechanisms may exist in the crack tip and the crack wake process zones for concrete, the fictitious crack approach is only an approximation for fracturing of concrete. Some actual features associated with crack propagation, such as the crack profile, computed based on the fictitious crack approach may not match with those experimentally measured. This is illustrated in Fig. 5-10, where the crack profile for a center-cracked plate under tension, experimentally measured using laser holography interferometry, is compared with that predicted by the fictitious crack model.[25] It is found that the predicted crack profile deviates from that experimentally measured. The actual crack profile at the tip is more similar to that obtained from LEFM, instead of the fictitious crack model.[28] To match the experimentally measured crack profile, the Griffith–Irwin energy dissipation mechanism should be included as in Eq. (5.3). Until cohesive fracture parameters G_F, the tensile strength f_t, and the shape of $\sigma(w)$ are further identified to be geometry and size independent, the fictitious crack approach requires some empirical fitting. It has also been indicated by Cox and Marshall[7] that the fictitious crack models are incapable of representing some essential characteristics of fracture and fatigue that are commonly observed in many other quasi-brittle materials, particularly for materials with some reinforcing phases.

Another major problem with the fictitious crack approach is the need to determine, experimentally, a size-independent value of the tensile strength that can be measured consistently and easily.

5.3 EFFECTIVE-ELASTIC CRACK APPROACH

5.3.1 Energy Dissipation for Effective-Elastic Crack

The fracture process zone in concrete can also be modeled by a single Griffith–Irwin energy dissipation mechanism by assuming $\sigma(w) = 0$. The effective-elastic crack approach models the fracture process zone by using an equivalent, traction-free elastic crack. The effective-elastic crack is governed by the criterion from LEFM, and the equivalence between the actual and the corresponding effective crack is prescribed explicitly in each model. As a result, the energy release rate for an effective-elastic, mode I crack is

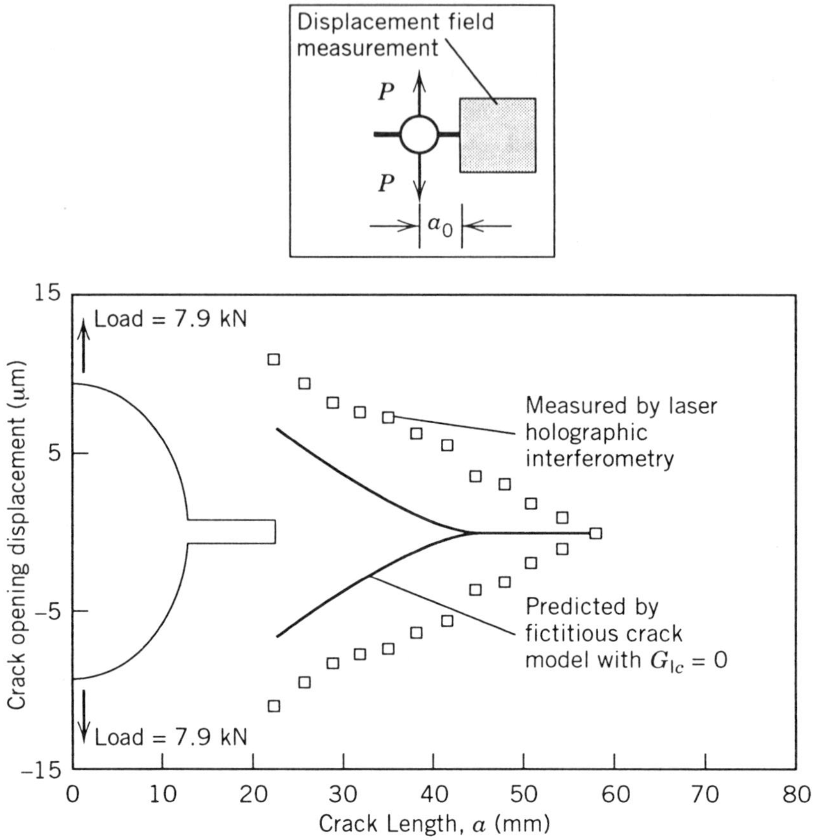

Figure 5-10 Crack profiles measured by laser holographic interferometry and predicted by fictitious crack model ($G_{Ic} = 0$).[25]

$$G_q = G_{Ic} \tag{5.19}$$

Equation (5.19) governs propagation of the effective-elastic crack due to change of applied load. In the application of Eq. (5.19), G_q is a function of structural size and geometry and applied load as well as length of the effective-elastic crack, whereas G_{Ic} is the critical energy release rate (i.e., the material fracture toughness). Since the crack length will increase with change of applied load for the case of stable crack propagation, an additional equation should be provided to determine the crack length before Eq. (5.19) can be used. The length of the effective-elastic crack has been empirically found to be structural size and geometry dependent so that it cannot be used directly as an independent fracture criterion. Therefore, another quantity should be introduced as the fracture criterion. Most of the effective-elastic crack models use two fracture parameters to define the inelastic fracture process and to govern crack propagation. Some effective-elastic crack models are presented below. For simplicity, all expressions presented are valid for the plane-stress condition. For the plane-strain condition, E in the corresponding equations should be replaced by $E/(1 - \mu^2)$, although for the normal value of Poisson's ratio μ for concrete, the expressions differ by no more than about 4%.

5.3.2 Two-Parameter Fracture Model by Jenq and Shah

Jenq and Shah[10] have proposed a two-parameter fracture model based on elastic fracture response of structures. To separate the elastic and plastic fracture responses of a given specimen as shown in Fig. 5-11*a*, the specimen may be loaded up to the maximum stress, and an unloading and reloading cycle performed as shown in Fig. 5-11*b*. Based on the obtained unloading compliance, the value of CMOD at the peak load, CMOD_c, can be divided into an elastic component and an inelastic component as (see Fig. 5-11*b*)

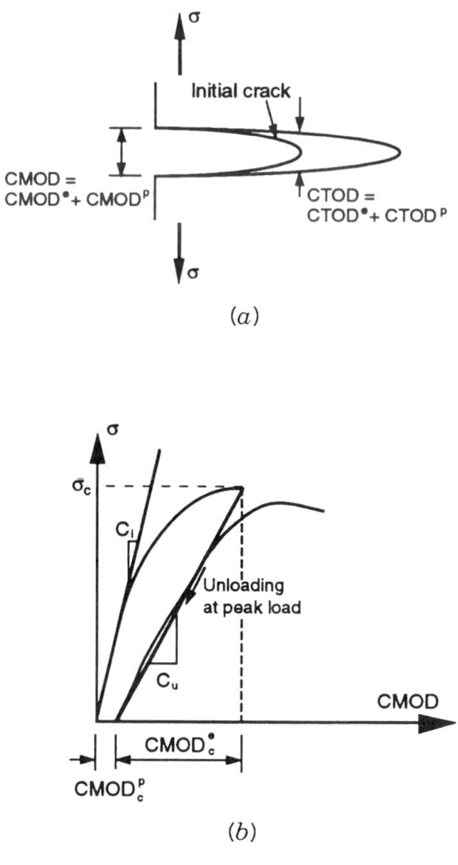

Figure 5-11 Testing specimen and procedure to determine values of K_{Ic}^s and CTOD_c: (*a*) elastic and plastic fracture responses and (*b*) loading and unloading procedure.

$$\mathrm{CMOD}_c = \mathrm{CMOD}_c^e + \mathrm{CMOD}_c^p \tag{5.20}$$

where CMOD_c^e and CMOD_c^p are the elastic and inelastic components of CMOD_c, respectively. The measured value of CMOD_c^e as well as the maximum stress σ_c are further substituted into the following LEFM equations to calculate the critical stress intensity factor K_{Ic}^s and the critical effective elastic crack length a_c:

$$K_{\mathrm{Ic}}^s = \sigma_c \sqrt{\pi a_c}\, g_1\left(\frac{a_c}{b}\right) \tag{5.21}$$

and

$$\mathrm{CMOD}_c^e = \frac{4\sigma_c a_c}{E}\, g_2\left(\frac{a_c}{b}\right) \tag{5.22}$$

The value of the critical crack tip opening displacement CTOD_c^e is then determined based on the obtained values of CMOD_c^e, σ_c, and a_c using the equation

$$\mathrm{CTOD}_c^e = \mathrm{CMOD}_c^e\, g_3\left(\frac{a_c}{b}, \frac{a_0}{a_c}\right) \tag{5.23}$$

It is noted that since CMOD_c^e is the elastic component, Eqs. (5.21)–(5.23) are based on LEFM, and geometric functions g_1, g_2, and g_3 for different specimens can be found from an LEFM handbook.[29] Geometric functions g_1, g_2, and g_3 for some conventional specimens have been given in Chapter 2.

From the experimental results obtained by using three-point bend beams, Jenq and Shah[10] found that for the beams with different sizes but made of the same material the values of K_{Ic}^s and CTOD_c^e are basically constant. As a result, they proposed that the critical fracture property of a quasi-brittle material may be characterized by the values of K_{Ic}^s and CTOD_c^e. For a given material, structures with different geometries and sizes when subjected to the critical fracture load (the peak load) will satisfy the following two conditions, as shown in Fig. 5-12:

$$\begin{aligned} K_{\mathrm{I}} &= K_{\mathrm{Ic}}^s \\ \mathrm{CTOD} &= \mathrm{CTOD}_c \end{aligned} \tag{5.24}$$

where K_{I} and CTOD are the stress intensity factor and the crack tip displacement, respectively, which can be calculated based on LEFM.[29] It is noted in Eq. (5.24) that CTOD_c^e is simply denoted as CTOD_c. Equation (5.24) is the so-called two-parameter fracture model. In Eq. (5.24) the values of K_{I} and

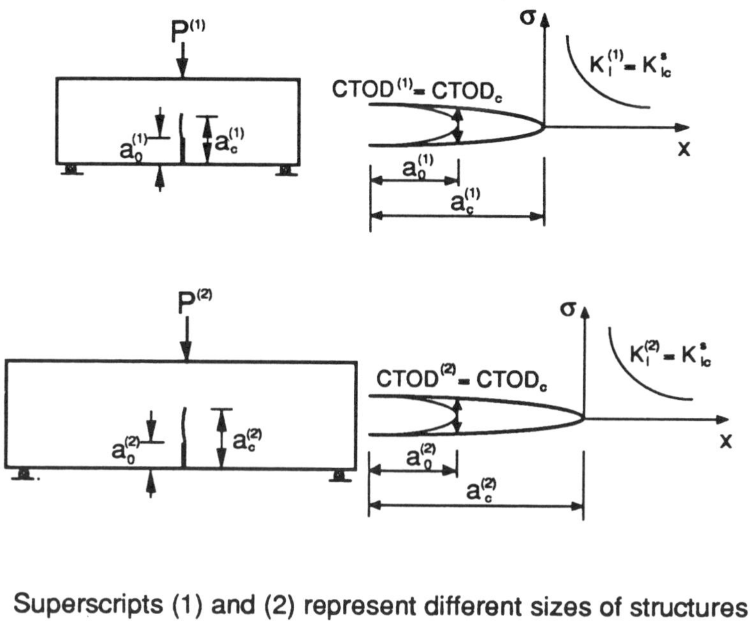

Figure 5-12 Two-parameter fracture model, fracture criteria: $K_I = K_{Ic}^s$ and CTOD = $CTOD_c$.

CTOD are functions of the applied load, structural geometry and size, and the crack length, whereas the critical stress intensity factor K_{Ic}^s and the critical crack tip opening displacement $CTOD_c$ are fracture parameters depending only on the material. As a result, Eq. (5.24) provides the two conditions needed to determine the critical fracture load and the critical crack length of a structure. It is seen from the aforementioned procedure of determining the values of K_{Ic}^s and $CTOD_c$ that in the two-parameter fracture model the effective-elastic crack exhibits a compliance equal to the unloading compliance of the actual structure. Therefore, the two-parameter fracture model determines the critical fracture state of a structure based on its elastic response.

When the values of K_{Ic}^s and $CTOD_c$ are given for a material, some of the material characteristic properties can be calculated. Jenq and Shah[10] introduced a material length, Q, which is proportional to the size of the fracture process zone for the same material. The value of Q is expressed as

$$Q = \left(\frac{E\,CTOD_c}{K_{Ic}^s}\right)^2 \tag{5.25}$$

It was found that the values of Q were in the range from 12.5–50 mm (0.5–2 in.) for hardened cement paste, 50–150 mm (2–6 in.) for mortar, and 150–350 mm (4–14 in.) for concrete. The material length Q can be used as a brittleness index for the material. The larger the value of Q, the more ductile the material. Furthermore, the tensile strength of a material may be evaluated from the failure stress of an infinitely-large uniaxial tensile plate with double-edge crack. Therefore, the theoretical tensile strength of a material, f_t, can be calculated by substituting the values of K_{Ic}^s and CTOD_c into the expressions for the stress intensity factor and the crack opening displacement given in Chapter 2 [i.e., Eqs. (2.27)–(2.29)] for the plate. This leads to

$$f_t = 1.4705 \frac{(K_{Ic}^s)^2}{E\ \text{CTOD}_c} \tag{5.26}$$

The two-parameter fracture model can also be used for an unnotched structure such as shown in Fig. 5-13. By substituting $a_0 = 0$ into LEFM expressions for K_I, CMOD, and CTOD for a three-point bend beam (see Chapter 2) and noting CTOD = CMOD and $a_c = \Delta a_c$ for an unnotched beam, one can calculate the value of the MOR based on the two-parameter fracture model. By assuming that the material length Q is equal to 20 mm (8 in.), which is the average material length of concrete and mortar, the influence of the structural size on the MOR of unnotched three-point bend beams predicted by the two-

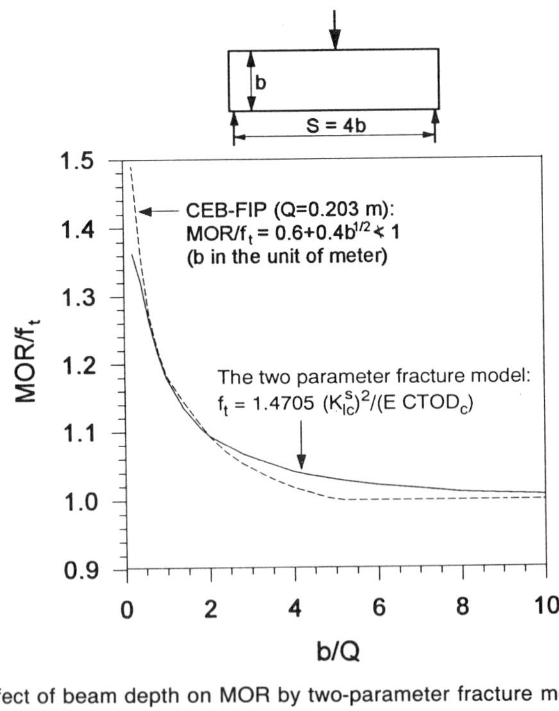

Figure 5-13 Effect of beam depth on MOR by two-parameter fracture model and CEB-FIP Model Code.[30]

parameter fracture model is plotted in Fig. 5-13. The CEB-FIP specification for the MOR is also given in Fig. 5-13.[30] It is seen that the CEB-FIP size effect MOR compares favorably with the prediction from the two-parameter fracture model.

John and Shah[31] found that both K_{Ic}^s and CTOD_c are related to the compressive strength of concrete, f_c'. Based on experimental observations, they proposed the following empirical equations for K_{Ic}^s, CTOD_c, and E in terms of f_c':

$$K_{Ic}^s = 0.06(f_c')^{0.75} \qquad \text{CTOD}_c = 0.00602(f_c')^{0.13} \qquad E = 4785\sqrt{f_c'} \tag{5.27}$$

where K_{Ic}^s is in $\text{MPa}\sqrt{\text{m}}$, CTOD_c is in millimeters, and E and f_c' are in megapascals.

It may be worthwhile to explain the physical significance of CTOD_c. The existence of the CTOD_c can be justified from the fact that all materials have some sort of initial flaws. This is especially true for concrete. When the material is subjected to an external load, these flaws will open, propagate, and sometimes coalesce with other cracks. The opening displacement of these flaws can be directly linked to the value of CTOD. At the critical fracture load, the crack opening displacement at the initial crack tip can be defined as CTOD_c. For quasi-brittle materials such as concrete, stable crack extension occurs before the critical fracture load. Since the CTOD_c is defined at the tip of the initial crack, it may primarily account for the growing size of the wake process zone. It is noted that CTOD_c is used together with K_{Ic}^s to determine the critical crack propagation for quasi-brittle materials. However, for ductile fracture in metals CTOD_c is used as a single fracture parameter.[32]

For a material obeying LEFM one parameter (K_{Ic}^s or CTOD_c or G_{Ic}) is sufficient to describe its fracture toughness. For example, for some metals increasing strain rate increases their yield strength but decreases their fracture toughness, as shown in Fig. 5-14.[33] For concrete if only one parameter (e.g., K_{Ic} or G_F) is considered, then one observes that the fracture toughness increases with increasing compressive strength or increasing strain rate. Such a one-parameter representation is misleading since concrete in fact becomes more brittle as its compressive strength increases. The importance of a two-parameter representation for high-strength concrete is shown in Fig. 5-15. In this figure the critical crack extension, which was calculated by substituting Eq. (5.27) into the two-parameter fracture model, is plotted against compressive strength. It is seen that the critical crack extension decreases with increasing compressive strength. For an ideally brittle material critical crack extension is zero. Thus the two-parameter fracture model correctly simulates the observed brittle response of high-strength concrete.

A large round-robin test program was conducted by the RILEM Committee TC89.[34] It was shown that the critical stress intensity factor evaluated at the

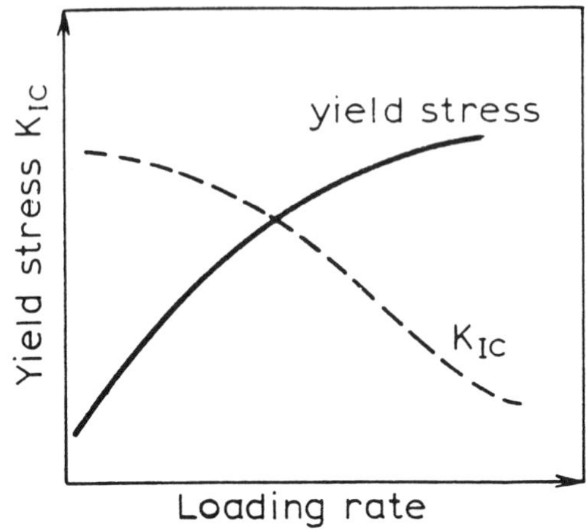

Figure 5-14 Effect of loading rate on critical stress intensity factor K_{Ic} and yield stress for steel.

tip of the effective crack was essentially independent of the dimensions of notched beams as well as independent of different methods used to estimate the effective crack length. A similar extensive comparison was not done for the $CTOD_c$. Since the $CTOD_c$ is a small quantity and difficult to measure, additional verification is needed to confirm that it is indeed a material property.

Equation (5.24) can always be used as fracture criteria for quasi-brittle materials, and a constant value of $CTOD_c$ may be used as first-order approxi-

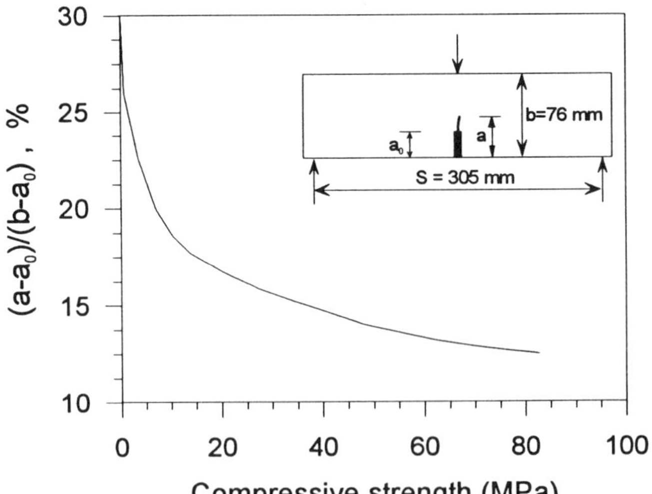

Figure 5-15 Effect of compressive strength on crack extension.[31]

mation of the fracture parameter for quasi-brittle materials. It is noted that introducing CTOD_c as a material constant makes Eq. (5.24) have a minimum number of material parameters for governing fracture of quasi-brittle materials. It has been found that when two fracture constants $K_{\mathrm{I}c}^{s}$ and CTOD_c are given for a material, the two-parameter fracture model can predict the critical fracture loads for structures with different sizes and geometries. These have been done for a three-point bend beam,[35] single-edged notched tensile specimen,[35] center-cracked plate specimen,[36] compact tension specimen,[37] and split-tension cylinder.[38] The fracture criteria given in Eq. (5.24) can easily be implemented in a finite element method.[38] John and Shah[31] applied the two-parameter fracture model to high-strength concrete. They found that the variation of uniaxial tensile strength, split-tension cylinder, and MOR with change of the compressive strength can be predicted satisfactorily for high-strength concrete using the two-parameter fracture model. John et al.[39] also found that if both E and $K_{\mathrm{I}c}^{s}$ are assumed to be rate independent while CTOD_e is assumed to decrease exponentially with the logarithm of the relative strain rate, the two-parameter fracture model is capable of predicting the rate sensitivity of mode I fracture of concrete. The effective crack length was reported to decrease with increasing rate from 10^{-6} to 1 per second. This would explain the observed increased brittleness with the increased strain rate.

5.3.3 Size Effect Model by Bazant and Kazemi

Bazant and Kazemi[11] have simulated fracture of quasi-brittle materials by an effective-elastic crack. They considered a series of geometrically similar structures. For two-dimensional similarity as shown in Fig. 5-16, these structures may have different sizes, but the ratio of initial crack length a_0 to characteristic dimension of a structure, D, is kept a constant (i.e., a_0/D = const), where D represents the depth of a beam, or the width of a tensile plate, and so on. It is, however, an important requirement that the third dimension (thickness) be kept constant. For these geometrically similar structures, the nominal stress at failure (or the nominal strength) is described by

$$\sigma_{Nc} = \frac{c_n P_c}{tD} \tag{5.28}$$

where P_c is the critical fracture load (or the peak load), t is the thickness of a structure, and c_n is a coefficient representing different types of structures, c_n = 1 for a tensile plate and $c_n = 1.5S/D$ for a beam (S is the span and D the depth). It is noted that the value of c_n is constant for a series of geometrically similar structures.

It may be assumed that the fracture energy dissipated at failure is a smooth function of structural dimensions and size of the fracture process zone. Based on this assumption, Bazant[40] has shown by dimensional analysis and similitude

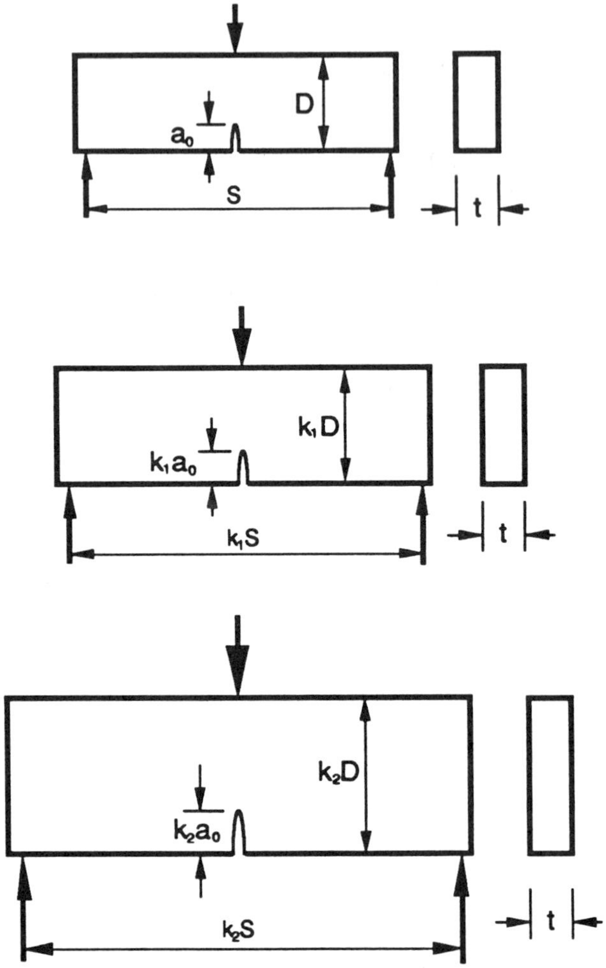

Figure 5-16 Series of geometrically similar structures (k_1 and k_2 are constant, two-dimensional similarity).

arguments that the failure stress of a series of geometrically similar structures of different sizes can be expressed by the infinite series

$$\sigma_{Nc} = Bf_t\left[\frac{D}{D_0} + 1 + L_1\left(\frac{D}{D_0}\right)^{-1} + L_2\left(\frac{D}{D_0}\right)^{-2} + \cdots\right]^{-1/2} \tag{5.29}$$

where B, D_0, L_1, L_2, $\cdots$ are constants and f_t is the size-independent tensile strength of a material. Equation (5.29) represents an asymptotic expansion of the failure stress of the geometrically similar structures with respect to the relative structure size, D/D_0. It was further demonstrated that for the size range up to $D/D_0 = \frac{1}{20}$, the asymptotic series can be truncated after the linear term. As a result, Eq. (5.29) reduces to[41]

$$\sigma_{Nc} = \frac{Bf_t}{\sqrt{1 + D/D_0}} \tag{5.30}$$

Equation (5.30) represents the Bazant size effect model for the failure stress of geometrically similar structures. Equation (5.30) is similar to Eq. (5.10), which was derived based on the crack band model. Below it will be demonstrated how to determine the two constants, B and D_0, based on the effective-elastic crack approach.

Since an effective-elastic crack is used, based on formulations from LEFM and using Eq. (5.28), the critical energy release rate for a certain size of structure, $G_{\mathrm{I}c}$, can be written as

$$G_{\mathrm{I}c} = \frac{K_{\mathrm{I}c}^2}{E} = \frac{\sigma_{Nc}^2 \pi a_c}{E} g_1^2\left(\frac{a_c}{D}\right) = \frac{P_c^2}{Et^2D} g\left(\frac{a_c}{D}\right) = \frac{\sigma_{Nc}^2 D}{Ec_n^2} g\left(\frac{a_c}{D}\right) \tag{5.31}$$

where $a_c = a_0 + \Delta a_c$ is the critical crack length, which is expressed as the sum of the initial crack length a_0 and the effective crack extension Δa_c for a given size structure, and the geometric function $g(a_c/D)$ is given by

$$g\left(\frac{a_c}{D}\right) = \frac{c_n^2 \pi a_c}{D} g_1^2\left(\frac{a_c}{D}\right) \tag{5.32}$$

Bazant and Kazemi[11] proposed using the critical energy release rate and the critical crack extension for an infinitely large structure as fracture parameters of a quasi-brittle material. Therefore, the two fracture parameters for mode I quasi-brittle fracture are defined as

$$G_f = \lim_{D \to \infty} G_{\mathrm{I}c} \tag{5.33}$$

and

$$c_f = \lim_{D \to \infty} \Delta a_c \tag{5.34}$$

where G_f and c_f are the critical energy release rate and the critical crack extension for an infinitely large structure, respectively. Substituting Eq. (5.31) into Eq. (5.33) and using Eq. (5.30) results in

$$G_f = \lim_{D \to \infty} \frac{\sigma_{Nc}^2 D g(a_c/D)}{E c_n^2} = \lim_{D \to \infty} \frac{B^2 f_t^2 D_0}{E c_n^2} \frac{D}{D + D_0} g\left(\frac{a_c}{D}\right) \tag{5.35}$$

By using the relationships

$$\lim_{D \to \infty} \frac{D}{D + D_0} = 1 \tag{5.36}$$

and

$$\lim_{D \to \infty} g\left(\frac{a_c}{D}\right) = g\left(\frac{a_0}{D}\right) \tag{5.37}$$

Eq. (5.35) becomes

$$G_f = \frac{B^2 f_t^2 D_0}{E c_n^2} g\left(\frac{a_0}{D}\right) \tag{5.38}$$

Equations (5.36) and (5.37) result from the fact that the length of the crack extension is negligible compared to the initial crack length when the structural size approaches infinity. Equation (5.38) relates the material fracture parameter G_f to a finite-size structure with a characteristic size of D_0.

To obtain the value of c_f, the geometric function $g(a_f/D)$ for an infinitely large structure may be expanded into a Taylor series at the value of a_0/D. If only up to the first-order terms are considered, this Taylor series is approximated as

$$g\left(\frac{a_f}{D}\right) \approx g\left(\frac{a_0}{D}\right) + g'\left(\frac{a_0}{D}\right)\left(\frac{a_f}{D} - \frac{a_0}{D}\right) = g\left(\frac{a_0}{D}\right) + g'\left(\frac{a_0}{D}\right)\frac{c_f}{D} \tag{5.39}$$

where $a_f = a_0 + c_f$ is the effective crack length for a large structure and $g'(a_0/D) = dg/d(a/D)$ for $a = a_0$. By combining Eqs. (5.36)–(5.39), when D approaches infinity, one obtains

$$g\left(\frac{a_f}{D}\right) = g\left(\frac{a_0}{D}\right)\frac{D + D_0}{D} \tag{5.40}$$

By substituting Eq. (5.40) into Eq. (5.39), c_f can be determined as

$$c_f = \frac{g(a_0/D)}{g'(a_0/D)} D_0 \tag{5.41}$$

Equation (5.41) represents a relationship between c_f and D_0. Substituting the value of D_0 obtained from Eq. (5.41) into Eq. (5.38), the value of Bf_t can be determined in terms of G_f and c_f as

$$Bf_t = c_n \left[\frac{EG_f}{c_f g'(a_0/D)} \right]^{1/2} \tag{5.42}$$

From Eqs. (5.41) and (5.42) the values of D_0 and Bf_t are related to the material fracture parameters G_f and c_f. Substituting Eqs. (5.40) and (5.42) into Eq. (5.30), the nominal strength of structures can be expressed as

$$\sigma_{Nc} = c_n \left[\frac{EG_f}{g'(a_0/D)\, c_f + g(a_0/D) D} \right]^{1/2} \tag{5.43}$$

Equation (5.43) describes the nominal strengths for a series of geometrically similar structures made of the same quasi-brittle material. Once the fracture parameters G_f and c_f are known for a given material, the nominal strength σ_{Nc} for a series of geometrically similar structures with different sizes can be determined from Eq. (5.43). The influence of structural geometry is taken into account by the functions $g(a_0/D)$ and $g'(a_0/D)$. Since the values of $g(a_0/D)$ and $g'(a_0/D)$ are constants for geometrically similar structures, the nominal strength σ_{Nc} decreases with increasing structural size D, as shown in Fig. 5-17. Experimental measurement of G_f and c_f will be discussed in Chapter 6.

Equation (5.43) for the size effect on the nominal strength of a series of geometrically similar structures is illustrated in Fig. 5-17. The nominal strength of structures is conventionally described by a strength criterion, which is a constant regardless of structural size and may only be used for relatively small size structures. On the other hand, the single-parameter failure criterion based on LEFM predicts a linear curve for the nominal strength, and this criterion may only be used for relatively large size structures. Equation (5.43) based on NLFM provides a smooth curve for the nominal strength of various sizes of

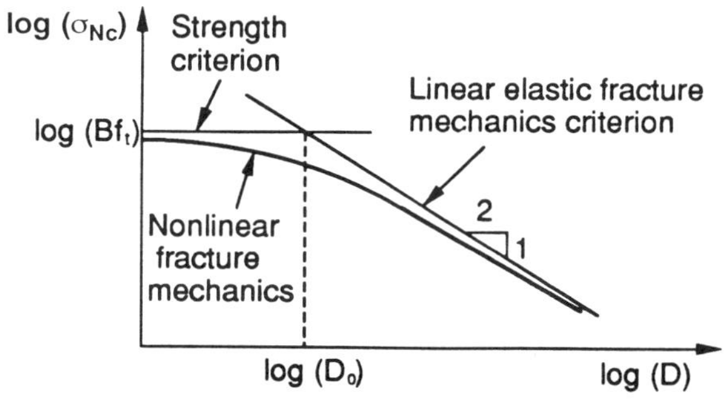

Figure 5-17 Size effect on nominal failure stress by Bazant.[41]

structures. The parameter D_0 may be physically interpreted as the size that characterizes the intersection of the horizontal strength criterion and the inclined criterion from LEFM, as shown in Fig. 5-17. If the value of D_0 is determined in such a way, the brittleness number β_b introduced by Bazant and Kazemi[11] can be expressed as

$$\beta_b = \frac{D}{D_0} = \frac{B^2 f_t g(a_0/D)}{c_n E G_f} D \tag{5.44}$$

It is noted that the brittleness number β_b depends on not only the material fracture parameter G_f but also the structural geometric function $g(a_0/D)$. Since the value of $B = \sigma_{Nc}/f_t$ relates to the nominal strength of a very small structure on the basis of the strength criterion, Eq. (5.44) may be suitable for relatively small size structures with D approaching zero. For relatively large size structures with D approaching infinity, the brittleness number β_b may be defined as[42]

$$\beta_b = \frac{D}{c_f} \frac{g(a_0/D)}{g'(a_0/D)} \tag{5.45}$$

It is noted that Eqs. (5.44) and (5.45) are based on different asymptotic approximations. Based on these definitions, brittleness of structures relates to not only material but also geometry and size of structures. The greater the value of β_b, the more brittle the structure. For $\beta_b < 0.1$, the strength criterion may be used. For $0.1 < \beta_b < 10$, the size effect model [Eq. (5.43)] should be used. For $\beta_b > 10$, the criterion based on LEFM may be used.

5.3.4 Effective Crack Model by Karihaloo and Nallathambi

Karihaloo and Nallathambi[43,44] have proposed an effective crack model based on the effective-elastic crack approach. In their model, the fracture toughness of concrete is determined using three-point bend beams. The basic concept of their effective crack model is similar to the two-parameter fracture model. However, a secant compliance at the maximum load, rather than the elastic part of the unloading compliance as in the two-parameter fracture model, is used to determine the effective-elastic crack length in the effective crack model. This is illustrated in Fig. 5-18, where a load–deflection curve of a three-point bend beam is shown. The initial linear part of the load–deflection curve is followed by a monotonic change in the slope. By connecting the maximum load to the origin, a secant compliance corresponding to the maximum load can be obtained from the load–displacement curve, as shown in Fig. 5-18. The critical effective crack length, assumed to correspond to the maximum load, can then be obtained using the secant compliance based on LEFM. By using

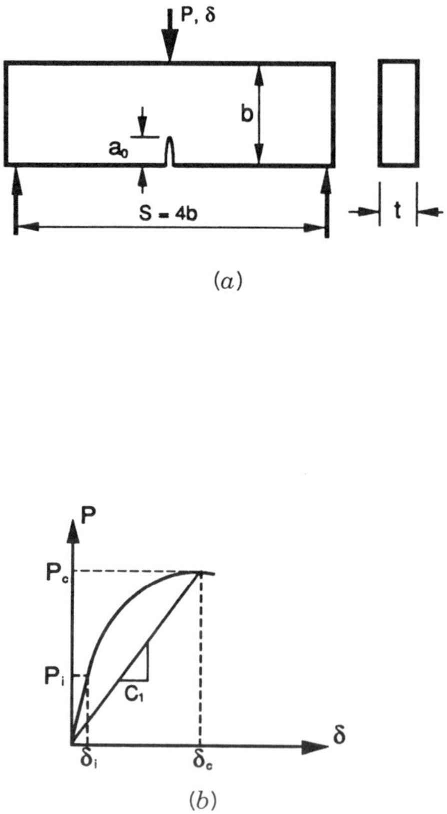

Figure 5-18 Effective crack model by Nallathambi and Karihaloo: (*a*) a three-point bend beam and (*b*) loading and load-point displacement curve.[43]

the critical crack length and the maximum load, the value of the critical stress intensity factor K_{Ic}^e can be calculated. Since the secant compliance at the maximum load includes effects of both elastic and inelastic deformations, the critical effective crack length obtained in such a way is usually greater than that obtained from the two-parameter fracture model, where the latter only includes effect of elastic deformation. The procedure used to obtain values of K_{Ic}^e will be discussed in detail in Chapter 6.

5.3.5 Effective Crack Model by Refai and Swartz

This model[45,46] is based upon observations obtained by the use of dye, of the development of the process zone in notched beams in three-point-bending, which has been described in Chapter 4. The use of precracked specimens following the procedure described in the next chapter is required. For such specimens, if the ratio of the crack length to the beam depth is less than 0.65 and provided the beam depth is greater than 203 mm (8 in.), it may be assumed, based on experimental evidence, that the process zone is fully developed and its shape stays relatively constant.

Using these ideas, a curve of crack length versus maximum load P_c is shown in Fig. 5-19 for 8-in-deep beams. Each data point is obtained after failure by measuring the dyed surface to obtain the average crack length. Since the process zone is completely developed, this average length is also considered to be effective crack length. For beams tested subsequently the calibration curve is used to obtain the effective crack length $a_c = a$, which is then used in an LEFM formula to obtain K_{Ic}^e. This procedure will be described in more detail in Chapter 6.

5.3.6 Some Comments on Effective-Elastic Crack Approach

For quasi-brittle materials, an initial flaw extends in a stable manner before it propagates critically at maximum load. Note that for negative-geometry structures for which the stress intensity factor initially decreases and then increases with the crack propagation, the maximum load and critical crack extension may not coincide. The energy absorbed in this stable crack extension can be accurately simulated by defining an equivalent-elastic crack. The critical stress intensity factor at the tip of the effective crack can be considered a material property similar to LEFM. However, one needs one more parameter to define the equivalent-elastic crack length. In the two-parameter fracture model and the size effect model, the values of CTOD_c and c_f are introduced as additional material parameters, respectively. The value of CTOD_c corresponds to the elastic response of a structure since it is determined based on the elastic un-

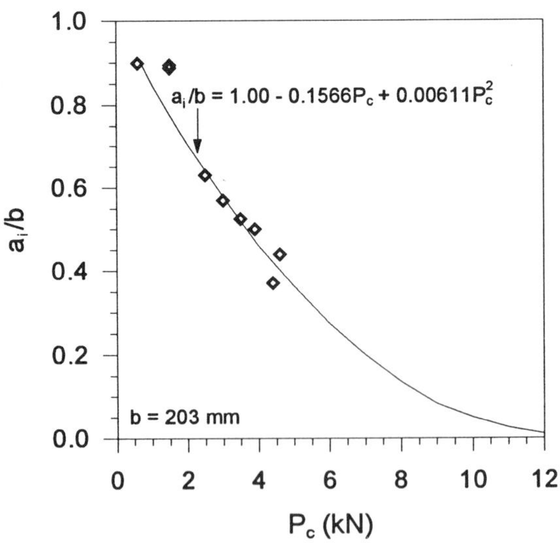

Figure 5-19 Maximum-load calibration curve for beams with depth of 203 mm (8 in.).[45]

loading compliance at the maximum load, whereas c_f is interpreted as the critical crack extension for an infinitely large structure. Therefore, the influence of the fracture process zone on the material fracture parameter is eliminated either by taking only the elastic response in the two-parameter fracture model or by using an infinitely large structure in the size effect model. It is noted that besides using three-point bend beams, the fracture parameters $K_{\mathrm{I}c}^s$, CTOD_c or G_f, c_f can also be determined using structures with other geometries. By using two fracture parameters, either $K_{\mathrm{I}c}^s$ and CTOD_c or G_f and c_f, the effective-elastic crack approach is capable of predicting failure (i.e., the maximum load and the corresponding effective crack length) of structures with different sizes and geometries. In the effective crack models proposed by Karihaloo and Nallathambi[43,44] and by Swartz and Refai,[45,46] some regression curves based on experimental observations were introduced to determine the critical effective crack length for notched beams. Although the critical stress intensity factor for a given material can be determined from these later effective crack models, regression analysis is needed for each different geometry (even for beams with proportional dimensions) to obtain $K_{\mathrm{I}c}^e$.

In the size effect model, a series of geometrically similar structures with a constant value of a_0/D was introduced. For real materials, however, the value of the initial flaw a_0 depends on the microstructure of the material. Then it is unlikely that the value of a_0 should linearly increase with increasing dimension of structures (provided the fabrication processes that may alter the microstructure are essentially size independent).

Since the material fracture parameters $K_{\mathrm{I}c}^s$ and CTOD_c, or G_f and c_f are defined for the critical fracture of a structure, the effective-elastic crack approach can only predict the maximum load and the corresponding displacement (or crack length) of a structure. In order to predict the entire stress–strain response of a structure, corresponding R-curves should be introduced, which will be discussed in Chapter 7. Since the equivalence between the actual crack and the effective-elastic crack is based on the critical state of a structure, the effective-elastic crack approach may predict the peak load and displacement well. Since the prepeak stable crack growth can be well controlled by the critical fracture, it may be reasonably predicted by using the effective-elastic crack approach. However, some discrepancy between experimental measurements and theoretical prediction may occur when the crack propagates unstably (i.e., after the peak load). This discrepancy increases as the fracture state is far away from the peak load. Generally, the effective-elastic crack approach may underestimate the displacement after the peak load because inelastic response is ignored. It is noted that although the two-parameter fracture model and the size effect model are capable of predicting failure of different structures, more studies should be conducted to identify $K_{\mathrm{I}c}^s$ and CTOD_c, or G_f and c_f as material fracture parameters. This will be discussed in Chapter 6. As noted earlier, the same is true for the three parameters for the Hillerborg et al.[8] fictitious crack model: f_t, G_F, and w_c (or the shape of the stress–separation curve).

5.4 COMPARISON BETWEEN FICTITIOUS AND EFFECTIVE-ELASTIC CRACK APPROACHES

Fracture of quasi-brittle materials is a complex process and many basic phenomena associated with the fracture process still remain unclear. However, the fracture process in quasi-brittle materials may be reasonably approximated by a combination of the Dugdale–Barenblatt energy dissipation mechanism and the Griffith–Irwin energy dissipation mechanism. The fictitious crack approach models material fracture using only the single Dugdale–Barenblatt energy dissipation mechanism, whereas the effective-elastic crack approach simulates material fracture by the single Griffith–Irwin energy dissipation mechanism. Both approaches are based on some parts of physical observations and are not as appropriate as the combined approach.

Planas and Elices[37,47] have compared the fictitious crack model by Hillerborg, the size effect model by Bazant, and the two-parameter fracture model by Jenq and Shah. Comparisons were made for three types of geometries: the three-point bend beam, the compact tension specimen, and the single edge-notched tensile specimen. They used the fictitious crack model with an exponential stress–separation curve as a reference, and the corresponding fracture parameters for the fictitious crack model examples are $G_F = 103$ N/m, $f_t = 3.21$ MPa, and $E = 30.0$ GPa. They plotted a relationship between two nondimensional parameters, l_{ch}/D and EG_F/K_{c0}^2. Here l_{ch} is the characteristic length of a material defined in Eq. (5-6), D is a characteristic size of the structure (e.g., the depth of the three-point beam), and K_{c0} is the stress intensity factor calculated from the actual peak load and the initial crack length based on LEFM. For a series of geometrically similar structures, the value of K_{c0} increases with increasing structural size because a large structure has a high value of the actual peak load. Some of their results for the three-point bend beam are shown in Fig. 5-20. In Fig. 5-20 the value of G_f or G_{Ic}^s obtained from the size effect model and the two-parameter model are expressed as G_F/G_f or G_F/G_{Ic}^s, respectively, for comparisons. For a given reference value of G_F, the values of G_f and G_{Ic}^s are determined in such a way that both the size effect model and the two-parameter fracture model achieve best fits to the fictitious crack model in the range of practical experimental sizes of structures, as shown in Figs. 5-20*a* and 5-20*b*, respectively. As a result, for the material parameters selected they obtained $G_f = 0.52G_F$ and $G_{Ic}^s = 0.48G_F$ as D approaches infinity. The values of G_f and G_{Ic}^s are comparable. However, the value of G_F is approximately twice as great as the values of G_f and G_{Ic}^s. It was also reported that the peak loads for an infinitely large structure predicted by the size effect model and the two-parameter fracture model are 28.1 and 30.8% lower than that predicted by the fictitious crack model, respectively. They concluded that by appropriately selecting material fracture parameters all three models are able to predict accurately the maximum loading capacity of concrete specimens in the practical range of sizes used in laboratory. However, these three models

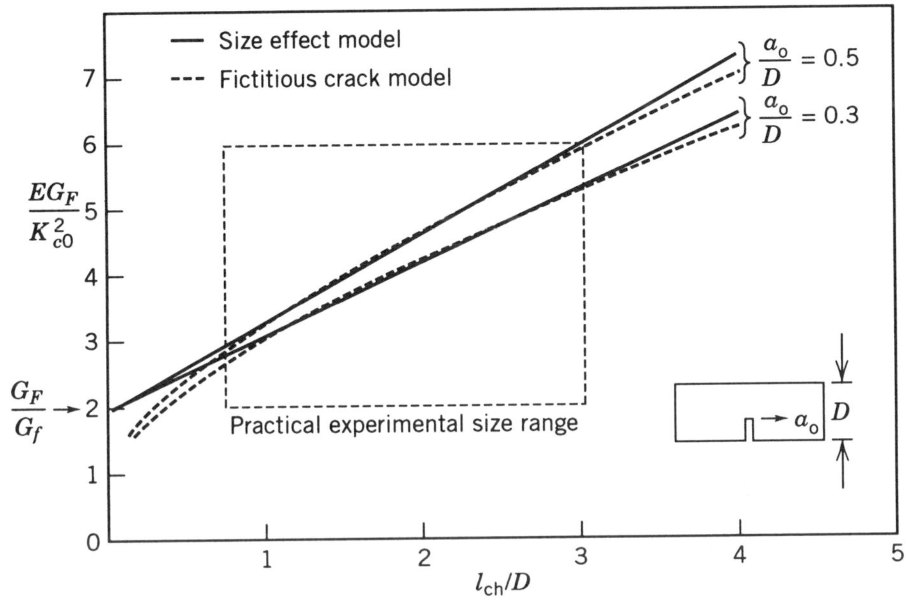

(a)

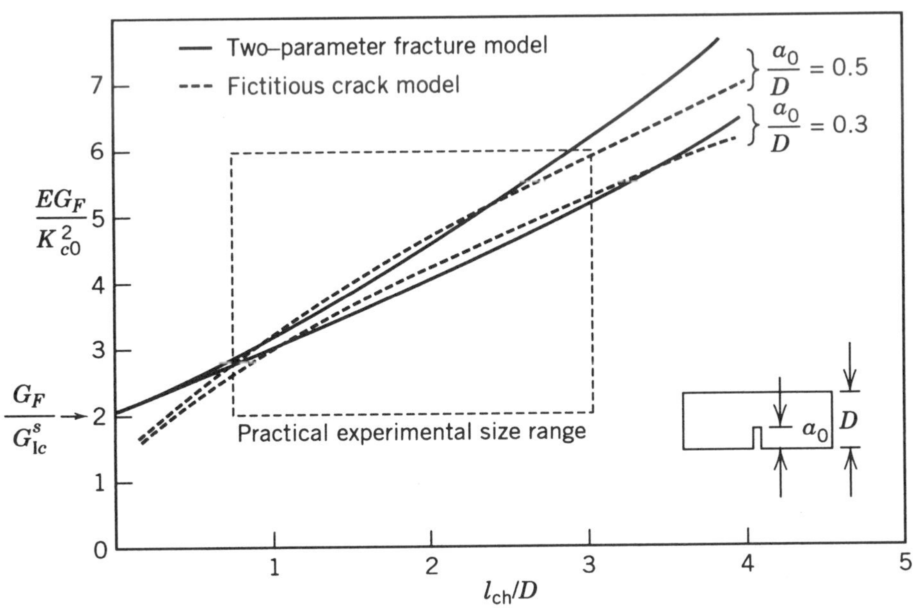

(b)

Figure 5-20 Comparison between cohesive crack approach and effective-elastic crack approach: comparison between (*a*) size effect and fictitious crack models and (*b*) two-parameter fracture and fictitious crack models.[47]

may exhibit differences in asymptotic behavior when they are used in the analysis of concrete structures much larger than those used for determining their parameters in laboratory. Planas and Elices[37,47] argued that the reason for the discrepancy among the three models at large structural sizes may be attributed to the inconsistency in the definition of the size effect in the different models. It is noted that G_f or G^s_{Ic} is the rate of strain energy required to extend a crack beyond a given point. On the other hand, G_F is the energy per unit area to completely separate a fracturing material.

5.5 FINITE ELEMENT ANALYSIS

The basic concept of the FE method is to replace the solid body with an interconnected system of subelements selected in such a way as to provide a reasonable approximation to the response of the body. The elements normally are connected to each other at nodes located at the corners and sides of each element. At these connections it is required that conditions of equilibrium of tractions (forces and couples) and compatibility of deformations (displacements and rotations) be satisfied.

Generally, for a given element, the following is assumed to be known: the material constitutive properties and functions describing the distributions of deformations throughout the interior of the element. The deformation distribution functions are correlated to the nodal deformations. Based upon these assumptions, it is possible to obtain explicit relationships between nodal tractions and displacements as well as internal distributions of stresses and strains throughout the element. A detailed description of the FE method is beyond the scope of this book. Standard textbooks are available as, for instance, Zienkiewicz and Taylor.[48] An excellent overview of the state of the art of applications of this method to problems of fracture in concrete is given by the Fracture Committee of the American Concrete Institute.[49] An earlier report on general applications to concrete structures is in the reference.[50]

The earliest published work in which a cracked, reinforced concrete structure was modeled by this method appears to be that of Ngo and Scordelis.[51] In this, the cracks were modeled as a release of the connectivity requirement between elements adjacent to the crack so that stress distributions in the connected material in the region of the crack tip could be obtained. However, at that time a crack propagation model was not developed.

In general, the basic objective of a FE analysis of a reinforced concrete structure is to obtain a picture of the load–deformation response at any desired location up to any level of distress. To do this, the method should be able to identify locations where cracks will initiate; should be able to model crack growth (propagation); should be able to model crack coalescence, which leads to the localization phenomenon; and should be able to model this process up to collapse of the structure.

As shown in Chapter 4, even the simplest physical problem (e.g., a plate in tension or a beam in bending) exhibits nonuniform effects throughout the volume in the region of the crack due to crack bridging and other process zone effects. As a result, the FE method would be most realistic if it was to model the solid with three-dimensional elements. Although this is an area of current research activity, problems of cost, computing facilities needed, and proven models have prevented general widespread usage of three-dimensional FE modeling. Instead, the developments, starting with Ngo and Scordelis,[51] have utilized two-dimensional elements of a wide variety. Thus, the approximations of either plane-strain or plane-stress are adopted, and further, the process zone shape and influence are assumed to be constant throughout the thickness. Thus, applications of these FE models are limited to appropriate physical problems such as beams, plates, slabs or walls, and thin shells. In this context, the physical crack models described earlier in this chapter have their FE counterparts. These are to model cracks discretely or to model cracks individually or in groups in a ''smeared'' fashion using a continuum mechanics approach.

5.5.1 Discrete Crack Approach

In this simplest application of this method the crack path is assumed a priori and a mesh is arranged so that the path coincides with boundaries between elements. An example of this approach was described earlier with the simple beam in three-point bending shown in Fig. 5-6. There are basically three problems to be solved: determining the location and initiation direction of the start of the crack, determining how the crack extends, and determining the direction of crack extension. The first problem is usually solved on the basis of maximum tensile stress. Once the crack location is identified, it may be necessary to redefine the mesh. In the beam example of Fig. 5-6, this was not necessary because of symmetry and the presence of the notch. Along the crack line it is necessary to either completely eliminate the nodal forces across the crack at the interface elements or to use a softening rule. At the crack tip, if LEFM is assumed, it is necessary to have a ''singular'' or ''super'' element that includes a crack tip stress function. For the beam shown in Fig. 5-21*a*, the notch location is again known; however, the starting direction is not known. This may be found from evaluating the state of stress in the element at the notch tip to determine the direction of the maximum tensile stress. Crack extension may be determined for LEFM problems by considering the stress intensity factor associated with the stress state in the notch tip element. This is usually the mode I stress intensity factor and it is compared to the material fracture toughness. When this exceeds the fracture toughness, the crack propagates. Alternatively, the critical energy release rate criterion may be used. If either of these approaches is not used for problems with open cracks (no process zone effects), then the solution is nonobjective. That is, the results change with mesh size, as first pointed out by Bazant.[52] This is because the maximum stresses are unbounded as the mesh size decreases at the crack tip. Hence, a strength criterion for propagation of a traction-free crack is not suitable. Some energy

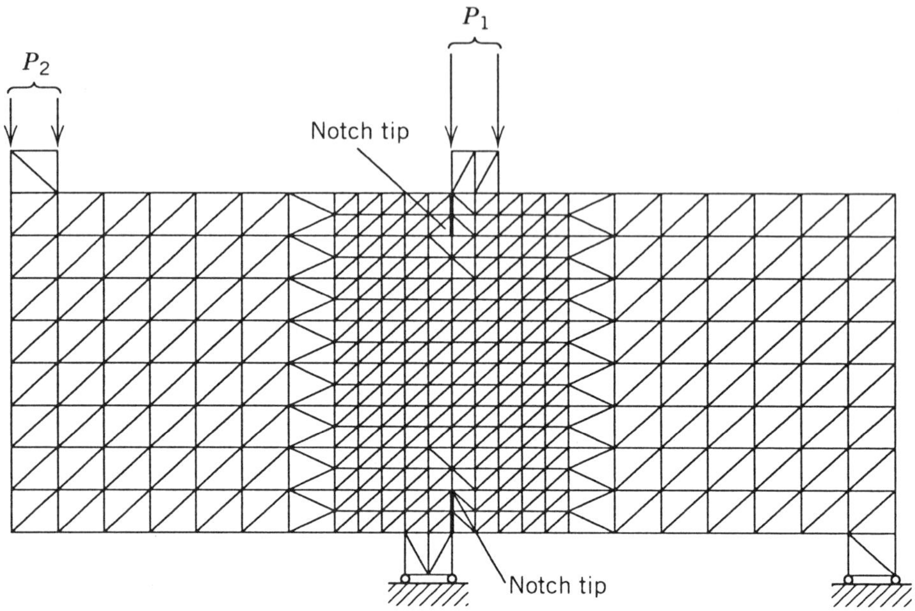

(*a*)

Figure 5-21 Finite element analysis of beam with four-point loading: (*a*) initial mesh, (*b*) two cracks and associated mesh, and (*c*) crack pattern.[57]

criterion is required to be used for crack propagation if the process zone effects are considered as, for instance, in the fictitious crack model.[8] But if the crack propagation is not self-similar that is in the same straight line with which it—for the notch—started, then the FE mesh must be redefined to incorporate the new crack direction. Ideally, this should be done in an objective and automated manner. This approach was first developed successfully by Ingraffea and his co-workers.[53–56] The application of this approach using the CRACKER FE program[55] to the beam in Fig. 5-21*a* was done by Taha.[57] The resulting mesh shown in Fig. 5-21*b* corresponds to the crack pattern given in Fig. 5-21*c*. In this LEFM was assumed. However, the problem was also solved using the fictitious crack model augmented with shearing stresses which were permitted to exist along the crack faces. The results obtained were virtually identical to those from LEFM.

The advantages of using the discrete crack model with interface elements and a singular crack tip element have been summarized in the ACI 446 report[49]:

1. Interface elements are a natural way of describing the crack and are consistent with the fictitious crack model.
2. Interface elements are economical in that the crack is modeled with the minimum number of degrees of freedom.

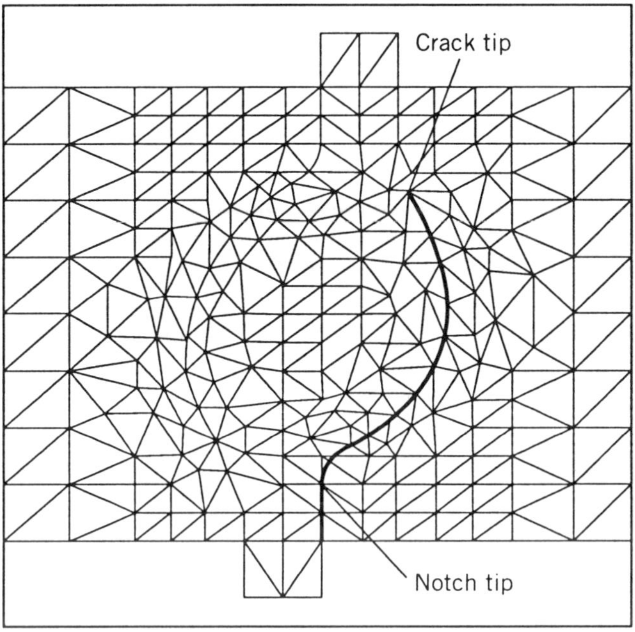

(*b*)

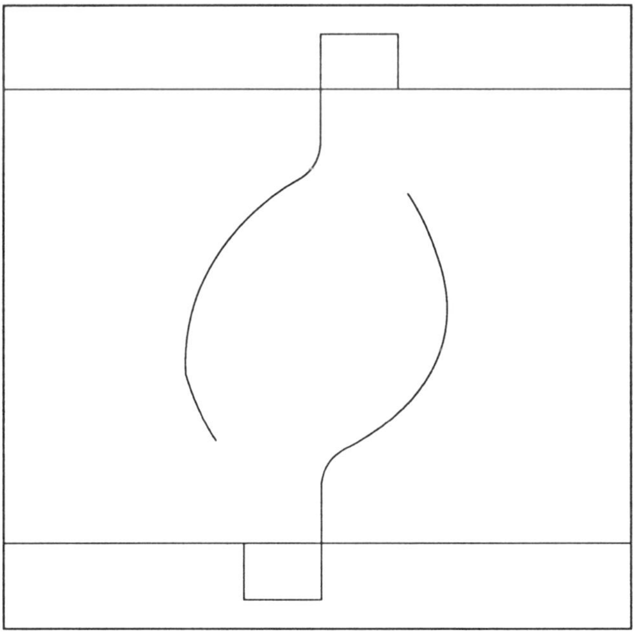

(*c*)

Figure 5-21 (*Continued*)

3. Mesh regeneration associated with crack propagation, and direction change, can be accomplished automatically. This eliminates one of the greatest criticisms of the discrete crack approach.
4. The formation of a discrete crack is therefore viewed as a problem of changing geometry rather than, as in the smeared crack approach, a change in properties.
5. As the interface forms, a graphical representation of the geometric discontinuity that develops is automatically obtained by the altered mesh.

There are, however, some disadvantages to the discrete crack model. The method is very computing intensive since large numbers of degrees of freedom are involved. This becomes particularly serious when many cracks are present, which frequently occurs in shear wall panels or concrete shells. In such cases of distributed damage, in which thousands of cracks may be involved, the smeared crack approach is more feasible.

5.5.2 Smeared Crack Approach

This approach, which is based on the concept of replacing the crack by a continuous medium with altered physical properties, was first introduced by Rashid.[58] The physical concept of the smeared crack approach is similar to the crack band model of Bazant and Oh.[9,14] An illustration of this concept leading to the development of a FE model presented by Bazant[59] is shown in Fig. 5-22. In this figure, the width of the crack band, w_c, which includes the actual crack or groups of cracks, must be several times larger than the largest inhomogeneity in the structure, that is, the aggregate size. Within this band, average properties are used based on orthotopic behavior that is a function of the crack direction. It is also shown that the bandwidth must be limited to assure stability in the solution. This width, w_c, is considered to be a material property.

Four reasons given for adapting the smeared crack model are as follows[49]:

1. There is computational convenience, and remeshing is not required.
2. Distributed damage and cracking have been observed in reinforced concrete structures.
3. The crack generally is not straight but tortuous. Such tortuosity can be modeled more readily using the smeared crack approach.
4. In the case where parallel cracks in concrete are densely and uniformly distributed, such cracks are well represented by the smeared crack model with a minimum crack band width equal to the actual spacing of the parallel cracks.

The most serious disadvantage of this model is the spurious mesh sensitivity.[52,60] This may be avoided by the use of the localization limiter. However, this also has a disadvantage and that is that the convergence of the solution for

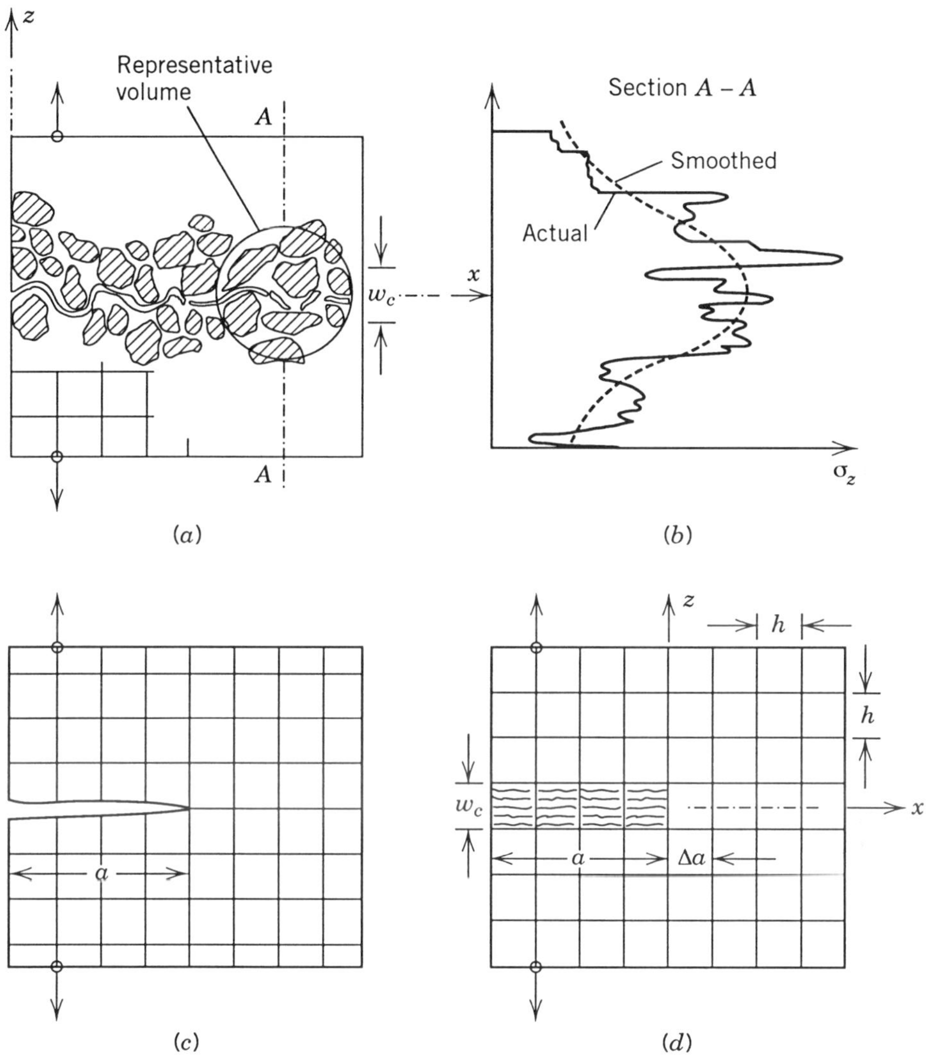

Figure 5-22 Random microstructure, scatter of microstresses, and crack band or sharp crack model.[59]

decreasing mesh sizes cannot be checked because the minimum size cannot be smaller than w_c.

An illustration of this method for the analysis of a beam in four-point bending by Rots and Blaauwendraad[61] is shown in Fig. 5-23. The initial mesh is shown in Fig. 5-23*a* and the final deformed mesh is shown in Fig. 5-23*c*. Note that the curved crack path was modeled without remeshing. The implementation of this example was performed using the computer program DIANA.[62]

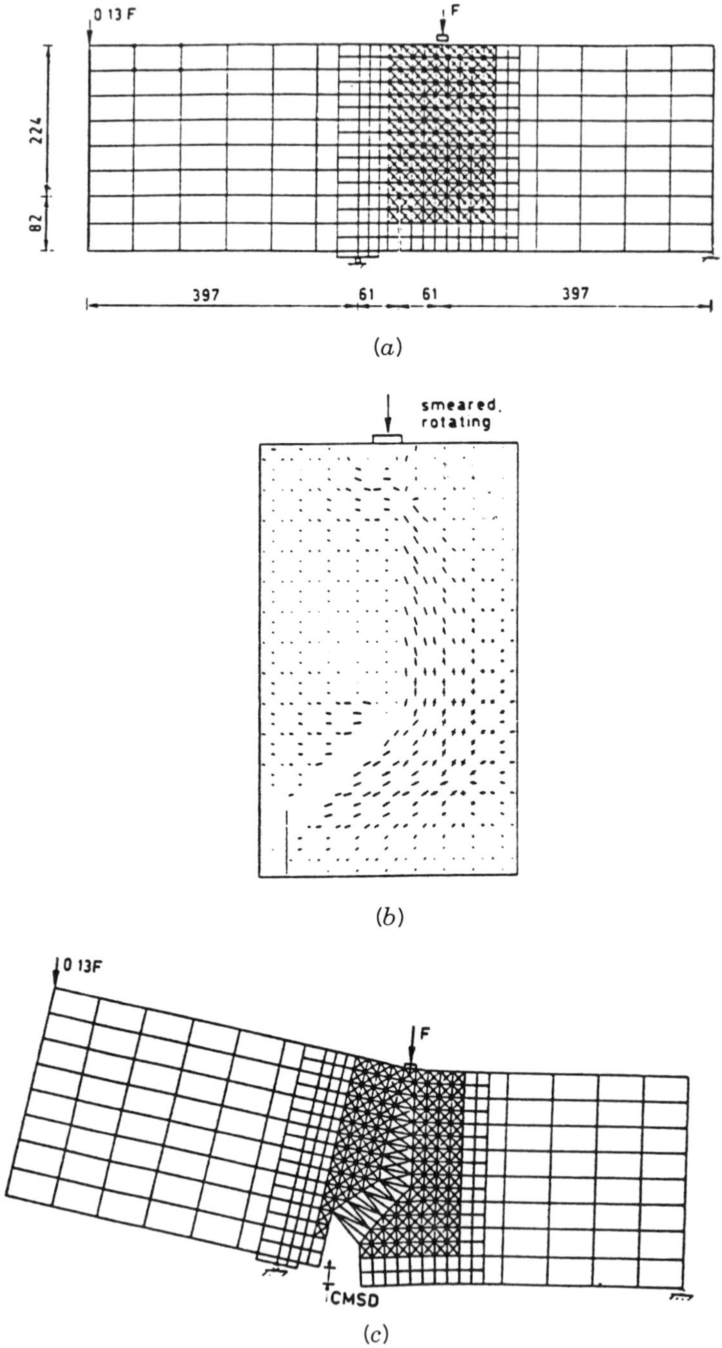

Figure 5-23 Beam in four-point bending analyzed by Rots: (*a*) element mesh for smeared crack analysis; (*b*) principal tensile stresses at final stage, stress locking for smeared cracks (coaxial, rotating); and (*c*) deformed at final stage.[61]

The smeared crack analysis was used by Sitaram[63] to model the cracking behavior of a reinforced concrete, folded plate shell to obtain load–displacement response to collapse. The analysis was made on a model constructed of microconcrete and loaded to failure by Pudhaphongsiriporn.[64] The tensile stress–strain curve of Figueiras and Owen[23] given in Table 5-1 was used. Among other objectives of the study was to determine the influence of the shape of the strain-softening curve and the influence of fracture energy on the predicted structural response. These results are shown respectively in Figs. 5-24 and 5-25. The analysis used one-quarter of the model and employed 49 elements using three layers for concrete and four layers for steel. It is seen from Fig. 5-24 that the main effect of changing the shape of the softening portion of the tensile stress–strain curve is to change the ductility of response. The effect of fracture energy shown in Fig. 5-25 is on both peak load and ductility.

5.5.3 Software Available

Many researchers prepare their own software packages, which usually are intended for specific and limited applications of fracture mechanics methods. Major commercial FE packages such as NASTRAN or ANSYS or structural analysis and design programs such as STRUDL have many of the features needed for a fracture analysis. These are basically adaptable to discrete crack modeling. In recent years, four fracture analysis FE programs have become available. Two of these were developed at Cornell University and are essentially in the public domain. These are CRACKER[55] and FRANC[56].

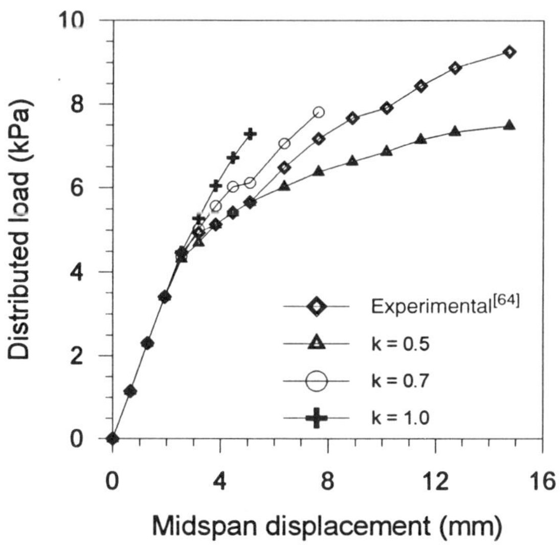

Figure 5-24 Effect of tension-softening parameter k, G_F = 200 N/m.[63]

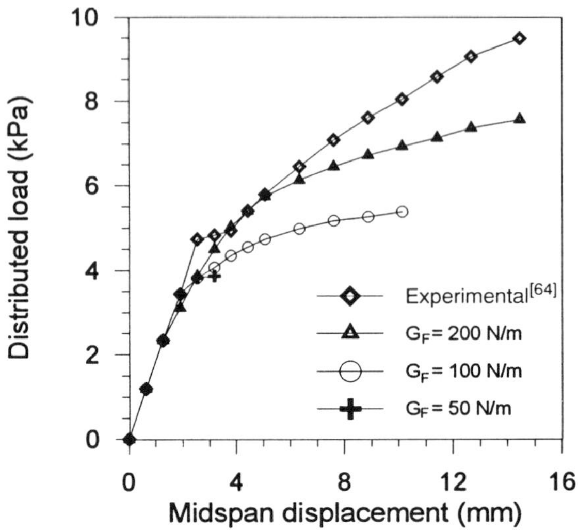

Figure 5-25 Effect of fracture energy G_F, $k = 0.5$.[63]

CRACKER is based on the discrete crack approach and is intended for applications of dynamic crack propagation. Cracks can be initiated at any desired location during an analysis either by direct input from the user or by specifying some criterion such as maximum stress. The crack can then propagate in directions that are not specified a priori. Significant features include the use of explicit time integration, triangular elements with quadratic shape functions, arbitrary numbering of nodes and elements, automated remeshing during crack propagation (although user interaction is required), and singular crack tip elements. Fracture may be modeled with LEFM concepts using stress intensity factors. At any time, the current stress intensity factor is compared to the critical stress intensity factor of the material at the current crack velocity. If the current value lies above the curve of the critical stress intensity factor, the crack is accelerated; whereas if it lies below the curve, the crack is decelerated. Crack arrest occurs when the velocity reaches zero. The direction of propagation is determined using the maximum circumferential stress criterion. For nonlinear fracture modeling the Dugdale approach is used where a stress–crack-opening displacement relation is specified. This is basically identical to the fictitious crack concept.[8] Quasi-static crack growth may also be investigated using this program. Software and documentation currently are available from Danial V. Swenson, Department of Mechanical Engineering, Kansas State University, Durland Hall, Manhattan, KS 66506-5106. It is implemented there on SUN workstations.

FRANC, a FRacture ANalysis Code, it also based on the discrete crack model and is available from Anthony R. Ingraffea, Department of Structural Engineering, Cornell University, Hollister Hall, Ithaca, NY 14853. It is an interactive workstation based program designed to perform modeling of two-dimensional fracture processes. Integral remeshing routines allow the FE mesh to be modified semiautomatically to represent new crack configurations. The program is run by means of the DEC Command Definition Utility (VAX/VMS) and is implemented in the VAX-II FORTRAN language. It is oriented to static crack propagation problems. A crack can be nucleated either as an edge crack or as an internal crack. Crack propagation involves user interaction. The user identifies the tip and the direction is determined by the program. Remeshing requires user interaction and generally is the same as in CRACKER. Both LEFM and nonlinear fracture models may be used with the crack interface elements.

The other two computer programs described here are based on the smeared crack concept and are available on a commercial basis,

DIANA is a trademark of TNO-IBBC and is available from the Institute TNO for Building Materials and Structures, 2600 AA Delft, the Netherlands. This is a general-purpose, FE code for the analysis of three-dimensional structures.[65] Material models for "concrete" elements may include effects of cracks, embedded bars, or combinations of these. Complete bond between steel and concrete may be assumed or a bond-slip relationship may be used. The concrete incorporates a crack model that is valid for loading combinations in which at least one tensile component is involved and a plasticity model to describe the behavior under compressive stress states. For a nonlinear analysis of unreinforced concrete, the fracture energy concept is used to obtain an appropriate stress–strain diagram for the element(s) representing the smeared crack. Crack initiation, however, is based on the maximum tensile strength criterion. For reinforced concrete elements the tension-stiffening concept is used. As with other commercial packages, a variety of element types are available for the user; preprocessing and postprocessing capabilities are extensive and user interactive.

The computer program SBETA[66] was developed jointly at the University of Stuttgart, Germany, and the Czech Technical University, Prague, Czech Republic, by V. Cervenka, R. Eligehausen, and R. Pukl. It is designed for use on personal computers using the DOS operating system with IBM-compatible machines. Processors should be at least Intel 80386. The program is also available on Apollo workstations. It is available from the Institute for Material Science, University of Stuttgart, Pfaffenwaldring 4, D 7000 Stuttgart 80, Germany. A two-dimensional, continuum mechanics model is used that incorporates nonlinear behavior under static loading conditions. The analysis includes all stages of structural response: service loading, failure (peak) loading, and postpeak response. Again, the smeared crack approach is used for both cracks and reinforcement. The minimum volume for a smeared region is one element. The following effects of concrete behavior are modeled[66]:

- nonlinear behavior in compression, including softening,
- tensile fracture in concrete based on the fracture energy criterion,

- biaxial stress failure criterion,
- tension-stiffening effect,
- shear stiffness retention after cracking (in case of the fixed crack model), and
- two crack models: fixed crack direction and rotated crack direction.

Bond slip is modeled only in the tension-stiffening model. The width of the crack band is approximated as $h_c = A^{1/2}$, where A is the area of the FE. The concrete continuum, cracked or uncracked, is represented by quadrilateral elements, each of which is made up of two four-node triangular elements. This results in an element with four corner nodes and one interior node. Reinforcement is modeled with bar elements that pass through the quadrilateral elements. As in the other programs, extensive preprocessing and postprocessing capabilities exist and the software interacts with the user in a ''friendly'' manner.

5.6 SUMMARY

Based on experimental observations, many theoretical models for predicting fracture of quasi-brittle materials have been proposed. These theoretical models can be categorized as the fictitious crack approach based on the Dugdale–Barenblatt energy dissipation mechanism and the effective-elastic crack approach based on the Griffith–Irwin energy dissipation mechanism. For the fictitious crack approach, both the fictitious crack model by Hillerborg et al. and the crack band model by Bazant have been presented. For the effective-elastic crack approach, the two-parameter fracture model by Jenq and Shah, the size effect model by Bazant and Kazemi, and effective crack models by Karihaloo and Nallathambi and by Refai and Swartz have been addressed. Two approaches based on FE analysis, the discrete crack approach and the smeared crack approach, have been discussed. Some FE softwares have been briefly presented. Since fracture behavior for quasi-brittle materials such as concrete is a complex phenomenon, many basic mechanisms associated with fracture process still remain unclear. Different models may catch some of the actual physical mechanisms. Since very intensive studies are currently ongoing in this area, many developments can be expected in the future.

REFERENCES

1. Kaplan, M. E., ''Crack Propagation and Fracture of Concrete,'' *Journal of the American Concrete Institute*, Vol. 58, No. 5, 1961, pp. 591–610.

2. Shah, S. P., and McGarry, F. J., "Griffith Fracture Criterion and Concrete," *Journal of Engineering Mechanics Division, ASCE*, Vol. 97, No. EM6, 1971, pp. 1663–1676.
3. Jenq, Y.-S., and Shah, S. P., "Fracture Toughness Criterion for Concrete," *Engineering Fracture Mechanics*, Vol. 21, 1985, pp. 1055–1069.
4. Rice, J. R., "Mathematical Analysis in the Mechanics of Fracture," in *Fracture: An Advanced Treatise*, edited by Liebowitz, H., Academic, New York, 1968, pp. 191–311.
5. Kobayashi, A. S., Hawkins, N. M., and Bradt, R. C., "Fracture Process Zone in Concrete and Ceramics: A Matter of Scaling," in *Toughening Mechanisms in Quasi-Brittle Materials*, edited by Shah, S. P., Kluwer Academic, Dordrecht, 1991, pp. 35–46.
6. Cook, R. F., Fairbanks, C. J., Lawn, B. R., and Mai, Y.-W., "Crack Resistance by Interfacial Bridging: Its Role in Determining Strength Characterization," *Journal of Materials Research*, Vol. 2, No. 3, 1987, pp. 345–356.
7. Cox, B. N., and Marshall, D. B., "Concepts in the Fracture and Fatigue of Bridged Cracks," *ACTA Metallurgica et Materialia*, 1994.
8. Hillerborg, A., Modeer, M., and Petersson, P.-E., "Analysis of Crack Formation and Crack Growth in Concrete by Means of Fracture Mechanics and Finite Elements," *Cement and Concrete Research*, Vol. 6, No. 6, 1976, pp. 773–782.
9. Bazant, Z. P., and Oh, B. H., "Crack Band Theory for Fracture of Concrete," *Materials and Structures, RILEM*, Vol. 16, 1983, pp. 155–177.
10. Jenq, Y. S., and Shah, S. P., "A Two Parameter Fracture Model for Concrete," *Journal of Engineering Mechanics*, Vol. 111, No. 4, 1985, pp. 1227–1241.
11. Bazant, Z. P., and Kazemi, M. T., "Determination of Fracture Energy, Process Zone Length and Brittleness Number from Size Effect, with Application to Rock and Concrete," *International Journal of Fracture*, Vol. 44, 1990, pp. 111–131.
12. Hilsdorf, H. K., and Brameshuber, W., "Code-Type Formulation of Fracture Mechanics Concepts for Concrete," *International Journal of Fracture*, Vol. 51, 1991, pp. 61–72.
13. Hillerborg, A., "Analysis of One Single Crack," in *Fracture Mechanics of Concrete*, edited by Wittmann, F. H., Elsevier Science, Amsterdam, 1983, pp. 223–249.
14. Bazant, Z. P., "Size Effect in Blunt Fracture: Concrete, Rock, Metal," *Journal of Engineering Mechanics, ASCE*, Vol. 110, No. 4, 1984, pp. 518–535.
15. Roelfstra, R. E., and Wittmann, F. H., "A Numerical Method to Link Strain Softening with Fracture in Concrete," in *Fracture Toughness and Fracture Energy in Concrete*, edited by Wittmann, F. H., Elsevier Science, Amsterdam, 1986, pp. 163–175.
16. Alvarado, A. M., and Torrent, R. J., "The Effect of the Shape of the Strain-Softening Diagram on the Bearing Capacity of Concrete Beam," *Materials and Structures*, Vol. 20, 1987, pp. 448–457.
17. Cedolin, L., DeiPoli, S., and Iori, I., "Tensile Behavior of Concrete," *Journal of Engineering Mechanics, ASCE*, Vol. 113, No. 3, 1987, pp. 431–449.
18. CEB-FIP Model Code 1990, First Predraft 1988, Bulletin d'Information No. 190a, 190b, Comite Euro-International du Beton (CEB), Lausanne.

19. Cho, K. Z., Kobayashi, A. S., Hawkins, N. M., Barker, D. B., and Jeang, F. L., "Fracture Process Zone of Concrete Cracks," *Journal of Engineering Mechanics, ASCE*, Vol. 110, No. 8, 1984, pp. 1174–1184.

20. Liaw, B. M., Jeang, F. L., Du, J. J., Hawkins, N. M., and Kobayashi, A. S., "Improved Non-linear Model for Concrete Fracture," *Journal of Engineering Mechanics, ASCE*, Vol. 116, No. 2, 1990, pp. 429–445.

21. Gopalaratnam, V. S., and Shah, S. P., "Softetning Response of Plain Concrete in Direct Tension," *ACI Journal*, Vol. 82, No. 3, 1985, pp. 310–323.

22. Du, J., Yon, J. H., Hawkins, N. M., Arakawa, K., and Kobayashi, A. S., "Fracture Process Zone for Concrete for Dynamic Loading," *ACI Materials Journal*, Vol. 89, No. 3, 1992, pp. 252–258.

23. Figueiras, J. A., and Owen, D. R. J. "Nonlinear Analysis of Reinforced Concrete Shell Structures," in *International Conference on Computer-Aided Analysis and Design of Concrete Structures, Part I*, Split, Yugoslavia, 1984, pp. 509–532.

24. Li, V. C., Chan, C.-M., and Leung, C. K. Y., "Experimental Determination of the Tension-Softening Relations for Cementitious Composites," *Cement and Concrete Research*, Vol. 17, 1987, pp. 441–452.

25. Miller, R. A., Castro-Montero, A., and Shah, S. P., "Cohesive Crack Models for Cement Mortar Examined Using Finite Element Analysis and Laser Holographic Measurement," *Journal of the American Ceramic Society*, Vol. 74, 1991, pp. 130–138.

26. Yon, J.-H., Hawkins, N. W., and Kobayashi, A. S., "Numerical Simulation of Mode I Dynamic Fracture of Concrete," *Journal of Engineering Mechanics, ASCE*, Vol. 117, No. 7, 1991, pp. 1595–1610.

27. Du, J., Yon, J. H., Hawkins, N. M., and Kobayashi, A. S., "Analysis of the Fracture Process Zone of a Propagating Concrete Crack Using Moire Interferometry," in *Micromechanics of Failure of Quasi-Brittle Materials*, edited by Shah, S. P., Swartz, S. E., and Wang, M. L., Elsevier Applied Science, London, 1990, pp. 146–155.

28. Shah, S. P., and Ouyang, C., "Toughening Mechanisms in Quasi-Brittle Materials," *Journal of Materials and Technologies, ASME*, Vol. 115, No. 3, 1993, pp. 300–307.

29. Tada, H., Paris, P. C., and Irwin, G. R., *The Stress Analysis of Cracks Handbook*, 2nd ed., Paris Productions, St. Louis, MO, 1985.

30. Jenq, T. S., and Shah, S. P., "Features of Mechanics of Quasi-Brittle Crack Propagation in Concrete," *International Journal of Fracture*, Vol. 51, 1991, pp. 103–120.

31. John, R., and Shah, S. P., "Fracture Mechanics Analysis of High-Strength Concrete," *Journal of Materials in Civil Engineering, ASCE*, Vol. 1, No. 4, 1989, pp. 185–198.

32. Broek, D., *Elementary Engineering Fracture Mechanics*, 4th ed., Martinus Nijhoff, Hingham, 1987.

33. Barson, J. M., and Rolfe, S. T., *Fracture and Fatigue Control in Structures*, Prentice-Hall, Englewood Cliffs, NJ, 1987.

34. Karihaloo, B. L., and Nallathambi, P., ''Notched Beam Test: Mode I Fracture Toughness,'' in RILEM Report 5, *Fracture Mechanics Test Methods for Concrete*, edited by Shah, S. P., and Carpinteri, A., Chapman & Hall, London, 1991, pp. 1–86.
35. Ouyang, C., and Shah, S. P., ''Geometry-Dependent *R*-Curve for Quasi-Brittle Materials,'' *Journal of the American Ceramic Society*, Vol. 74, No. 11, 1991, pp. 2831–2836.
36. Ouyang, C., Mobasher, B., and Shah, S. P., ''An *R*-Curve Approach for Fracture of Quasi-Brittle Materials,'' *Engineering Fracture Mechanics*, Vol. 37, No. 4, 1990, pp. 901–913.
37. Planas, J., and Elices, M., ''Fracture Criteria for Concrete: Mathematical Approximations and Experimental Validation,'' *Engineering Fracture Mechanics*, Vol. 35, No. 1/2/3, 1990, pp. 87–94.
38. Tang, T., Shah, S. P., and Ouyang, C., ''Fracture Mechanics and Size Effect of Concrete in Tension,'' *Journal of Structural Engineering, ASCE*, Vol. 118, No. 11, 1992, pp. 3169–3185.
39. John, R., Shah, S. P., and Jenq, Y.-S., ''A Fracture Mechanics Model to Predict the Rate Sensitivity of Mode I Fracture of Concrete,'' *Cement and Concrete Research*, Vol. 17, 1987, pp. 249–262.
40. Bazant, Z. P., ''Fracture Mechanics and Strain-Softening in Concrete,'' in *Proceedings of U.S.–Japan Seminar on Finite Element Analysis of Reinforced Concrete Structures*, American Society of Civil Engineers, New York, 1986, pp. 121–150.
41. Bazant, Z. P., ''Fracture Energy of Heterogeneous Materials and Similitude,'' in *Fracture of Concrete and Rock*, edited by Shah, S. P., and Swartz, S. E., Springer-Verlag, New York, 1989, pp. 229–241.
42. Bazant, Z. P., and Pfeiffer, P. A., ''Determination of Fracture Energy from Size Effect and Brittleness Number,'' *ACI Materials Journal*, Vol. 84, No. 6, 1987, pp. 463–480.
43. Nallathambi, P., and Karihaloo, B. L., ''Determination of Specimen-Size Independent Fracture Toughness of Plain Concrete,'' *Magazine of Concrete Research*, Vol. 38, No. 135, 1986, pp. 67–76.
44. Karihaloo, B. L., and Nallathambi, P., ''An Improved Effective Crack Model for the Determination of Fracture Toughness of Concrete,'' *Cement and Concrete Research*, Vol. 19, 1989, pp. 603–610.
45. Swartz, S. E., and Refai, T. M. E., ''Influence of Size Effects on Opening Mode Fracture Parameters for Precracked Concrete Beams in Bending,'' in *Fracture of Concrete and Rock*, Springer-Verlag, New York, 1988, pp. 243–254.
46. Swartz, S. E., and Refai, T., ''Cracked Surface Revealed by Dye and Its Utility in Determining Fracture Parameters,'' in *Fracture Toughness and Fracture Energy*, edited by Mihashi, H., et al., Balkema, The Netherlands, 1989, pp. 509–520.
47. Elices, M., and Planas, J., ''Size Effect and Experimental Validation of Fracture Models,'' in *Analysis of Concrete Structres by Fracture Mechanics*, edited by Elfgren, L., and Shah, S. P., Chapman & Hall, London, 1991, pp. 99–127.

48. Zienkiewicz, O. C., and Taylor, R. L., *The Finite Element Method*, Vol. 1: *Basic Formulations and Linear Problems*, 4th edition, McGraw-Hill, London, 1989.

49. ACI Committee 446, ''Finite Element Analysis of Fracture in Concrete Structures,'' Report of Subcommittee 3, American Concrete Institute, Detroit, 1995 (expected).

50. ASCE, *Finite Element Analysis of Reinforced Concrete Structures: State-of-the-Art Report*, Committee on Concrete and Masonry Structures, American Society of Civil Engineers, New York, 1982.

51. Ngo, D., and Scordelis, A. C., ''Finite Element Analysis of Reinforced Concrete Beams,'' *ACI Journal*, Vol. 64, No. 3, 1967, pp. 152–163.

52. Bazant, Z. P., ''Instability, Ductility and Size Effect in Strain-Softening Concrete,'' *Journal of Engineering Mechanics, ASCE*, Vol. 102, 1976, pp. 331–344.

53. Ingraffea, A. R., ''Discrete Fracture Propagation in Rock: Laboratory Tests and Finite Element Analysis,'' Ph.D. Dissertation, University of Colorado, Boulder, CO, 1977.

54. Ingraffea, A. R., and Saouma, V., ''Numerical Modeling of Discrete Crack Propagation in Reinforced and Plain Concrete,'' in *Fracture Mechanics of Concrete: Structural Application and Numerical Calculation*, edited by Sih, G., and Tommaso, A. D., Martinus Nijhoff, Dordrecht, 1984, pp. 171–225.

55. Swenson, D. V., ''Modeling Mixed-Mode Dynamic Crack Propagation Using Finite Elements: Theory and Applications,'' Ph.D. Dissertation, Cornell University, Ithaca, NY, 1986.

56. Wawrzynek, P. A., and Ingraffea, A. R., ''Fracture Analysis Code (*FRANC*),'' Cornell University, Ithaca, NY, 1987.

57. Taha, N. M., ''Mixed Mode Fracture of Four-Point Loaded Concrete Beams,'' Ph.D. Dissertation, Kansas State University, Manhattan, KS, 1988.

58. Rashid, Y. R., ''Ultimate Strength Analysis of Prestressed Concrete Pressure Vessels,'' *Nuclear Engineering and Design*, Vol. 7, No. 4, 1968, pp. 334–344.

59. Bazant, Z. P., ''Mechanics of Fracture and Progressive Cracking in Concrete Structures,'' in *Fracture Mechanics of Concrete*, edited by Sih, G. C., and Tommase, A. D., Martinus Nijhoff, Dordrecht, 1984, pp. 1–94.

60. Bazant, Z. P., and Cedolin, L., ''Blunt Crack Bond Propagation in Finite Element Analysis,'' *Journal of Engineering Mechanics, ASCE*, Vol. 105, No. 2, 1979, pp. 297–315.

61. Rots, J. G., and Blaauwendraad, J., ''Crack Models for Concrete: Discrete or Smeared? Fixed, Multi-Directional or Rotating?'' *Heron*, Vol. 34, No. 1, 1989, 59 pp.

62. Deborst, R., Kusters, G. M. A., Nauta, P., and deWitte, F. C., ''*DIANA:* A Comprehensive But Flexible Finite Element System,'' in *Finite Element Systems Handbook*, 3rd ed., Springer-Verlag, Berlin, 1983.

63. Sitaram, P., ''Ultimate Load Analysis of Reinforced Concrete Folded Plate Structures,'' Ph.D. Dissertation, Kansas State University, Manhattan, KS, 1993.

64. Pudhaphongsiriporn, P., ''An Experimental Study of a Reinforced Concrete Folded Plate Structure,'' Ph.D. Dissertation, Purdue University, West Lafayette, IN, 1978.

65. van Mier, J. G. M., "Examples of Non-Linear Analysis of Reinforced Concrete Structures with *DIANA*," *Heron*, Vol. 32, No. 3, 1987, 147 pp.

66. Cervenka, V., Eligehausen, R., and Pukl, R., "*SBETA* Computer Program for Nonlinear Finite Element Analysis of Reinforced Concrete Structures," Report No. 4/9-89/22, University of Stuttgart, November 1989.

6

TEST METHODS TO DETERMINE MODE I FRACTURE PROPERTIES FOR CONCRETE AND ROCK

6.1 INTRODUCTION

Several NLFM models for describing failure of quasi-brittle materials have been presented in the previous chapter. Each of these models introduces some material fracture parameters, which are expected to depend only on material fracture properties regardless of structural geometry and size. Unless these fracture parameters are determined, these NLFM models cannot be used for quasi-brittle materials such as concrete. Due to the presence of a large size of fracture process zone in such materials, conventional test methods based on LEFM cannot be used directly to measure these material fracture parameters. Test methods to determine fracture parameters for concrete and rock based on different fracture models are discussed in this chapter.

Three draft recommendations to measure concrete fracture parameters proposed by RILEM committees on fracture mechanics of concrete (Committees 50-FMC and TC89-FMT) will be first presented. These three draft recommendations are based on the fictitious crack model by Hillerborg et al., the two-parameter fracture model by Jenq and Shah, and the size effect model by Bazant et al., respectively. Test methods for determining fracture parameters of rock, recommended by the ISRM, will be discussed. Some other test methods for fracture properties of concrete and rock will also be introduced.

6.2 RILEM METHOD OF HILLERBORG ET AL., G_F

6.2.1 Specimens and Test Procedure

The RILEM Technical Committee 50-FMC on Fracture Mechanics of Concrete—Test Methods proposed a draft recommendation to measure the material fracture energy G_F using a three-point bend beam in 1985.[1] This method was based on the fictitious crack model by Hillerborg et al.,[2,3] which has been presented in detail in the previous chapter. It is noted that the fictitious crack model includes three fracture parameters: the fracture energy G_F, the material tensile strength f_t, and the critical crack separation displacement w_c. However, the draft test recommendation provided only for determination of G_F. This test method will be briefly presented below.

The three-point bend beam shown in Fig. 6-1*a* was recommended to be used. The size of the beam depends on the maximum size of aggregate, d_a, according to Table 6-1. The notch depth is equal to half the beam depth $\pm$ 5 mm, and the notch width at the tip should be less than 10 mm. It is recom-

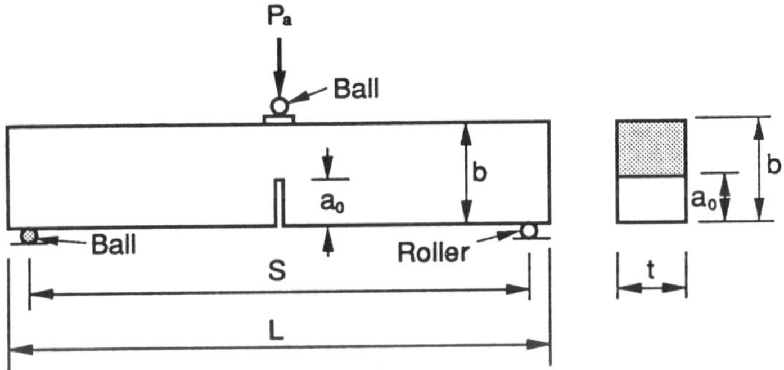

(*a*)

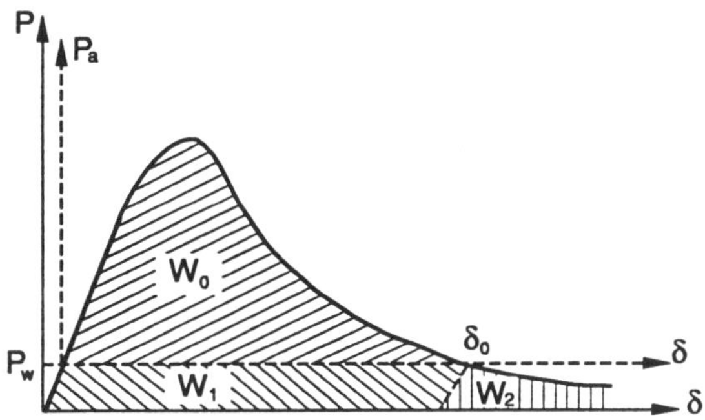

(*b*)

Figure 6-1 Determination of fracture energy G_F based on RILEM proposal: (*a*) a three-point bend beam and (*b*) load displacement curve for evaluation of the fracture energy (downward load and displacement as shown in (*a*)).

Table 6-1 Recommended Sizes of Beams for Measuring G_F

Maximum sizes of Aggregates, d_a (mm)	Depth, b (mm)	Width, t (mm)	Length, L (mm)	Span, S (mm)
1.0–16	100 ± 5	100 ± 5	840 ± 10	800 ± 5
16.1–32	200 ± 5	100 ± 5	1,190 ± 10	1,130 ± 5
32.1–48	300 ± 5	150 ± 5	1,450 ± 10	1,385 ± 5
48.1–64	400 ± 5	200 ± 5	1,640 ± 10	1,600 ± 5

Source: From ref. 1.

mended that the notch be sawn under wet conditions at least one day before the test. During the period of curing the ambient temperature shall be 20 ± 2°C. After removal from the mold, the beam shall be stored in lime-saturated water until less than 30 min before testing. The supports and loading arrangements shall be such that the forces acting on the beam are statically determinate, such as shown in Fig. 6-1*a*.

The test is performed with an approximately constant rate of deformation, which allows the maximum load to be reached within about 30–60 s after the start of the test. A relationship between the load and the load point displacement should be recorded during testing. Deformation should be measured with an accuracy of at least 0.01 mm, and the load should be measured with an accuracy of at least 2% of the maximum value in the test. Possible nonelastic deformation at the supports and the loading point should be eliminated.

In order to obtain a complete load and load point displacement curve, a closed-loop servo control testing machine is recommended. Otherwise the testing machine should be stiff enough to allow one to obtain the softening part of the curve in a stable manner. In the latter case, to obtain a stable test, the stiffness of the testing machine must be greater than the steepest slope of the descending part of the load–displacement curve. The requirements on the stiffness of the testing machine have been discussed in detail by Petersson.[4]

6.2.2 Test Results and Calculations

Based on the obtained load and load point displacement curve, the fracture energy G_F can be calculated. Since the self-weight of the beam has an appreciable effect in the test according to Fig. 6-1*a*, its influence on the value of G_F may be estimated as follows.[4] The weight of the beam between supports is equal to $Sbt\rho g$, where S, b, and t are the span, the depth, and the thickness of the beam, respectively, ρ is the density of the concrete, and g is the acceleration due to gravity and its value is equal to 9.8 m/s^2 (32.2 ft/s^2). The corresponding net bending stress σ_w above the notch due to the self-weight of the beam is

$$\sigma_w = \frac{btS^2\rho g}{8} \frac{6}{t(b - a_0)^2} \tag{6.1}$$

where a_0 is the initial crack length. For beams made of normal concrete and with $a_0/b = 0.5$, Eq. (6.1) becomes

$$\sigma_w \approx \frac{7000S^2}{b} \tag{6.2}$$

where S and b are in meters and σ_w is in pascals. As a result, $\sigma_w \approx 0.4$ MPa is obtained due to $S^2/b = 6.4$ for all beams listed in Table 6-1. This value of σ_w is not small compared to the tensile strength of normal concrete (about 4 MPa) and should be taken into account when calculating the value of G_F.

The influence of the self-weight may be represented by an additional equivalent force P_w, which supplies the same amount of energy based on the load–displacement curve, as shown in Fig. 6-1*b*. Therefore, the total load P on the beam is represented by $P = P_w + P_a$, where P_a is the load applied by the testing machine, provided that load and displacement are both downward. It is noted that the weight of test equipment for load distribution supported by the beam should also be included in P_w. The energy absorbed by the beam is represented by the area under the load–displacement curve, the P–δ curve, as shown in Fig. 6-1*b*, where δ is the load point displacement. In the test, however, only the P_a–δ curve can be measured. The displacement corresponding to $P_a = 0$ is termed as δ_0. The total area under the P–δ curve is referred to as W, which may be divided into three parts, W_0, W_1, and W_2, as shown in Fig. 6-1*b*. Here W_0 is the area below the measured P_a–δ curve, and $W_1 = P_w\delta_0$. Thus, both values of W_0 and W_1 can be determined from the measured P_a–δ curve. It has been demonstrated by Petersson[4] and Swartz and Yap[5] that the value of W_2 is approximately equal to the value of W_1. Therefore, the total fracture energy of the beam, W_t, is

$$W_t = W_0 + 2P_w\delta_0 \tag{6.3}$$

By assuming that energy absorption takes place only in the fracture zone, the fracture energy per unit projected area can be calculated by

$$G_F = \frac{W_t}{(b - a_0)t} = \frac{W_0 + 2P_w\delta_0}{(b - a_0)t} \tag{6.4}$$

The assumption that energy absorption takes place only in the fracture zone implies that all deformations outside this zone are purely elastic. This may be approximately true if the stresses are low in the major part of the beam. A rather deep notch has been used in order to keep the maximum bending moment low for the test.

This specification assumes that loading, beam self-weight, and deflection all act in the same direction—downward. Two other situations are possible:

1. The beam is loaded on its side. In this case, as pointed out by Petersson,[4] the scalar product of self-weight and displacement is zero. In this case, Eq. (6.4) becomes

$$G_F = \frac{W_0}{(b - a_0)t} \tag{6.5}$$

   Note again that W_0 is the area under the *measured* load–displacement diagram.

2. The beam applied load and measured deflection are upward. In this case the scalar product of self-weight and deflection is negative while the scalar product of applied load P_a and deflection is still positive. It has been demonstrated by Swartz and Yap[5] that for this case

$$G_F = \frac{W_0 - P_w\delta_0}{(b - a_0)t} \tag{6.6}$$

   Note that in Eq. (6.6) the sign of the second term in the numerator is negative and the value is one-half that in Eq. (6.4). This need for modification was first observed by Hilsdorf and Brameshuber.[6]

A round-robin test among different laboratories was performed to measure the value of G_F using different sizes of beams.[7] The round-robin test included three comparative test series. Since only the second test series gave values of all three parameters G_F, f_t, and w_c, the results obtained from this test series are summarized in Table 6-2. If G_F is a material fracture parameter, its value should be independent of sizes of the beams. The values of G_F measured from different sizes of the beams are shown in Fig. 6-2.[7] Generally, the values of G_F are specimen size dependent. As an average, the values of G_F increased by 20% when the beam depth increased by a factor of 2 and by 30% when the depth increased by a factor of 3. This beam size dependence on the value of G_F may partially be due to the unwanted energy absorption outside the fracture zone. This unwanted energy increases when the beam size increases. Guinea et al.[8-10] have recently studied the influence of experimental procedures, bulk energy dissipation, and cutting the tail of the P–δ curve on the measured value of G_F. They reported that if extraneous energy dissipations from hysteresis in the testing equipment, friction of supports, and deformation of bulk material are taken into account and the effect of cutting the tail of the P–δ curve is correctly estimated, the measured value of G_F may be more or less size inde-

Table 6-2 Values of G_F from Three-Point Bend Beams from RILEM Round-Robin Test

Series Number	Dimensions (mm) S	b	t	w/c	Cement Content (kg/m³)	d_a (mm)	Age (days)	Curing[a]	Number of Beams Tested	a_0/b	E (GPa)	f_t (MPa)	G_F (N/m)
1	600	100	100	0.59	—	16	100	B	5	0.5	33.5	3.02	178
	600	100	150	0.59	—	16	100	B	4	0.5	33.5	3.02	173
	600	150	100	0.59	—	16	100	B	4	0.5	33.5	3.02	184
2	800	100	100	0.50	250	13	100	C	3	0.2	24.6	2.10	113
	800	100	100	0.50	250	13	100	C	3	0.3	24.6	2.10	133
	800	100	100	0.50	250	13	100	C	3	0.4	24.6	2.10	113
	800	100	100	0.50	250	13	100	C	3	0.5	24.6	2.10	156
	450	150	150	0.50	250	13	100	C	3	0.2	24.6	2.10	176
	450	150	150	0.50	250	13	100	C	3	0.3	24.6	2.10	170
	450	150	150	0.50	250	13	100	C	3	0.4	24.6	2.10	174
	450	150	150	0.50	250	13	100	C	3	0.5	24.6	2.10	202
	800	150	150	0.50	250	13	100	C	3	0.2	24.6	2.10	164
	800	150	150	0.50	250	13	100	C	3	0.3	24.6	2.10	170
	800	150	150	0.50	250	13	100	C	3	0.4	24.6	2.10	178
	800	150	150	0.50	250	13	100	C	3	0.5	24.6	2.10	170
3	800	100	100	0.40	400	16	28	A	10	0.5	36.7	3.37	152
	800	100	100	0.58	315	32	28	A	10	0.5	30.4	2.34	168
4	800	100	100	0.45	350	30	7	A	8	0.5	29.6	3.13	112
	800	100	100	0.58	300	30	7	A	6	0.5	24.7	2.43	81
	800	100	100	0.50	400	8	7	A	8	0.5	22.4	3.32	85

Table 6-2 (*Continued*)

Series Number	Dimensions (mm) S	b	t	w/c	Cement Content (kg/m³)	d_a (mm)	Age (days)	Curing[a]	Number of Beams Tested	a_0/b	E (GPa)	f_t (MPa)	G_F (N/m)
4	800	100	100	0.50	400	3	7	A	7	0.5	20.4	3.04	62
(cont.)	800	100	100	0.40	400	3	7	A	3	0.5	24.4	3.55	79
	800	100	100	0.40	400	8	7	A	3	0.5	26.5	4.10	78
	800	100	100	0.58	300	30	34	A	6	0.5	29.1	3.81	105
	800	100	100	0.48	350	30	33	A	8	0.5	32.7	4.27	120
	800	100	100	0.40	400	30	30	A	9	0.5	35.1	4.41	146
5	800	100	100	0.48	423	9.5	30	A	7	0.5	33.9	2.96	99
	800	100	100	0.58	350	9.5	30	A	7	0.5	31.0	2.34	109
	800	100	100	0.68	300	9.5	30	A	7	0.5	29.9	1.93	101
	800	100	100	0.48	364	19	30	A	13	0.5	42.2	3.79	188
	800	100	100	0.58	302	19	30	A	14	0.5	39.9	2.45	200
	800	100	100	0.68	259	19	30	A	14	0.5	35.7	2.27	193
6	450	100	100	0.50	375	8	28	D	5	0.5	30.0	2.90	113
7	800	100	100	0.40	524	25	28	A	13	0.5	26.3	4.03	112

Source: From ref. 7.

[a] A = moist cured 1 day, then in lime-saturated water; B = in water 8 days, then in 65% RH; C = covered with wet burlap for 1 day, then exposed to air; D = demolding after 3 days, then in 95% RH.

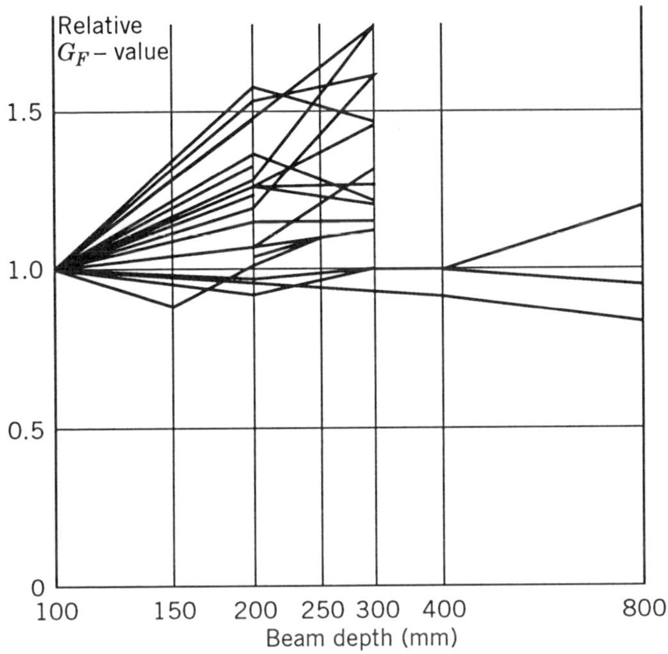

Figure 6-2 Variation of relative G_F values with beam depth.[7]

pendent. Obviously, more studies are needed to access and evaluate these requirements. Hu and Wittmann[11] have discussed the effect of width of the fracture process zone on the value of G_F. They introduced a concept of a local fracture energy that varies with the width of the fracture process zone. As a crack approaches the back end of a specimen, the fracture process zone becomes more and more confined and hence the local fracture energy decreases. This decrease of the local fracture energy may partly account for the observed size effect on the value of G_F.

The RILEM recommendation does not address the question of how to make the displacement measurement. As stated above, errors can be introduced into displacement measurements related to beam settlement on the supports, or "punching." This is illustrated in Fig. 6-3, where three possible displacement measurements are indicated: δ_1, δ_2, and δ_3. (This setup is for a beam loaded upward. The same conclusions apply to a beam loaded downward.) As presented by Swartz and Kan,[12,13] the greater error occurs if δ_1 is measured. This is simply the measurement of piston movement. More accurate but still in error is δ_2 because load point punching is included in the measurement. Most accurate is δ_3, and this is the recommended measurement method. Care must be taken that the reference bar support at pins 1 and 2 is free to rotate relative to the pins.

Since G_F is the value somewhat easiest to use, researchers often use G_F as

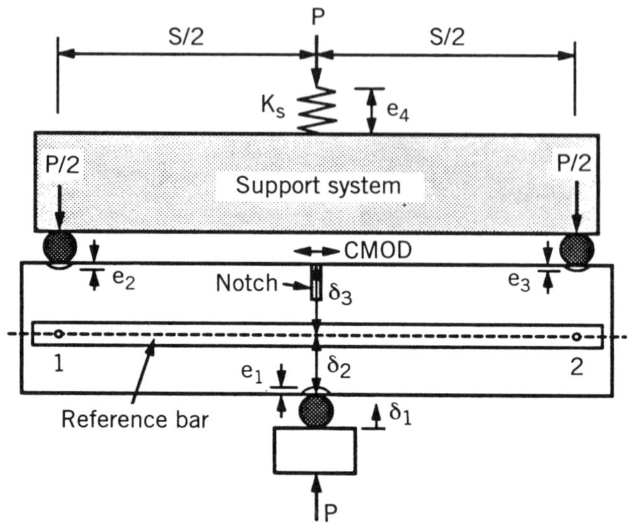

Figure 6-3 Schematic of test setup(s).

a sole parameter to quantify fracture toughness. However, the model proposed by Hillerborg et al.[2] is based on determining three parameters, G_F, f_t, and w_c. If only a single parameter is used, then erroneous conclusions can be drawn. This is illustrated in Fig. 6-4, where the value of G_F is plotted as the function of compressive strength.[14] It is observed that G_F increases with increasing compressive strength. From this it should not be inferred that concrete is tougher the higher the compressive strength. In fact, concrete becomes more brittle with an increasing value of the compressive strength. If a brittleness index, which combines the three fracture parameters, is introduced, then it can be shown that such an index increases with the compressive strength.

6.2.3 Example

An example of measuring fracture toughness G_F was reported by Petersson.[15] The mix proportions of the concrete were 1.0 : 2.04 : 2.71 of cement, sand, and aggregate by weight. A crushed quartzite aggregate was used. Particle sizes of the sand used were up to 4 mm (0.16 in.), whereas particle sizes for the aggregate used ranged from 4 to 12 mm (0.16 to 0.5 in.). The water-to-cement ratio used was 0.5. The specimen geometry is shown in Fig. 6-5*a*. The specimens were stored in 100% relative humidity (RH) and at a temperature of 20°C during the first 24 h after casting and then stored in water until testing. The specimen was tested at an age of 28 days. A notch of 25 mm (1 in.) depth was sawn by a diamond saw at 1 day before testing.

The testing arrangement is shown in Fig. 6-5*a*. The loading rate was chosen

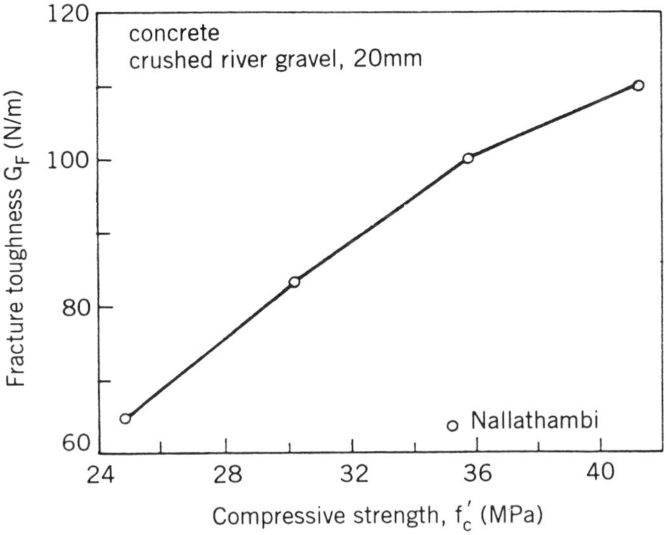

Figure 6-4 Effect of compressive strength of concrete on fracture toughness G_F (aggregate is crushed river gravel with maximum size of 20 mm).[14]

in such a way that the maximum load was reached in about 30 s after the loading started. The load–displacement curve obtained is shown in Fig. 6-5*b*. The area under the load–displacement curve can be measured, and a value of $G_F = 90$ N/m was calculated based on Eq. (6.4). In addition, the value of $f_t = 3.4$ MPa was obtained from a direct tensile test, and a value of $E = 42.0$ GPa was estimated.

6.3 RILEM METHOD OF JENQ AND SHAH, K_{Ic}^s AND $CTOD_c$

6.3.1 Specimen and Test Procedure

The RILEM Technical Committee 89-FMT on Fracture Mechanics of Concrete—Test Methods proposed in 1990 a recommendation to measure the material fracture parameters K_{Ic}^s and $CTOD_c$ using a three-point bend beam.[16] This method is based on the two-parameter fracture model by Jenq and Shah,[17] which has been presented in detail in the previous chapter. This test method will be briefly presented below.

The three-point bend beam with $S/b = 4$ as shown in Fig. 6-6*a* was recommended. The size of the beam depends on the maximum size of aggregate, d_a, according to Table 6-3. The initial notch-to-depth ratio should be $\frac{1}{3}$, and the notch width should be less than 5 mm (0.2 in.). After casting, the speci-

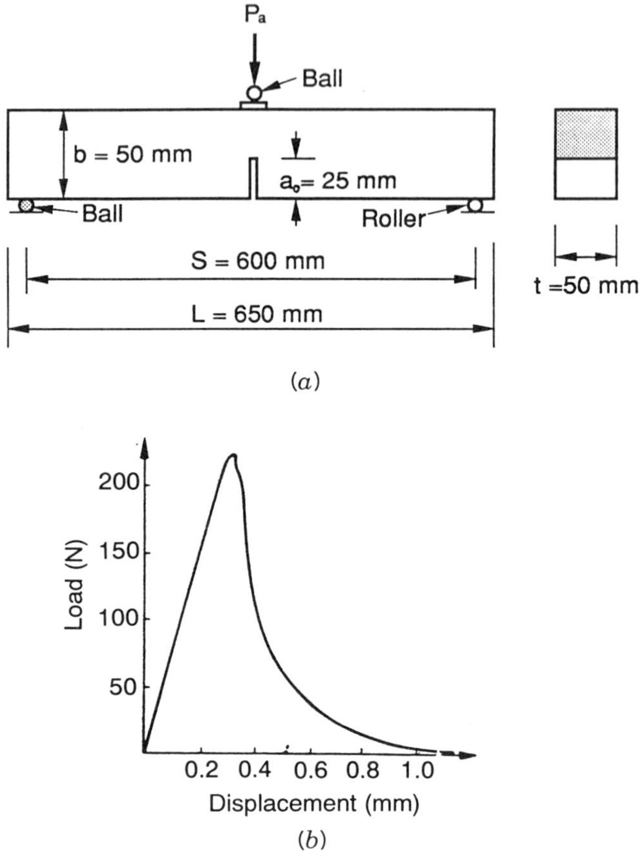

Figure 6-5 Example of measurement of the G_F-value by three-point bend beam: (*a*) the three-point bend beam and (*b*) the measured load-displacement curve.[15]

mens should be covered with wet burlap or kept in the curing room with 100% RH at 23 $\pm$ 2°C for the first 24 h. On the second day, all the specimens should be transferred to the curing room until 4 h before testing. A minimum of four specimens are required for each type of material.

A closed-loop testing machine with the CMOD as feedback signal or a relatively stiff machine is required to achieve a stable failure. The CMOD and the applied load shall be recorded continuously during the test. A clip gauge is recommended to measure the CMOD. However, if a clip gauge is not available, an LVDT can be used as a replacement. The rate of loading shall be controlled by a constant rate of increment of CMOD such that the peak load is reached in about 5 min.

The beam shall be monotonically loaded up to the maximum load. The applied load is then manually reduced after the load passes the maximum load and within 95% of the maximum load, as shown in Fig. 6-6*b*. When the ap-

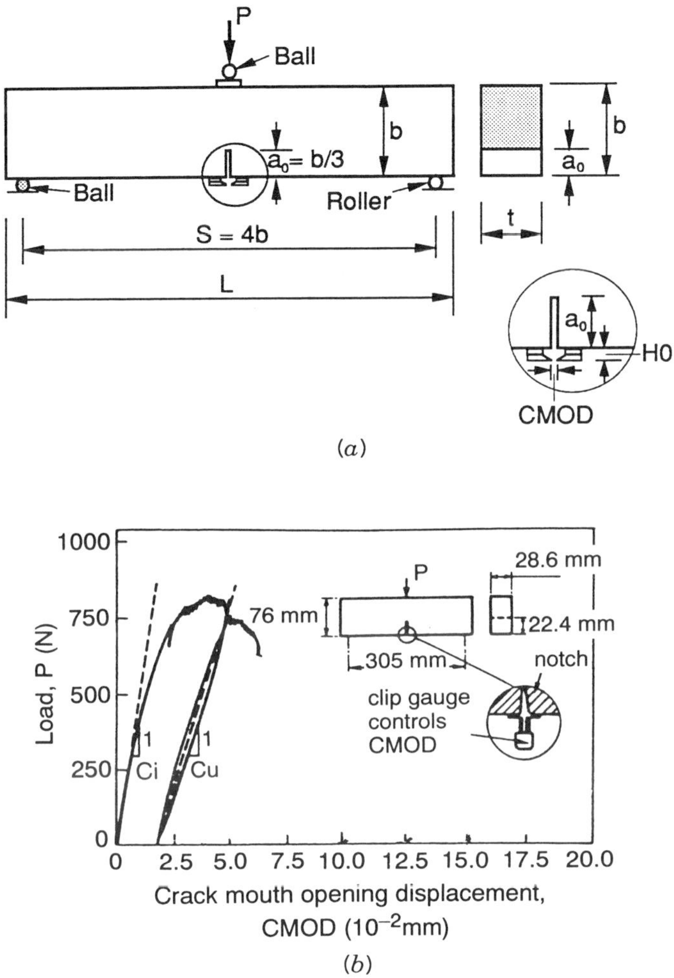

Figure 6-6 Determination of K_{Ic}^s and CTOD$_c$ according to RILEM draft recommendation: (*a*) testing configuration and specimen geometry and (*b*) a typical load-CMOD curve measured.[16]

Table 6-3 Sizes of Beams for Measuring K_{Ic}^s and CTOD$_c$

Maximum Sizes of Aggregates, d_a (mm)	Depth, b (mm)	Width, t (mm)	Length, L (mm)	Span, S (mm)
1.0–25	150 ± 5	80 ± 5	700 ± 10	600 ± 5
25.1–50	250 ± 5	150 ± 5	1,100 ± 10	1,000 ± 5

Source: From ref. 16.

plied load is reduced to zero, reloading might be applied. The rate of unloading and reloading may not be the same as the initial loading rate. The loading and unloading cycle might be finished in about 1 min.

6.3.2 Test Results and Calculations

A typical load–CMOD curve with loading–unloading cycle is shown in Fig. 6-6*b*. The modulus of elasticity E, the critical stress intensity factor K_{Ic}^{s}, and the critical crack tip opening displacement $CTOD_c$ can be determined from the obtained load–CMOD curve as below.

The modulus of elasticity E is calculated from

$$E = \frac{6Sa_0 g_2(\alpha_0)}{C_i b^2 t} \tag{6.7}$$

where C_i is the initial compliance calculated from the load–CMOD curve, as shown in Fig. 6-6*b*, and the geometric function $g_2(\alpha_0)$ is calculated as

$$g_2(\alpha_0) = 0.76 - 2.28\alpha_0 + 3.87\alpha_0^2 - 2.04\alpha_0^3 + \frac{0.66}{(1 - \alpha_0)^2} \tag{6.8}$$

where $\alpha_0 = (a_0 + HO)/(b + HO)$ and S, a_0, HO, b, and t are defined in Fig. 6-6*a*.

An effective-elastic critical crack length a_c may be defined in such a way that it results in an unloading compliance C_u within 95% of the peak load based on LEFM. Therefore, by using this, the modulus of elasticity E can also be calculated from

$$E = \frac{6Sa_c g_2(\alpha_c)}{C_u b^2 t} \tag{6.9}$$

where C_u is the unloading compliance within 95% of the peak load calculated from the load–CMOD curve, as shown in Fig. 6-6*b*. The geometric function $g_2(\alpha_c)$ is given by Eq. (6.8) but $\alpha_c = (a_c + HO)/(b + HO)$, instead of α_0, should be used.

By equating Eqs. (6.7) and (6.9), the value of the effective-elastic critical crack length a_c for the tested beam can be solved. This results in

$$a_c = a_0 \frac{C_u}{C_i} \frac{g_2(\alpha_0)}{g_2(\alpha_c)} \tag{6.10}$$

This can be solved fairly readily by trial and error. Then the critical stress intensity factor is calculated using the equation

$$K_{Ic}^s = 3(P_c + 0.5\, W_h)\, \frac{S\sqrt{\pi a_c}\, g_1(a_c/b)}{2b^2 t} \tag{6.11}$$

where P_c is the peak load, $W_h = W_{h0}S/L$, W_{h0} is the self-weight of the beam, and

$$g_1\left(\frac{a_c}{b}\right) = \frac{1.99 - (a_c/b)\,(1 - a_c/b)\,[2.15 - 3.93\, a_c/b + 2.70\,(a_c/b)^2]}{\sqrt{\pi}\,(1 + 2\, a_c/b)\,(1 - a_c/b)^{3/2}} \tag{6.12}$$

The critical crack tip opening displacement is then calculated using the equation

$$\mathrm{CTOD}_c = \frac{6(P_c + 0.5W_h)Sa_c g_2(a_c/b)}{Eb^2 t} \cdot \left[(1 - \beta_0)^2 + \left(1.081 - 1.149\,\frac{a_c}{b}\right)(\beta_0 - \beta_0^2)\right]^{1/2} \tag{6.13}$$

where $\beta_0 = a_0/a_c$ and $g_2(a_c/b)$ is based on Eq. (6.8) but α_0 should be replaced by a_c/b.

Refai and Swartz[18] have tested a series of concrete three-point bend beams with different sizes and initial notch lengths. Their results and corresponding values of K_{Ic}^s and CTOD_c are listed in Table 6-4. Although these beams have a span-to-depth ratio of 3.75 (i.e., $S/b = 3.75$), Eqs. (6.7)–(6.13) are approximately used to evaluate the values of K_{Ic}^s and CTOD_c in Table 6-4. Karihaloo and Nallathambi[19] have summarized results of three-point bend beams tested by Catalano and Ingraffea,[20] Jenq and Shah,[17] and John and Shah.[21] These results are listed in Tables 6-5 to 6-7. To demonstrate the variation of the value of K_{Ic}^s with different beam sizes, some relative values of K_{Ic}^s obtained from these tables are plotted as functions of the beam depth in Fig. 6-7. It is seen from the figure that when the beam depth varies from 50 to 800 mm, the value of K_{Ic}^s changes only approximately 30%. Note that the value of K_{Ic}^s is fairly constant if the beam depth is greater than 400 mm. It is generally agreed that the value of K_{Ic}^s is essentially specimen size independent.

It is seen from Tables 6-4 to 6-7 that the values of CTOD_c have much greater scatter than the values of K_{Ic}^s. Several possible explanations may be offered for some reported discrepancies on the values of CTOD_c. The determination of CTOD_c is very sensitive to the precision of the measurements. For example, when a large-size specimen is tested, the testing machine may not be stiff enough to attain a stable failure even under a closed-loop controlled testing system. As a result, unloading at the peak load (or within 95% of the postpeak region) as required in the proposed procedure may be violated, which

Table 6-4 Values of K_{Ic}^s and $CTOD_c$ from Three-Point Bend Beams Tested by Refai and Swartz

Test Number[a]	Dimensions (mm)			f'_c (MPa)	w/c	d_a (mm)	E (GPa)	a_0/b	a_c/b	K_{Ic}^s (MPa m$^{1/2}$)	$CTOD_c$ (mm)	G_{Ic}^s (N/m)[b]
	S	b	t									
1	381	102	76	55.8	0.5	19	36.8	0.290	0.550	1.153	0.0219	36.32
2	381	102	76	55.8	0.5	19	36.8	0.320	0.560	1.184	0.0214	38.30
3	381	102	76	55.8	0.5	19	36.8	0.460	0.680	1.158	0.0217	36.64
4	381	102	76	55.8	0.5	19	36.8	0.520	0.690	0.939	0.0084	24.09
5	381	102	76	55.8	0.5	19	36.8	0.620	0.720	0.594	0.0067	9.64
6	381	102	76	55.8	0.5	19	36.8	0.670	0.810	0.560	0.0086	8.57
7	762	203	76	53.1	0.5	19	38.4	0.290	0.410	1.054	0.0153	28.93
8	1143	305	76	54.4	0.5	19	39.3	0.300	0.420	1.378	0.0240	48.32
9	1143	305	76	54.4	0.5	19	39.3	0.500	0.630	1.250	0.0257	39.76
10	1143	305	76	54.4	0.5	19	39.3	0.670	0.730	0.755	0.0102	14.5
11	762	203	76	53.1	0.5	19	38.4	0.250	0.440	1.142	0.0222	33.96
12	762	203	76	53.1	0.5	19	38.4	0.275	0.480	1.160	0.0243	35.04
13	762	203	76	53.1	0.5	19	38.4	0.293	0.620	1.124	0.0360	32.90
14	762	203	76	53.1	0.5	19	38.4	0.454	0.720	1.156	0.0350	34.80
15	762	203	76	53.1	0.5	19	38.4	0.563	0.780	1.212	0.0337	38.25
16	762	203	76	53.1	0.5	19	38.4	0.608	0.830	1.264	0.0385	41.61
17	1143	305	76	54.4	0.5	19	39.3	0.158	0.540	1.251	0.0500	39.76
18	1143	305	76	54.4	0.5	19	39.3	0.187	0.400	1.319	0.0323	44.27
19	1143	305	76	54.4	0.5	19	39.3	0.208	0.550	1.228	0.0457	38.37
20	1143	305	76	54.4	0.5	19	39.3	0.221	0.430	1.224	0.0301	38.12
21	1143	305	76	54.4	0.5	19	39.3	0.250	0.490	1.337	0.0373	45.49
22	1143	305	76	54.4	0.5	19	39.3	0.258	0.460	1.243	0.0305	39.31
23	1143	305	76	54.4	0.5	19	39.3	0.266	0.420	1.094	0.0222	30.45
24	1143	305	76	54.4	0.5	19	39.3	0.411	0.580	1.309	0.0309	43.60
25	1143	305	76	54.4	0.5	19	39.3	0.454	0.600	1.312	0.0285	43.80
26	1143	305	76	54.4	0.5	19	39.3	0.479	0.620	1.138	0.0245	32.96
27	1143	305	76	54.4	0.5	19	39.3	0.500	0.650	1.052	0.0240	28.16
28	1143	305	76	54.4	0.5	19	39.3	0.545	0.710	1.138	0.0289	32.95
29	1143	305	76	54.4	0.5	19	39.3	0.570	0.690	1.033	0.0209	27.15

Source: From ref. 18.

[a]Beams from tests 1–10 have a precast notch, and beams from 11–29 have an initial notch created by load recycling the specimen. There is no noticeable difference in values of K_{Ic}^s and $CTOD_c$ for two types of beams. [b]$G_{Ic}^s = (K_{Ic}^s)^2/E$

Table 6-5 Values of K^s_{Ic} and $CTOD_c$ from Three-Point Bend Beams Tested by Bascoul

Test Number	Dimensions (mm) S	b	t	f'_c (MPa)	w/c	d_a [a] (mm)	E (GPa)	a_0/b	a_c/b	K^s_{Ic} (MPa m$^{1/2}$)	$CTOD_c$ (mm)	G^s_{Ic} (N/m) [b]
1	160	40	24	—	0.53	3.15	36.8	0.250	0.323	0.982	0.0037	26.20
2	160	39	24	—	0.53	3.15	36.8	0.263	0.307	0.854	0.0025	19.82
3	160	40	24	—	0.53	3.15	36.8	0.274	0.301	0.792	0.0018	17.05
4	320	80	50	—	0.53	3.15	36.8	0.512	0.550	0.931	0.0050	23.55
5	320	80	50	—	0.53	3.15	36.8	0.500	0.545	0.914	0.0053	22.70
6	320	80	50	—	0.53	3.15	36.8	0.512	0.562	0.923	0.0057	23.15
7	320	80	49	—	0.53	3.15	36.8	0.262	0.310	1.025	0.0057	28.55
8	320	83	50	—	0.53	3.15	36.8	0.263	0.308	0.899	0.0048	21.96
9	320	83	50	—	0.53	3.15	36.8	0.264	0.310	0.869	0.0047	20.52
10	320	80	50	—	0.53	3.15	36.8	0.125	0.154	0.779	0.0032	16.49
11	320	83	50	—	0.53	3.15	36.8	0.156	0.191	0.843	0.0038	19.31
12	320	83	50	—	0.53	3.15	36.8	0.144	0.176	0.911	0.0039	22.55

Source: From ref. 19.

[a] Aggregate used is crushed marble.
[b] $G^s_{Is} = (K^s_{Ic})^2/E$.

Table 6-6 Values of K_{Ic}^s and $CTOD_c$ from Three-Point Bend Beams Tested by Catalano and Ingraffea

Test Number	Dimensions (mm) S	b	t	f'_c (MPa)	w/c	d_a (mm)	E (GPa)	a_0/b	a_c/b	K_{Ic}^s (MPa m$^{1/2}$)	$CTOD_c$ (mm)	G_{Ic}^s (N/m)[a]
1	1220	305	102	60.7	0.44	6	33.5	0.197	0.278	1.236	0.0104	45.60
2	1220	305	102	60.7	0.44	6	33.5	0.211	0.248	1.046	0.0185	32.66
3	1220	305	152	45.5	0.52	12.7	31.0	0.205	0.261	1.286	0.0131	53.35
4	1220	305	152	45.5	0.52	12.7	31.0	0.198	0.352	1.736	0.0227	97.22
5	1220	305	152	45.5	0.52	12.7	31.0	0.202	0.252	1.403	0.0302	63.50
6	1220	305	152	43.4	0.52	12.7	31.0	0.211	0.267	1.529	0.0196	75.41
7	1220	305	152	43.4	0.52	12.7	31.0	0.202	0.244	1.504	0.0156	72.97
8	1220	305	152	43.4	0.52	12.7	31.0	0.205	0.275	1.558	0.0155	78.30

Source: From ref. 20.

[a] $G_{Ic}^s = (K_{Ic}^s)^2/E$.

Table 6-7 Values of K^s_{Ic} and $CTOD_c$ from Three-Point Bend Beams Tested by Jenq and Shah and John and Shah

Test Number[a]	Dimensions (mm) S	b	t	f'_c (MPa)	w/c	d_a (mm)	E (GPa)	a_0/b	a_c/b	K^s_{Ic} (MPa m$^{1/2}$)	$CTOD_c$ (mm)[c]	G^s_{Ic} (N/m)[b]
1	305	76	29	25.2	0.65	19	33.6	0.293	0.531	1.127	0.0231	46.63
2	305	76	29	25.2	0.65	19	33.6	0.293	0.535	0.989	0.0198	35.91
3	305	76	29	25.2	0.65	19	33.6	0.293	0.548	0.891	0.0180	29.14
4	305	76	29	25.2	0.65	19	33.6	0.293	0.498	0.968	0.0196	34.40
5	609	152	57	25.2	0.65	19	33.6	0.324	0.511	1.001	0.0269	36.78
6	609	152	57	25.2	0.65	19	33.6	0.318	0.377	0.787	0.0089	22.74
7	914	229	86	25.2	0.65	19	33.6	0.333	0.385	0.940	0.0122	32.43
8	914	229	86	25.2	0.65	19	33.6	0.333	0.347	1.106	0.0076	44.91
9	305	76	29	27.2	0.65	4.8	25.4	0.293	—	0.708	0.0077	19.74
10	609	152	57	27.2	0.65	4.8	25.4	0.324	—	0.719	0.0118	20.35
11	914	229	86	27.2	0.65	4.8	25.4	0.333	—	0.693	0.0083	18.95
12	305	76	29	39.4	0.45	4.8	32.4	0.293	—	0.897	0.0084	24.77
13	609	152	57	39.4	0.45	4.8	32.4	0.324	—	0.966	0.0085	28.75
14	914	229	86	39.4	0.45	4.8	32.4	0.333	—	1.010	0.0121	31.39
15	305	76	29	27.6	0.45	Paste	20.8	0.293	—	0.601	0.0076	17.41
16	609	152	57	27.6	0.45	Paste	20.8	0.324	—	0.599	0.0062	17.27
17	914	229	86	27.6	0.45	Paste	20.8	0.333	—	0.654	0.0068	20.63
18	203	76	25	110	0.22	8	56.6	0.329	—	2.130	0.0112	80.23

Source: From refs. 17 and 21.

[a] Values of K^s_{Ic} and $CTOD_c$ for tests 9–17 are average values of several beams.

[b] $G^s_{Ic} = (K^s_{Ic})^2/E$.

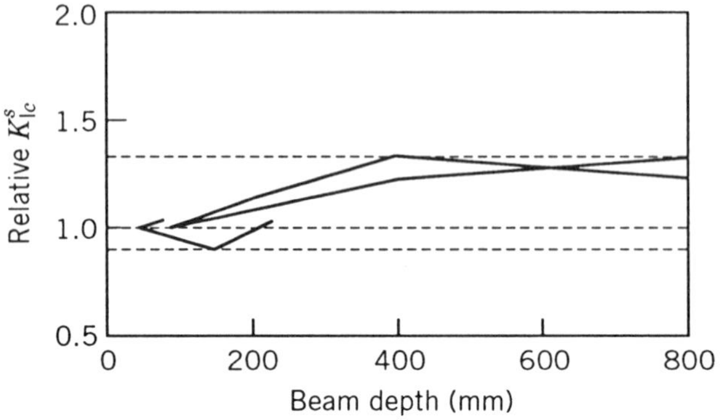

Figure 6-7 Variation of relative K^s_{Ic} values with beam depth.[19]

will result in extra postpeak crack extension. Also, for large specimens, compensation for the self-weight during specimen handling is critical because the self-weight of the beam may already preload and damage the beam before applying external load. The effects of the machine stability and the self-weight were not addressed in the proposed recommendation, since the proposed beams were relatively small and these effects are, in general, negligible. Therefore, more accurate measurements on $CTOD_c$ are needed to appraise its size independence. Furthermore, according to the two-parameter fracture model, the value of $CTOD_c$ is determined based on the elastic (unloading) compliance at peak load. It is noted that the total crack tip opening displacement, $CTOD_t$, at the peak load consists of the part from elastic deformation, $CTOD^e_c$, which is simplified as $CTOD_c$ here, and the part from inelastic deformation, $CTOD^p_c$. The unloading procedure was introduced in the two-parameter fracture model so that $CTOD^p_c$ can be subtracted from $CTOD_t$. Theoretically, specimens with any size and geometry can be used to determine the value of $CTOD_c$. However, since $CTOD^p_c$ has a greater portion in the $CTOD_t$ for a small-size specimen (i.e., the fracture process zone is larger compared to specimen size), use of too small a size of specimen may easily result in error in determining $CTOD_c$. As a result, one should be very careful when using small-size specimens to determine $CTOD_c$.

Several conflicting results on measuring $CTOD_c$ have been recently reported. Zhao et al.[22] measured $CTOD_c$ from wedge-splitting specimens. They found that the values of $CTOD_c$ are basically independent of specimen size. On the other hand, Brühwiler et al.[23] concluded that $CTOD_c$ is size dependent from their study using similar wedge-splitting specimens. However, Brühwiler et al. used finite element analysis to calculate $CTOD_c$, and their method of measurement was such that it included other extraneous deformations beside CTOD.

To overcome these difficulties in measuring $CTOD_c$, Tang et al.[24] have proposed a method to determine K^s_{Ic} and $CTOD_c$. This method only needs to measure peak loads of at least three different specimens and will be discussed later.

6.3.3 Example

An example of measuring fracture parameters of K_{Ic}^{s} and CTOD$_c$ presented by Jenq and Shah[25] is given. The mix proportion of the mortar tested was 1.0 : 2.6 of cement and sand. The water-to-cement ratio was 0.65, and the maximum particle size of the sand was 4.76 mm. The span, depth, and width of the beam were 304.8 mm (12 in.), 76.2 mm (3 in.), and 28.6 mm (1.125 in.), respectively. The beam had an initial notch of 22.4 mm (0.88 in.), and the notch was precast with a thin steel plate with the thickness of 0.8 mm ($\frac{1}{32}$ in.). The beam was tested after curing for about 90 days in an environment with 96% RH and 26.7°C of temperature.

The beam was tested in a servo-controlled closed-loop testing machine. The peak load was reached in about 10 min. The obtained load–CMOD curve is shown in Fig. 6-6*b*. As measured from the load–CMOD curve, values of the initial loading compliance C_i and the peak load unloading compliance C_u are equal to 0.00002015 and 0.00003570 mm/N, respectively. The maximum load P_c is 795 N. The values of $S = 304.8$ mm, $b = 76.2$ mm, $t = 28.6$ mm, $a_0 = 22.4$ mm, $HO = 1$ mm, $C_i = 0.00002015$ mm/N, $C_u = 0.00003570$ mm/N, $P_c = 795$ N, and $W_h = 10$ N will be used in the following calculation.

By substituting $\alpha_0 = (a_0 + HO)/(b + HO) = (22.4 + 1)/(76.2 + 1) = 0.303$ into Eq. (6.8), the value of $g_2(\alpha_0)$ is obtained as

$$g_2(\alpha_0) = 0.76 - 2.28 \times 0.303 + 3.87 \times 0.3032$$
$$- 2.04 \times 0.3033 + \frac{0.66}{(1 - 0.303)^2} = 1.727$$

The value of the modulus of elasticity E is determined from Eq. (6.7):

$$E - \frac{6 \times 304.8 \times 22.4 \times 1.727}{0.00002015 \times 76.22 \times 28.6} - 21128.9 \text{ MPa}$$

The effective critical crack length a_c is then calculated using Eq. (6.10):

$$\frac{6 \times 304.8 a_c g_2(\alpha_c)}{0.0000357 \times 76.22 \times 28.6} = 21128.9$$

where the geometric function $g_2(\alpha_c)$ is based on Eq. (6.8), but $\alpha_c = (a_c + HO)/(b + HO)$, instead of α_0, should be used. The value of $a_c = 30.68$ mm can be solved from this nonlinear equation.

Substituting $a_c/b = 30.68/76.2 = 0.403$ into Eq. (6.12) leads to

$$g_1\left(\frac{a_c}{b}\right)$$

$$= \frac{1.99 - 0.403\ (1 - 0.403)\ (2.15 - 3.93 \times 0.403 + 2.7 \times 0.4032)}{3.1421/2 \times (1 + 2 \times 0.403)\ (1 - 0.403)3/2}$$

$$= 1.183$$

The value of K_{Ic}^s can then be determined from Eq. (6.11) as

$$K_{Ic}^s = \frac{3(795 + 5) \times 304.8 \times (3.142 \times 30.68)^{1/2} \times 1.183}{2 \times 76.22 \times 28.6}$$

$$= 25.5 \text{ MPa mm}^{1/2}$$

The geometric factor $g_2(a_c/b)$ is evaluated using Eq. (6.8), but $a_c/b = 30.68/76.2 = 0.403$, instead of α_0, should be used. This results in

$$g_2(a_c/b) = 0.76 - 2.28 \times 0.403 + 3.87 \times 0.403^2$$

$$- 2.04 \times 0.403^3 + \frac{0.66}{(1 - 0.403)^2} = 2.186$$

The value of CTOD_c is then obtained by substituting $\beta_0 = a_0/a_c = 22.4/30.68 = 0.73$ into Eq. (6.13) as

$$\text{CTOD}_c$$

$$= \frac{6 \times (795 + 5) \times 304.8 \times 30.68 \times 2.186 \times [(1 - 0.73)^2 + (1.081 - 1.149 \times 0.73)\ (0.73 - 0.732)]^{1/2}}{21128.9 \times 76.2^2 \times 28.6}$$

$$= 0.0123 \text{ mm}$$

The material fracture parameters K_{Ic}^s and CTOD_c as obtained can be used to predict failure of structures made of the same material.

6.4 RILEM METHOD OF BAZANT G_f AND c_f

6.4.1 Specimen and Test Procedure

The RILEM Technical Committee 89-FMT on Fracture Mechanics of Concrete—Test Methods has proposed in 1990 a draft recommendation for measuring the material fracture energy G_f using three-point bend beams.[26] This method is based on the size effect model by Bazant et al.,[27,28] which has been presented in detail in the previous chapter. This test method will be presented as follows.

Three-point bend beams used should have span-to-depth ratio not less than 2.5, as shown in Fig 6-8. The initial notch-to-depth ratio should be between 0.15 and 0.5. The notch width should be as small as possible and should not exceed 0.5 times the maximum aggregate size d_a. The width and the depth of the beam, b, must not be less than $3d_a$.

Specimens of at least three different sizes, characterized by beam depths $b = b_1, b_2, \ldots, b_n$ and spans $S = S_1, S_2, \ldots, S_n$, must be tested (see Fig. 6-8). The smallest depth b_1 must not be larger than $5d_a$, and the largest depth b_n must not be smaller than $10d_a$. The ratio of b_n to b_1 must be at least 4. The ratios of the adjacent sizes should be approximately constant. The choice of $b_j/d_a = 3, 6, 12, 24$ is preferable. At least three identical specimens should be tested for each beam size. The ratios of S/b, a_0/b, and L/b should be the same for all specimens. It is important that thickness t be kept constant.

Specimens should be loaded at a constant (or almost constant) displacement rate. The loading rates should be such that the maximum load is reached in about 5 min. Although an ordinary uniaxial testing machine may be used, a machine with high-strength and/or closed-loop control is better. All tests should be performed on the same machine and in an identical manner.

6.4.2 Test Results and Calculations

Only the maximum loads $P_1, P_2, \ldots, P_n$ for specimens of various sizes of $b_1, b_2, \ldots, b_n$ and the modulus of elasticity E are needed to determine the

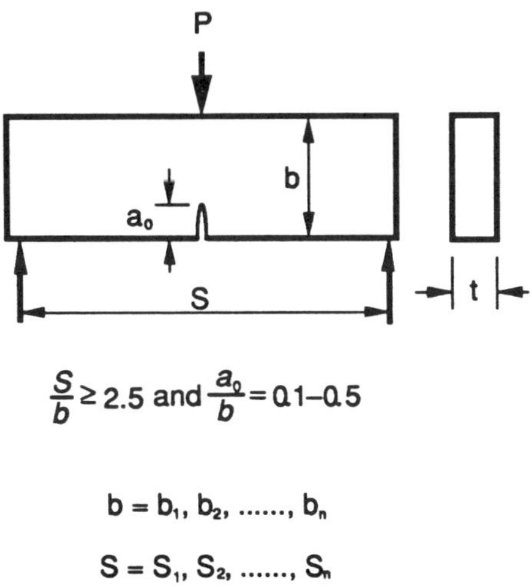

Figure 6-8 Three-point bend beams for determining G_f.

values of fracture parameters G_f. The calculation procedure for the value of G_f is presented below.

The corrected maximum loads $P_1^0, P_2^0, \ldots, P_n^0$, which take the weight of the specimen into account, should be first calculated as

$$P_j^0 = P_j + \frac{2S_j - L_j}{2S_j} gm_j \qquad (j = 1, 2, \ldots, n) \tag{6.14}$$

where m_j is the mass of specimen j, $g = 9.8$ m/s^2 (32.2 ft/s^2) is the acceleration due to gravity, and n is the number of tests conducted.

By introducing

$$Y_j = \left(\frac{b_j t}{P_j^0}\right)^2 \qquad X_j = b_j \tag{6.15}$$

for $j = 1, 2, \ldots, n$, a linear regression $Y = A_B X + C_B$ can be plotted. The values of the slope A_B and the intercept C_B can be obtained as below:

$$A_B = \frac{\sum_{j=1}^{n} (X_j - \overline{X})(Y_j - \overline{Y})}{\sum_{j=1}^{n} (X_j - \overline{X})^2} \tag{6.16}$$

and

$$C_B = \overline{Y} - A_B \overline{X} \tag{6.17}$$

where

$$\overline{X} = \frac{1}{n} \sum_{j=1}^{n} X_j \qquad \overline{Y} = \frac{1}{n} \sum_{j=1}^{n} Y_j \tag{6.18}$$

are the centroid of all data points.

The geometric factor $g(\alpha_0)$ is calculated as

$$g(\alpha_0) = \left(\frac{S}{b}\right)^2 \pi\alpha_0[1.5g_1(\alpha_0)]^2 \tag{6.19}$$

where $\alpha_0 = a_0/b$. The value of $g_1(\alpha_0)$ is determined based on ratios of S/b. For $S/b = 2.5$,

$$g_1(\alpha_0) = \frac{1.0 - 2.5\alpha_0 + 4.49\alpha_0^2 - 3.98\alpha_0^3 + 1.33\alpha_0^4}{(1 - \alpha_0)^{3/2}} \tag{6.20}$$

For $S/b = 4$

$$g_1(\alpha_0) = \frac{1.99 - \alpha_0(1 - \alpha_0)\,(2.15 - 3.93\alpha_0 + 2.70\alpha_0^2)}{\sqrt{\pi}\,(1 + 2\alpha_0)\,(1 - \alpha_0)^{3/2}} \tag{6.21}$$

For $S/b = 8$

$$g_1(\alpha_0) = 1.11 - 1.552\alpha_0 + 7.71\alpha_0^2 - 13.55\alpha_0^3 + 14.25\alpha_0^4 \tag{6.22}$$

Linear interpolation may be used to obtain $g_1(\alpha_0)$ for other values of S/b.

The values of the material fracture energy G_f can then be determined from

$$G_f = \frac{g(\alpha_0)}{EA_B} \tag{6.23}$$

The length of the fracture process zone for an infinitely large specimen c_f can be further obtained from

$$c_f = \frac{g(\alpha_0)}{g'(\alpha_0)}\left(\frac{C_B}{A_B}\right) \tag{6.24}$$

where $g'(\alpha_0)$ is the value of the first derivative of $g(a/b)$ with respect to a/b at $a = a_0$.

To verify the linear regression, one needs to calculate statistics:

$$s_X^2 = \frac{1}{n-1}\sum_{j=1}^{n}(X_j - \overline{X})^2 \qquad s_Y^2 = \frac{1}{n-1}\sum_{j=1}^{n}(Y_j - \overline{Y})^2 \tag{6.25}$$

$$s_{X|Y}^2 = \frac{1}{n-2}\sum_{j=1}^{n}(Y_j - Y_j^*)^2 = \frac{n-1}{n-2}\sum_{j=1}^{n}(s_Y^2 - A_B^2 s_X^2) \tag{6.26}$$

$$\omega_{X|Y} = \frac{s_{X|Y}}{\overline{Y}} \qquad \omega_X = \frac{s_X}{\overline{X}} \tag{6.27}$$

$$\omega_A = \frac{s_{Y|X}}{A_B s_X \sqrt{n-1}} \qquad \omega_C = \frac{s_{Y|X}}{C_B\sqrt{n-1}}\sqrt{1 + \frac{1}{\omega_X^2}} \qquad m = \frac{\omega_{Y|X}}{\omega_X} \tag{6.28}$$

and the approximation

$$\omega_G^2 \approx \omega_A^2 + \omega_E^2 \tag{6.29}$$

where X_j are the specimen sizes, Y_j are measured data points (not averages of some groups of data points), $Y_j - Y_j^*$ are the vertical deviations of these points from the regression line, ω_X is the coefficient of variation of the sizes chosen,

$\omega_{Y|X}$ is the coefficient of variation of the errors (vertical deviations from the regression line), ω_A is the coefficient of variation of the slope of the regression line, ω_C is the coefficient of variation of the intercept, m is the relative width of the scatterband, and ω_E is the coefficient of variation of values of the modulus of elasticity E used in Eq. (6.23). The coefficient of variation of G_f is approximately equal to ω_G if the values of E and G_f are assumed to be uncorrelated and is approximately equal to ω_A if the values of E and G_f are assumed to be perfectly correlated. In reality, the coefficient of variation of G_f is between ω_A and ω_G for concrete.

The value of ω_A should not exceed about 0.10 and the value of ω_C and m about 0.20. These conditions prevent situations in which the size range used is insufficient compared with the scatter of results. Such a situation is illustrated in Fig. 6-9*a*, in which a unique regression slope A_B is obtained but is highly uncertain. The case where the large scatter of test results necessitates the use of a very broad size range is illustrated in Fig. 6-9*b*, whereas the case where a small scatter of test results permits the use of a narrow size range is shown in Fig. 6-9*c*. Obviously, the necessary size range can be reduced by carefully controlled testing that results in a low scatter.

It is noted that since the value of G_f obtained corresponds to the fracture energy dissipated in an infinitely large specimen, it is independent of specimen size. However, since the value of G_f is obtained from an extrapolation of the peak loads of limited numbers of finite-size specimens by statistical regression, a slight error in the calculation of the slope A_B can significantly alter the value of G_f. Therefore, extreme care should be exercised in determining the slope A_B of the regression line.[25] On the other hand, to obtain an accurate estimate of fracture toughness according to the size effect model, it would seem essential that the test specimen dimensions meet all the size requirements listed in the proposal. It may not always be easy to meet these requirements, especially if they lead to testing very large specimens, for example, when large-size coarse aggregate is used.[19] This is evident in Tables 6-8 and 6-9, where only one set of experimental results satisfies the requirement on ω_A and m. It is noted that

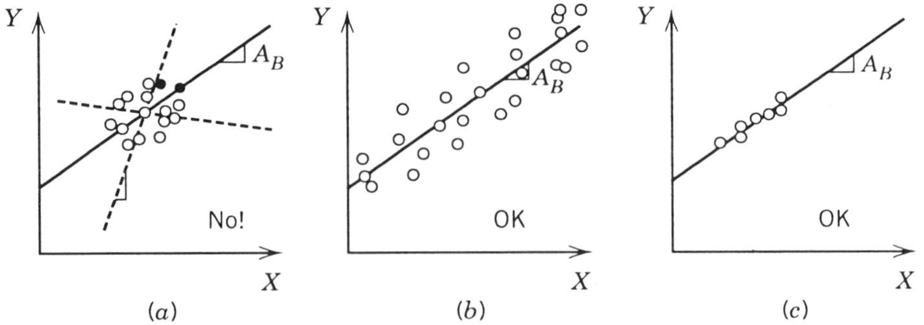

Figure 6-9 Examples to construct regression plot.

Table 6-8 Values of G_f from Three-Point Bend Beams Tested by Bazant and Pfeiffer[a]

Series Number[b]	Dimensions (mm)			f'_c (MPa)	w/c	d_a (mm)	E (GPa)	a_0/b	A_B (mm^{-1} MPa^{-2})	G_f (N/m)	ω_A	m
	S	b	t									
1	95	38	38	34.1	0.60	13	27.7	0.167	0.00598	36.60	0.067	0.144
2	191	76	38	34.1	0.60	13	27.7	0.167	0.00598	36.60	0.067	0.144
3	381	152	38	34.1	0.60	13	27.7	0.167	0.00598	36.60	0.067	0.144
4	762	305	38	34.1	0.60	13	27.7	0.167	0.00598	36.60	0.067	0.144
5	95	38	38	47.6	0.50	5	32.9	0.167	0.00873	23.67	0.031	0.048
6	191	76	38	47.6	0.50	5	32.9	0.167	0.00873	23.67	0.031	0.048
7	381	152	38	47.6	0.50	5	32.9	0.167	0.00873	23.67	0.031	0.048
8	762	305	38	47.6	0.50	5	32.9	0.167	0.00873	23.67	0.031	0.048

Source: From ref. 27.

[a] For a valid test, the value of ω_A should not exceed about 0.10 and m should be about 0.20 according to the RILEM proposal.

[b] Three beams were tested for each series.

Table 6-9 Values of G_f from Three-Point Bend Beams Tested by Alexander[a]

Series Number[b]	Dimensions (mm)			f'_c (MPa)	w/c	d_a (mm)	E (GPa)	a_0/b	A_B (mm^{-1} MPa^{-2})	G_f (N/m)	ω_A	m
	S	b	t									
1 (3)	400	100	100	29.0	0.53	19	32.5	0.40	0.02700	71.06	0.537	0.585
2 (2)	800	200	100	29.0	0.53	19	32.5	0.40	0.02700	71.06	0.537	0.585
3 (2)	2000	500	100	29.0	0.53	19	32.5	0.40	0.02700	71.06	0.537	0.585
4 (8)	400	100	100	26.3	0.53	19	32.0	0.20	0.00709	94.06	0.331	0.435
5 (2)	800	200	100	26.3	0.53	19	32.0	0.20	0.00709	94.06	0.331	0.435
6 (2)	1200	300	100	26.3	0.53	19	32.0	0.20	0.00709	94.06	0.331	0.435
7 (1)	2000	500	100	26.3	0.53	19	32.0	0.20	0.00709	94.06	0.331	0.435
8 (1)	3200	800	100	26.3	0.53	19	32.0	0.20	0.00709	94.06	0.331	0.435

Source: From ref. 19.

[a] For a valid test, the value of ω_A should not exceed about 0.10 and m should be about 0.20 according to the RILEM proposal.
[b] Numbers in parentheses indicate number of beams tested.

the great values of ω_A and m for the experimental results given in Table 6-9 lead to unrealistic G_f values of 71.06 and 94.06 N/m for the normal-strength concrete.

The value of c_f can be obtained from Eq. (6.24). For two sets of specimens listed in Table 6-8, Bazant and Pfeiffer[27] have reported c_f = 13.5 mm for a concrete with a maximum aggregate size of 13 mm and c_f = 1.90 mm for a mortar with a maximum aggregate size of 5 mm.

6.4.3 Example[27]

The mix proportion of the mortar used to measure fracture toughness G_f was 1 : 2 : 2 of cement–sand–aggregate ratio by weight. Crushed limestone and siliceous river sand were used. The maximum aggregate size was 12.7 mm (0.5 in.), whereas the maximum sand size was 4.83 mm (0.19 in.). The water-to-cement ratio used was 0.6. The specimens were cured for 27 days until testing in a moist room of 95% RH and at temperature of 25.6°C (78°F). A notch of 25 mm (1 in.) was sawn by a diamond saw at one day before testing. The three-point bend beams used had $S/b = 2.5$, $t = 38$ mm (1.5 in.), and $a_0 = b/6$. Four different sizes of beams with b values of 38 mm (1.5 in.), 76 mm (3 in.), 152 mm (6 in.), and 304 mm (12 in.) were tested. Three identical specimens were tested for each size. Although a closed-loop testing machine is not required, all specimens were tested using a servo-controlled closed-loop testing machine. The loading rate was chosen in such a way that the maximum load was reached in about 5 min.

The maximum loads obtained for different beams are listed in Table 6-10, and the values of X_j and Y_j for different specimens calculated using Eq. (6.15) are given in Table 6-11. The values of E = 27.7 GPa and ω_E = 0.02 were obtained from separate testing. By applying a linear regression to the values of X_j and Y_j as shown in Fig. 6-10, regression constants for $Y = A_B X + C_B$ were obtained as

$$A_B = 0.00598 \text{ mm}^{-1} \text{ MPa}^{-2} \quad \text{and} \quad C_B = 0.465 \text{ MPa}^{-2}$$

Based on Eqs. (6.24) and (6.25), (6.26) and (6.37), and (6.27), regression coefficients were obtained as

Table 6-10 Maximum Loads Measured

Depth, b (mm)	Maximum Loads, P_j^0 (N)			
	Beam 1	Beam 2	Beam 3	Mean
38	1800	1810	1850	1820
76	3010	3140	3160	3103
152	4400	4630	4880	4637
305	7730	7740	7890	7787

Table 6-11 Data for Regression

$X_j = b$ (mm)	$Y_j = (bt/P_j^0)^2$ (MPa^{-2}) Beam 1	Beam 2	Beam 3	Mean
38	0.644	0.636	0.609	0.630
76	0.921	0.846	0.835	0.864
152	1.723	1.556	1.401	1.560
305	2.248	2.242	2.158	2.216

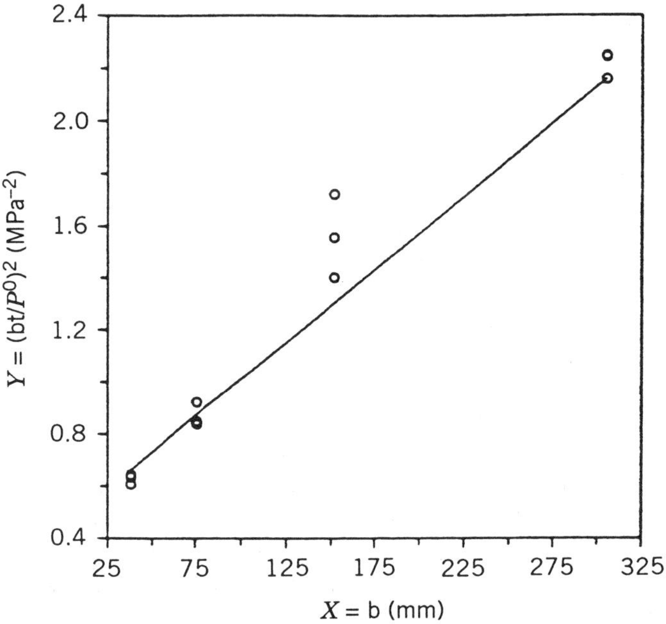

Figure 6-10 Linear regression plot constructed from maximum loads measured.[27]

$$s_{X|Y} = 0.142 \text{ MPa}^{-2} \qquad s_X = 118.0 \text{ mm} \qquad \omega_{X|Y} = 0.108 \qquad \omega_X = 0.8526$$

The value of $g(\alpha_0) = 6.07$ can be estimated for the given geometry.

The values of G_f, ω_A, ω_C, and m are determined from Eqs. (6.23) and (6.28), (6.41), and (6.46) as

$$G_f = \frac{6.07}{(27.7 \times 10^3 \text{ MPa})\ (5.98 \times 10^{-3} \text{ mm}^{-1} \text{ MPa}^{-2})} = 36.6 \text{ N/m}$$

$$\omega_A = 0.0672 \qquad \omega_C = 0.154 \qquad m = 0.144$$

Since $\omega_A < 0.10$ and $m < 0.20$, the value of G_f obtained is valid.

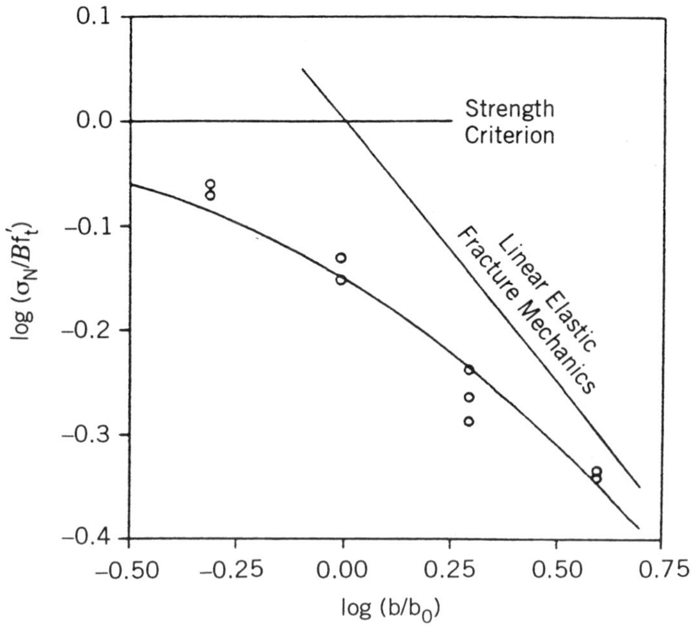

Figure 6-11 Size effect plot constructed from maximum loads measured ($\sigma_N = P^0/bt$, $b_0 =$ 77.8 mm, and $Bf_t' = 1.47$ MPa).[27]

The corresponding size effect on nominal strength of the beams tested is shown in Fig. 6-11. It is seen that the nominal strength of the beams decreases and approaches the limit of LEFM with increasing beam size.

6.5 SOME COMMENTS ON RILEM METHODS

Three RILEM draft methods based on the fictitious crack model by Hillerborg et al.,[2,3] the two-parameter fracture model by Jenq and Shah,[17] and the size effect model by Bazant et al.,[27,28] respectively, have been presented. All three draft recommendations are used to measure values of fracture toughness, which are represented by G_F, $G_{\text{I}c}^s$, and G_f, respectively. Here, G_f is really based on the size effect model. However, $G_{\text{I}c}^s$ is obtained indirectly from $K_{\text{I}c}^s$ and E as $G_{\text{I}c}^s = (K_{\text{I}c}^s)^2/E$. Both $G_{\text{I}c}^s$ and G_f are based on the effective-elastic crack approach, and it has been proven that both values of $G_{\text{I}c}^s$ and G_f are comparable.[19] On the other hand, G_F is based on the fictitious crack approach. It was reported that the value of G_F was approximately twice as great as $G_{\text{I}c}^s$ (or G_f).[29] This may be due to fact that G_F is a parameter based on a global load and load point displacement curve. According to the RILEM draft recommendation for measuring G_F, not only the true surface separation energy but also possible spurious energy and energy dissipated outside the fracture process zone are used to calculate G_F. In addition to this, energy required to form the fracture process zone is also included. As a result, the value of G_F thus obtained is certainly higher than the true fracture energy. Furthermore, the value of G_F is the fracture energy averaged by the whole ligament area at a crack front, which includes both areas within and outside the fracture process zone. But G_{Ic}^s (or G_f) is based on the effective-elastic crack approach and represents the fracture energy dissipated on a unit crack area in the fracture process zone. The value of G_{Ic}^s (or G_f) is determined based on the applied critical load and the critical effective-elastic crack length, and the spurious energy and the energy dissipated outside the fracture process zone are not included when calculating the value of G_{Ic}^s (or G_f). As discussed in Chapter 5, both the fictitious crack approach and the effective-elastic crack approach are based on experimental observations. Although the three proposed RILEM draft recommendations rely on different fracture energy dissipation mechanisms, they do provide some common grounds to evaluate fracture parameters. There are certainly no conflicts between the three draft recommendations. Current research is directed toward developing a unified approach for the fracture of concrete.

6.6 OTHER METHODS FOR MEASURING FRACTURE TOUGHNESS OF CONCRETE

6.6.1 Karihaloo and Nallathambi Method for K_{Ic}^e

This method[30-32] is also based on the concept of the effective-elastic crack. Through finite element analysis and other numerical modeling, expressions have been developed to compute E and K_{Ic}^e from beam specimens in three-point bending with specific geometries. The loading apparatus and testing arrangement are shown in Fig. 6-12. The measured displacement is

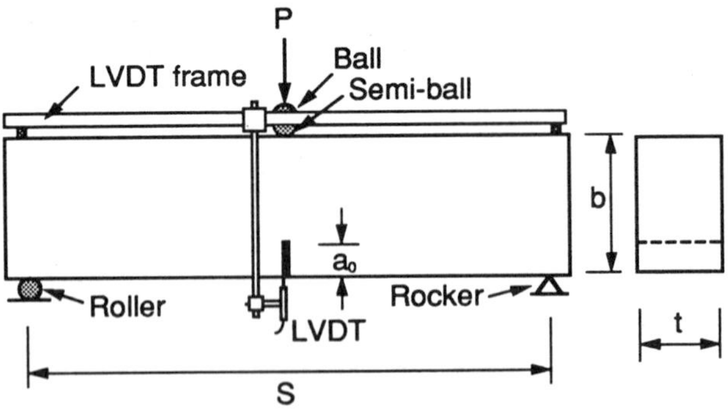

Figure 6-12 Loading apparatus and fixing arrangement for LVDT.[31]

at the load point and this testing arrangement corresponds to measuring δ_3 in Fig. 6-3.

6.6.1.1 *Test Specimen*

The following specimen dimension ranges are suggested. No critical linear dimension should be less than five times the maximum aggregate size d_a used, which may be in the range of 5–25 mm. For instance, the depth b of the specimen for 20 mm maximum aggregate size mix should be such that the uncracked ligament size is at least 100 mm. The width t may be between 40 and 100 mm, and the loading span–depth ratio S/b may be between 4 and 8. The notch–depth ratio a_0/b may be between 0.2 and 0.6 but preferably is either 0.3 or 0.4.

At least four specimens are recommended for each type of material tested. Sawcut notch or precast notch (using a well-greased wedge) of width not greater than 3 mm and milled to a sharp edge may be used.

After casting, specimens should be covered with wet burlap or kept in a curing room with 100% RH and 23 $\pm$ 2°C temperature for the first 24 h. On the second day all the specimens should be stripped from their molds (and any wedges used for precast notches gently removed) and transferred to the curing room until about an hour before testing. The specimens should be measured before testing.

6.6.1.2 *Requirements of Test Setup*

Any relatively stiff machine may be used to load the specimens incrementally while simultaneously measuring the deflection adjacent to the notch. While measuring the deflection at the midspan, care should be exercised to avoid any support disturbances. To achieve this, an additional fixture may be necessary to attach the measuring device, such as a transducer to the specimen, as shown in Fig. 6-12. Plot load values against midspan deflection values up to the peak load. Alternatively, if a more sophisticated machine is available allowing a closed-loop setup, then the CMOD (measured with a clip gauge) or the load point displacement (measured with a LVDT) may be used as a feedback signal to achieve a stable failure. It is, however, emphasized that the present method does not require any load–deflection information past the peak load. The load should be incremented so that the peak load is attained in 1–10 min.

6.6.1.3 *Calculation of Material Properties*

The modulus of elasticity E of the mix may be determined in accordance with ACI Code requirements. It is however preferable to measure strains using two electrical strain gauge lengths of 50 mm glued to diametrically opposite sides at midheight of a standard cylinder with a diameter of 150 mm. Note that the determination of E using the cylindrical specimen is required only when a continuous record of loads and deflections up to the maximum load cannot be kept. However, if a continuous load–deflection record is available, record (P_i, δ_i) corresponding to the proportional limit. Then, the modulus of elasticity may be calculated from

$$E = \frac{P_i}{4t\delta_i}\left(\frac{S}{b}\right)^3\left[1 + \frac{5SW_L}{8P_i} + \left(\frac{b}{S}\right)^2\left(2.7 + 1.35\,\frac{SW_L}{P_i}\right) - 0.84\left(\frac{b}{S}\right)^3\right]$$
$$+ \frac{9P_i}{2t\delta_i}\left(1 + \frac{SW_L}{2P_i}\right)\left(\frac{S}{b}\right)^2 g_4\left(\frac{a_0}{b}\right) \tag{6.30}$$

where P_i is the load corresponding to the deflection; E is the modulus of elasticity; S, b, and t are the span, depth, and thickness of the beam, respectively; and W_L is the self-weight of the beam per unit length. The function $g_4(a_0/b)$ is given by

$$g_4\left(\frac{a_0}{b}\right) = \int_0^{a_0} \frac{\pi a}{b^2}\, g_1^2\left(\frac{a}{b}\right) da \tag{6.31}$$

where the function $g_1(a/b)$ is given in Eq. (6.12) but a_c should be replaced by a. A slightly less accurate expression (error $\leq 1\%$) for $g_1(a/b)$ is available in the range $0.1 \leq a/b \leq 0.6$ for two span-to-depth ratios ($S/b = 4, 8$) with linear interpolation permitted with and outside of these ratios,

$$g_1\left(\frac{a}{b}\right) = \frac{1}{\sqrt{\pi}}\left[B_0 + B_1\left(\frac{a}{b}\right) + B_2\left(\frac{a}{b}\right)^2 + B_3\left(\frac{a}{b}\right)^3 + B_4\left(\frac{a}{b}\right)^4\right] \tag{6.32}$$

where

$$\begin{aligned} B_0 &= +0.0075\,\frac{S}{b} + 1.90 \\ B_1 &= +0.0800\,\frac{S}{b} - 3.39 \\ B_2 &= -0.2175\,\frac{S}{b} + 15.40 \\ B_3 &= +0.2825\,\frac{S}{b} - 26.24 \\ B_4 &= -0.1450\,\frac{S}{b} + 26.38 \end{aligned} \tag{6.33}$$

It is assumed that the critical notch depth a_c may be calculated by introducing a fictitious beam containing a notch a_c, whose stiffness (proportional to E) would be equal to the reduced stiffness of the real beam containing a notch of depth a_0, that is,

$$\delta_c = \frac{P_c}{4tE}\left(\frac{S}{b}\right)^3\left[1 + \frac{5SW_L}{8P_c} + \left(\frac{b}{S}\right)^2\left(2.7 + 1.35\frac{SW_L}{P_c}\right) - 0.84\left(\frac{b}{S}\right)^3\right]$$
$$+ \frac{9P_c}{2tE}\left(1 + \frac{SW_L}{2P_c}\right)\left(\frac{S}{b}\right)^2 g_4\left(\frac{a_c}{b}\right) \tag{6.34}$$

where P_c and δ_c are the peak load and the corresponding load point displacement and the function $g_4(a_c/b)$ is given by

$$g_4\left(\frac{a_c}{b}\right) = \int_0^{a_c} \frac{\pi a}{b^2} g_1^2\left(\frac{a}{b}\right) da \tag{6.35}$$

From Eq. (6.34) the value of a_c is calculated by a trial-and-error procedure as follows. For a given notched beam, the initial modulus of elasticity E is calculated from Eq. (6.30). Using a_0 of this beam as a starting value, $g_4(a_c/b)$ is calculated from Eq. (6.35) and substituted into Eq. (6.34) together with the corresponding P_c and δ_c to calculate a fresh value of E, say E^* (in the first iteration, E^* will naturally be less than E). The notch depth is incremented by $\Delta a = 0.001b$, and the procedure is repeated until $E^* = E \pm \epsilon\ (= 0.5\%)$. The value of a at this final step is taken equal to a_c.

After applying the above procedure to a large number of the beams tested, Karihaloo and Nallathambi[30] proposed a regression equation to estimate the values of a_c for beams with different sizes and mix proportions:

$$\frac{a_c}{b} = \gamma_1\left(\frac{\sigma_c}{E}\right)^{\gamma_2}\left(\frac{a_0}{b}\right)^{\gamma_3}\left(1 + \frac{d_a}{b}\right)^{\gamma_4} \tag{6.36}$$

where the nominal strength $\sigma_c = 6M_c/(b^2t)$, $M_c = (P_c + SW_L/2)S/4$, d_a is the maximum size of aggregate, and constants $\gamma_1 = 0.088$, $\gamma_2 = -0.208$, $\gamma_3 = 0.451$, $\gamma_4 = 1.653$.

This procedure, although somewhat complex, may be easily programmed for implementation by computer to obtain the value of a_c. Then, the critical stress intensity factor may be obtained from

$$K_{Ic}^e = \sigma_c\sqrt{\pi a_c}\, g_1\left(\frac{a_c}{b}\right) \tag{6.37}$$

where $\sigma_c = 3P_{c1}S/(2b^2t)$ is the nominal strength and $P_{c1} = P_c + SW_L/2$.

Karihaloo and Nallathambi[31] have shown that values of K_{Ic}^e from this method are close to those of K_{Ic}^s from the two-parameter fracture model.

6.6.2 Swartz and Refai Method for K_{Ic}^{e}

As described in the previous chapter, this test method uses precracked beams for which the fracture process zone is fully developed. The effective crack length for each beam is considered to be the average length of the fracture surface shown by dye. Based on test results[33] the following limits are recommended: crack length–depth ratio ≤ 0.65 and beam depth ≥ 203 mm (8 in.). Testing requires at least 12 replicate specimens that are used to establish a calibration curve of crack length versus maximum load covering a range of crack–depth ratio from 0.2 to 0.8. All specimens are precracked and dye inserted according to the following procedure.[34]

The test specimen is shown in Fig. 4-11 in which it is seen that the beam is loaded upward and a starter notch of depth $a_0 = 0.1b$ is made at the top. This arrangement is necessary to allow the dye to flow into the beam, but it also is convenient for testing large beams in order to prevent self-weight effects from influencing the stability of the cracking process.

After the beam is placed in the testing machine the procedure for precracking is the following:

1. Using a ramping function and strain control, the beam is loaded beyond peak load into the softening regime. Notice the plotted curve $Y = 2.5$ in Fig. 6-13*a*. The load is removed at a level about one-half the peak load and the process restarted with $Y = 1.0$ (scale change only for convenience). It is seen that the initial slope of the restarted plot ($Y = 1.0$) changed due to crack closure effects. The load is again removed at about $P_c/2$ on the softening branch of the

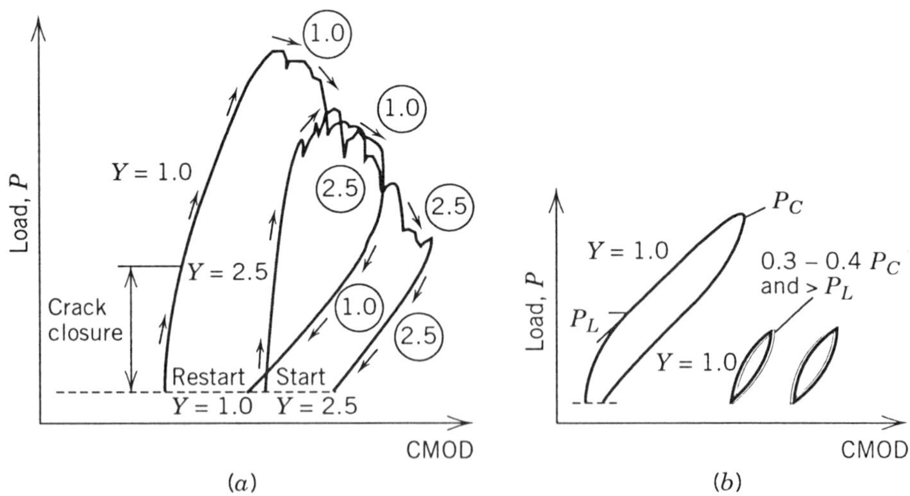

Figure 6-13 Specimen precracking and dye application: (*a*) precracking process, dye not applied, and (*b*) load cycling, with dye applied.[33]

second cycle (or about one-fourth of the original maximum load). At what point to remove the load depends on what crack length is desired.

2. Referring to the left curve in Fig. 6-13*b*, a load–displacement plot is made for reference purposes. The load is then removed.

3. The dye is inserted. Then the beam is load cycled, as shown to the right in Fig. 6-13*b*, between zero and P_p (either strain control or load control) such that $P_L \leq P_p \leq$ about $P_c/3$, where P_p is the peak load and P_L is the smallest load at which the crack is completely open. Thus, the crack is opened to allow dye penetration, but it does not grow during this cycling process. This procedure is done slowly and takes about 30 min.

4. The dye is removed, the plates are removed and the beam is dried. The beam is now precracked; the surface after failure will be shown by the dye. The earlier load–displacement curves are not used subsequently because of possible stiffening effects from the reservoir. The specimen is now ready for testing to determine fracture parameters.

5. The beam is tested to failure and continuous plots of P–CMOD and P–LPD (load point displacement) made. These are not required but may be used with other methods.

6. After failing the beam, the dyed surface area is measured. The initial, equivalent crack length is

$$a_i = \frac{\text{area of dyed surface}}{t} \tag{6.38}$$

7. This process is repeated for all beams to obtain a curve of a_i versus P_c. Such a curve was presented in Fig. 5-19 for 203-mm- (8-in.-) deep beams. For this curve it is convenient, although not required, to obtain a fitting equation. This will, of course, vary with beam sizes and mix designs. As given for the beams in Fig. 5-19,[33] a fitting formula is

$$\frac{a_i}{b} = 1.00 - 0.1566P_c + 0.0061103P_c^2 \tag{6.39}$$

where P_c is the maximum load in units of kilonewtons. This curve is used to obtain a_c by replacing P_c by $0.95P_c$ in Eq. (6.39).

8. The value of K_{Ic}^e from an LEFM formula using P_c and a_c is calculated. As an example, in reference,[33] the fitting formula obtained for 305-mm- (12-in.-) deep beam was

$$\frac{a_i}{b} = 0.90 - 0.09393P_c + 0.002459P_c^2 \tag{6.40}$$

For these beams, $S/b = 3.75$ and an LEFM formula for the critical stress intensity factor can be calculated from[35]

$$K^e_{\mathrm{I}c} = \frac{P_c S}{4tb^{3/2}} [-0.065B_c^2 - 3.483B_c - 0.120 + 5.706B_c^{-1} + 0.166B_c^{-2}] \tag{6.41}$$

where $B_c = 1 - a_c/b$ and a_c is obtained from Eq. (6.40) using $0.95P_c$. By combining Eqs. (6.40) and (6.41) one obtains the curve in Fig. 6-14*a*. This process was followed to obtain the curve in Fig. 6-14*b* for 203-mm- (8-in.-) deep beams.

It is seen in Fig. 6-14 that values of $K^e_{\mathrm{I}c}$ from this method are comparable with $K^s_{\mathrm{I}c}$ from the two-parameter fracture model and $K^e_{\mathrm{I}c}$ from the Karihaloo and Nallathambi[30–32] method. Values of the critical stress intensity factor obtained from these three models are generally comparable. However, the two-parameter fracture model predicts slightly higher values of the critical stress intensity factor, whereas the effective crack model by Karihaloo and Nallathambi predicts somehow lower values of the critical stress intensity factor. It is noted that since some regression equations were introduced in both the Karihaloo and Nallathambi method, and the Swartz and Refai[33,34] method, it is convenient to obtain the critical stress intensity factor using both methods. However, since the regression equations are valid only for three-point bend beams with certain span-to-depth ratios, both methods cannot apply to specimens of other geometries. In contrast, the two RILEM methods previously presented, the two-parameter fracture model and the size effect model, may in principle be extended to specimens of other geometries instead of the three-point bend beam.

6.6.3 Li et al. Method for Stress–Separation Curve of Material

As discussed previously, the cohesive crack approach requires a unique stress–separation curve $\sigma(w)$ that is determined by the fracture toughness G_F, the tensile strength f_t, the critical crack separation displacement w_c, and an assumed functional of $\sigma(w)$. The first RILEM draft recommendation[1] only proposed a method to measure the value of G_F for a given material. Therefore, unique determinations of the values of f_t and w_c are crucial for implementing the fictitious crack approach. Li et al.[36] have proposed an experimental method capable of determining the entire softening stress–separation curve $\sigma(w)$. Their method will be presented below.

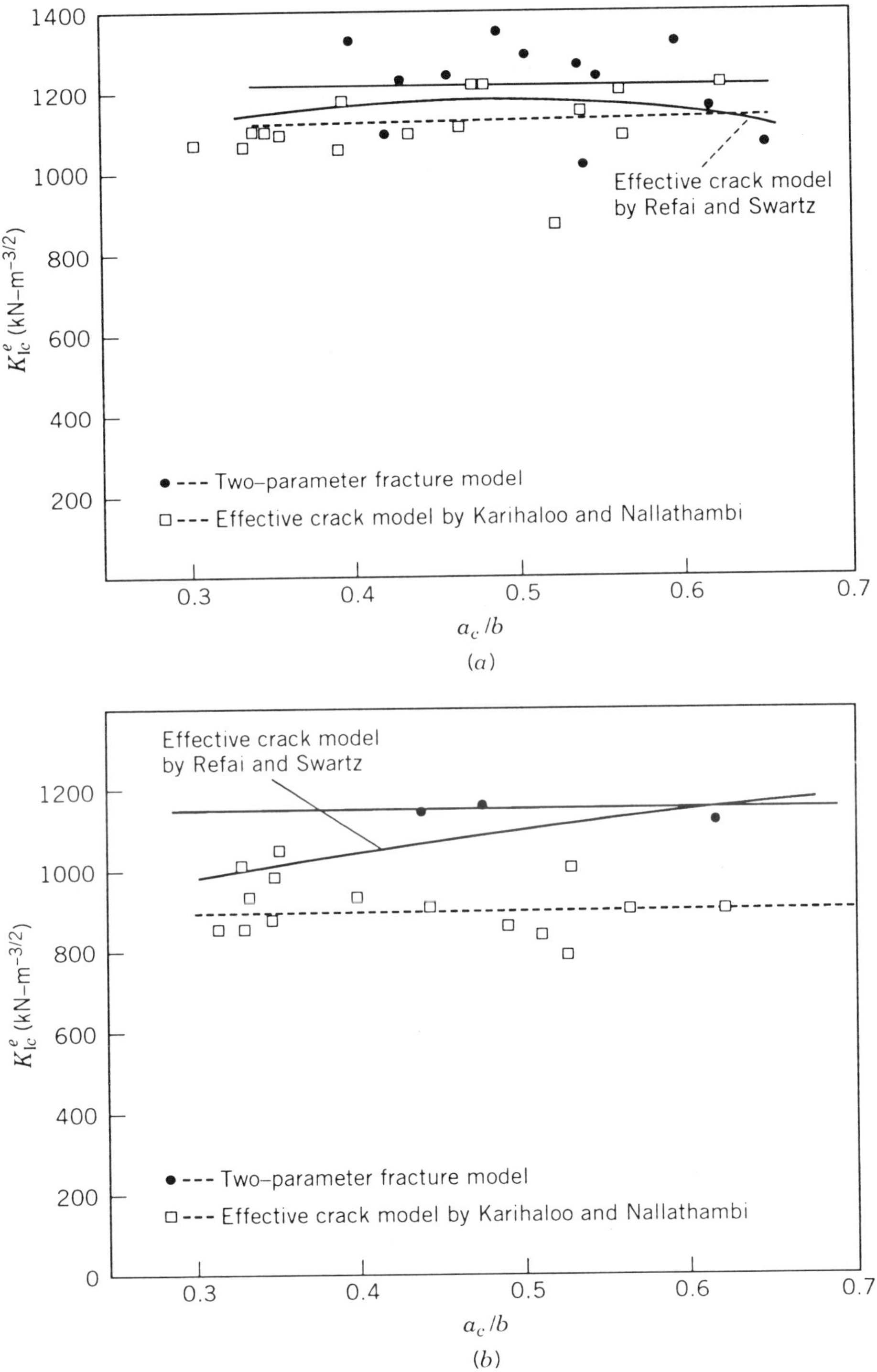

Figure 6-14 Critical stress intensity factors based on different effective crack models: beam series with (*a*) b = 305 mm and (*b*) b = 203 mm.[33]

Based on the energy principles discussed in Chapter 3, the energy release rate G_q can be expressed as a function of the strain energy $U(a, u)$ as

$$G_q = -\frac{1}{t}\left[\frac{\partial U(a, u)}{\partial a}\right]_u \tag{6.42}$$

where t is the thickness of the specimen and u is the load point displacement. It is noted that subscript u indicates that displacement keeps constant during infinitesimal crack extension. On the other hand, according to the fictitious crack approach, in which only the Dugdale–Barenblatt energy dissipation mechanism is considered, the energy release rate can be written as

$$G_q = \int_0^w \sigma(w^*)\, dw^* \tag{6.43}$$

Since a change of the strain energy may be obtained from an increment of area under a global load–displacement curve, Eq. (6.42) can be rewritten as

$$G_q = -\frac{1}{t}\int_0^u \left[\frac{\partial P(a, u^*)}{\partial a}\right]_{u^*} du^* \tag{6.44}$$

where P is the applied load and is a function of the crack length a and the displacement u. Based on Eq. (6.44) the energy release rate G_q can be determined from the load–displacement curve. Since it is difficult to measure accurately crack extension at each load increment, Eq. (6.44) cannot be directly used to evaluate the value of G_q. An approximate procedure using two identical specimens with different notch lengths a_1 and a_2, as shown in Fig. 6-15, was introduced by Li et al.[36] Based on their procedure, if the difference $a_2 - a_1$ is small, Eq. (6.44) may be approximated by

$$G_q = \frac{1}{t(a_2 - a_1)}\int_0^u [P_2(u^*) - P_1(u^*)]\, du^* = \frac{\Delta A}{t(a_2 - a_1)} \tag{6.45}$$

The integral in Eq. (6.45) indicates difference in areas of load–displacement curves of two specimens with notch lengths a_1 and a_2, respectively, and is represented by ΔA here. As a result, the stress–separation curve $\sigma(w)$ can be determined by comparing Eqs. (6.44) and (6.45). This leads to

$$\sigma(w) = \frac{\partial G_q}{\partial w} = \frac{1}{t(a_2 - a_1)}\frac{\partial[\Delta A(u)]}{\partial w} \tag{6.46}$$

Equation (6.46) can be approximately used to determine the stress–separation

Figure 6-15 Two identical specimens with different initial notches: (*a*) shorter notch specimen and (*b*) longer notch specimen.

curve $\sigma(w)$. It is noted that ΔA is expressed as a function of the load point displacement u rather than the crack tip separation w in Eq. (6.46). Therefore, w should be expressed as a function of u before Eq. (6.46) can be implemented. Such a relationship may be obtained by recording both w and u during loading. Thus the experimental outputs required to obtain the $\sigma(w)$ curve are a set of P–u and P–w relationships for each type of specimen with different notch lengths. The value of $a_2 - a_1$ should be as small as possible. However, this value depends on experimental accuracy and material variability.

To demonstrate the validity of their approach, Li et al.[36] have presented a result using compact tension specimens. The geometry of the compact tension specimen tested is shown in Fig. 6-15. Two specimens with initial notches of $a_1 = 35.56$ mm and $a_2 = 43.18$ mm are tested. The measured load and load point displacement curves for the two specimens are shown in Fig. 6-16*a*. The measured w–u curves for the two specimens are indicated in Fig. 6-16*b*, and the average of two w–u relationships was used to determine $\sigma(w)$. The $\Delta A(u)$

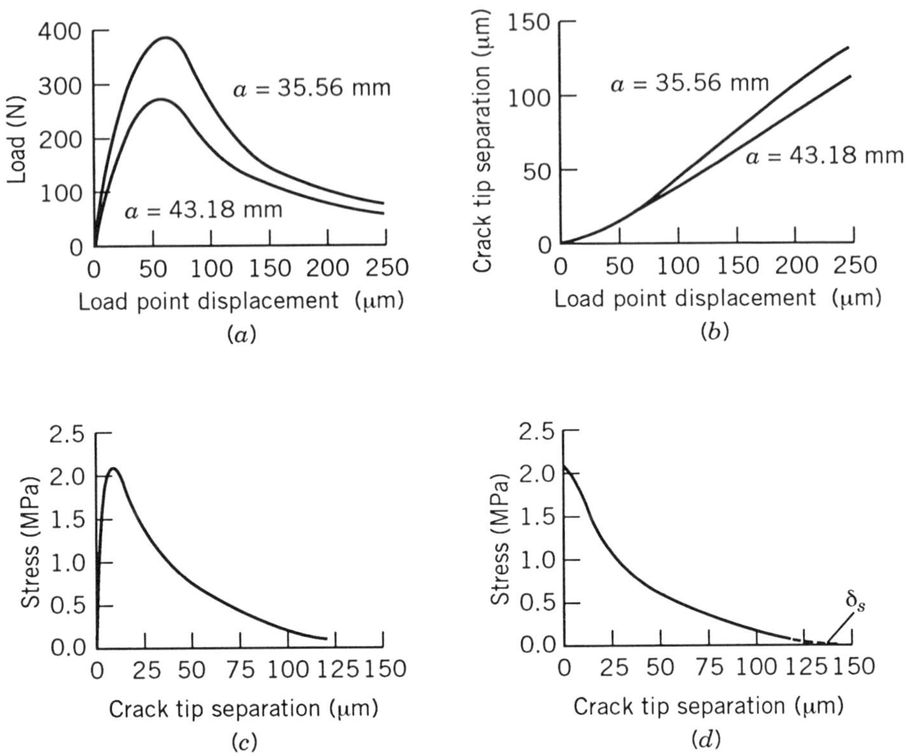

Figure 6-16 Measurement of stress and crack tip separation curve from two identical specimens with different notches: (*a*) load–load point displacement, (*b*) crack tip separation–load point displacement, (*c*) deduced tension softening, and (*d*) stress–crack tip separation.[36]

relationship can be obtained by subtracting the two P–u curves given in Fig. 6-16*a*. The stress–separation curve can then be numerically deduced by substituting the $\Delta A(u)$ relationship and the average w–u curve shown in Fig. 6-16*b* into Eq. (6.46). The deduced stress–separation curve, which includes an ascending part and a descending part, is shown in Fig. 6-16*c*. It is noted that $\sigma(w)$ is only the descending part of the deduced curve, as shown in Fig. 6-16*d*.

6.6.4 Brühwiler et al. Method Using Wedge Splitting

As discussed earlier, measurements of material fracture parameters using three-point bend beams have been proposed by RILEM. A possible problem associated with the use of the three-point bend beam is that when the size of the beam tested is relatively large, the effect of self-weight of the beam on material fracture parameters should be very carefully evaluated. Additionally, although the three-point bend beam can easily be handled in laboratories, it may not lend itself either to fabrication at a building site or to the use of material drilled

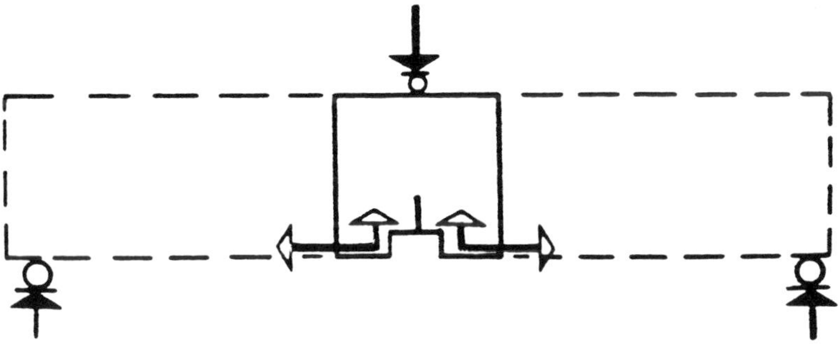

Figure 6-17 Wedge-splitting specimen as "compact" three-point bend beam.

(cored) from existing structures. To overcome these problems, an alternative method using a wedge-splitting test, which may be regarded as a "compact" three-point bend beam as shown in Fig. 6-17, has been proposed.[37,38]

The wedge-splitting specimen, previously proposed by Linsbauer and Tschegg,[39] is characterized by a groove and a starter notch that can be either molded or saw cut. Cubical and cylindrical specimens as shown in Fig. 6-18 are recommended. The principle of the wedge-splitting test is schematically presented in Fig. 6-19. The specimen is placed on a line support fixed to the lower plate of the testing machine (Fig. 6-19*a*). Two massive steel loading devices equipped with roller or needle bearings on each side are placed on the top of the specimen (Fig. 6-19*b*). A steel profile with two identical wedges is fixed at the upper plate of the testing machine. The actuator of the testing machine is moved so that the wedge enters between the bearings, which results in a horizontal splitting force component (Fig. 6-19*c*). The fracture section of the specimen is essentially subjected to a bending moment. The load in the vertical direction, F_v, and the COD are monitored during testing. The splitting force F_{sp} is the horizontal component of the force acting on the bearings and

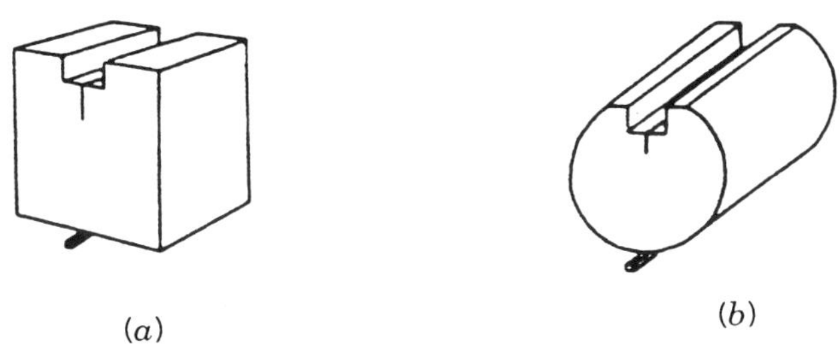

(*a*) (*b*)

Figure 6-18 Geometry of wedge-splitting specimens.

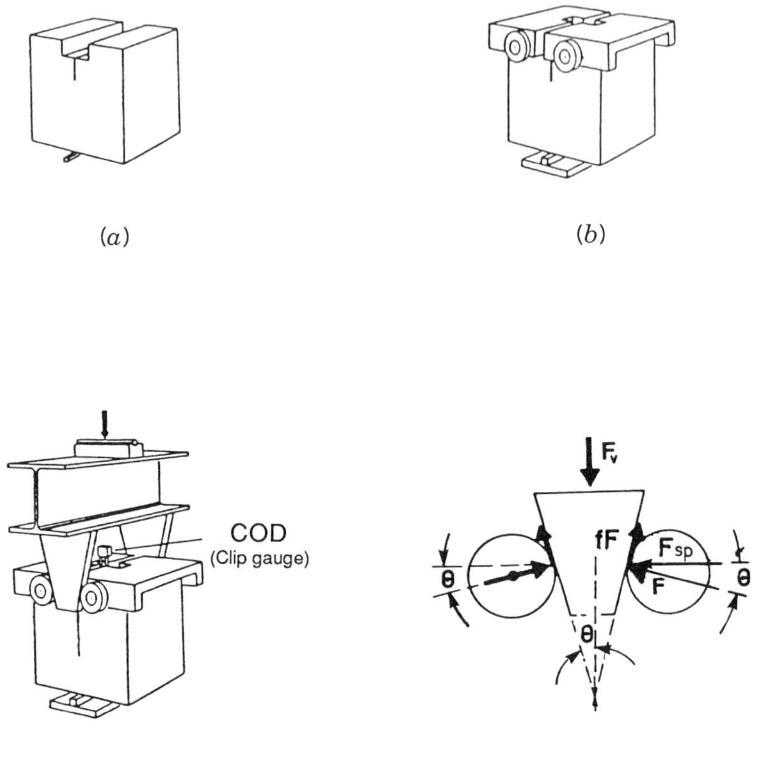

Figure 6-19 Principle of wedge-splitting test: (*a*) a test specimen on a linear support, (*b*) placing of two load devices with roller bearing, (*c*) pressing a wedge between bearings, and (*d*) forces acting on the wedge.[38]

can be determined in terms of the wedge angle and frictional forces. Equilibrium of forces on the wedge results in (Fig. 6-19*d*)

$$F_{sp} = \frac{F_v}{2 \tan \theta} \frac{1 - f \tan \theta}{1 + f \operatorname{cotan} \theta} \approx \frac{F_v}{2(1 + f \operatorname{cotan} \theta) \tan \theta} \tag{6.47}$$

where θ is the wedge angle indicated in Fig. 6-19*d* and f is the coefficient of friction ranging from 0.1 to 0.5% for the roller bearings used. In Eq. (6.47), the relationship $f \tan \theta << 1$ has been used.

Typical load–COD curves obtained from the wedge-splitting test using the cubical and cylindrical specimens, respectively, are shown in Fig. 6-20.[37] The values of G_F were determined from the area under the obtained load–COD curve. On the other hand, the values of K_{Ic}^s and $CTOD_c$ were calculated from

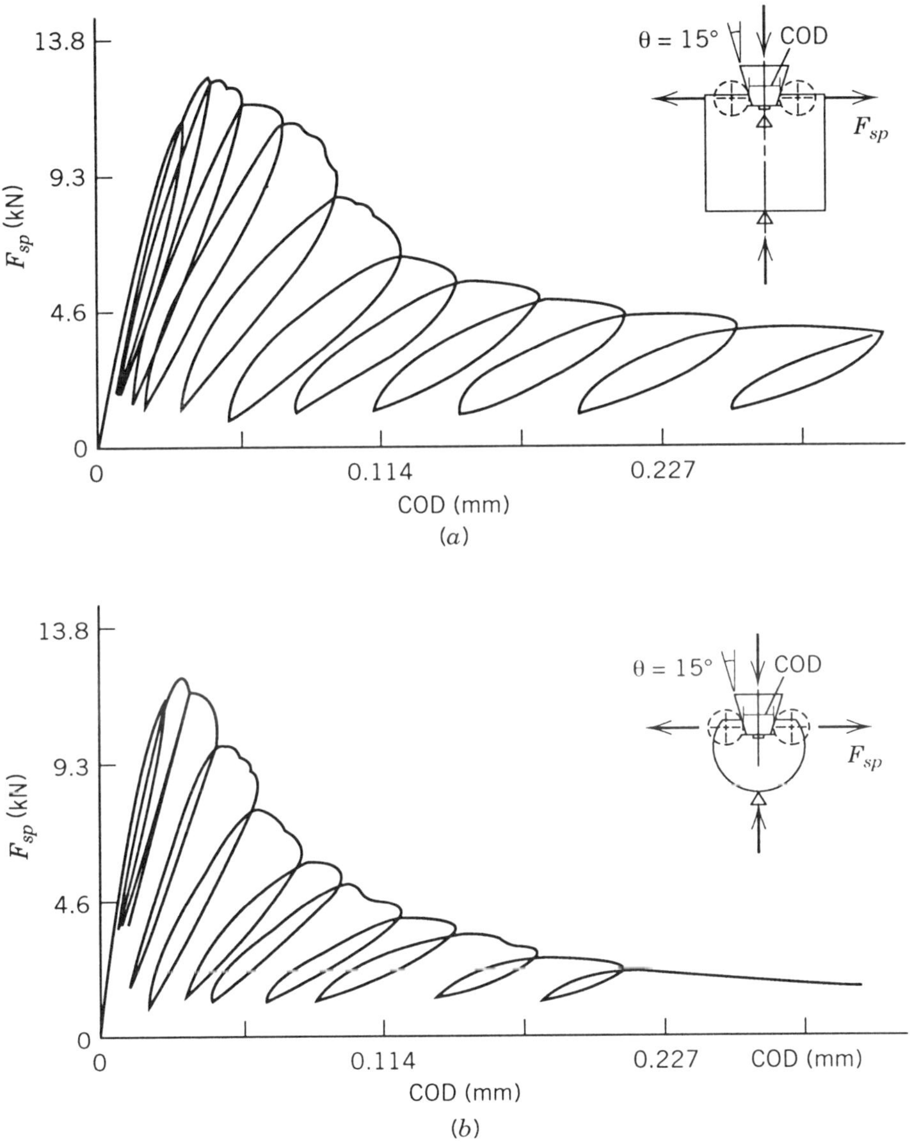

Figure 6-20 Measured splitting force and crack opening displacement curves for different specimens: (*a*) cylinder-type and (*b*) cube type geometry.[38]

the initial compliance and the unloading compliance at the peak load by the finite element method. These values of G_F, K^s_{Ic}, and $CTOD_c$ for two concrete mixes are listed in Table 6-12.[38] It is seen that values of G_F and K^s_{Ic} are basically independent of the specimen geometry. But the values of $CTOD_c$ obtained from the cylindrical specimens are greater than those obtained from the cubical specimens. This may be due to the fact that the values of $CTOD_c$ for concrete are rather small, and large errors may occur unless the specimens are carefully instrumented.

6.6.5 Tang et al. Method for K^s_{Ic} and $CTOD_c$ Based on Peak-Load Measurement

The RILEM procedure for measuring K^s_{Ic} and $CTOD_c$ requires an unloading at a load past the maximum value but not less than 95% of the peak load. To achieve a stable unloading after the peak load, a closed-loop testing system is usually required. As previously discussed, the difficulty involved in the unloading also results in some deviation in the $CTOD_c$ value. To overcome these problems, Tang et al.[24] have proposed a simple method for determining K^s_{Ic} and $CTOD_c$ using peak loads. This method is presented in the following.

As mentioned before, Eq. (5.24) can be used to determine the nominal failure stress and corresponding crack extension for any structure given the material fracture parameters K^s_{Ic} and $CTOD_c$. If one applies Eq. (5.24) to two different specimens, the following four equations are obtained:

$$K^1_I(\sigma^1_{Nc}, a^1_c) = K^s_{Ic} \tag{6.48}$$

$$CTOD^1(\sigma^1_{Nc}, a^1_c) = CTOD_c \tag{6.49}$$

$$K^2_I(\sigma^2_{Nc}, a^2_c) = K^s_{Ic} \tag{6.50}$$

$$CTOD^2(\sigma^2_{Nc}, a^2_c) = CTOD_c \tag{6.51}$$

where superscripts 1 and 2 designate two different specimens. If nominal strengths (or nominal failure stresses) σ^1_{Nc} and σ^2_{Nc} for the two different specimens (either two specimens with different sizes or two specimens with the same size but different initial crack lengths) are known, Eqs. (6.48)–(6.51) provide four simultaneous equations for the four unknowns a^1_c, a^2_c, K^s_{Ic}, and $CTOD_c$. This is the theoretical basis of the proposed peak-load method. To avoid numerical difficulties involved in solving four simultaneous nonlinear equations, an efficient procedure for determining K^s_{Ic} and $CTOD_c$, in conjunction with a^1_c and a^2_c, based on Eqs. (6.48)–(6.51) is suggested in the following.

With σ^1_{Nc} known, Eqs. (6.48) and (6.49) can be written as

$$CTOD_c = CTOD_c(a^1_c) \tag{6.52}$$

$$K^s_{Ic} = K^s_{Ic}(a^1_c) \tag{6.53}$$

Table 6-12 Values of G_F, K_{Ic}^s, and $CTOD_c$ from Wedge-Spitting Tests

Specimen Geometry	w/c	Cement Content (kg/m^3)	d_a (mm)	Age Tested (days)	Number of Beams Tested	E (GPa)	f_c (MPa)	f_t (MPa)	G_F (N/m)	K_{Ic}^s (MPa m$^{1/2}$)	$CTOD_c$ (mm)
Cubic	0.47	400	20.0	45	4	36.6	51	3.7	68.7	0.92	0.0062
	0.48	400	12.5	150	4	36.0	55	4.5	68.0	0.11	0.0081
Cylinder	0.47	400	20.0	45	5	36.6	51	3.7	87.5	1.10	0.0120
	0.48	400	12.5	150	4	36.0	55	4.5	75.9	1.37	0.0131

Source: From ref. 38.

Note: f_c = compressive strength determined from cylinders with diameter of 160 mm and length of 320 mm; f_t = tensile strength determined using Brazilian splitting test. Values of E, f_c, and f_t were determined at the age of 28 days. Values of G_F, K_{Ic}^s, and $CTOD_c$ are average of at least four specimens.

The combination of these two equations can be considered as the *parametric form* of a function with a_c^1 as the parameter. This function can be expressed as

$$\text{CTOD}_c = f^1(K_{\text{Ic}}^s) \tag{6.54}$$

for specimen 1. Although the explicit form of CTOD_c in terms of K_{Ic}^s is difficult to obtain, the relationship between CTOD_c and K_{Ic}^s for specimen 1 can be established through the parameter a_c^1 for specimen 1 [Eqs. (6.48) and (6.49)]. For a series of given values of a_c^1, corresponding values of CTOD_c and K_{Ic}^s are calculated using Eqs. (6.52) and (6.53), and the relationship between CTOD_c and K_{Ic}^s as given by Eq. (6.54) is thus determined. A similar equation can be obtained for specimen 2. If tests of two distinct specimens of a given material are conducted with "exact" measurement, the point of intersection of the two K_{Ic}^s–CTOD_c curves should be the solutions of K_{Ic}^s and CTOD_c.

Since random errors always exist in measured values of σ_{Nc}^1 and σ_{Nc}^2, more than two different specimens may be tested for better estimation of K_{Ic}^s and CTOD_c. It is suggested that at least three distinct specimens should be used. Distinct specimens can be either geometrically similar specimens with different sizes or specimens with the same size and geometry but different initial crack lengths. A procedure based on statistics is used to determine values of K_{Ic}^s and CTOD_c. When peak loads of several geometrically similar specimens are used to determine values of K_{Ic}^s and CTOD_c, the K_{Ic}^s–CTOD_c curves obtained for different specimens are in a manner as given in Fig. 6-21. In this figure, for a given value of CTOD_c there are several corresponding values of K_{Ic}^s for different specimens, and an average value of K_{Ic}^s over all specimens can be calculated. As a result, the average K_{Ic}^s–CTOD_c curves is obtained for these specimens. The *sample deviation* of values of K_{Ic}^s on all the K_{Ic}^s–CTOD_c curves at the same value of CTOD_c needs to be calculated. The sample standard deviation is defined as

$$s(\text{CTOD}_c) = \sqrt{\frac{\sum_{i=1}^{n} [(K_{\text{Ic}}^s)^{\text{ave}} - (K_{\text{Ic}}^s)^i]^2}{n-1}} \tag{6.55}$$

where n is the total number of groups of specimens ($n = 4$ in Fig. 6-21), $(K_{\text{Ic}}^s)^{\text{ave}}$ is the average value of K_{Ic}^s for all n specimens, and $(K_{\text{Ic}}^s)^i$ is the K_{Ic}^s value for the ith specimen. It is noted that the sample standard deviation can also be given in terms of K_{Ic}^s, instead of CTOD_c as in Eq. (6.55). This has been discussed in detail by Tang et al.[24]

When values of s are plotted against CTOD_c as given in Fig. 6-22, the value of CTOD_c corresponding to the minimum value of s may be taken as the ma-

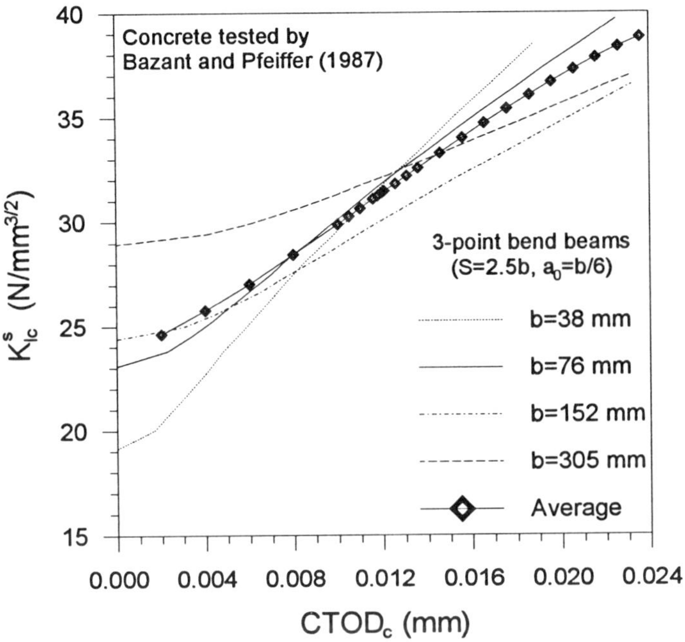

Figure 6-21 The K^s_{Ic}–CTOD$_c$ curves for concrete tested by Bazant and Pfeiffer.[24]

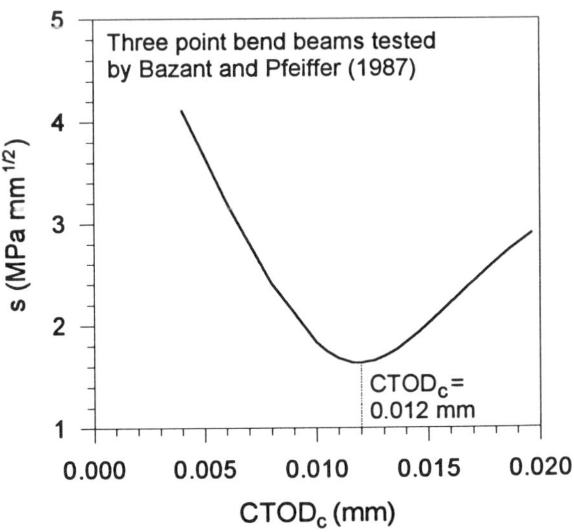

Figure 6-22 Relationship between sample standard deviation and CTOD$_c$.

terial parameter $CTOD_c$. By substituting this value of $CTOD_c$ into the $(K_{Ic}^s)^{ave}$–$CTOD_c$ curve (see Fig. 6-21), another fracture parmaeter K_{Ic}^s is determined.

Tang et al.[24] have tested a series of three-point bend beams with the same size but with different initial crack lengths. The beams tested have $S = 203$ mm, $b = 50.8$ mm, and $t = 25.4$ mm. Values of a_0 for the beams vary from 10.2 to 25.4 mm. Tang et al. reported that values of K_{Ic}^s and $CTOD_c$ determined by the proposed peak-load method are in good agreement with those obtained by the RILEM method.[16] They also found that values of K_{Ic}^s and $CTOD_c$ obtained using the three-point bend beams are close to those obtained using split-tension cylinders with different lengths of center notch.

6.7 RELATIONSHIP BETWEEN (K_{Ic}^s, $CTOD_c$) AND (G_f, c_f)

As previously discussed, both the two-parameter fracture model and the size effect model assume an equivalent elastic crack. As a result, one may expect a relationship between (K_{Ic}^s, $CTOD_c$) and (G_f, c_f). This relationship has been studied by Ouyang et al.[40] and is presented below.

6.7.1. Equivalency between (K_{Ic}^s, $CTOD_c$) and (G_f, c_f)

As indicated by the RILEM procedure[16] for determining the values of K_{Ic}^s and $CTOD_c$, the effective-elastic crack length in the two-parameter fracture model is defined by a compliance equal to the unloading compliance of the actual structure as shown in Fig. 5-11. In other words, the two-parameter fracture model determines the critical fracture state of a structure based on its elastic response. On the other hand, G_f and c_f of the size effect model are based on an infinitely large specimen. It is noted that when specimen size tends to infinity, inelastic response of the structure vanishes since the fracture process zone is negligible compared to the uncracked segment. Accordingly, an equivalency between (K_{Ic}^s, $CTOD_c$) and (G_f, c_f) can be established based on infinitely large specimens. This equivalency can be used to establish relationships between (K_{Ic}^s, $CTOD_c$) and (G_f, c_f) using LEFM concepts based on an infinite large specimen.

6.7.2 Relationship between K_{Ic}^s and G_f

Parameters K_{Ic}^s and G_f can be directly related to each other for the infinitely large specimen based on LEFM,

$$G_f = \frac{(K_{If})^2}{E'} = \frac{(K_{Ic}^s)^2}{E'} \tag{6.56}$$

where $E' = E$ for the plane-stress condition and $E' = E/(1 - \mu^2)$ for the plane-strain condition, E is the modulus of elasticity, and μ is Poisson's ratio of a material. Equation (6.56) has been confirmed by Planas and Elices[29] and Karihaloo and Nallathambi.[32]

According to the size effect model, G_f is obtained by testing a series of geometrically similar specimens for which a_0/b is constant. This implies $a_0 \rightarrow \infty$ when $b \rightarrow \infty$ (infinitely large specimen). On the other hand, K_{Ic}^s can be related to an infinitely large specimen with either an infinite or a finite value of a_0. Despite this difference in the value of a_0 when determining G_f and K_{Ic}^s, Eq. (6.56) is valid regardless of specimen geometry and initial crack length. This is because Eq. (6.56) is based on LEFM, which is applicable for quasi-brittle materials when the size of the fracture process zone is much smaller than the size of the specimen. For an infinitely large specimen, the length of the fracture process zone is always negligible compared to the specimen size regardless of the length of the initial crack and the specimen geometry.

6.7.3 Relationship between $CTOD_c$ and c_f

Based on LEFM, the COD near the crack tip can be evaluated from the stress intensity factor K_I:

$$\text{COD} = \frac{K_I}{E'}\sqrt{\frac{32r}{\pi}} \tag{6.57}$$

where r is the distance from the crack tip and a is the crack length. Equation (6.57) only includes the term with the first order of r and only can be used when $r << a$. Otherwise, terms associated with the second order of r should be considered. The dependence of COD on specimen geometry and size is implied in K_I.

Equation (6.57) can be applied to an infinitely large specimen. Since COD $= \text{CTOD}_c$ when $r = c_f = a_c - a_0$, Eq. (6.57) becomes

$$\text{CTOD}_c = \frac{K_{If}}{E'}\sqrt{\frac{32c_f}{\pi}} = \sqrt{\frac{32G_f\, c_f}{E'\pi}} \tag{6.58}$$

It is noted that Eq. (6.58) is only valid when $a_0 >> c_f$. This equation was reported by Bazant et al.[41] for an infinitely large specimen with an infinitely long initial crack (this is due to different sizes of geometrically similar specimens). Since the value of a_0 usually relates to the material defect, the length of the fracture process zone (the value of c_f) may not be very small compared to the material defect even for a very large size specimen. Therefore, a more general relationship between CTOD_c and c_f should be considered.

The relationship between CTOD_c and c_f generally depends not only on the type of the material but also on the specimen geometry. For infinitely large two-dimensional specimens, the initial crack can be located either inside the specimen or at the edge of the specimen. The former is represented by the tensile plate with a center notch, whereas the latter is represented by the tensile plate with a single-edge notch or the three-point bend beam with a single-edge notch. Crack opening displacements along the cracked surface (crack profiles) obtained from LEFM for these three types of geometries are shown in Fig. 6-23. The crack profiles given in Fig. 6-23 are not restricted to the case of $a_0 >> c_f$ and are valid for COD at both near and far from the crack tip. It is seen that regardless of loading configuration, these specimens with single-edge notch (the tensile plate with a single-edge notch and the three-point bend beam) have an almost identical crack profile. On the other hand, the tensile plate with a center notch has a crack profile distinct from the other two. Therefore, only the infinitely large three-point bend beam and tensile plate with a center notch are studied in the following. The relationship between CTOD_c and c_f depends on crack profile, as shown in Fig. 6-23. It is noted that c_f represents the length of the fracture process zone in an infinitely large specimen with either a finite or an infinite value of a_0 in the following. This is slightly different from the original definition of c_f introduced by Bazant and Kazemi,[28] where c_f is based on infinitely large specimens with an infinite value of a_0.

6.7.3.1 Tensile Specimen with Center Crack Based on LEFM, the strain energy release rate G and the COD for an infinitely large tensile plate (unit thickness) with a center crack are given by

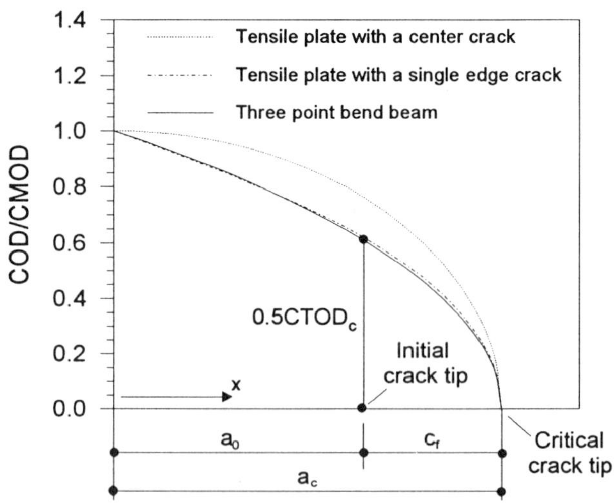

Figure 6-23 Effect of crack profile on relationship between CTOD_c and c_f.

$$G = \frac{\sigma^2 \pi a}{E} \tag{6.59}$$

and

$$\text{COD} = \frac{4\sigma}{E} \sqrt{a^2 - x^2} \tag{6.60}$$

where σ is the applied tensile stress. It is noted that for the tensile plate with a center crack Eq. (6.60) is valid for any value of a and x. For other geometries, Eq. (6.60) may be approximately used for the crack profile near the crack tip.

Since $a = a_c = a_0 + c_f$, $G = G_f$, and $(\text{COD})_{a=a_c, x=a_0} = \text{CTOD}_c$ when $\sigma = \sigma_f$, where σ_f is the critical applied stress, the following relationships are obtained:

$$G_f = \frac{\sigma_f^2 \pi (a_0 + c_f)}{E} \tag{6.61}$$

and

$$\text{CTOD}_c = \frac{4\sigma_f}{E} \sqrt{2a_0 c_f + c_f^2} \tag{6.62}$$

Combining Eqs. (6.61) and (6.62) leads to

$$\text{CTOD}_c = \sqrt{\frac{16 G_f (2a_0 + c_f) c_f}{E\pi (a_0 + c_f)}} \tag{6.63}$$

Equation (6.63) can be solved for c_f. This results in

$$c_f = \frac{E\pi \ \text{CTOD}_c^2}{32 G_f} - a_0 + \sqrt{a_0^2 + \left(\frac{E\pi \ \text{CTOD}_c^2}{32 G_f}\right)^2} \tag{6.64}$$

Equations (6.63) and (6.64) describe the relationship between CTOD$_c$ and c_f. This relationship depends on the initial crack length and is valid for COD at both near and far from the crack tip. Equation (6.58) can be directly obtained from Eq. (6.63) if only the case $a_0 >> c_f$ (the near-field COD) is considered.

6.7.3.2 Three-Point Bend Beam For an infinitely large three-point bend beam (b tends to infinity) of a unit thickness, G_f and CTOD$_c$ can be respectively expressed as

$$G_f = \frac{1.261\pi\sigma_f^2\,(a_0 + c_f)}{E} \tag{6.65}$$

and

$$\text{CTOD}_c = \text{CMOD}_c\left[1 - 0.92\left(\frac{a_0}{a_0 + c_f}\right) - 0.08\left(\frac{a_0}{a_0 + c_f}\right)^2\right]^{1/2} \tag{6.66}$$

where the nominal strength $\sigma_f = 3P_f S/(2b^2)$, P_f is the peak load, and the critical crack mouth displacement (CMOD_c) is

$$\text{CMOD}_c = \frac{5.68\sigma_f(a_0 + c_f)}{E} \tag{6.67}$$

It is noted that the crack profile given by Eq. (6.66) is a regression expression based on numerical analysis of the crack opening displacement.[17]

Substituting Eqs. (6.65) and (6.67) into Eq. (6.66) results in

$$\text{CTOD}_c = 2.854\sqrt{\frac{G_f}{E}\left(0.081a_0 + c_f - \frac{0.081a_0^2}{a_0 + c_f}\right)} \tag{6.68}$$

Equation (6.68) can be solved for c_f. This leads to

$$c_f = \frac{0.061\ \text{CTOD}_c^2\ E}{G_f} - 0.54a_0 + \sqrt{0.292a_0^2 + 0.057\frac{\text{CTOD}_c^2\ Ea_0}{G_f} + 0.0038\left(\frac{\text{CTOD}_c^2\ E}{G_f}\right)^2} \tag{6.69}$$

Equations (6.68) and (6.69) give the relationship between CTOD_c and c_f for an infinitely large three-point bend beam. This relationship also theoretically depends on the initial crack length. It is noted that when $c_f/a_0 \to 0$, Eq. (6.68) does not approach Eq. (6.58) because $(32/\pi)^{1/2} = 3.192 \neq 2.854$. This may be due to the fact that the regression relation (6.66) has a slight deviation from the exact values for very small values of COD while it describes the whole crack profile. Since the far-field COD may be several orders greater than the near-field COD, this slight deviation does not influence the generality of Eq. (6.68).

Equations (6.63) and (6.68) [or Eqs. (6.64) and (6.69)] are valid for both the near-field and the far-field crack profiles for the tensile plate with a center notch and the three-point bend beam (i.e., $a_0 >> c_f$ is not required).

6.7.4 Results and Discussions

Fracture properties of a series of geometrically similar concrete beams were reported by Bazant and Pfeiffer.[27] These three-point bend beams had the values $S = 2.5b$, $a_0 = b/6$, and $t = 38$ mm. Four different sizes of beams with b values of 38, 76, 152, and 304 mm were tested. Three identical specimens were tested for each size. Average peak loads were 1820 N for $b = 38$ mm, 3103 N for $b = 76$ mm, 4637 N for $b = 152$ mm, and 7787 N for $b = 304$ mm. The value of $E = 27{,}700$ MPa was measured. Using the size effect model, values of $G_f = 0.037$ N/mm and $c_f = 13.47$ mm were obtained.[27]

To obtain values of K_{Ic}^s and CTOD$_c$ using the aforementioned peak-load method, the obtained values of peak loads for different sizes of beams are substituted into Eqs. (6.65)–(6.67). It is assumed that infinitely large three-point bend beams with $S = 4b$ and $S = 2.5b$ have the same crack profile here. It is noted that $\sigma_f = 3P_f S/(2b^2)$ and $G_f = (K_{Ic}^s)^2/E$ here. By substituting Eq. (6.65) into Eq. (6.66), the term $(a_o + c_f)$ can be eliminated. This results in K_{Ic}^s–CTOD$_c$ curves for four different sizes of the beams as shown in Fig. 6-21. The average K_{Ic}^s–CTOD$_c$ curve is also shown in Fig. 6-21. By substituting these curves in Eq. (6.55), the sample standard deviation s is obtained as indicated in Fig. 6-22. The material fracture parameter CTOD$_c$ = 0.012 mm for the concrete is found from the minimum s value. Substituting this CTOD$_c$ value into the average K_{Ic}^s–CTOD$_c$ curve results in the other fracture parameter $K_{Ic}^s = 31.2$ N/mm$^{3/2}$ for the concrete.

These independently obtained values of (K_{Ic}^s, CTOD$_c$) and (G_f, c_f) are used to check the proposed relationships. Substituting $E = 27{,}700$ MPa and $K_{Ic}^s = 31.2$ N/mm$^{3/2}$ obtained from the two-parameter model into Eq. (6.56) leads to the value of $G_f = 0.035$ N/mm, which is close to the value of $G_f = 0.037$ N/mm predicted by the size effect model. Since the relationship between CTOD$_c$ and c_f depends on specimen geometry and initial crack length, the value of CTOD$_c$ = 0.012 mm from the two-parameter fracture model is substituted into Eqs. (6.64) and (6.69) to calculate the value of c_f. The obtained c_f values are plotted as functions of the initial crack length a_0 in Fig. 6-24 (values of a_0 ranged from 6.33 to 50.83 mm for the four different sizes of geometrically similar beams tested). Although c_f is a function of a_0 based on Eq. (6.69), the influence of a_0 is numerically negligible for the three-point bend beam. Values of c_f predicted by Eq. (6.69) using the result from the two-parameter fracture model are almost identical to the value of $c_f = 13.47$ mm determined by the size effect model. The value of $c_f = 11.12$ mm is obtained from CTOD$_c$ using Eq. (6.58). These results may explain why both (K_{Ic}^s, CTOD$_c$) and (G_f, c_f) can independently serve as material fracture parameters. These results also provide a unified approach to measure (K_{Ic}^s, CTOD$_c$) and (G_f, c_f), instead of two separate RILEM methods.[16,26]

The proposed relationships are also checked using the experimental results reported by Bazant et al.[41] for an Indiana limestone. The three-point bend beams tested had $S = 4b$, $a_0 = 0.4b$, and $t = 13$ mm. Four sizes of beams with b

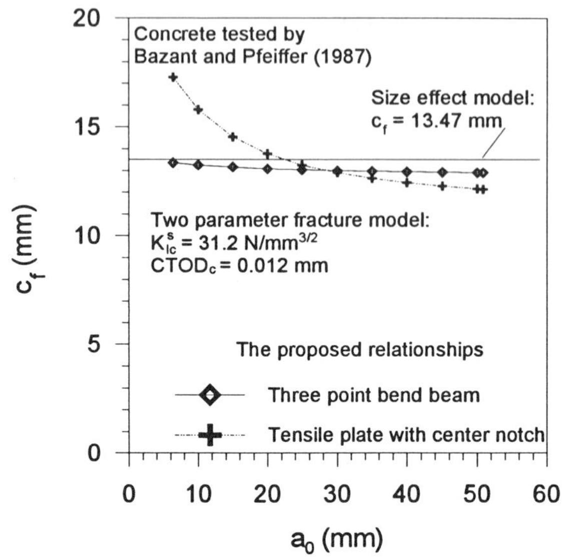

Figure 6-24 Values of c_f by proposed relationship and by size effect model for concrete.

values of 13, 25, 51, and 102 mm were tested. Three identical specimens were tested for each size. Average peak loads were 1820 N for $b = 38$ mm, 3103 N for $b = 76$ mm, 4637 N for $b = 152$ mm, and 7787 N for $b = 304$ mm. The value of $E = 15{,}300$ MPa was measured. Using the size effect model, values of $G_f = 0.061$ N/mm and $c_f = 9.3$ mm were obtained by Bazant et al.[41]

Values of $K_{\text{Ic}}^s = 27.11$ N/mm$^{3/2}$ and CTOD$_c$ = 0.0143 mm are obtained when the aforementioned procedure based on peak-load measurement is used (see Fig. 6-25). Substituting $E = 15{,}300$ MPa and $G_f = 0.061$ N/mm from the size effect model into Eq. (6.56) leads to the value of $K_{\text{Ic}}^s = 30.55$ N/mm$^{3/2}$. On the other hand, values of c_f for the three-point bend beam and for the tensile plate with a center notch were respectively predicted by Eqs. (6.69) and (6.64) for different a_0 values. This is shown in Fig. 6-25, which exhibits a trend similar to that of the concrete. It is noted that the value of $c_f = 6.563$ mm was obtained from Eq. (6.58). These results further confirm that the obtained relationships for (K_{Ic}^s, CTOD$_c$) and (G_f, c_f) are applicable for quasi-brittle materials.

As indicated by Planas and Elices,[29] both the two-parameter fracture model and the size effect model are capable of predicting accurately the maximum loading capacity of concrete specimens in the practical range of sizes used in laboratory. However, it is noted that the size effect model implies that an infinitely large specimen has an infinite initial crack length due to geometric similarity in specimen dimensions. Although this may not be true if the initial crack length represents a material defect, the results given in Figs. 6-24 and 6-25 indicate that generally the size effect model can also be applied for an

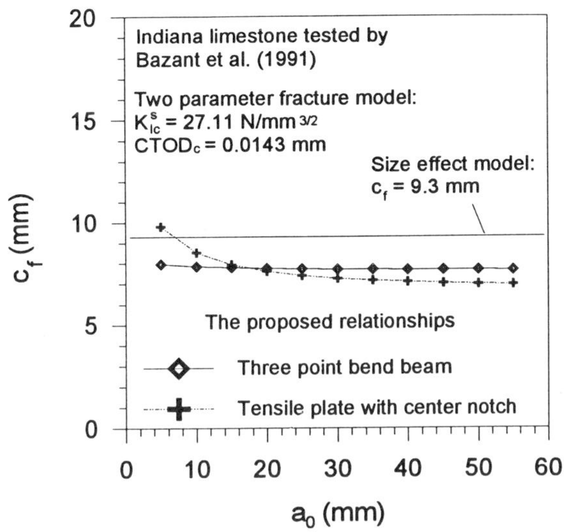

Figure 6-25 Values of c_f by proposed relationship and by size effect model for rock.

infinitely large specimen with a finite crack length, except for the tensile plate with short initial crack ($a_0 < 20$ mm).

6.8 ISRM METHODS FOR ROCK

Fracture testing for rock bears some similarities to that for concrete. However, there are at least two major differences. Unlike concrete test specimens fabricated in the laboratory or job site from a plastic mixture placed in molds, the rock test specimens must be obtained from the in situ mass by cutting. The resulting test specimens will typically be cylindrical cores, although sometimes sawed rectangular prisms are used. A second difference is that many rock formations have bedding planes. Thus, the test samples may not be isotropic and also may have nonhomogeneities. In Fig. 6-26 one sees how these bedding planes may be oriented relative to notches used in chevron bend (CB) specimens and short-rod (SR) specimens. A more general situation of notch orientation and notations is shown in Fig. 6-27.

There are similarities with respect to concrete testing also. The crack development and growth is very much like that for concrete except at a different scale that is related to the grain size of the rock vis-à-vis the maximum aggregate size in the concrete. For the most part, LEFM cannot be applied directly, although historically many attempts were made to do this in a manner conforming to ASTM standards for metals.[42,43] This included the use of fatigue precracking to grow natural cracks prior to determining fracture properties. This in turn led to chevron-notched specimens in which slow, controlled crack

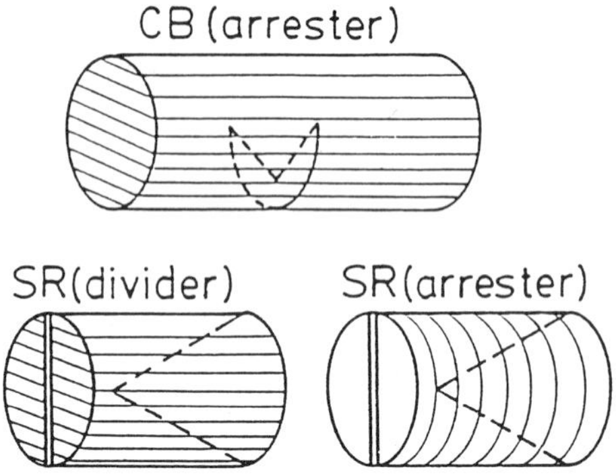

Figure 6-26 Coring direction and notch orientation relative to bedding planes in specimen of Inada granite, Ogino tuff, and Tohoku marble.[42]

growth may take place in specimens subjected to static loads. Out of this has emerged proposed testing methods by the ISRM.[44] These are directed to obtaining the material fracture toughness.

Two testing geometries are proposed: the CB specimen as shown in Fig. 6-28 and the SR specimen as shown in Fig. 6-29. It is noted that the dimensions of the CB specimen are such that, after testing to failure, one can prepare SR specimens from the unbroken halves.

The specimen dimensions are as follows: (i) method 1 for the CB specimen: $S = 3.33D \pm 0.02D$, or $|\Delta S| \leq 0.02D$, $a_0 = 0.15D \pm 0.10D$, $2\theta = 90.0° \pm 1.0°$, and $t \leq 0.03D$ or 1 mm, whichever is greater; (ii) method 2 for the SR specimen: $L = 1.45D \pm 0.02D$, or $|\Delta L| \leq 0.02D$, $a_0 = 0.48D \pm 0.02D$, $a_1 - a_0 = 0.97D \pm 0.02D$, $2\theta = 54.6° \pm 1.0°$, and $t \leq 0.03D$ or 1 mm, whichever is greater. In these, the minimum value of the dimension D will be given later. Testing may be done at two levels of accuracy (and complexity).

The testing concept is described in the following.[43] A set of characteristic stress intensity factor curves for the SR is shown in Fig. 6-30 together with a material curve showing the crack extension resistance (fracture toughness) as having reached a constant, representative level. The former curves have a minimum at a specific crack length, $a = a_m$ say. Up to this point the crack extension is stable due to the chevron notch and is achieved by a steadily increasing load. During this period a sharp natural crack is created in the specimen. Beyond a_m the crack extension tends to become unstable. Hence, ideally, all one would have to do during a test is to read off the failure load $F_{\max}$ corresponding to the tangency condition and determine the fracture toughness from

$$K_{SR} = K_I(a_m, F_{\max}) = K_{I\min}(F_{\max}) = \frac{A_{\min} F_{\max}}{D^{3/2}} \tag{6.70}$$

The brittle (level I) fracture toughness value is calculated from

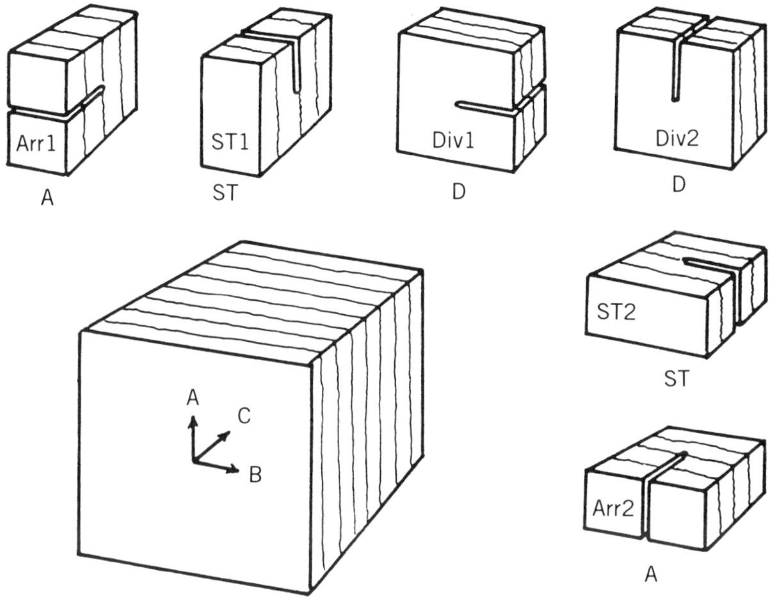

A//B: Specimen cored in A direction, with notch parallel to core axis (i.e a SR specimen) and crack front in B direction.

C⊥A: Specimen cored in C direction, with notch perpendicular to core axis (i.e a CB specimen) and crack front in A direction

CB notation	Extended transverse isotropic notation	SR notation
A⊥B	Arrester 1 (Arr1)	C//B
B⊥A	Arrester 2 (Arr2)	C//A
A⊥C	Divider 1 (Div1)	B//C
B⊥C	Divider 2 (Div2)	A//C
C⊥B	Short transverse 1 (ST1)	A//B
C⊥A	Short transverse 2 (ST2)	B//A

Figure 6-27 Notation used to denote coring direction and crack orientation in specimen of anisotropic rock [42]

$$K_{\text{ISRM}} = K_{\text{CB}} \quad \text{or} \quad K_{\text{SR}} = \frac{A_{\min} F_{\max}}{D^{3/2}} \tag{6.71}$$

where the failure load $F_{\max}$ is measured during the test. The nondimensional stress intensity factor $A_{\min}$ is given by

$$\text{CB:} \qquad A_{\min} = \left[1.835 + 7.15 \frac{a_0}{D} + 9.85 \left(\frac{a_0}{D} \right)^2 \right] \frac{S}{D} \tag{6.72}$$

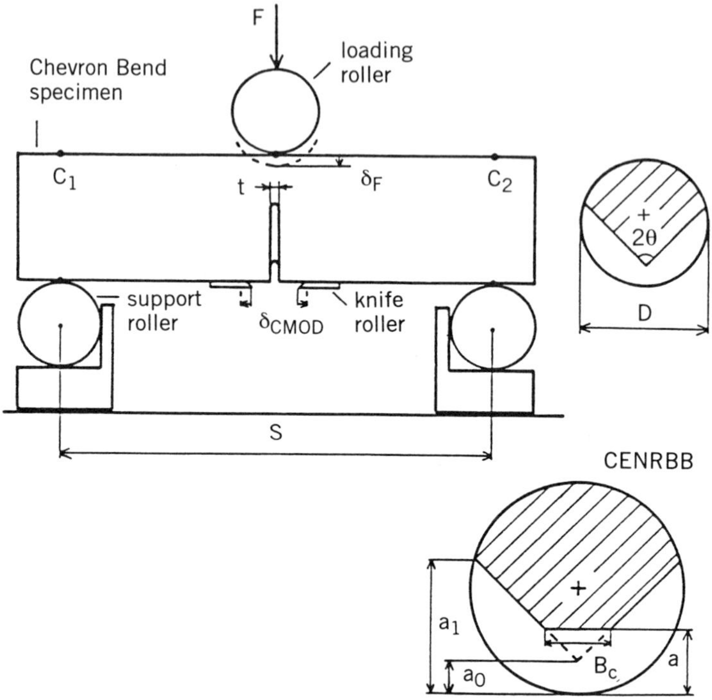

Basic notation:
D = diameter of specimen
S = distance between support points, 3.33D
2θ = chevron angle, 90°
a_0 = chevron tip distance from specimen surface, 0.015D
a_1 = maximum depth of chevron flanks
t = notch width
a = crack length
B_c = crack front length
δ_F = deflection of load point relative to support points, LPD
δ_{CMOD} = relative opening of knifes edges, CMOD

Figure 6-28 The CB specimen with basic notation.

and

$$\text{SR:} \qquad A_{\min} = 24.0\left[1 - 0.6\frac{\Delta w}{D} + 1.4\frac{\Delta a_0}{D} - 0.01\ \Delta(2\theta)\right] \tag{6.73}$$

where Δ denotes deviation from nominal proportions. This presumes that the crack in the specimen has progressed from the apex a_0 to a predetermined material independent evaluation point a_c at the failure load,

$$\text{CB:} \qquad a_c = a_0 + 0.15D \tag{6.74}$$

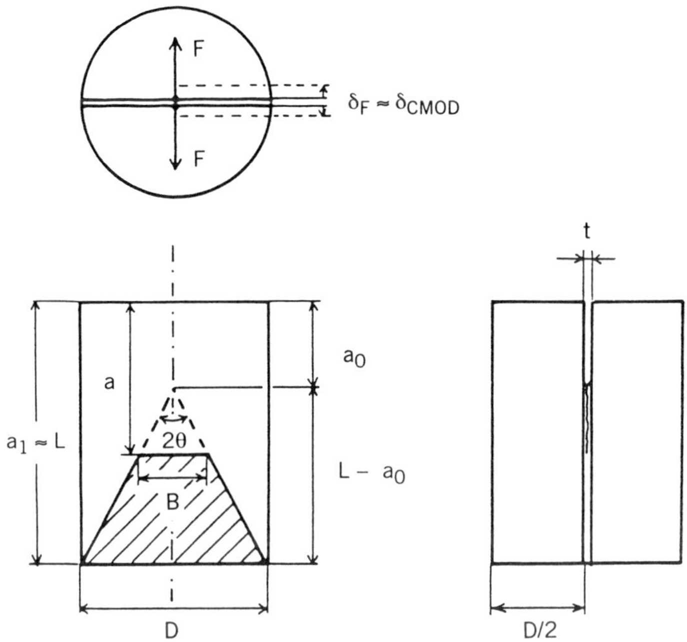

Basic notation:
D = diameter of short rod specimen
L = length of specimen, 1.45D
2θ = chevron angle, 54.6°
a_0 = chevron tip distance from load line, 0.481D
a_1 = maximum depth of chevron flanks
t = notch width
a = crack length
B = crack front length
F = load line displacement, LPD
δ_{CMOD} = crack mouth opening displacement, CMOD ≈ LPD

Figure 6-29 The SR specimen with basic notation.

or

$$\text{SR:} \qquad a_c = a_0 + 0.342D \tag{6.75}$$

A second level of testing is offered in each method that requires continuous load and displacement measurements to be made during the test. The reason is that level I testing of normal core sizes tends to yield fracture toughness values that are lower than values obtained from larger specimens. Since level II testing in some cases yields values that are independent of core size and that correlate well with results from other tests methods, level II testing is sometimes justified.

Where level I testing is simple and cheap, yielding indexlike results, level

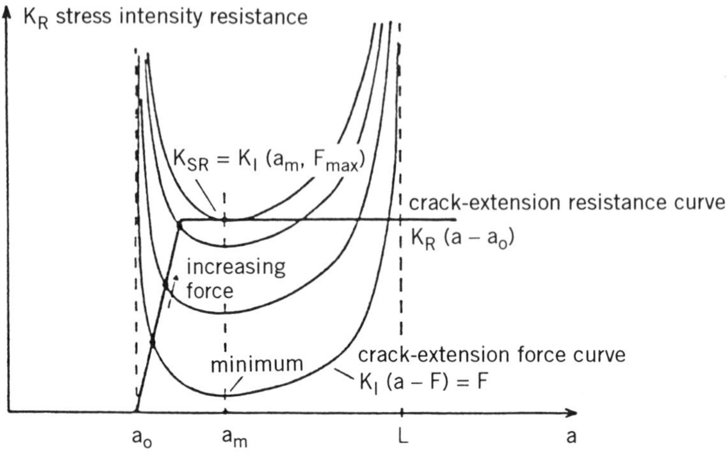

Figure 6-30 Principal appearance of stress intensity factor $K_I(a, F)$ curve of SR specimen. Crack extension resistance $K_R(a - a_0)$ with flat part (representative toughness level) assumed in K_{SR} evaluation in Eq. (6.70).[43]

II testing is more complicated but the prospective user is given a choice between accuracy and cost. In order to conduct level II testing, a closed-loop system is required and the LPD is plotted continuously with the load. Note that for the CB specimen in Fig. 6-28 spurious deflections may be recorded as described earlier.[12,13] A trace similar to that shown in Fig. 6-31 is obtained. Actual test data for Iidate granite are shown in Fig. 6-32.[42] When level II testing is made, Young's modulus E of the rock material is conveniently determined from the initial tangent slope S_{F0} of low-amplitude load cycling, re-

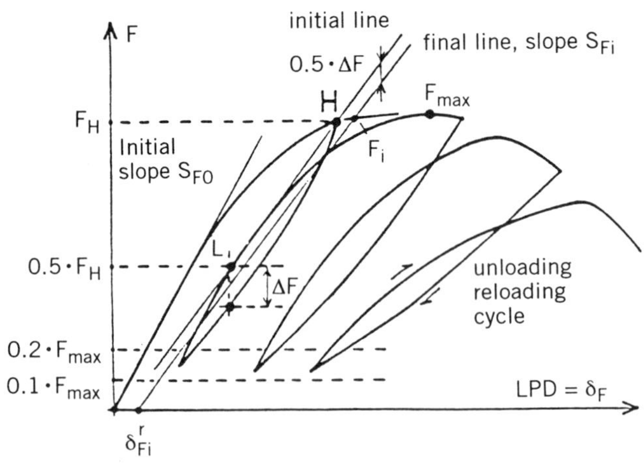

Figure 6-31 Construction of linear unloading in F–LPD curve.

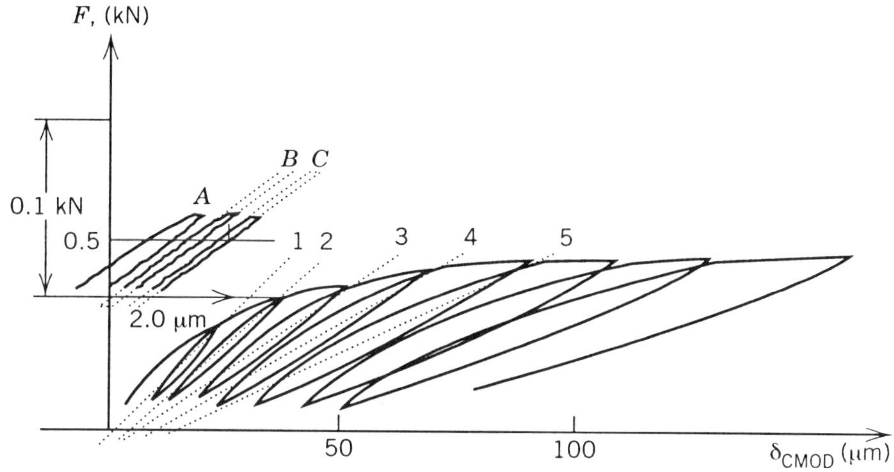

Figure 6-32 Test record from SR specimen of lidate granite. Notice low-amplitude load cycling used for Young's modulus determination (left side).[42]

corded at high resolution before the actual fracture testing takes place, as shown in Figs. 6-31 and 6-32. It is then given by

$$E = \frac{g_0 S_{F0}}{D} \tag{6.76}$$

where

$$\text{CB:} \qquad g_0 = 20.8 + 19.4\frac{a_0}{D} + 142.3\left(\frac{a_0}{D}\right)^2 \tag{6.77}$$

or

$$\text{SR:} \qquad g_0 = 84.5\left[1 + 2.9\frac{\Delta a_0}{D} + 2.5\left(\frac{t}{D} - 0.012\right)\right] \tag{6.78}$$

The level II fracture toughness takes nonideal fracture behavior into account through a nonlinearity correction factor p_F,

$$K^c_{\text{ISRM}} = \left(\frac{1 + p_F}{1 - p_F}\right)^{1/2} K_{\text{ISRM}} = \left(\frac{1 + p_F}{1 - p_F}\right)^{1/2} A_{\min} F_m D^{3/2} \tag{6.79}$$

if the evaluation load F_m does not coincide with $F_{\max}$. Determination of the value of p_F will be discussed later. Note that K_{ISRM} is taken to mean either K_{CB}

or K_{SR}. This evaluation procedure assumes that the specimen compliance is a good measure of crack length and it involves three steps.

In the first step, the crack length during testing is monitored by determining both the unloading compliance or stiffness at a number of points and the associated residual displacement δ_F^r at complete unloading. The compliance λ_F or stiffness $S_F = 1/\lambda_F$ is to be determined at a number of points on the load and LPD curve. Barker's[45] linearization procedure is specified (see Fig. 6-31). It assumes that a partial unloading cycle down to $0.1F_{max}$ to $0.2F_{max}$ is made for each point.

The final line of cycle i has slope S_{Fi}, its intersection with the smoothed load–LPD curve defines an evaluation point F_i, and its intersection with the LPD axis defines the residual LPD, δ_{Fi}^r. Note that this line lies basically halfway between the unload curve and the reload curve of cycle i. This step is repeated a number of times; cf. Fig. 6-32.

The second step is finding the desired evaluation point F_m corresponding to $a = a_m$, the critical crack length. For this point, the slope of the final line is S_{Fm}. The value of S_{Fm} is equal to $S_{F0}s_m$, where

$$\text{CB:} \qquad s_m = 1.05 - 2.15\frac{a_0}{D} + 4.21\left(\frac{a_0}{D}\right)^2 \tag{6.80}$$

or

$$\text{SR:} \qquad s_m = 0.5 \tag{6.81}$$

Then, referring to Fig. 6-33, first find the two points with slope values that span S_{Fm}, say F_{i-1} and F_i. Then find point F_l on the final line through F_{i-1}. It shall have the same amount of displacement recovery as F_i, that is,

$$l_e = \delta_{F_i} - \delta_{F_i}^r = \frac{F_i}{S_{F_i}} = \frac{F_l}{S_{F_{i-1}}} \tag{6.82}$$

Then draw a straight line through F_i and F_l and find the interpolation point whose load is

$$F_e = S_{Fm} l_e = \frac{F_i\, S_{Fm}}{S_{F_i}} \tag{6.83}$$

The corresponding displacement is δ_{Fe}. Next find the point $(\delta_{Fe} - l_e, 0)$ on the LPD axis. Draw a straight line through it and F_e. It will automatically have the slope S_{Fm}. Let its intersection with a load–LPD curve, which has been smoothed to remove any dips due to the unloading and reloading cycles, define the evaluation point F_m with coordinates (δ_{Fm}, F_m). If $F_m \approx F_{max}$, then F_{max} is used instead.

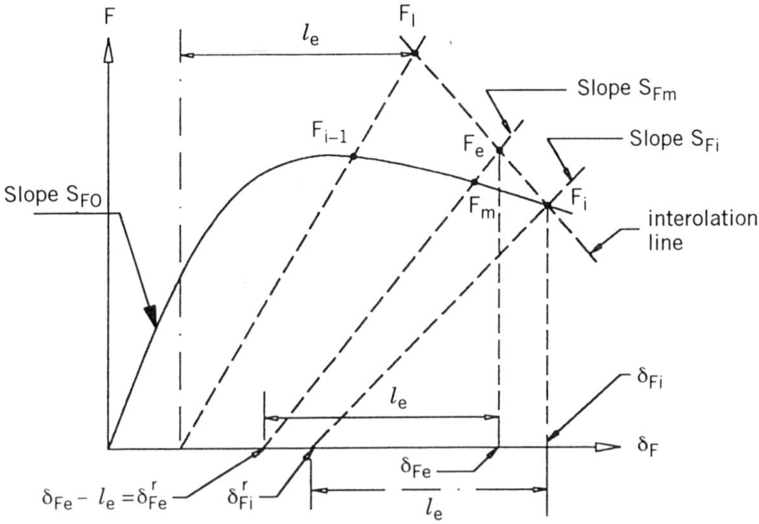

Figure 6-33 Derivation at point F at which fracture toughness is evaluated.

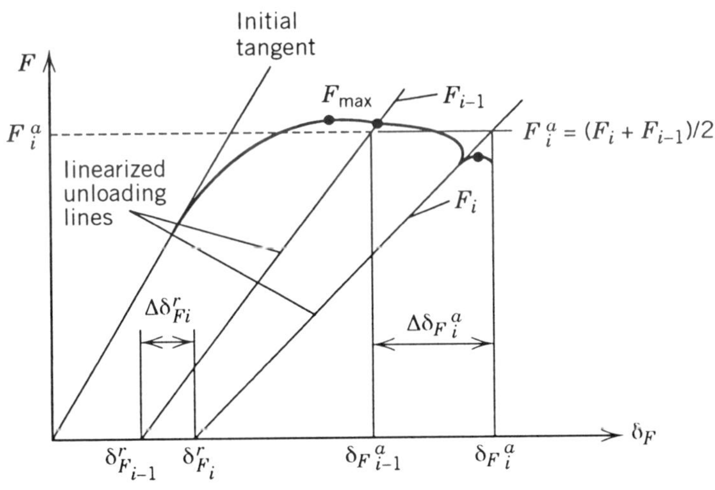

Figure 6-34 Construction of nonlinearity correction factor p_F.

The third step concerns the nonlinearity correction. To this end define first the average load $F^a_i = (F_i + F_{i-1})/2$ as shown in Fig. 6-34. Next let this load define two matching displacements along the unloading lines through F_{i-1} and F_i. Their differences,

$$\Delta\delta_{F^a_i} = \delta_{F^a_i} - \delta_{F^a_{i-1}} \quad \text{and} \quad \Delta\delta^r_{F_i} = \delta^r_{F_i} - \delta^r_{F_{i-1}} \tag{6.84}$$

are used to define p_F:

$$p_F = \frac{\Delta \delta^r_{F_i}}{\Delta \delta_{F^a_i}} \tag{6.85}$$

cf. reference 46. This value of p_F is used to determine $K^c_{\text{ISRM}}$from Eq. (6.79).

The minimum value of D for the test specimen has to be determined. Based upon extensive testing reported in reference 42, it is recommended that a minimum size to obtain a toughness value as a representative material property, that is, $K^c_{\text{ISRM}} = K_{\text{I}c}$, is

$$D_{\min} = \alpha_D \left(\frac{K_{\text{I}c}}{\sigma_t} \right)^2 \tag{6.86}$$

where $\alpha_D = 0.71$ for the SR specimen or 1.6 for the CB specimen and σ_t is the tensile strength. If $K_{\text{I}c}$ is not known, an estimate of $D_{\min}$ can be made by calculating

$$K^e_{\text{I}c} = 0.23E^{0.65} \tag{6.87}$$

and inserting $1.2K^e_{\text{I}c}$ into Eq. (6.86) in place of $K_{\text{I}c}$. In Eq. (6.87), the units are metric (i.e., $K^e_{\text{I}c}$ is in MPa$\sqrt{\text{m}}$ and the Young's modulus E is in GPa).

The determination of the three major testing quantities E, K_{ISRM}, and K^c_{ISRM} involves errors, especially the steps that give F_m and p_F as part of K^c_{ISRM}. The modulus determination has proven to be accurate, both in comparison with strain gauge values and in terms of a small scatter and compatibility between CB and SR values for the same rock.

A recent compilation[43] of fracture toughness values obtained by the ISRM working group members and their colleagues is listed in Table 6-13 where testing was made essentially according to the recommended procedures. Beside the range of values to be expected as well as the variability and anisotropy present in rock material, the scatter indicates that a reproducibility of 5–10% can be expected in level I testing. The frequently larger scatter of level II results illustrates the larger errors involved.

That K^c_{CB} is sometimes less than K_{CB} is an indication that material R-curve effects may be present. A procedure on how to modify the CB results to obtain crack lengths and K_{CB} values corresponding to those from the SR specimens is presented in reference 42.

Reference 42 contains a number of recommendations to establish the validity of the calculated fracture toughness. A flow chart presented in Fig. 6-35 shows whether the measured K^c_{ISRM} value can be considered a material property, $K_{\text{I}c}$, or not. Sometimes the testing according to the suggested methods fails; then K_R testing to measure the crack resistance curve can be made and a $K_{\text{I}c}$ value can possibly be obtained.

Table 6-13 Mean Values of Fracture Toughness from Core Specimens

Type of Rock	Chevron Bend Specimens				Short-Rod Specimen			
	Test Number	K_{CB} (MPa $m^{1/2}$)	Test Number	K_{CB}^c (MPa $m^{1/2}$)	Test Number	K_{CB} (MPa $m^{1/2}$)	Test Number	K_{CB}^c (MPa $m^{1/2}$)
Tampomas andesite	7	1.50 ± 0.12	7	1.68 ± 0.15				
Tampomas andesite	5	1.26 ± 0.10	5	1.26 ± 0.26				
Whin Sill dolerite	—	—	—	—	6	2.86 ± 0.12	6	3.26 ± 0.09
Kallax gabbro	—	—	—	—	13	2.22 ± 0.11	18	2.58 ± 0.22
Kallax gabbro	—	—	—	—	10	2.86 ± 0.14	13	3.23 ± 0.34
Bohus granite	5	1.46 ± 0.07	—	—	3	1.8	3	2.4
Bohus granite	4	1.69 ± 0.04	4	1.42 ± 0.14				
Cornwall granite	4	1.32 ± 0.10	—					
Epprechtstein granite	8	1.74 ± 0.18	—					
Falkenberg granite	4	0.65 ± 0.14	—					
Falkenberg granite	—	—	5	1.52 ± 0.20				
Iidate granite	5	1.09 ± 0.13	5	1.73 ± 0.21				
Iidate granite	3	1.37 ± 0.13	3	2.26 ± 0.65	4	1.01 ± 0.18	4	1.12 ± 0.35
Iidate granite D, A	3	1.43 ± 0.01	3	0.83 ± 0.13	3	1.58 ± 0.08	3	1.85 ± 0.06
Iidate granite ST	—	—	—	—	3	1.11 ± 0.12	3	1.26 ± 0.18
Krakemala granite	3	1.64 ± 0.04	3	2.16 ± 0.23	6	1.69 ± 0.17	5	2.22 ± 0.24
Merrivale granite	—	—	—	—	24	1.50 ± 0.10	24	1.80 ± 0.13
Pink granite	—	—	—	—	4	1.58 ± 0.04	4	2.03 ± 0.08
Rasjö granite	—	—	—	—	6	2.37 ± 0.32	6	2.80 ± 0.33
Strain hallafale granite	—	—	—	—	11	1.80 ± 0.10	11	2.19 ± 0.11
Stripa granite	—	—	—	—	5	2.01 ± 0.14	11	2.36 ± 0.13
Stripa granite	—	—	—	—	9	2.37 ± 0.15	9	2.70 ± 0.27
Westerly granite	—	—	—	—	9	1.64 ± 0.03	9	1.82 ± 0.07
Westerly granite	—	—	—	—	17	2.28 ± 0.19		
Westerly granite	—	—	—	—	4	2.04 ± 0.05	4	2.27 ± 0.03

Table 6-13 (*Continued*)

Type of Rock	Chevron Bend Specimens				Short-Rod Specimen			
	Test Number	K_{CB} (MPa m$^{1/2}$)	Test Number	K^c_{CB} (MPa m$^{1/2}$)	Test Number	K_{CB} (MPa m$^{1/2}$)	Test Number	K^c_{CB} (MPa m$^{1/2}$)
Finnsjön granodiorite	—	—	—	—	3	2.95 ± 0.11	3	3.35 ± 0.08
Grey norite	—	—	—	—	11	2.23 ± 0.11	6	2.69 ± 0.16
Ogino tuff	17	1.05 ± 0.11	9	1.08 ± 0.10	19	1.02 ± 0.05	19	1.06 ± 0.05
Pennant sandstone	—	—	—	—	6	1.98 ± 0.06	6	2.56 ± 0.07
Ruhr sandstone	10	1.03 ± 0.04						
Älvdalen sandstone	6	1.51 ± 0.08	6	0.73 ± 0.08	11	1.54 ± 0.08	11	1.91 ± 0.14
Indiana limestone	—	—	—	—	7	0.90 ± 0.11	7	1.13 ± 0.09
Indiana limestone	—	—	—	—	9	1.05 ± 0.06		
Klinthagen limestone	2	1.31	—	—	8	1.41 ± 0.19	8	1.87 ± 0.25
Shelly limestone	—	—	—	—	5	1.40 ± 0.03	5	1.44 ± 0.04
Anvil Points oil shale D	—	—	—	—	5	0.56 ± 0.09	3	1.02 ± 0.14
Anvil Points oil shale ST	—	—	—	—	4	0.25 ± 0.04	2	0.47 ± 0.07
Carrara marble	5	1.26 ± 0.08	5	1.38 ± 0.09				
Ekeberg marble	16	1.89 ± 0.12	—	—	9	1.83 ± 0.35	9	2.25 ± 0.36
Ekeberg marble ST	—	—	—	—	3	1.48 ± 0.16	3	1.82 ± 0.10
Ekeberg marble D	—	—	—	—	2	2.28 ± 0.01	2	2.62 ± 0.05
Ekeberg marble	4	1.79 ± 0.07	4	1.76 ± 0.21				
Treuchtlingen marble	6	1.26 ± 0.07	6	1.70 ± 0.09			—	
Yizhang marble	7	1.49 ± 0.09	7	1.83 ± 0.18				

Source: From ref. 43.

Notes: A, D, and ST mean arrester, divider, and short transverse orientations of crack growth with respect to discernible structure in rock (see Fig. 6-27).

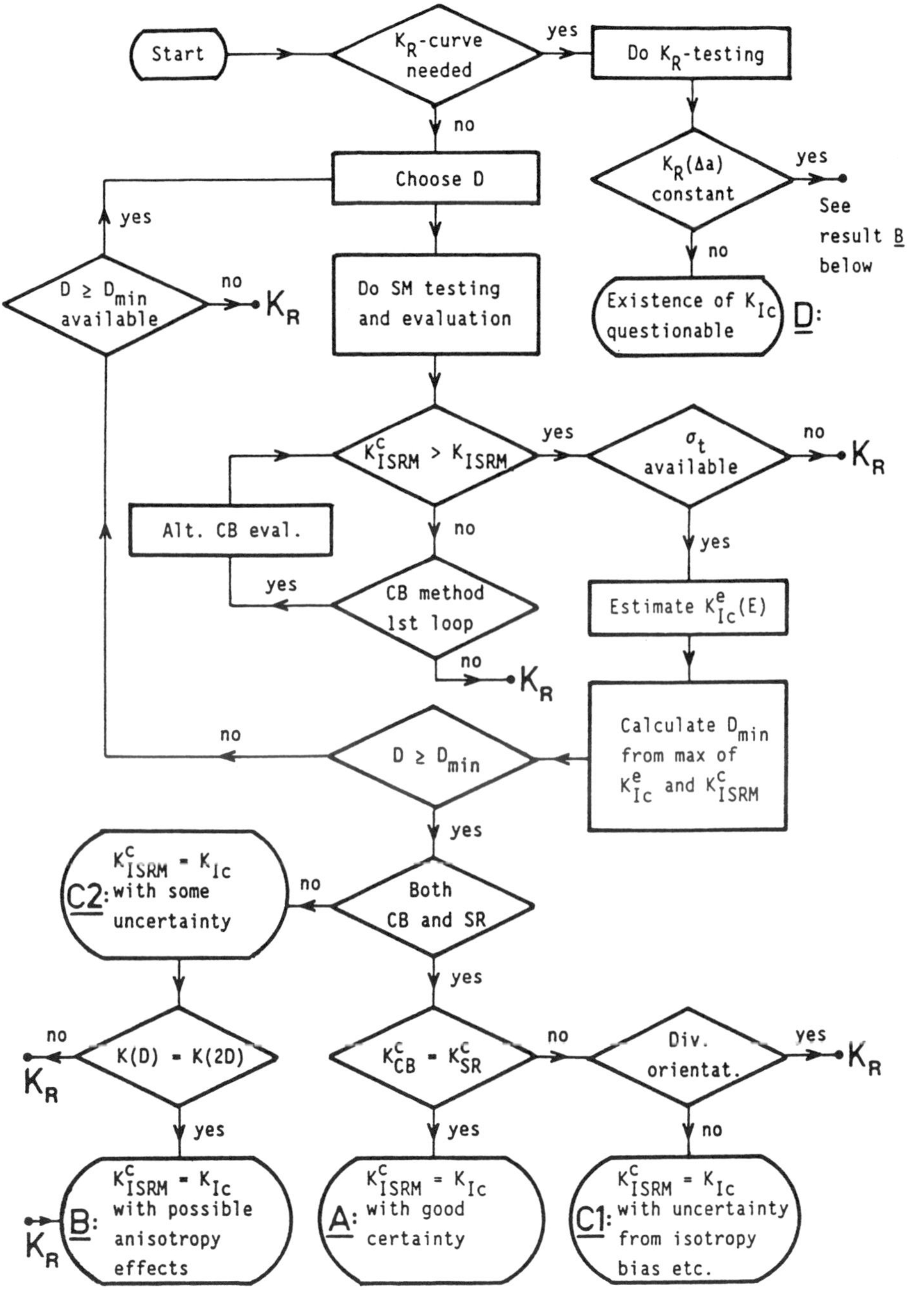

Figure 6-35 Flow chart for fracture toughness testing of rock that shows level of accuracy of result obtained.[42]

The central validity recommendation of the suggested test method is obviously $D \geq D_{\min}$, which requires knowledge of σ_t. The accuracy of the $K_{\mathrm{I}c}$ statement becomes better if one or several of the following points are investigated further: (i) specimen size effects, (ii) specimen geometry effects, and (iii) anisotropic effects.

The following levels of accuracy can be reached using the flow chart (Fig. 6-35). In this, SM is the proposed standard method of test and K_R is resistance curve testing.

A. $K_{\mathrm{I}c} = K^c_{\mathrm{ISRM}}$ *with good accuracy.* This occurs for ideal, that is, linearly elastic, and brittle rocks for which $p_F \approx 0$ and

$$K_{\mathrm{CB}} = K_{\mathrm{SR}} \approx K^c_{\mathrm{CB}} = K^c_{\mathrm{SR}} = K_{\mathrm{I}c} \tag{6.88}$$

One example is Ogino tuff. It also occurs for nonbrittle rocks for which $p_F \neq 0$ and

$$K_{\mathrm{CB}} < K_{\mathrm{SR}} \leq K^c_{\mathrm{CB}} \leq K^c_{\mathrm{SR}} \approx K_{\mathrm{I}c} \tag{6.89}$$

which is valid if $D \geq D_{\min}$. The rocks need to be isotropic or to have the crack in the divider orientation to satisfy Eqs. (6.88) and (6.89).

B. $K_{\mathrm{I}c} = K^c_{\mathrm{ISRM}}$ *with a possible inaccuracy due to isotropy bias of evaluation formulas.* This occurs for the nonbrittle rocks under A above when testing:

$$\begin{aligned} \text{SM:} \quad & K_{\mathrm{I}c} = K^c_{\mathrm{ISRM}}(2D) = K^c_{\mathrm{ISRM}}(D) \\ K_R\text{:} \quad & K_{\mathrm{I}c} = K_R(\Delta a) \quad \text{when } \Delta a \geq \Delta a_{\min} \end{aligned} \tag{6.90}$$

C1: $K_{\mathrm{I}c} \approx K^c_{\mathrm{ISRM}}$ *with some unknown inaccuracy due to isotropy bias of the evaluation formulas.* This occurs for a nonbrittle anisotropic rock with a crack in *Arr* or *ST* orientation if

$$K^c_{\mathrm{CB}} \neq K^c_{\mathrm{SR}} \quad \text{when } D \geq D_{\min} \tag{6.91}$$

A reasonable $K_{\mathrm{I}c}$ estimate can probably be obtained from

$$K_{\mathrm{I}c} \approx \tfrac{1}{2}(K^c_{\mathrm{SR}} + K^c_{\mathrm{CB}}) \tag{6.92}$$

C2. $K_{\mathrm{I}c} \approx K^c_{\mathrm{ISRM}}$ *with engineering accuracy.* This is the basic statement possible when $D \geq D_{\min}$ but specimen size, specimen geometry, or anisotropy effects have not been investigated further.

D. $K_{\mathrm{I}c} \neq K^c_{\mathrm{ISRM}}$ *and existence of* $K_{\mathrm{I}c}$ *not proven by experiments.* The testing could have been made with $D \leq D_{\min}$, σ_t not available or $\Delta a \leq \Delta a_{\min}(K_R)$. Here, $K_{\mathrm{I}c}$ may even be nonexistent.

6.9 OTHER METHODS FOR ROCK

A number of other testing methods for determining fracture mechanics properties for rock specimens have been proposed. The most promising of these are probably the semicircular specimen developed by Chong and his colleagues[47–50] and the axial loaded specimen of Luong.[51,52]

6.9.1 Semicircular Core Specimen

This specimen is obtained from a cylindrical core by slicing the core in half, thus obtaining two specimens, and cutting a central, vertical notch, as shown in Fig. 6-36.[47] This is called the semicircular bend (SCB) fracture specimen and is used to obtain fracture toughness, fracture energy, and stress-softening curves.[47–50] This specimen is considered to be especially suitable for short cores and materials with foliation because the fracture mechanism involves the average fracture behavior. This of course depends on the specimen size relative to the larger thicknesses. Typical dimensions are 50–100 mm radius and 10–35 mm thickness. The ratio of S to r_d is 0.8 (Fig. 6-36).

Testing is performed using a closed-loop setup, shown schematically in Fig. 6-37 with the feedback control from a strain-gauge extensometer. The early work done using this specimen included growing a crack to a depth of about $r_d/2$ by subjecting the specimen to fatigue loading (5 Hz cycle with $P = 0.60P_{max}$). Subsequently, the precracked specimen was loaded to failure statically and a load displacement trace plotted. Basically, the ASTM standard

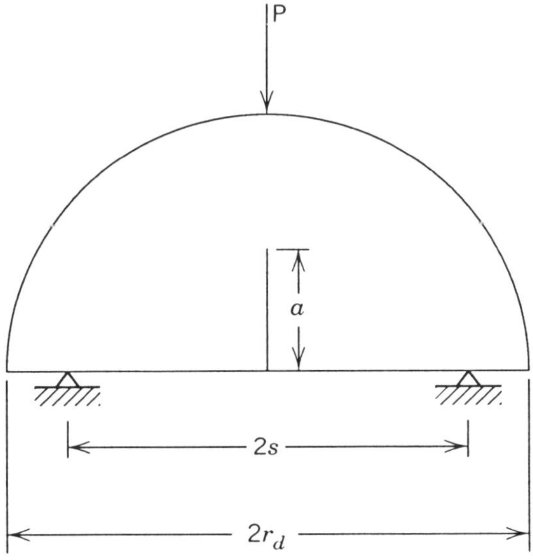

Figure 6-36 An SCB fracture specimen.[48]

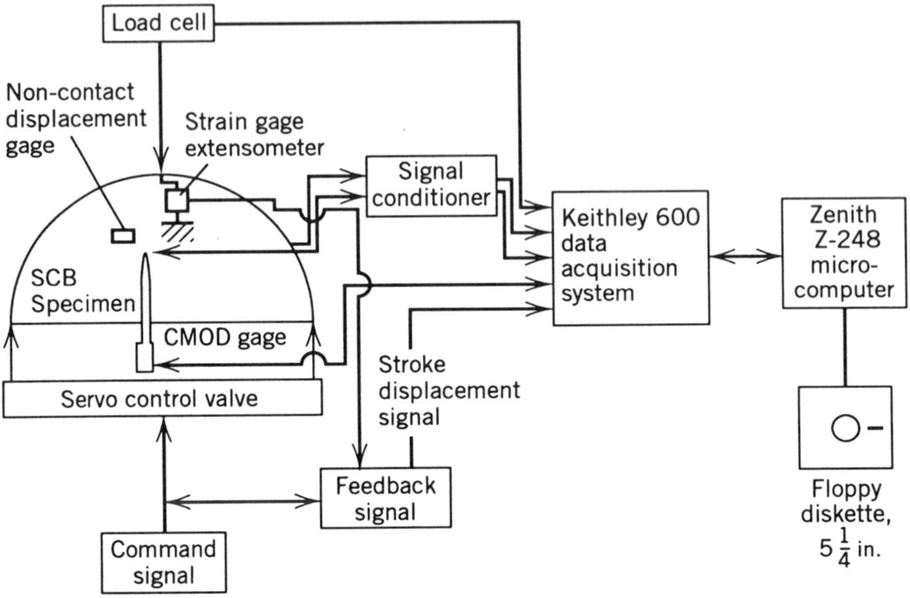

Figure 6-37 Instrumentation scheme for fracture toughness testing.[49]

procedure for metals was followed to identify the load corresponding to a 2% increment of crack extension and the apparent fracture toughness K_Q, computed from an approximate stress intensity formula.[47,50]

This specimen also lends itself very well to the tension-softening approach of Li et al.[36] In this method, two specimens with slightly different notch lengths are used. With two SCB specimens cut from one core, specimen variability is kept to a minimum. Let the two specimens be denoted 1 and 2 with notch lengths a_1 and a_2, $a_1 > a_2$. The load–LPD curves obtained from the tests on the two sandstone specimens are shown in Fig. 6-38*a*. The following definitions and procedures are given in reference 49.

The critical value of the J-integral obtained from the difference between the two load–LPD curves measured is

$$J_c(u) = \left(\frac{A_1}{t_1} - \frac{A_2}{t_2}\right)\frac{1}{a_2 - a_1} \tag{6.93}$$

where u is the LPD, A_1 and A_2 are areas under the load–LPD curves for specimens 1 and 2, respectively, and t_1 and t_2 are thicknesses of specimens 1 and 2 (frequently $t_1 = t_2$).

On the other hand, the value of J_c can also be evaluated in terms of crack tip separation w as

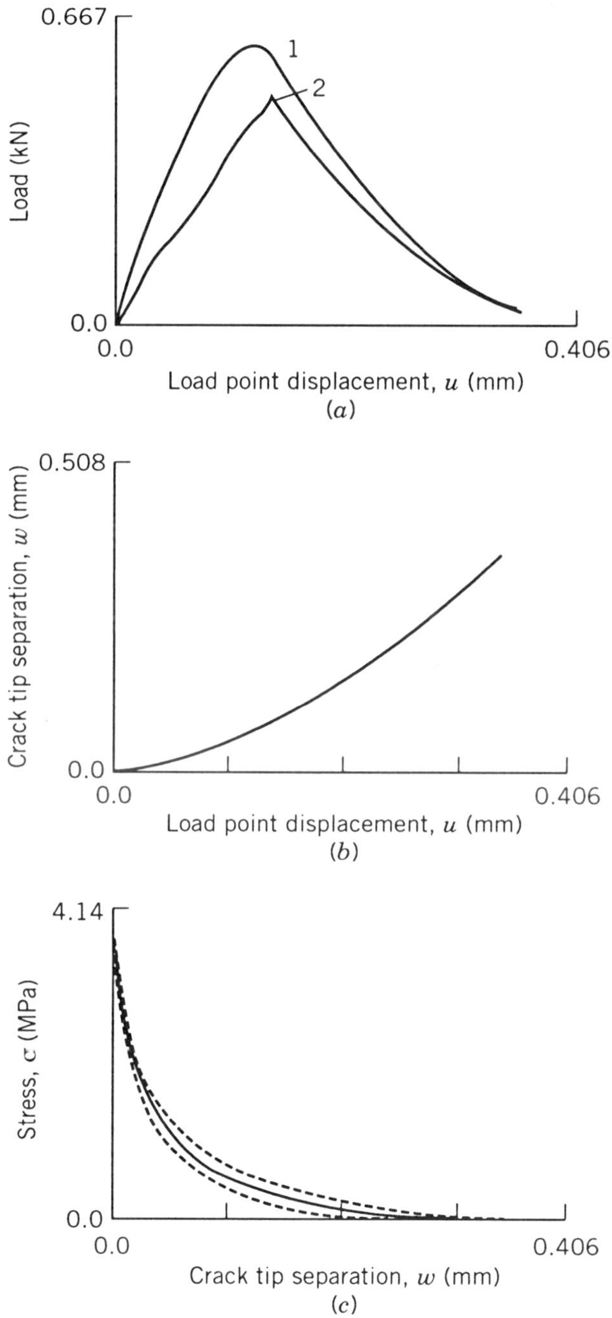

Figure 6-38 The SCB test results for sandstone: (*a*) typical load–LPD curve, (*b*) typical crack tip separation–LPD regression curve, and (*c*) mean stress–separation curve with upper and lower bounds.[49]

$$J_c(w) = \int_0^{w_c} \sigma(w)\, dw \tag{6.94}$$

where σ is the tensile stress at a crack tip and w_c is the critical crack tip separation. If $w < w_c$ (a noncritical case), Eq. (6.94) becomes

$$J(w) = \int_0^{w} \sigma(w)\, dw \tag{6.95}$$

By taking the derivative with respect to w, Eq. (6.95) becomes

$$\sigma(w) = \frac{\partial J(w)}{\partial w} = \frac{\partial J(u)}{\partial u}\frac{\partial u}{\partial w} \tag{6.96}$$

Hence, the stress–separation curve may be obtained indirectly from knowing the change of J with respect to the crack tip separation. For $J < J_c$, combining Eqs. (6.93) and (6.96) yields

$$\sigma(w) = \frac{1}{a_2 - a_1}\left(\frac{\partial A_1}{\partial w}\frac{1}{t_1} - \frac{\partial A_2}{\partial w}\frac{1}{t_2}\right) \tag{6.97}$$

Since A_1 and A_2 are defined with respect to u_1 and u_2, a direct evaluation of Eq. (6.97) is not possible. A correlation between the load point displacement u and the crack tip separation w must be established. This can be done by measuring both values of u and w through the complete loading process. Such a relation is shown in Fig. 6-38*b*.

This indirect method of obtaining the stress–separation curve is summarized in the following steps[49]:

1. Two identical specimens with slightly different notch lengths, a_1 and a_2 ($a_2 > a_1$), are loaded to failure.
2. Load P, load point displacement u, and crack tip separation w are monitored and recorded continuously. The feedback control signal should be the load point displacement.
3. From the recorded data the following relationships for both specimens 1 and 2 can be established: (i) load versus load point displacement (Fig. 6-38*a*), (ii) load versus crack tip separation, and (iii) crack tip separation versus load point displacement (Fig. 6-38*b*).
4. Based on Eq. (6.96) the σ–w relationship can be determined by substituting the integral form of A_1 and A_2 (areas under the P–u curves for specimens 1 and 2, respectively) into Eq. (6.93) and differentiating with respect to u. This modification results in

$$\sigma_i(w_i) = \frac{\partial J_i(u_i)}{\partial u_i}\frac{\partial u_i}{\partial w_i} = \frac{1}{a_2 - a_1}\left[\frac{P_1(u_i)}{t_1} - \frac{P_2(u_i)}{t_2}\right]\frac{\partial u_i}{\partial w_i} \tag{6.98}$$

where $P_1(u_i)$ and $P_2(u_i)$ are loads corresponding to the value of u_i for specimens 1 and 2, respectively, and u_i with $i = 1, 2, \ldots, n$ are values of the load point displacement at different intervals. Note that with the use of Eq. (6.98), which is the modification of Eq. (6.96), $\sigma(w)$ can be computed more efficiently because $\partial J(u)/\partial u$ and $\partial u/\partial w$ can easily be obtained directly from the curves of P–u and u–w defined in previous steps, respectively. This modification [Eq. (6.98)] enables one to avoid defining the $J(u)$ curve by computing the area between two $P(u)$ curves based on Eq. (6.93) and by differentiating the polynomial regression function of $J(u)$ originally proposed by Li et al.[36]

5. A corrected $\sigma(w)$ relationship is obtained by subtracting the inelastic damaged portion from the deduced $\sigma(w)$ curve. The corrected curve is shown in Fig. 6-38*c*.

6. Finally, J_c can be computed from Eq. (6.94), and $K_{Ic} = (EJ_c)^{1/2}$ can then be obtained.

Results for two types of rock are presented in Table 6-14. It is of interest to note that the value of K_{Ic} computed from the stress intensity factor method based on LEFM agrees well with the one derived from the tension-softening measurement $J_c(u)$ for the basalt rock, whereas the stress intensity factor method gives a much lower value for the sandstone. The explanation may be that the basalt specimens are larger than the sandstone specimens. Thus, the size of the basalt SCB specimen appears to be adequate so that LEFM concepts yield relatively "true" or size-independent fracture toughness values. It is also noted that the basalt has a much higher tensile strength than the sandstone and is therefore more brittle.

Finally, another useful feature of this specimen is that mixed-mode behavior

Table 6-14 Summary of Results for SCB Test

Type of Rock	Mean Radius (mm)	Young's Modulus E (GPa)	Derived f_t (MPa)	Measured[a] f_t (MPa)	K_{Ic} (MPa m^{-1}) SIF[b]	$J_c(D)$ Based	l_{ch}[c] (mm)
Sandstone	73.1	10.34	3.84	3.62	0.400	1.264	132.6
Basalt	95.8	15.11	18.89	19.17	3.36	3.46	33.6

Source: From ref. 49.

[a] Split-cylinder test.

[b] Stress intersity factor method by Chong and Kuruppu (ref. 47).

[c] Characteristic length $l_{ch} = EG_c/(f_t)^2$, where f_t is the derived value and $G_c = J_c(d)$ for a linear elastic material.

can also be investigated easily by either cutting the notch at an angle to the vertical to making it off center.

6.9.2 Direct-Load Core

A test specimen to determine the tensile properties of rock and concrete has been proposed by Luong.[51,52] As shown in Fig. 6-39*a*, the specimen has some distinct advantages over other tensile specimens: (i) it provides direct tensile stresses, (ii) it is loaded in compression so no special gripping fixtures are

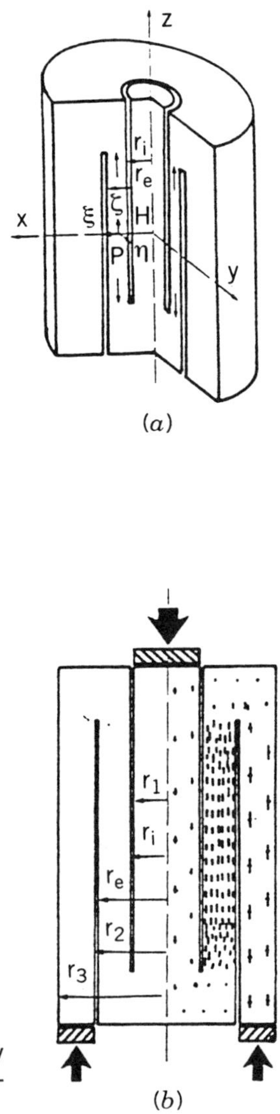

Figure 6-39 Direct-load core specimen: (*a*) specimen geometry and (*b*) finite element calculation of test specimen for linear elastic material.[52]

required, and (iii) it is relatively easy to align the load to obtain uniaxial tension.

The finite element calculation for determination of the distribution of internal stresses, shown in Fig. 6-39*b*, indicates a uniform distribution of tensile stresses throughout the test volume. This would be valid for a perfect geometry. Sources of error that would perturb these results include imperfect cutting of core and imperfect cutting of the slots to create the cylinders. It is stated that usually the tensile failures occur in the central part of the specimen. This is because, in the vicinity of the bored notch ends, fracture occurs in a mixed mode that causes a higher tensile resistance than the direct-tension mode. Forty five tests on sandstone specimens with $r_e = 25$ mm and $r_i = 13$ mm were conducted. The tensile stress was computed as

$$\sigma_z = \frac{F}{\pi(r_e^2 - r_i^2)} \tag{6.99}$$

A histogram of the test results is presented in Fig. 6-40*a*. It is seen that the average tensile strength is 1.92 MPa with a standard deviation of 0.22 MPa or a coefficient of variation of 11.45%. Stress–strain curves are shown in Fig. 6-40*b*.

Finally, it is noted that this geometry may be easily modified for use in a mixed-mode tests (modes II and III). This will be described further in Chapter 8.

6.10 SUMMARY

Use of NLFM models is necessary in order to determine material fracture parameters. Different methods for measuring fracture parameters of concrete have been presented in this chapter. These include three RILEM draft recommendations for measuring G_F, K_{Ic}^s and CTOD$_c$, and G_f, respectively. In addition, Karihaloo and Nallathambi[30–32] and Swartz and Refai[33,34] have proposed methods that are very similar to the two-parameter fracture model by Jenq and Shah,[16] respectively. The *J*-integral method by Li et al.[36] for obtaining the stress–separation curve $\sigma(w)$ and the wedge-splitting test by Brühwiler et al.[37,38] were also presented. The peak-load method by Tang et al.[24] can be easily used to determine values of K_{Ic}^s and CTOD$_c$ of the two-parameter fracture model. These test methods are associated with different NLFM models that are categorized into the fictitious crack approach and the equivalent-elastic crack approach. Since the two-parameter fracture model and the size effect model are based on the equivalent-elastic approach, the relationship between (K_{Ic}^s, CTOD$_c$) and (G_f, c_f) has been derived. Although some fracture parameters from different methods have the same physical meaning, parameters measured based on the fictitious crack approach may not be used for the equivalent-

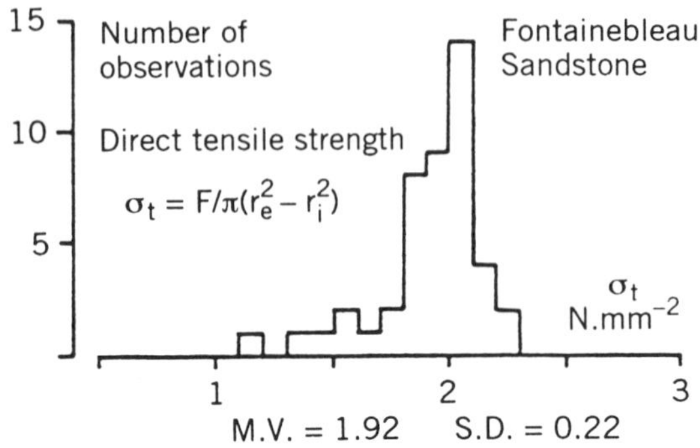

(*a*)

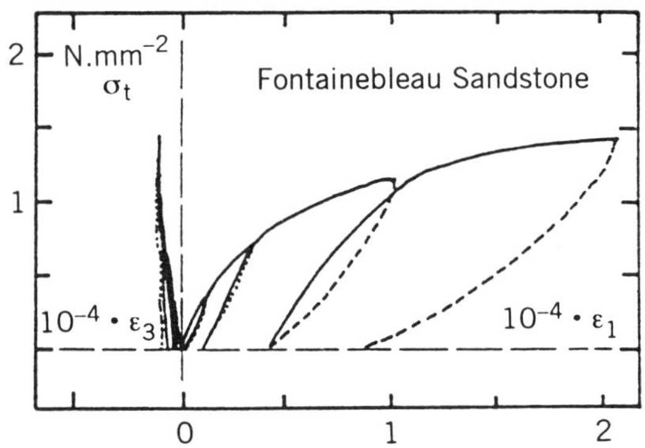

(*b*)

Figure 6-40 Direct-tension tests on sandstone: (*a*) histogram of 45 direct tensile test observations and (*b*) stress–strain curve of Fontainbleau sandstone under tension.[52]

elastic crack approach. For example, the value of G_F based on the fictitious crack model cannot as present be used as G_{Ic}^s (or G_f) based on the equivalent-elastic crack approach. However, research continues to establish some sort of equivalency.

Different methods for measuring fracture toughness of rock have been also presented. These include methods recommended by ISRM and methods proposed by Chong et al.[47–50] and Luong.[51,52] It is noted that the Chong et al.[47–50] method is an extension of the *J*-integral proposed by Li et al.[36]

REFERENCES

1. RILEM Committee on Fracture Mechanics of Concrete—Test Methods, "Determination of the Fracture Energy of Mortar and Concrete by Means of Three-Point

Bend Tests on Notched Beams,'' *Materials and Structures,* Vol. 18, No. 106, 1985, pp. 285–290.

2. Hillerborg, A., Modeer, M., and Petersson, P.-E., ''Analysis of Crack Formation and Crack Growth in Concrete by Means of Fracture Mechanics and Finite Elements,'' *Cement and Concrete Research,* Vol. 6, No. 6, 1976, pp. 773–782.
3. Hillerborg, A., ''Existing Methods to Determine and Evaluate Fracture Toughness of Aggregative Materials: RILEM Recommendation on Concrete,'' in *Fracture Toughness and Fracture Energy, Test Methods for Concrete and Rock,* edited by Mihashi, H., et al., Balkema, Rotterdam, 1989, pp. 145–151.
4. Petersson, P. E., ''Crack Growth and Development of Fracture Zones in Plain Concrete and Similar Materials,'' Report TVBM-1006, Division of Building Materials, Lund Institute of Technology, 1981.
5. Swartz, S. E., and Yap, S. T., ''The Influence of Dead Load on Fracture Energy Measurements Using the *RILEM* Method,'' *Materials and Structures,* Vol. 21, 1988, pp. 410–415.
6. Hilsdorf, H. K., and Brameshuber, W., ''Size Effects in the Experimental Determination of Fracture Mechanics Parameters,'' in *Application of Fracture Mechanics to Cementitious Composites,* edited by Shah, S. P., Martinus Nijhoff, Dordrecht, 1985, pp. 319–360.
7. Hillerborg, A., ''Results of Three Comparative Test Series for Determining the Fracture Energy G_F of Concrete,'' *Materials and Structures,* Vol. 18, No. 107, 1985, pp. 407–413.
8. Guinea, G. V., Planas, J., and Elices, M., ''Measurement of the Fracture Energy Using Three-Point Bend Tests: Part I. Influence of Experimental Procedures,'' *Materials and Structures,* Vol. 25, 1992, pp. 212–218.
9. Planas, J., Elices, M., and Guinea, G. V., ''Measurement of the Fracture Energy Using Three-Point Bend Tests: Part II. Influence of Bulk Energy Dissipation,'' *Materials and Structures,* Vol. 25, 1992, pp. 305 312.
10. Elices, M., Guinea, G. V., and Planas, J., ''Measurement of the Fracture Energy Using Three-Point Bend Tests: Part III. Influence of Cutting the P–δ Tail,'' *Materials and Structures,* Vol. 25, 1992, pp. 327–334.
11. Hu, X.-Z., and Wittmann, F. H., ''Fracture Energy and Fracture Process Zone,'' *Materials and Structures,* Vol. 25, 1992, pp. 319–326.
12. Swartz, S. E., and Kan, Y.-C., ''Effect of Support Conditions on Fracture Energy Measurements for Concrete Beams,'' in *Fracture Behavior and Design of Materials and Structures,* edited by Firao, D., Engineering Materials Advising Services, Ltd., London, 1990, pp. 660–666.
13. Swartz, S. E., and Kan, Y.-C., ''On the Validity of Indirect Measurement of the LPD for SEN Concrete Beams,'' in *Fracture Processes in Concrete, Rock and Ceramics,* edited by van Mier, J. G. M., Rots, J. G., and Bakker, A., E & FN SPON, London, 1991, pp. 771–778.
14. Shah, S. P., ''Fracture Toughness of Cement-Based Materials,'' *Materials and Structures,* Vol. 21, No. 122, 1988, pp. 145–150.
15. Petersson, P. E., ''Fracture Energy of Concrete: Practical Performance and Experimental Results,'' *Cement and Concrete Research,* Vol. 10, 1980, pp. 91–101.

16. RILEM Committee on Fracture Mechanics of Concrete—Test Methods, ''Determination of the Fracture Parameters (K^s_{Ic} and $CTOD_c$) of Plain Concrete Using Three-Point Bend Tests,'' *Materials and Structures,* Vol. 23, 1990, pp. 457–460.

17. Jenq, Y. S., and Shah, S. P., ''A Two Parameter Fracture Model for Concrete,'' *Journal of Engineering Mechanics,* Vol. 111, No. 4, 1985, pp. 1227–1241.

18. Refai, M. E., and Swartz, S. E., ''Fracture Behavior of Concrete Beams in Three-Point Bending Considering the Influence of Size Effects,'' Report No. 190, Engineering Experimental Station, Kansas State University, Manhattan, KS, 1987.

19. Karihaloo, B. L., and Nallathambi, P., ''Notched Beam Test: Mode I Fracture Toughness,'' in *Fracture Mechanics Test Methods for Concrete,* RILEM Report 5, edited by Shah, S. P., and Carpinteri, A., Chapman & Hall, London, 1991, pp. 1–86.

20. Catalano, D. M., and Ingraffea, A. R., ''Concrete Fracture: A Linear Elastic Fracture Mechanics Approach,'' Report No. 82-1, Department of Structural Engineering, Cornell University, Ithaca, NY, 1982.

21. John, R., and Shah, S. P., ''Fracture Mechanics Analysis and High Strength Concrete,'' *Journal of Materials in Civil Engineering, ASCE,* Vol. 1, No. 4, 1989, pp. 185–198.

22. Zhao, G., Jiao, H., and Xu, S., ''Study on Fracture Behavior with Wedge Splitting Test Method,'' in *Fracture Processes in Concrete, Rock and Ceramics,* edited by van Mier, J. G. M., et al., E & FN SPON, London, 1991, pp. 789–799.

23. Brühwiler, E., Broz, J. J., and Saouma, V. E., ''Fracture Model Evaluation of Dam Concrete,'' *Journal of Materials in Civil Engineering, ASCE,* Vol. 3, 1991, pp. 235–251.

24. Tang, T., Ouyang, C., and Shah, S. P., ''A Simple Method for Determining Material Fracture Parameters from Peak Loads,'' *ACI Materials* Journal, 1995, In press.

25. Jenq, Y.-S., and Shah, S. P., ''A Fracture Toughness Criterion for Concrete,'' *Engineering Fracture Mechanics,* Vol. 21, No. 5, 1985, pp. 1055–1069.

26. RILEM Committee on Fracture Mechanics of Concrete—Test Methods, ''Size-Effect Method for Determining Fracture Energy and Process Zone Size of Concrete,'' *Materials and Structures,* Vol. 23, 1990, pp. 461–465.

27. Bazant, Z. P., and Pfeiffer, P. A., ''Determination of Fracture Energy from Size Effect and Brittleness Number,'' *ACI Materials Journal,* Vol. 84, No. 6, 1987, pp. 463–480.

28. Bazant, Z. P., and Kazemi, M. T., ''Determination of Fracture Energy, Process Zone Length and Brittleness Number from Size Effect, with Application to Rock and Concrete,'' *International Journal of Fracture,* Vol. 44, 1990, pp. 111–131.

29. Planas, J., and Elices, M., ''Fracture Criteria for Concrete: Mathematical Approximations and Experimental Validation,'' *Engineering Fracture Mechanics,* Vol. 35, No. 1/2/3, 1989, pp. 87–94.

30. Nallathambi, P., and Karihaloo, B. L., ''Determination of Specimen-Size Independent Fracture Toughness of Plain Concrete,'' *Magazine of Concrete Research,* Vol. 38, No. 135, 1986, pp. 67–76.

31. Karihaloo, B. L., and Nallathambi, P., ''An Improved Effective Crack Model for

the Determination of Fracture Toughness of Concrete,'' *Cement and Concrete Research,* Vol. 19, 1989, pp. 603–610.

32. Karihaloo, B. L., and Nallathambi, P., ''Notched Beam Test: Mode I Fracture Toughness,'' in *Fracture Mechanics Test Methods for Concrete,* RILEM Report 5, edited by Shah, S. P., and Carpinteri, A., Chapman & Hall, London, 1991, pp. 1–86.
33. Swartz, S. E., and Refai, T. M. E., ''Influence of Size Effects on Opening Mode Fracture Parameters for Precracked Concrete Beams in Bending,'' in *Fracture of Concrete and Rock,* edited by Shah, S. P., and Swartz, S. E., Springer-Verlag, New York, 1988, pp. 243–254.
34. Swartz, S. E., and Refai, T., ''Cracked Surface Revealed by Dye and Its Utility in Determining Fracture Parameters,'' in *Fracture Toughness and Fracture Energy,* edited by Mihashi, et al., Balkema, The Netherlands, 1989, pp. 509–520.
35. Go, C. G., Swartz, S. E., and Hu, K. K., ''Stress Intensity Factors for Single-Edge-Notch Beam,'' *Journal of Engineering Mechanics, ASCE,* Vol. 110, No. 4, 1984, pp. 629–633.
36. Li, V. C., Chan, C.-M., and Leung, C. K. Y., ''Experimental Determination of the Tension-Softening Relations for Cementitious Composites,'' *Cement and Concrete Research,* Vol. 17, 1987, pp. 441–452.
37. Brühwiler, E., and Wittmann, F. H., ''The Wedge Splitting Test, a New Method of Performing Stable Fracture Mechanics Tests,'' *Engineering Fracture Mechanics,* Vol. 35, No. 1/2/3, 1990, pp. 117–125.
38. Rossi, P., Brühwiler, E., Chhuy, S., Jenq, Y.-S., and Shah, S. P., ''Fracture Properties of Concrete as Determined by Means of Wedge Splitting Tests and Tapered Double Cantilever Beam Tests,'' in *Fracture Mechanics Test Methods for Concrete,* RILEM Report 5, edited by Shah, S. P., and Carpinteri, A., Chapman & Hall, London, 1991, pp. 87–128.
39. Linsbauer, H. N., and Tschegg, E. K., ''Fracture Energy Determination of Concrete with Cube-Shaped Specimens,'' *Zement and Beton,* Vol. 31, 1986, pp. 38–40 (in German).
40. Ouyang, C., Tang, T., and Shah, S. P., ''Relationship between Fracture Parameters from Two Parameter Fracture Model and from Size Effect Model, *Materials and Structures, RILEM,* 1995, In press.
41. Bazant, Z. P., Gettu, R., and Kazemi, M. T., ''Identification of Nonlinear Fracture Properties from Size Effects Tests and Structural Analysis Based on Geometry-Dependent *R*-Curves,'' *International Journal of Rock Mechanics and Mineral Science,* Vol. 28, No. 1, 1991, pp. 43–51.
42. Ouchterlony, F., Takahashi, H., Matsuki, K., and Hashida, T., ''Experiences from Fracture Toughness Testing of Rock According to the ISRM Suggested Methods,'' Report DS 1991: 2, Swedish Detonic Research Foundation, Stockholm, Sweden, 1991.
43. Ouchterlony, F., ''Fracture Toughness Testing of Rock with Core Based Specimens, the Development of an ISRM Standard,'' in *Fracture Toughness and Fracture Energy, Test Methods for Concrete and Rock,* edited by Mihashi, H., et al., Balkema, Rotterdam, 1989, pp. 231–251.
44. ISRM, ''Suggested Methods for Determining the Fracture Toughness of Rock,''

Ouchterlony, F., Working Group Coordinator, *International Journal of Rock Mechanics and Mineral Science,* Vol. 25, 1988, pp. 71–96.

45. Barker, L. M., "Data Analysis Methods for Short Rod and Short Bar Fracture Toughness Tests of Metallic Materials," Report TR80-12, Terra Tek, Salt Lake City, UT, 1980.
46. Barker, K. M., "Theory for Determining K_{Ic} from Small, Non-LEFM Specimens, Supported by Experiments on Aluminum," *International Journal of Fracture,* Vol. 15, 1979, pp. 515–536.
47. Chong, K. P., Kuruppu, M. D., and Kuszmaul, J. S., "Fracture Toughness Determination of Rocks with Core-Based Specimens," Keynote Paper, in *Fracture of Concrete and Rock,* edited by Shah, S. P., and Swartz, S. E., Springer-Verlag, New York, 1989, pp. 13–25.
48. Chong, K. P., Basham, K. D., and Wang, D. Q., "Fracture Parameters Derived from Tension-Softening Measurement Using Semi-Circular Specimens," in *Fracture Toughness and Fracture Energy,* edited by Mihashi, H., et al., Balkema, Rotterdam, 1989, pp. 279–286.
49. Basham, K. D., Chong, K. P., and Boresi, A. P., "A New Development in Devising Tension-Softening Curves for Brittle Materials," in *Proceedings of the Ninth International Conference on Experimental Mechanics,* Copenhagen, Denmark, 1990, pp. 1423–1432.
50. Chong, K. P., and Kuruppu, M. D., "New Specimen for Fracture Toughness Determination of Rock and Other Materials," *International Journal of Fracture,* Vol. 26, 1984, pp. R59–R62.
51. Luong, M. P., "Un Nouvel Essai pour la Mesure de la Resistance a la Traction," *Revue Française de Geotechnique,* No. 34, 1986, pp. 69–74.
52. Luong, M. P., "Tensile and Shear Strengths of Concrete and Rock," *Engineering Fracture Mechanics,* Vol. 35, No. 1/2/3, 1990, pp. 127–135.

7

FRACTURE RESISTANCE CURVES (R-CURVES) FOR QUASI-BRITTLE MATERIALS

7.1 GENERAL DESCRIPTION OF R-CURVES

According to the energy principles discussed in Chapter 3, the necessary condition for a structure in an equilibrium state during crack propagation is that the first-order derivative of the potential energy, Π, be equal to zero:

$$\frac{\partial \Pi}{\partial a} = \frac{\partial}{\partial a}(U - F + W) = 0 \quad \text{or} \quad \frac{\partial}{\partial a}(F - U) = \frac{\partial W}{\partial a} \tag{7.1}$$

where U is the strain energy of the structure, F is the work done by the applied force, and W is the energy required for crack formation. Equation (7.1) can be rewritten as

$$G_q = R \tag{7.2}$$

where $G_q = 1/t\ \partial(F - U)/\partial a$ is the energy release rate for propagation of a unit length of crack in a structure with a unit thickness, $R = 1/t\ \partial W/\partial a$ is the fracture resistance of the material, and t is the thickness of the structure. For linearly elastic materials (perfectly brittle) R is a material constant, as shown in Fig. 7-1, and any extension of the initial crack means catastrophic failure of the structure. Therefore, Eq. (7.2) can serve as a fracture criterion for linearly elastic materials.

For quasi-brittle materials, R is a rising function with increasing crack length before material failure, as shown in Fig. 7-1. When the value of G_q increases due to increasing applied load, the value of R also increases. Therefore Eq.

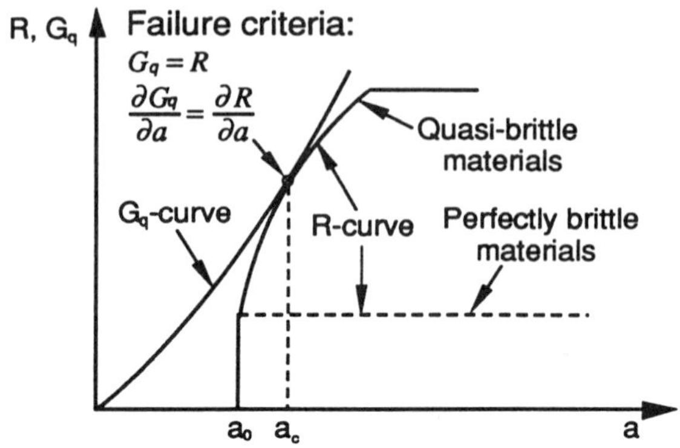

Figure 7-1 The *R*-curves for quasi-brittle materials.

(7.2) can only serve as the necessary condition for failure of a structure for quasi-brittle materials. Crack propagation may be stable or unstable when Eq. (7.2) is satisfied. Since G_q generally increases at a faster rate than R does, a second condition given below can be satisfied at a certain crack length as

$$\frac{\partial G_q}{\partial a} = \frac{\partial R}{\partial a} \tag{7.3}$$

Equation (7.3) can be regarded as a sufficient condition for the onset of unstable crack propagation (failure) in quasi-brittle materials. The combination of Eqs. (7.2) and (7.3) provides the necessary and sufficient conditions for failure of a structure made of a quasi-brittle material. Both Eqs. (7.2) and (7.3) are illustrated in Fig. 7-1. In order to use Eqs. (7.2) and (7.3) to describe the failure of a structure, both G_q and R should be known. Generally, the energy release rate G_q is a function of applied load, structural geometry, and boundary conditions and can be obtained from analytical or numerical stress analysis.[1] However, the fracture resistance R usually is an unknown function depending on the material fracture property and structural geometry. This chapter will discuss different approaches to construct and measure R-curves for quasi-brittle materials. It is noted that in some literature the term K_R is used to represent the material fracture resistance instead of R. The relationship between K_R and R for the plane-stress condition is

$$R = \frac{(K_R)^2}{E} \tag{7.4}$$

For the plane-strain condition, the modulus of elasticity E should be replaced by $E/(1 - \mu^2)$.

Since Eqs. (7.2) and (7.3) can serve as failure criteria for a quasi-brittle material, they may be regarded as the basic definition for *R*-curves. Equations (7.2) and (7.3) suggest that two methods may be used to measure or construct *R*-curves. First, since *R*-curves are characterized by a stable crack propagation prior to failure according to Eq. (7.2), one may experimentally measure *R*-curves by monitoring positions of crack length during loading for a specimen. The crack tip in quasi-brittle materials such as ceramics may be monitored by using a microscope. This method has been used by many investigators for ceramics.[2–4] However, since many toughening mechanisms exist in the fracture process zone and the size of the fracture process zone varies through the thickness (as discussed in Chapter 4), an effective crack length should be used for quasi-brittle materials when principles of fracture mechanics are applied. The crack tip observed on the specimen surface may not coincide with the tip of the effective crack. Use of the surface crack length in fracture mechanics for ceramics may not result in a great error due to the relatively small size of the fracture process zone in these materials. However, concrete is more anisotropic and heterogeneous than ceramics, and the effective crack length should be introduced in order to use the principles of fracture mechanics. As presented in Chapter 3, both ASTM E 1152-87 and ASTM E 1290-89 recommend the use of a compliance method to determine the effective crack length for both *J–R* curves and CTOD–*R* curves of metallic materials. Since G_q is specimen geometry dependent, *R*-curves based on Eqs. (7.2) and (7.3) generally also depend on structural geometry except for ideally brittle materials. The disadvantage of measuring an *R*-curve based on the concept of stable crack propagation is that the *R*-curve measured using one geometry may not apply to other geometries.

On the other hand, Eqs. (7.2) and (7.3) indicate that an *R*-curve is a tangent to a G_q-curve at the critical crack length for a given structure. Therefore if an *R*-curve is assumed to be invariant for a certain category of structures, it may be mathematically constructed as an envelope of energy release rates G_q for the given category of structures. Based on this interpretation, *R*-curves can be constructed in semianalytical and semiexperimental ways. Mathematical functions of *R*-curves containing some constants are first derived based on Eqs. (7.2) and (7.3). These unknown constants are then determined using material fracture parameters. It is possible to include some geometric coefficients to account for the geometric effect on *R*-curves in the functions of *R*-curves obtained.

The concept of an *R*-curve was first introduced using the energy criterion by Irwin for fracture of metals.[5] Krafft et al.[6] proposed that the *R*-curve is a unique function of crack extension for metals. This suggestion was used by Broek[7] to derive a semiempirical solution of an *R*-curve. His solution, however, is only valid for short cracks. The *R*-curve approach has been recently widely used for fracture of cementitious composites and ceramics.[8–10] Since the fracture behavior of quasi-brittle materials can be well described by an *R*-curve, many attempts have been made to theoretically construct and experimentally measure the *R*-curve. The major difficulty in the study of *R*-curve behavior is that *R*-curves depend on both specimen geometry and material parameters. Wecharatana and Shah[11,12] have used an *R*-curve to model the stable crack growth in concrete. Cook et al.[13] have developed an *R*-curve with four material constants based on a cohesive crack. Cook and Clarke[14] assumed that an *R*-curve is a power function with two geometry-dependent constants. Kim and Joe[15] have proposed a method that determines an *R*-curve from a series of load and load point displacement curves of specimens with different initial crack lengths. Faber et al.[16] have studied *R*-curves of ceramics. Microcrack shielding and bridging by whiskers were taken into account in their study. Becher[17] analyzed toughening by discontinuous bridging phases based on a cohesive fracture model. Bennison and Lawn[18] have realized that *R*-curves depend on the initial flaw size and found that if no toughening mechanism is active prior to crack extension, then *R*-curves can be approximately constructed as a function of the crack extension only. Bennison et al.[19] have discussed toughening mechanisms and microstructure associated with *R*-curve behavior for alumina-based materials. Zok[20] has used the *R*-curve to predict the failure strength of composite materials. Elangovan[21] has developed an *R*-curve as a nonlinear function of size of the crack tip plastic zone. Sakai and Ichikawa[22] have proposed an *R*-curve including energy dissipated during irreversible deformation. Li and Liang[23] have proposed a stability theory for crack propagation based on the fictitious crack approach. Since second-order derivatives of a potential energy were considered, their theory is equivalent to the *R*-curve approach.

As discussed in Chapter 3, since *R*-curves can be mathematically interpreted as an envelope of a series of G_q-curves for a certain type of structures, they may be categorized into three types of envelopes of G_q-curves for a given structural geometry and material fracture property: (i) an *R*-curve may be defined as an envelope of G_q-curves for a series of structures with the same size but different initial crack lengths, (ii) an *R*-curve may be defined as an envelope of G_q-curves for a series of structures with different sizes but the same initial crack length, and (iii) an *R*-curve may be defined as an envelope of G_q-curves for a series of geometrically similar structures with increasing sizes.

On the other hand, based on different energy dissipation mechanisms in quasi-brittle materials (see Chapter 5), *R*-curves can be classified based on the fictitious crack approach and the equivalent-elastic crack approach. In the fictitious crack approach, a cohesive force on the crack tip is used to model *R*-curves. In the equivalent-elastic crack approach, a traction-free, effective-elastic crack is assumed, and *R*-curves are derived based on LEFM with at least two fracture parameters. It is noted that the *R*-curve is not unique for a given structure due to its different definitions discussed above. In other words, different *R*-curves can be used to describe fracture behavior for a given structure. In this chapter, *R*-curves based on the fictitious crack approach and the equivalent-elastic crack approach, proposed by different investigators, will be presented.

7.2 *R*-CURVES BASED ON BRIDGING MECHANISMS

7.2.1 *R*-Curve Proposed by Foote et al.

By modeling the fracture process zone with a cohesive stress, Foote et al.[24] have proposed a fracture resistance curve in terms of K_R. For a structure subjected to an applied load and the cohesive stress as shown in Fig. 7-2, the stress intensity factor at a crack tip in a quasi-brittle material, K_q, can be divided into two parts: (i) the stress intensity to create two new cracked surfaces, K_{Ic}, and the stress intensity to overcome the cohesive pressure $\sigma(w)$ in separating the surfaces, K_σ. Therefore, K_q is written as

$$K_q = K_{Ic} + K_\sigma \tag{7.5}$$

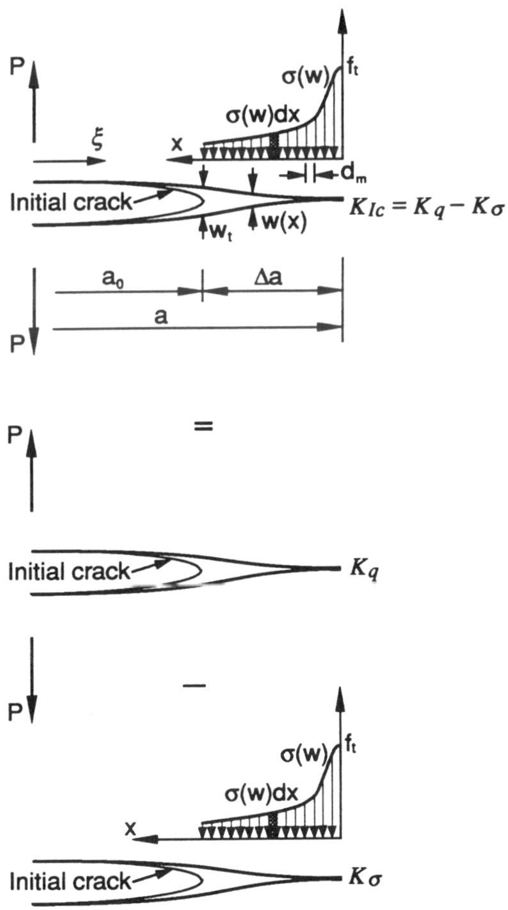

Figure 7-2 Superposition of stress intensity factors at crack tip.

Since $G = K^2/E$, K_q, K_{Ic}, and K_σ correspond to G_q, G_{Ic}, and G_σ, respectively, as mentioned in Chapter 5. The term K_σ is defined as a positive increasing function here. Since K_q should be balanced by K_R, the following expression is obtained:

$$K_R = K_{Ic} + K_\sigma \tag{7.6}$$

Both the Griffith–Irwin and the Dugdale–Barenblatt energy dissipation mechanisms are included in Eq. (7.6). The value of K_σ can be evaluated based on a Green's function as

$$K_\sigma = \int_{a0}^{a} \phi(a, x)\sigma(x)\, dx \tag{7.7}$$

where $\phi(a, x)$ is the appropriate Green's function that represents the stress intensity factor at the crack tip due to a unit force acting at a distance x from the crack tip, as shown in Fig. 7-3. Its value based on LEFM can be obtained from an appropriate handbook.[1] Foote et al.[24] suggested the following softening function for $\sigma(w)$:

$$\sigma(w) = f_t \left(1 - \frac{w}{w_c}\right)^m \tag{7.8}$$

where f_t and w_c are the tensile strength of a material and the crack separation when the tensile stress drops to zero, respectively. The parameter m is the softening index. It is noted that $m = 0$ for ideally brittle materials and $m = \infty$ for ideally plastic materials, as shown in Fig. 7-4.

The term $\sigma(w)$ given in Eq. (7.8) is a function of w and cannot be directly used in Eq. (7.7), which requires that σ is a function of position x. Therefore it is necessary to evaluate the value of w in terms of x. This can be done using Castigliano's theorem.[1] Consider a structure with a crack as shown in Fig. 7-3. To find $w(x)$ at a point x from the crack tip due to the applied load P acting

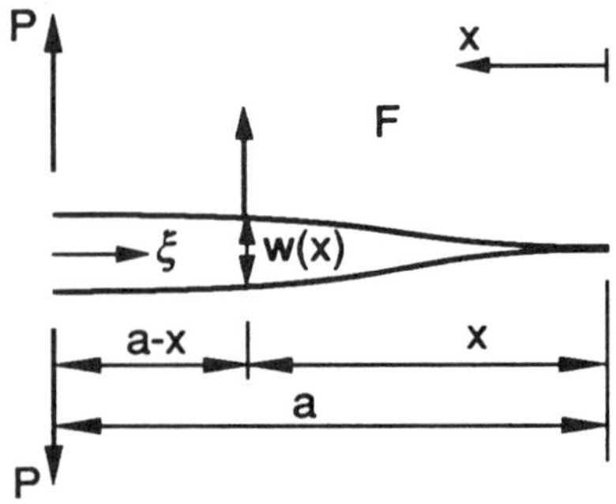

Figure 7-3 Estimation of crack opening displacement by Castigliano's theorem (F is a fictitious force).

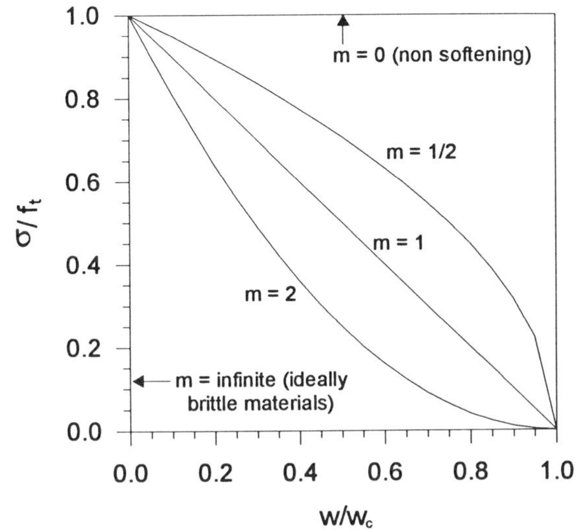

Figure 7-4 Relationship between cohesive stress and crack opening displacement.

at a point a away from the tip, one can apply a fictitious force F at x and determine the total strain energy U of the system as

$$U = U_0 + \Delta U_a = U_0 + \int_0^a \frac{(K_{IP} + K_{IF})^2}{E}\, d\xi \tag{7.9}$$

where U_0 is the strain energy in the uncracked specimen, ΔU_a is the strain energy change due to creation of the crack while holding forces P and F constant, and K_{IP} and K_{IF} are the stress intensity factors due to the applied force P and the fictitious force F, respectively. The term ΔU_a is evaluated by integrating the strain energy release rate of the structure over the crack length in Eq. (7.9).

Castigliano's theorem is now used to estimate the crack opening displacement $w(x)$, which is given by

$$w(x) = \left(\frac{\partial U}{\partial F}\right)_{F=0} = \left(\frac{\partial U_0}{\partial F}\right)_{F=0} + \frac{1}{E}\int_0^a \left(\frac{\partial K_{IP}^2}{\partial F} + 2K_{IP}\frac{\partial K_{IF}}{\partial F} + 2K_{IF}\frac{\partial K_{IF}}{\partial F}\right)_{F=0} d\xi \tag{7.10}$$

such that

$$\frac{\partial U_0}{\partial F} = 0 \qquad \frac{\partial K_{IP}}{\partial F} = 0 \qquad (K_{IF})_{F=0} = 0 \tag{7.11}$$

Moreover until the crack length ξ grows beyond the location of the forces F, it is seen that

$$\frac{\partial K_{IF}}{\partial F} = 0 \quad \text{for } \xi \le a - x \tag{7.12}$$

Substituting Eqs. (7.11) and (7.12) into Eq. (7.10) results in

$$w(x) = \frac{2}{E} \int_{a-x}^{a} K_{IP}(\xi) \left[\frac{\partial K_{IF}(\xi - a + x)}{\partial F} \right]_{F=0} d\xi \tag{7.13}$$

where ξ is the crack coordinate used for the integration. In Eq. (7.13), $K_{IP}(\xi)$ indicates that the stress intensity factor K_{IP} should be expressed as a function of the crack length ξ, whereas $K_{IF}(\xi - a + x)$ indicates that K_{IF} should be expressed as a function of the crack length $\xi - a + x$. Equation (7.13) is valid for the plane-stress condition. For the plane-strain condition, the modulus of elasticity E should be replaced by $E/(1 - \mu^2)$.

A K_R-curve is obtained by substituting Eqs. (7.7), (7.8), and (7.13) into Eq. (7.6). Since the crack opening displacement $w(x)$ determined from Eq. (7.13) is a nonlinear function of x, a numerical iterative scheme is usually needed to obtain the K_R-curve. The Green's function in Eq. (7.7) and the crack opening displacement calculated from Eq. (7.13) depend on specimen geometry and size. As a result, the K_R-curve obtained here are geometry and size dependent. Foote et al.[24] have calculated the approximated K_R-curve by assuming that $w(x)$ is a linear function of x. They have found that the approximated R-curve based on the linear crack opening profile is close to the K_R-curve in terms of the nonlinear crack opening profile for a double-cantilever beam.

7.2.2 Simplified *R*-Curve by Mai et al.

Mai et al.[25,26] have further simplified the above K_R-curve by using the Green's function for a semi-infinite plate with a line crack and by neglecting the effect of the closure stress on the crack opening profile. In their simplification, Mai et al. assumed that bridging is a major toughening mechanism for materials, and the bridging mechanism is characterized by the mean distance between bridges, d_m, as shown in Fig. 7-2. This simplified K_R-curve is presented below.

Substituting the Green's function for a semi-infinite plate with a line crack into Eq. (7.7) leads to

$$K_\sigma = \frac{2g_4\sqrt{a}}{\pi} \int_{a_0}^{a} \frac{\sigma(\xi)}{\sqrt{a^2 - \xi^2}} \, d\xi \tag{7.14}$$

where g_4 is a geometric factor depending on the ratio of ξ/a and approximately equal to $\sqrt{\pi}$. It is assumed that the value of $K_\sigma = 0$ when the crack length is smaller than the mean distance between bridges, d_m. One may assume that the crack opening displacement $w(\xi)$ is approximately determined from the condition of $K_q - K_\sigma = K_{Ic}$ for simplification.[25] This leads to

$$w(a, \xi) = \frac{\sqrt{8}g_4K_{Ic}}{E\pi} \sqrt{a - \xi} \tag{7.15}$$

Substituting Eqs. (7.8), (7.14), and (7.15) into Eq. (7.6) results in the following K_R-curve:

$$K_R = K_\infty - (K_\infty - K_{Ic}) \left[1 - \left(\frac{c - d_m}{c_F - d_m} \right)^{1/2} \right]^{m+1} \tag{7.16}$$

where the plateau K_∞ is the value of K_R at $w = w_c$, which is given by

$$K_\infty = K_{Ic} + \frac{Ef_tw_c}{(m + 1)K_{Ic}} \tag{7.17}$$

and c is the crack extension, and the value of c_F is the crack extension at $w = w_c$, which is given by

$$c_F = d_m + \left(\frac{\pi E w_c}{\sqrt{8}g_4K_{Ic}} \right)^2 \tag{7.18}$$

Equation (7.16) is valid for $d_m < c < c_F$. Here, $K_R = K_\infty$ for $c > c_F$.

Four materials parameters, f_t, w_c, m, and d_m, are required for the K_R-curve. Since Eqs. (7.14) and (7.15) are approximate, the K_R-curve given by Eq. (7.16) does not depend on the geometry and size of specimen.

A K_R-curve for a macro-defect-free cement paste predicted by Eq. (7.16) is illustrated in Fig. 7-5.[26] Values of $d_m = 50$ μm, $K_{Ic} = 1.40$ MPa $\sqrt{\text{m}}$, $K_\infty = 2.11$ MPa $\sqrt{\text{m}}$, and $c_F = 3.1$ mm were used, and the m values 0, 1, and 2, respectively, were used. The experimental data measured using a double-cantilever beam specimen are also given in the figure. It is seen that by appropriately selecting values of f_t, w_c, m, and d_m (or d_m, K_{Ic} K_∞, and m), a K_R-curve for a material can be calculated using Eq. (7.16).

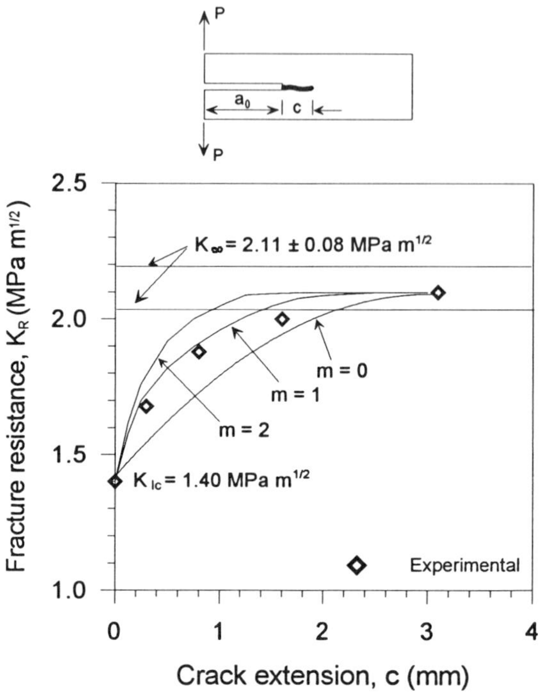

Figure 7-5 The *R*-curve for macro-defect-free cement paste predicted by Mai and Lawn's model and experimental results.[26]

7.2.3 *R*-Curves with $K_{Ic} = 0$

If the Griffith–Irwin energy dissipation mechanism is ignored [$K_{Ic} = 0$ in Eq. (7.6)], the fracture process in a quasi-brittle material may be fully characterized by the crack separation curve $\sigma(w)$. As a result, Eq. (7.6) can be rewritten into an equivalent form as[27]

$$R(w) = \frac{(K_R)^2}{E} = \int \sigma(w)\, dw \tag{7.19}$$

Equation (7.19) is valid for a plane-stress condition. For a plane-strain condition, the modulus of elasticity E should be replaced by $E/(1 - \mu^2)$. Based on the definition of $\sigma(w)$, the following relationships can be obtained from Eq. (7.19):

$$\frac{\partial R(w)}{\partial w} = \sigma(w) \tag{7.20}$$

and

$$\left[\frac{\partial R(w)}{\partial w}\right]_{w=0} = f_t \qquad [R(w)]_{w=w_c} = G_F \tag{7.21}$$

If $\sigma(w)$ is regarded as a material fracture property, the $R(w)$ curve given by Eq. (7.19) seems to be independent of geometry and size of structures. However, unless the unknown function w is first determined for a given structure, Eq. (7.19) cannot be directly used to predict fracture behavior of the structure. Since the value of w depends on the geometry and size of the structure as well as the loading configuration, this determination of the function w makes the $R(w)$ curve structural geometry and size dependent. Planas et al.[28] have applied the R-curve given by Eq. (7.19) to concrete structures.

7.3 *R*-CURVE PROPOSED BY OUYANG ET AL.

Ouyang et al.[29,30] have proposed an R-curve using the equivalent-elastic crack approach. Based on Eqs. (7.2) and (7.3), which can be regarded as the definition of the R-curve, a governing differential equation for R-curves can be derived. The R-curve obtained by solving this differential equation is interpreted as an envelope of the energy release rates for a series of structures with the same specimen geometry and initial crack length but different specimen sizes, as shown in Fig. 7-6, and can be classified in the second category of R-curves, as discussed in Chapter 3. Their R-curve will be presented below.

7.3.1 Differential Equation Governing *R*-Curves and Its Solution

The energy release rate G_q, as shown in Fig. 7-6, can be expanded as a Taylor series at a critical point,

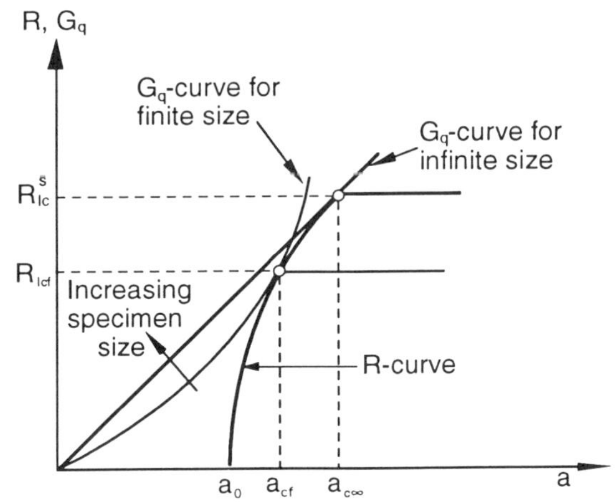

Figure 7-6 The R-curve as an envelope of G_q-curves for a series of structures with the same initial crack length but different sizes as proposed by Ouyang et al.[30]

$$G_q = (G_q)_c + \sum_{n=1}^{\infty} \frac{1}{n!} \left(\frac{d^n G_q}{da^n} \right)_c (a - a_c)^n \tag{7.22}$$

where a_c is the critical crack length, $(G_q)_c$ is the value of G_q at the critical point, and subscript c indicates the critical point. Using $G_q = 0$ at $a = 0$ results in the following:

$$(G_q)_c = -\sum_{n=1}^{\infty} \frac{1}{n!} \left(\frac{d^n G_q}{da^n} \right)_c (-a_c)^n \tag{7.23}$$

For a structure with precritical stable crack propagation, the critical crack length a_c can be regarded as proportional to the initial crack length a_0:

$$a_c = a_0 + \Delta a_c = \alpha a_0 \tag{7.24}$$

where α describes the precritical stable crack growth ($\alpha \geq 1$), and it may be regarded as a brittleness index.

Substituting Eq. (7.24) into Eq. (7.23) and noting $da = d(a_0 + \Delta a) = d\Delta a$, one obtains

$$(G_q)_c = -\sum_{n=1}^{\infty} \frac{1}{n!} \left(\frac{d^n G_q}{d\,\Delta a^n} \right)_c \left(\frac{\alpha}{\alpha - 1} \right)^n (-\Delta a_c)^n \tag{7.25}$$

since $(G_q)_c$ represents critical points on the G_q curves, and they can be replaced by R. Thus Eq. (7.25) can be rewritten as

$$R + \sum_{n=1}^{\infty} \frac{1}{n!} \frac{d^n R}{dc^n} \left(\frac{\alpha}{\alpha - 1} \right)^n (-c)^n = 0 \tag{7.26}$$

where $c = \Delta a_c$ is used for convenience. If the R-curve is an envelope of G-curves for a certain category of structures, each point on the R-curve corresponds to a critical point for some structure. Therefore, c is a variable in Eq. (7.26).

Since Eq. (7.26) must be valid for any crack extension c, it yields a differential equation governing R-curves for a quasi-brittle material. From the mathematical point of view, Eq. (7.26) can be classified as an *Euler* equation. By using the transformation

$$x = \ln(c) \tag{7.27}$$

Eq. (7.26) can be transformed into a linear differential equation with constant coefficients, with its auxiliary equation represented by

$$\sum_{n=1}^{\infty} d(d-1)(d-2)\cdots(d-n+1)\left(\frac{\alpha}{\alpha-1}\right)^n(-1)^n + 1 = 0 \tag{7.28}$$

The solution of Eq. (7.26) can be written as

$$R = \sum_{n=1}^{\infty} \beta_n c^{d_n} \tag{7.29}$$

where the β_n are constants to be determined and the d_n are roots of Eq. (7.28).

It can be shown that for $n = 1$, where G_q is a linear function of the crack length, the following R-curve can be solved from Eq. (7.26):

$$R = \beta_0(a - a_0)^{d_0} \qquad d_0 = \frac{\alpha - 1}{\alpha} \tag{7.30}$$

where α and β_0 are constants. Equation (7.30) is the R-curve obtained by Broek[7] for materials with a small crack extension, such as metals, and cannot be used for quasi-brittle materials.

If only terms up to $n = 2$ are considered, which indicates that G_q is approximated by a quadratic expression of crack length, the following R-curve can be solved from Eq. (7.26):

$$R = \beta_1(a - a_0)^{d_1} + \beta_2(a - a_0)^{d_2} \tag{7.31}$$

where

$$\begin{aligned} d_1 &= \frac{1}{2} + \frac{\alpha-1}{\alpha} + \left[\frac{1}{4} + \frac{\alpha-1}{\alpha} - \left(\frac{\alpha-1}{\alpha}\right)^2\right]^{1/2} \\ d_2 &= \frac{1}{2} + \frac{\alpha-1}{\alpha} - \left[\frac{1}{4} + \frac{\alpha-1}{\alpha} - \left(\frac{\alpha-1}{\alpha}\right)^2\right]^{1/2} \end{aligned} \tag{7.32}$$

Equations (7.31) and (7.32) are a general R-curve formulation with three parameters, α, β_1, and β_2. Once these parameters are determined, the R-curve is known.

7.3.2 Determination of Parameters α, β_1 and β_2

Values of α, β_1, and β_2 can theoretically be determined based on the geometric function and material fracture property of any given size of structures. The presence of the geometric function $g_1(a/b)$, which generally is a nonlinear function in terms of the relative crack length a/b, will result in a great complexity for determining parameters α, β_1, and β_2. Since $g_1(a/b)$ approaches a constant for an infinite-size specimen, the procedure to determine parameters

α, β_1, and β_2 will be greatly simplified if infinite-size specimens are used. As a result, values of α, β_1, and β_2 will be determined using infinite-size specimens in the following.

According to LEFM,[1] the strain energy release rate can be generally written as

$$G_q = \frac{\sigma^2 \pi a [g_1(a/b)]^2}{E} \tag{7.33}$$

where $g_1(a/b)$ is the geometry function on the stress intensity factor and σ is the nominal applied load.

For infinite-size specimens, Eq. (7.33) is reduced to

$$G_q = \frac{\sigma^2 \pi a f_1^2}{E} \tag{7.34}$$

where $f_1 = [g_1(a/b)]_{b\to\infty}$ is a geometric factor for the stress intensity factor, and its value is equal to 1 for the center-notch specimen and equal to 1.123 for the single-edge notch tensile specimen and for the three-point bend beam (with $S = 4b$).[1]

Substituting Eqs. (7.31), (7.32), and (7.34) into Eqs. (7.2) and (7.3), which is the basic definition of *R*-curves, a relationship between β_1 and β_2 is obtained, and the *R*-curve becomes

$$R = \beta_2 \psi_a (a - a_0)^{d_2} \tag{7.35}$$

where

$$\psi_a = 1 - \frac{d_2\alpha - \alpha + 1}{d_1\alpha - \alpha + 1}\left(\frac{\alpha a_0 - a_0}{a - a_0}\right)^{d_2 - d_1} \tag{7.36}$$

Only two parameters, α and β_2, need to be further determined. Based on the equivalent-elastic crack approach, the values of α and β_2 can be determined from K_{Ic}^s and $CTOD_c$, or G_f and c_f, which are fracture parameters for quasi-brittle materials and have been discussed in detail in Chapters 5 and 6.

In order to use the critical stress intensity factor K_{Ic}^s and the critical crack tip opening displacement $CTOD_c$ to determine values of α and β_2, one needs to know the opening profile of a crack. Similar to Cook et al.,[13] Sneddon's solution[31] for the near-field displacement of an equilibrium crack is used,

$$\text{COD} = \text{CMOD}\sqrt{1 - \left(\frac{x}{a}\right)^2} \tag{7.37}$$

where

$$\text{CMOD} = \frac{4\sigma a f_2}{E} \tag{7.38}$$

The term $f_2 = [g_2(a/b)]_{b\to\infty}$ is a geometry factor for the crack mouth opening displacement (CMOD), $f_2 = 1$ for the center-notch tensile specimen, $f_2 = 1.454$ for the single-edge notch tensile specimen, and $f_2 = 1.42$ for the three-point bend beam (with $S = 4b$).[1] It is noted that Sneddon's crack profile is an exact solution for the infinite-size, center-notch tensile specimen and is only an approximation for infinite-size specimens with other geometries. Using the critical conditions of $R = (K_{\text{Ic}}^s)^2/E$ and COD $= \text{CTOD}_c$ when $x = a_0$ and $a = a_c = \alpha a_0$, the values of α and β_2 can be determined as

$$\alpha = \frac{\pi E^2 f_1^2\,\text{CTOD}_c^2}{32 a_0\,(K_{\text{Ic}}^s)^2\,f_2^2} + \sqrt{\left[\frac{\pi E^2\,f_1^2\,\text{CTOD}_c^2}{32 a_0\,(K_{\text{Ic}}^s)^2\,f_2^2}\right]^2 + 1} \tag{7.39}$$

and

$$\beta_2 = \frac{(K_{\text{Ic}}^s)^2\,(d_1\alpha - \alpha + 1)}{E\alpha(d_1 - d_2)\,(\alpha a_0 - a_0)^{d_2}} \tag{7.40}$$

It is noted that since an infinitely large specimen is used for obtaining Eqs. (7.39) and (7.40), the value of α is equal to $a_{c\infty}/a_0$ hereafter, where $a_{c\infty}$ is the critical crack length of the infinitely large specimen.

On the other hand, parameters α and β_2 can be determined based on the fracture toughness G_f and the critical crack extension for infinite-size structure c_f. The condition of $R = G_f$ when $a = a_{c\infty} = \alpha a_0 = a_0 + c_f$ results in

$$\alpha = 1 + \frac{c_f}{a_0} \tag{7.41}$$

and

$$\beta_2 = \frac{G_f\,(d_1\alpha - \alpha + 1)}{\alpha(d_1 - d_2)\,(\alpha a_0 - a_0)^{d_2}} \tag{7.42}$$

To obtain a geometry-dependent R-curve for an infinite-size specimen, the material constants K_{Ic}^s and CTOD_c (or G_f and c_f) are first substituted into Eqs. (7.39) and (7.40) [or Eqs. (7.41) and (7.42)] to determine the values of α and β_2. The R-curve can then be determined from Eqs. (7.32), (7.35), and (7.36). It is noted that the value of R should not exceed its maximum value of R_{Ic}^s. This leads to

$$R = R_{\mathrm{I}c}^{s} = \frac{(K_{\mathrm{I}c}^{s})^2}{E} \quad \text{for } a > a_{c\infty} \tag{7.43}$$

The geometry effect on the R-curve is taken into account by factors f_1 and f_2, which can be obtained from a LEFM handbook.[1]

The R-curve obtained in such a way is regarded as an envelope of G_q-curves of a series of structures with the same geometry and initial crack length but different sizes. For a given geometry and a given initial crack length, different-size specimens follow the same R-curve with the exception of different plateau values, as shown in Fig. 7-6.

7.3.3 *R*-Curve for Finite-Size Specimen

For the same geometry and initial crack length, finite-size specimens follow the same R-curve as that for the infinite-size specimen. However, the critical crack length for the finite-size specimens is a_{cf} rather than $a_{c\infty}$ due to the size effect. The value of a_{cf} can be determined by substituting Eqs. (7.33) and (7.35) into Eqs. (7.2) and (7.3), where the geometry function $g_1(a/b)$ is not constant. This procedure results in the equation

$$2\frac{d[g_1(a/b)]}{da} - \frac{d\psi_a}{da}\frac{g_1(a/b)}{\psi_a} - \frac{d_2 g_1(a/b)}{a - a_0} + \frac{g_1(a/b)}{a} = 0 \tag{7.44}$$

Since both $g_1(a/b)$ and ψ_a are known functions of crack length, the only unknown in Eq. (7.44) is the crack length a. Therefore, Eq. (7.44) provides a nonlinear equation to determine the value of a_{cf}.

It is noted that the applied load reaches a maximum value when $a = a_{cf}$ for a given geometry and size. Therefore, by using the energy balance ($G_q = R$), the value of a_{cf} can be alternatively obtained by plotting the nominal applied stress σ as the function of the crack length a as below:

$$\sigma = \left[\frac{R}{(G_q)_{\sigma=1}}\right]^{1/2} \tag{7.45}$$

By increasing the crack length a, the nominal stress σ given by Eq. (7.45) initially increases and then reaches a maximum value. This is followed by a softening-type stress–crack length curve. The crack length corresponding to the maximum nominal applied stress is the critical crack length a_{cf} for the given finite-size specimen. As a result, the value of a_{cf} can be obtained by maximizing σ given by Eq. (7.45). The use of Eq. (7.45) can avoid solving the nonlinear equation to obtain the value of a_{cf}. A difference in the values of a_{cf} results in different plateau values of R-curves for the finite-size specimens. These values, termed $R_{\mathrm{I}cf}$ hereafter, can be determined by substituting different values of a_{cf} into Eq. (7.35).

Since Eqs. (7.2) and (7.3) have been used to determine parameter β_1 for infinite-size specimens and a_{cf} for finite-size specimens, the proposed *R*-curve automatically satisfies these two equations for specimens of any geometries and sizes.

To obtain an *R*-curve for an infinite-size specimen, the material constants K_{Ic}^s and $CTOD_c$ (or G_f and c_f) are first substituted into Eqs. (7.39) and (7.40) [or Eqs. (7.41) and (7.42)] to determine the values of α and β_2. The *R*-curve can then be determined from Eqs. (7.32), (7.35), and (7.36). The *R*-curve obtained here has a maximum value of $R_{Ic}^s = (K_{Ic}^s)^2/E$. The finite-size specimen with the same geometry and initial flaw size follows the same *R*-curve expression but with different plateau values, which can be determined from Eq. (7.35) using the value of a_{cf} obtained from Eq. (7.45). The *R*-curve obtained in this way is interpreted as an envelope of the energy release rates (G_q-curves) for a series of structures with the same geometry and the initial crack length but different sizes, as shown in Fig. 7-6, and can be classified in the second category of *R*-curve, as discussed in Chapter 3.

7.3.4 Fracture Response Predicted from *R*-Curves

Fracture responses of a specimen, including the load–CMOD and the load–displacement curves, can be predicted by the use of the proposed *R*-curve based on the equivalent-elastic crack approach.

The relationship between the nominal applied load σ and the effective crack length a is first calculated from Eq. (7.45). After knowing the value of σ for a given crack length, the CMOD can be computed based on corresponding formulations from an LEFM handbook.[1]

To obtain the load–load point displacement curve, it is necessary to know the specimen compliance. According to LEFM, the energy release rate can also be written as

$$G_q = \frac{\sigma^2 b^2}{2} \frac{\partial C}{\partial a} \tag{7.46}$$

where C is the compliance at the load point. Integrating Eq. (7.46) and using $G_q = R$ result in the compliance of a specimen with a different crack length:

$$C = C_0 + \int_{a_0}^{a} \frac{2R(a)}{b^2\sigma^2}\, da \tag{7.47}$$

where C_0 represents the compliance of specimen with the initial crack. After knowing the compliance, the load point displacement can be calculated.

7.3.5 Comparison and Discussion

The proposed *R*-curve approach was applied to predict the response of portland cement mortar subjected to uniaxial tension. The experimental results used were those reported by Maji and Shah.[32] The mortar used had a mix proportion of 1 : 2 : 0.5 cement–sand–water by weight. Specimens tested were 304.8 × 76.2 × 19 mm in size. A single-edge notch of 12 mm was used, and the rate of opening of the notch tip was used as the control parameter in loading the specimens. Since the required matrix properties for the mortar mix were not available, the average values of material properties (K_{Ic}^{s}, $CTOD_c$, and E) from two mortars, which have the mix proportions of 1 : 2.6 : 0.65 and 1 : 2.6 : 0.45 (cement–sand–water), respectively, and were obtained from a three-point bend beam by Jenq and Shah,[33] were used as an input for theoretical solutions. A good agreement between the experimental results and the theoretical predictions is indicated in Fig. 7-7.

As mentioned before, *R*-curves depend on specimen size, initial crack length, and geometry. The *R*-curves for different sizes, initial crack lengths, and geometries are discussed here. Values of $K_{Ic}^{s} = 31.4\ \text{N/mm}^{3/2}$, $CTOD_c = 0.02$ mm, and $E = 20$ GPa reported by Jenq and Shah[33] are used in the following discussion.

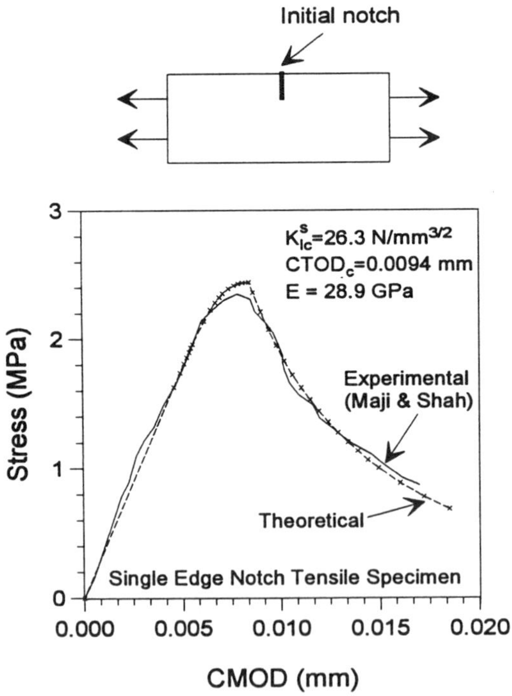

Figure 7-7 Theoretical and experimental load–CMOD curve for single-edge notch tensile specimen.[30]

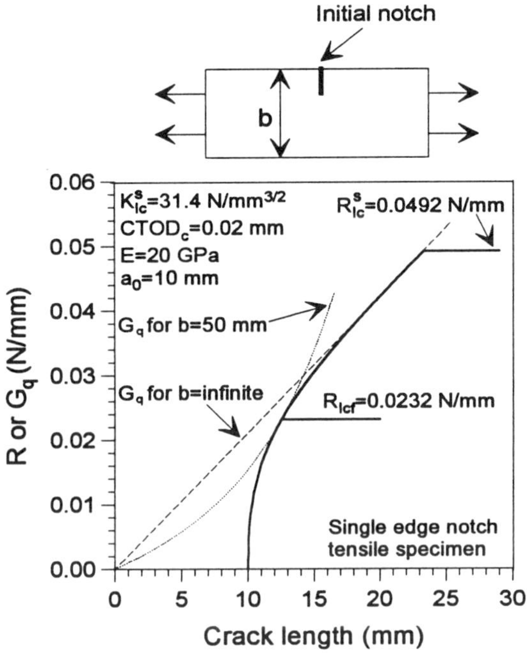

Figure 7-8 The G_q and *R*-curves for infinite- and finite-size specimens.[30]

The *R*-curves for different specimen sizes predicted by the proposed approach are indicated in Fig. 7-8, where single-edge notch tensile specimens are used. It is seen that the *R*-curve satisfies Eqs. (7.2) and (7.3) at critical points. The infinite-size and the finite-size specimens follow the same *R*-curve but with different plateau values.

The *R*-curves for different specimen geometries are shown in Fig. 7-9. These *R*-curves correspond to an infinite specimen size, and the value of a_0 is 10 mm for these *R*-curves. It is shown that these *R*-curves strongly depend on geometry.

The *R*-curves for different values of a_0 are given in Fig. 7-10, where the infinite-size, single-edge notch tensile specimens are used. According to the proposed *R*-curve formulation, the *R*-curve approaches that of a linear function when the value of a_0 tends to zero. With increasing values of a_0, the *R*-curve becomes steeper and more nonlinear. Note that a similar phenomenon has been experimentally observed for alumina by Steinbrech and Schmenkel.[34]

7.4 *R*-CURVE PROPOSED BY BAZANT ET AL.

Bazant et al.[35–37] have proposed an *R*-curve based on the size effect model. They studied the fracture behavior of a series of geometrically similar struc-

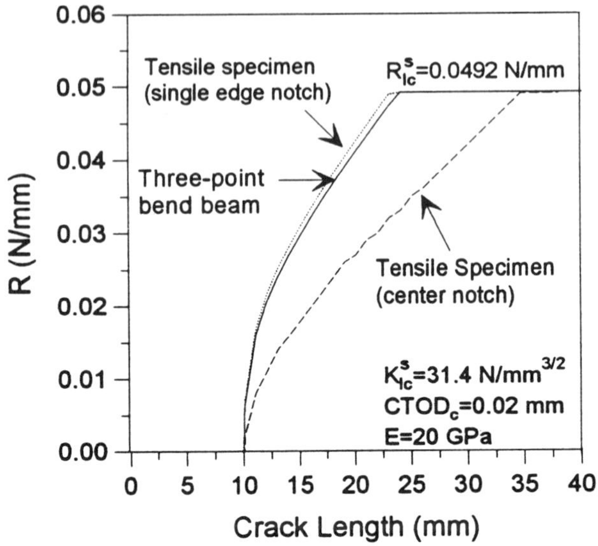

Figure 7-9 Predicted *R*-curves for infinite-size specimens with different geometries.[30]

tures that have different sizes but the same ratio of specimen size to initial notch. This has been discussed in detail in Chapter 5. An *R*-curve was proposed based on Eqs. (7.2) and (7.3). The *R*-curve obtained is interpreted as an envelope of G_q-curves for a series of geometrically similar structures with increasing sizes and is classified into the third category. Their *R*-curve will be presented in the following.

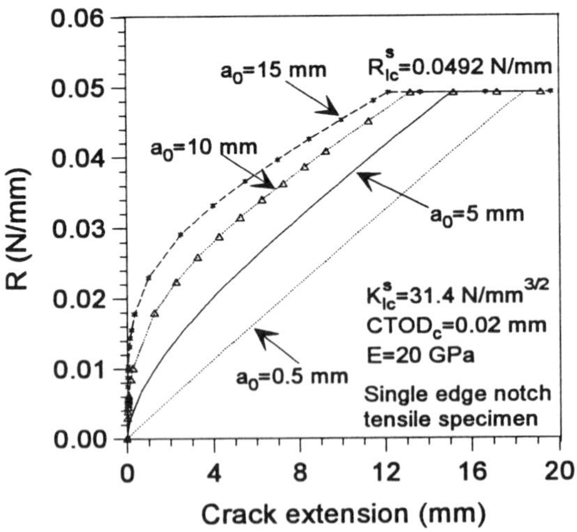

Figure 7-10 Predicted *R*-curves for infinite-size specimens with different initial crack lengths.[30]

7.4.1 *R*-Curve from Size Effect Model

Consider a series of geometrically similar structures with different sizes D. Based on LEFM, when a specimen fails the energy release rate, $(G_q)_c$ can be expressed as

$$(G_q)_c = \frac{P_c^2 g(a_c/D)}{Et^2 D} \tag{7.48}$$

where P_c is the maximum applied load, t is the thickness of specimens, and $g(a_c/D)$ is the geometric function, which has been discussed in Chapter 5. Both t and $g(a_c/D)$ are the same for all sizes of geometrically similar specimens. Since the R-curve proposed by Bazant et al.[35–37] consists of values of $(G_q)_c$ for different sizes of geometrically similar specimens as illustrated in Fig. 7-11, both $(G_q)_c$ and $g(a_c/D)$ will be replaced by G_q and $g(a/D)$, respectively, for convenience in the following discussion.

According to Bazant's size effect model, which has been discussed in Chapter 5 [see Eqs. (5.28), (5.30), (5.41), and (5.42)], the following relationships can be obtained:

$$P_c^2 = \left(\frac{\sigma_N\, Dt}{c_n}\right)^2 = \left(\frac{Bf_t\, Dt}{c_n}\right)^2 \frac{D_0}{D + D_0}$$

$$(Bf_t)^2 = \frac{c_n^2 E G_f}{D_0 g(a_0/D)} \tag{7.49}$$

Substituting Eq. (7.49) into Eq. (7.48) leads to

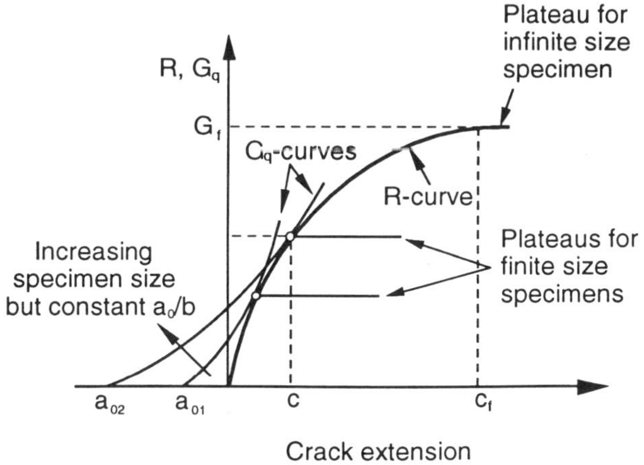

Figure 7-11 The *R*-curve as an envelope of G_q-curves for series of geometrically similar structures proposed by Bazant et al. (change specimen size but keep ratio a_0/b constant.[37]

$$G_q = G_f \frac{g(a/D)}{g(a_0/D)} \frac{D}{D + D_0} \tag{7.50}$$

As discussed previously, Eqs. (7.2) and (7.3) are regarded to be the basic definition of R-curves. It is noted from Eq. (7.48) that since $g(a/D)$ is a constant for a series of geometrically similar specimens, G_q can be expressed as a function of specimen size D rather than the crack length a. Therefore, Eq. (7.3) can be rewritten as

$$\frac{\partial G_q}{\partial D} - \frac{\partial R}{\partial D} = 0 \tag{7.51}$$

If the R-curve is interpreted as an envelope of G_q-curves for a series of geometrically similar specimens with different sizes, the different-size specimens follow the same R-curve but different plateau values. The R-curve expression defined in such a way is not a function of the specimen size, which means $\partial R/\partial D = 0$. As a result, Eq. (7.51) becomes

$$\frac{\partial G_q}{\partial D} = 0 \tag{7.52}$$

Substituting Eq. (7.50) into Eq. (7.52) results in

$$D + D_0 = \frac{D_0 g(a/D)}{(a/D - a_0/D)g'(a/D)} \tag{7.53}$$

where $g'(a/D) = \partial g/\partial(a/D)$. It is noted that to obtain Eq. (7.53), the following relationship has been used:

$$\frac{\partial}{\partial D}\left(\frac{a}{D}\right) = \frac{\partial}{\partial D}\left(\frac{a_0 + c}{D}\right) = -\frac{c}{D^2} = -\left(\frac{a}{D} - \frac{a_0}{D}\right)\frac{1}{D} \tag{7.54}$$

Also note that, by assumption of geometric similarity among different sizes of specimens, a_0/D = constant so $\partial(a_0/D)/\partial D = 0$. It is also assumed in Eq. (7.54) that the crack extension c and the specimen size D are two independent variables. On the other hand, the following relationships can be obtained from the size effect model [Eq. (5.41)]:

$$\frac{a}{D} - \frac{a_0}{D} = \frac{c}{D} \tag{7.55}$$

and

$$D_0 = c_f \frac{g'(a_0/b)}{g(a_0/D)} \tag{7.56}$$

Substituting Eq. (7.53) into Eq. (7.50) and using both Eqs. (7.55) and (7.56), one can obtain the R-curve

$$R(c) = G_f \frac{g'(a/D)}{g'(a_0/D)} \frac{c}{c_f} \tag{7.57}$$

It is seen from Eq. (7.57) that the proposed R-curve depends on specimen geometry and material fracture parameters G_f and c_f.

Since the three variables R, a, and c are included in Eq. (7.57), the R-curve cannot be directly determined using Eq. (7.57) alone. Two computational schemes can be used to calculate the R-curve mentioned above. The first scheme starts from selecting a series of values of a/D. For each of the selected value of a/D, the corresponding value of c/D can be calculated from Eq. (7.55). It is noted here that the value of a_0/D is a known constant for geometrically similar specimens with changing sizes. After the value of D_0 is calculated from Eq. (7.56), the corresponding value of D can be determined from Eq. (7.53). Then the value of c can be calculated from the previously known value of c/D. Since the values of c and the relative value of a/D have been determined at this point, the corresponding value of $R(c)$ can be computed using Eq. (7.57). The R-curve can be obtained by repeating this procedure for different given values of a/D. On the other hand, the second computational scheme starts from a given value of c. After the value of D_0 is calculated from Eq. (7.56), Newton iterations should be performed for Eq. (7.53) to determine the corresponding value of a/D. The value of $R(c)$ is then computed using Eq. (7.57).

Note that the R-curve given by Eq. (7.57) is for an infinite-size specimen. Geometrically similar finite-size specimens are assumed to follow the same R-curve with the infinite-size specimen but different plateau values. This will be further demonstrated in an example given below.

The R-curve obtained in such a way depends on the geometric function $g(a/D)$ and is interpreted as an envelope of G_q-curves for a series of geometrically similar structures with increasing sizes. These R-curves can be classified into the third category.

7.4.2 Comparison and Discussion

Bazant et al.[38] have tested a set of geometrically similar three-point bend beams of limestone. The obtained load and load point displacement curves are shown in Fig. 7-12. By applying the size effect model recommended by RILEM, which has been presented in Chapters 5 and 6, they calculated $G_f = 61$ N/m

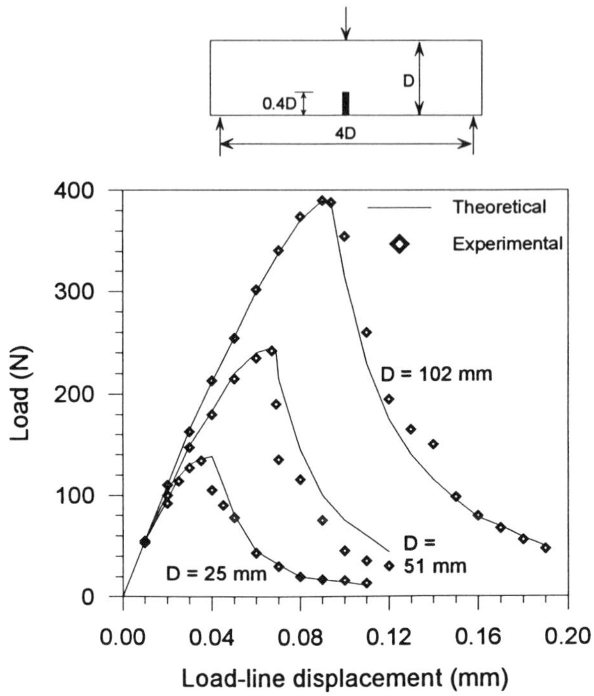

Figure 7-12 Load–load point displacement curves for limestone specimens.[38]

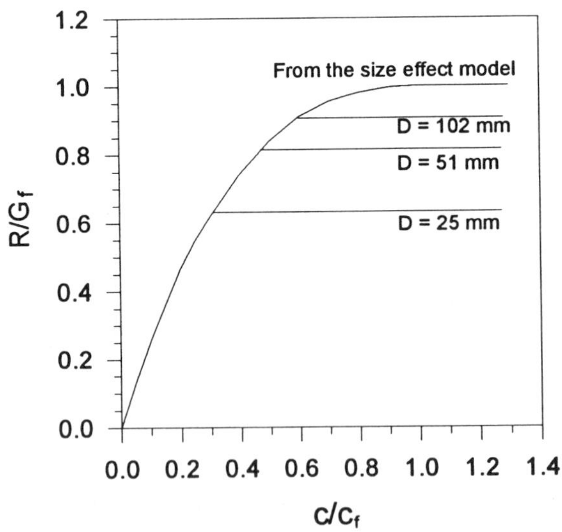

Figure 7-13 The *R*-curves proposed by Bazant et al. for three-point bending beams of limestone.[38]

(0.35 lb/in.), $c_f = 9.3$ mm (0.37 in.), and $E = 15.3$ GPa. Values of $g(a_0/D) = 62.84$ and $g'(a_0/D) = 347.7$ can be obtained from LEFM. By using the computational schemes previously presented, the R-curve shown in Fig. 7-13 is obtained. It is seen that different sizes of geometrically similar specimens follow the same R-curve but with different plateau values. Different from the R-curve proposed by Ouyang et al.,[29] the R-curve of Bazant et al.[38] smoothly approaches the line of $R = G_f$ at $c = c_f$ for infinite-size specimens. After the R-curves are obtained, the load and load point displacement can be predicted from the compliance given by Eq. (7.47). These predicted load and load point displacement curves are also shown in Fig. 7-12. The theoretical and experimental load and load point displacement curves match quite well.

7.5 SUMMARY

Two explanations of R-curves have been introduced in this chapter. The R-curves can be defined as energy release rates during stable crack propagation prior to failure of structures. The R-curves can also be interpreted as an envelope of G_q-curves for a certain category of structures which depends on both structural geometry and the material fracture property.

Based on these two explanations, three approaches to construct R-curves have been presented. The R-curve proposed by Foote et al.[24] includes both the Griffith–Irwin and the Dugdale-Barenblatt energy dissipation mechanisms. This R-curve requires one to solve an integral equation. Mai et al.[25,26] simplified this R-curve by introducing a Green's function for the semi-infinite plate and a certain crack opening profile. However, the simplified R-curve becomes geometry independent.

Ouyang et al.[29,30] have developed an R-curve using the equivalent-elastic crack approach. A governing differential equation for R-curves was derived based on the failure criteria of quasi-brittle materials. The R-curve was then obtained by solving this differential equation. This R-curve is interpreted as an envelope of the energy release rates for a series of structures with the same specimen geometry and the initial crack length but different specimen sizes.

Bazant et al.[35–37] have proposed an R-curve based on their size effect model. They considered the maximum loads of a series of geometrically similar structures with different size. The R-curve was established by using failure criteria of the structures with different size. The R-curve obtained is interpreted as an envelope of G_q-curves for a series of geometrically similar structures with increasing sizes.

REFERENCES

1. Tada, H., Paris, P. C., and Irwin, G. R., *The Stress Analysis of Cracks Handbook,* 2nd ed., Paris Productions, St. Louis, MO, 1985.

2. Sakai, M., Yoshimura, J., Goto, Y., and Inagaki, M., "*R*-Curve Behavior of a Polycrystalline Graphite: Microcracking and Grain Bridging in the Wake Region," *Journal of the American Ceramic Society,* Vol. 71, No. 8, 1986, pp. 609–616.
3. Marshall, D. B., and Swain, M. V., "Crack Resistance Curves in Magnesia-Partially-Stabilized Zirconia," *Journal of the American Ceramic Society,* Vol. 71, No. 6, 1988, pp. 399–407.
4. Anderson, R. M., and Braun, L. M., "Technique for the *R*-Curve Determination of Y-TZP Using Indentation-Produced Flaws," *Journal of the American Ceramic Society,* Vol. 73, No. 10, 1990, pp. 3059–3062.
5. McCabe D. E. (Ed.), "Fracture Toughness Evaluation by *R*-Curve Method," ASTM-STP 527, American Society for Testing and Materials, Philadelphia, PA, 1973.
6. Krafft, J. M., Sullivan, A. M., and Boyle, R. W., "Effect of Dimensions on Fast Fracture Instability of Notched Sheets," in *Proceedings of the Crack Propagation Symposium,* College of Aeronautics, Cranfield, England, Vol. I, 1961, pp. 8–28.
7. Broek, D., *Elementary Engineering Fracture Mechanics,* 4th ed., Martinus Nijhoff, Hingham, MA, 1987.
8. Jenq, Y. S., and Shah, S. P., "A Fracture Toughness Criterion for Concrete," *Engineering Fracture Mechanics,* Vol. 21, No. 5, 1985, pp. 1055–1069.
9. Hsueh, C. H., and Becher, P. F., "Evaluation of Bridging Stress from *R*-Curve Behavior for Nontransforming Ceramics," *Journal of American Ceramics Society,* Vol. 71, No. 5, 1988, pp. 234–237.
10. Homeny, J., and Vaughn, W. L., "*R*-Curve Behavior in a Silicon Carbide Whisker/Alumina Matrix Composite," *Journal of American Ceramics Society,* Vol. 73, No. 7, 1990, pp. 2060–2062.
11. Wecharatana, M., and Shah, S. P., "Slow Crack Growth in Cement Composites," *Journal of Structural Engineering,* Vol. 108, No. 6, 1982, pp. 1400–1413.
12. Wecharatana, M., and Shah, S. P., "Predictions of Nonlinear Fracture Process Zone in Concrete," *Journal of Engineering Mechanics,* Vol. 109, No. 5, 1983, pp. 1231–1246.
13. Cook, R. F., Fairbanks, C. J., Lawn, B. R., and Mai, Y.-W., "Crack Resistance by Interfacial Bridging: Its Role in Determining Strength Characteristics," *Journal of Materials Research,* Vol. 2, No. 3, 1987, pp. 345–356.
14. Cook, R. F., and Clarke, D. R., "Fracture Stability, *R*-Curves and Strength Variability," *Acta Metall.,* Vol. 36, No. 3, 1988, pp. 555–562.
15. Kim, B. H., and Joe, C. R., "On the Resistance to Crack Initiation and Growth," *Engineering Fracture Mechanics,* Vol. 34, No. 1, 1989, pp. 221–231.
16. Faber, K. T., Gu, W.-H., Cai, H., Winholtz, R. A., and Magley, D. J., "Fracture Properties in SiC-Based Particulate Composites," in *Toughening Mechanisms in Quasi-Brittle Materials,* edited by Shah, S. P., Kluwer Academic, Dordrecht, The Netherlands, 1991, pp. 3–18.
17. Becher, P. F., "Crack Bridging Process in Toughened Ceramics," in *Toughening Mechanisms in Quasi-Brittle Materials,* edited by Shah, S. P., Kluwer Academic, Dordrecht, The Netherlands, 1991, pp. 19–34.
18. Bennison, S. J., and Lawn, B. R., "Flaw Tolerance in Ceramics with Rising

Crack Resistance Characteristics,'' *Journal of Materials Science,* Vol. 24, 1989, pp. 3169–3175.

19. Bennison, S. J., Lathabai, S., Rodel, J., and Lawn, B. R., ''Microstructure, *R*-curve and Mechanical Properties of Alumina-Based Ceramics,'' in *Toughening Mechanisms in Quasi-Brittle Materials,* edited by Shah, S. P., Kluwer Academic, Dordrecht, The Netherlands, 1991, pp. 209–234.
20. Zok, F., ''The Fracture Resistance and Strength of Brittle Matrix Composites,'' in *Toughening Mechanisms in Quasi-Brittle Materials,* edited by Shah, S. P., Kluwer Academic, Dordrecht, The Netherlands, 1991, pp. 425–440.
21. Elangovan, P. T., ''Formulations for Fracture Stress, CTP Zone and *R*-curves,'' *Engineering Fracture Mechanics,* Vol. 40, No. 2, 1991, pp. 433–463.
22. Sakai, M., and Ichikawa, H., ''Work-of-Fracture of Brittle Materials with Microcracking and Crack Bridging,'' *International Journal of Fracture,* Vol. 55, 1992, pp. 65–79.
23. Li, Y. N., and Liang, R. Y., ''Stability Theory of Cohesive Crack Model,'' *Journal of Engineering Mechanics, ASCE,* Vol. 118, No. 3, 1992, pp. 587–603.
24. Foote, R. M. L., Mai, Y.-W., and Cotterell, B., ''Crack Growth Resistance Curves in Strain Softening Materials,'' *Journal of Mechanics and Physics of Solids,* Vol. 34, No. 6, 1986, pp. 593–607.
25. Mai, Y.-W., and Lawn, B. R., ''Crack-Interface Grain Bridging as a Fracture Resistance Mechanism in Ceramics: II. Theoretical Fracture Mechanics Model,'' *Journal of the American Ceramic Society,* Vol. 70, No. 4, 1987, pp. 289–294.
26. Mai, Y.-W., Barakat, B., Cotterell, B., and Swain, M. V., ''*R*-Curve Behavior in a Macro-Defect-Free Cement Paste,'' *Philosophical Magazine A,* Vol. 62, No. 3, 1990, pp. 347–361.
27. Rice, J. R., ''Mathematical Analysis in the Mechanics of Fracture,'' in *Fracture: Advanced Treatise,* Vol. II, edited by Liebowits, H., Academic, New York, 1968, pp. 191–311.
28. Planas, J., Eliccs, M., and Toribio, J., ''Approximation of Cohesive Crack Model by *R–CTOD* Curve,'' in *Fracture of Concrete and Rock: Recent Developments,* edited by Shah, S. P., Swartz, S. E., and Barr, B., Elsevier Applied Science, London, 1989, pp. 203–212.
29. Ouyang, C., Mobasher, B., and Shah, S. P., ''An *R*-Curve Approach for Fracture of Quasi-Brittle Materials,'' *Engineering Fracture Mechanics,* Vol. 37, No. 4, 1990, pp. 901 916.
30. Ouyang, C., and Shah, S. P., ''Geometry-Dependent *R*-Curve for Quasi-Brittle Materials,'' *Journal of the American Ceramic Society,* Vol. 74, No. 11, 1991, pp. 2831–2836.
31. Sneddon, I. N., ''The Distribution of Stress in the Neighborhood of a Crack in an Elastic Solid,'' *Proceedings of the Royal Society, Series A,* Vol. 187, 1946, pp. 229–260.
32. Maji, A. K., and Shah, S. P., ''Process Zone and Acoustic Emission Measurement in Concrete,'' *Experimental Mechanics,* Vol. 28, 1988, pp. 27–33.
33. Jenq, Y. S., and Shah, S. P., ''A Two Parameter Fracture Model for Concrete,'' *Journal of Engineering Mechanics,* Vol. 111, No. 4, 1985, pp. 1227–1241.
34. Steinbrech, R. W., and Schmenkel, O., ''Crack-Resistance Curves of Surface

Cracks in Alumina,'' *Journal of the American Ceramic Society,* Vol. 71, No. 5, 1988, pp. C271–C273.

35. Bazant, Z. P., and Cedolin, L., ''Approximate Linear Analysis of Concrete Fracture by *R*-Curves,'' *Journal of Structural Engineering, ASCE,* Vol. 110, No. 6, pp. 1336–1355.
36. Bazant, Z. P., Kim, J.-K., and Pfeiffer, P. A., ''Nonlinear Fracture Properties from Size Effect Tests,'' *Journal of Structural Engineering,* ASCE, Vol. 112, No. 2, 1986, pp. 289–307.
37. Bazant, Z. P., and Kazemi, M. T., ''Determination of Fracture Energy, Process Zone Length and Brittleness Number from Size Effect, with Application to Rock and Concrete,'' *International Journal of Fracture,* Vol. 44, 1990, pp. 111–131.
38. Bazant, Z. P., Gettu, R., and Kazemi, M. T., ''Identification of Nonlinear Fracture Properties from Size Effect Tests and Structural Analysis Based on Geometry-Dependent *R*-Curves,'' *International Journal of Rock Mechanics and Mineral Science,* Vol. 28, No. 1, 1991, pp. 43–51.

8

FRACTURE MECHANICS FOR OTHER DEFORMATION MODES

8.1 STRESS FIELDS AT TIP OF PLANAR MIXED-MODE CRACK

The influence of shearing tractions that may be present in the plane of the crack front needs to be assessed to provide a complete picture of general crack propagation and fracture. At the onset this will be discussed from the macrolevel viewpoint in terms of overall structural response while recognizing that this basically depends on the scale, or size, of observation. As mentioned in earlier chapters, the tortuosity of the cracking surface due to different grain sizes, relative strengths of matrix and aggregates, and other effects may be modeled on smaller scales.

In general, there are three components of deformation near the crack tip, as shown in Fig. 8-1. These are mode I, opening; mode II, in-plane sliding; and mode III, out-of-plane sliding or tearing. The combination of modes I and II will be discussed in the major portion of this chapter followed by a discussion of mode III and combinations of all three modes.

In the past few years, extensive efforts have been directed to modeling crack propagation and fracture in concrete and rock subjected to mixed-mode loading.[1] The term *mixed mode* implies the presence of in-plane, normal, and shearing tractions in the immediate region of the front of an existing crack and behind the front as well. Thus the crack and its "tip" are subjected to combined stresses. The deformation at the crack tip may have both mode I (opening) and mode II (sliding) components. This influence on the process of cracking may be of great importance in concrete structures.[1] And, as stated by Whittaker et al.[2] with respect to the fracture behavior of rock (in situ), studies of mode II and mixed-mode fracture have received little attention. Neverthe-

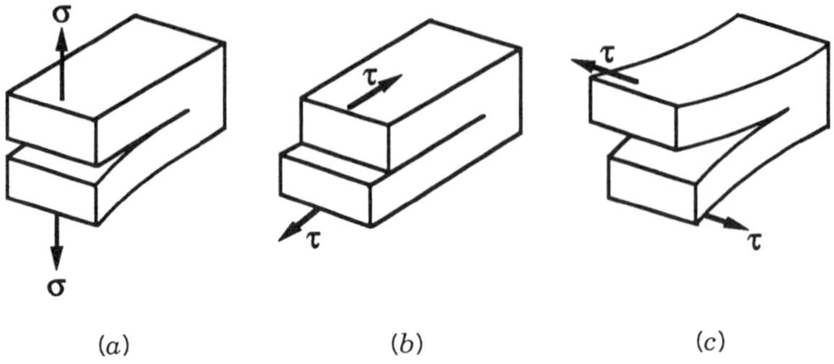

Figure 8-1 Deformation modes for crack propagation and fracture: (*a*) mode I, (*b*) mode II, and (*c*) mode III.

less, they believe that pure mode I loading seldom represents a realistic condition, and for many practical situations mode II fracture may dominate.[2]

At this point it may be appropriate to emphasize that normal loads—sometimes referred to as *mode I loading*—and shearing loads in the plane of the crack—sometimes referred to as *mode II loading*—do not necessarily yield mode I or mode II *deformations*. In the terminology of Fig. 8-1, the different modes refer to the deformation at the crack tip. In particular, mode II fracture implies a *self-similar* crack extension related solely to mode II deformation. Again, this deformation may be interpreted at different size scales.

As presented in Chapter 2, the cartesian stresses in the near region of the crack tip as shown in the Fig. 8-2*a* may be expressed in terms of the LEFM stress intensity factors K_I and K_{II} as[3]

$$\sigma_x = \frac{1}{\sqrt{2\pi r}}\left[K_I\cos\frac{\theta}{2}\left(1 - \sin\frac{\theta}{2}\sin\frac{3\theta}{2}\right) - K_{II}\sin\frac{\theta}{2}\left(2 + \cos\frac{\theta}{2}\cos\frac{3\theta}{2}\right)\right]$$

$$\sigma_y = \frac{1}{\sqrt{2\pi r}}\left[K_I\cos\frac{\theta}{2}\left(1 + \sin\frac{\theta}{2}\sin\frac{3\theta}{2}\right) + K_{II}\sin\frac{\theta}{2}\cos\frac{\theta}{2}\cos\frac{3\theta}{2}\right] \tag{8.1}$$

$$\tau_{xy} = \frac{1}{\sqrt{2\pi r}}\left[K_I\sin\frac{\theta}{2}\cos\frac{\theta}{2}\cos\frac{3\theta}{2} + K_{II}\left(1 - \sin\frac{\theta}{2}\sin\frac{3\theta}{2}\right)\right]$$

It is more convenient to express these in polar form, as shown in Fig. 8-2*b*,

$$\sigma_r = \frac{1}{\sqrt{2\pi r}}\left[K_I\left(\frac{5}{4}\cos\frac{\theta}{2} - \frac{1}{4}\cos\frac{3\theta}{2}\right) + K_{II}\left(-\frac{5}{4}\sin\frac{\theta}{2} + \frac{3}{4}\sin\frac{3\theta}{2}\right)\right]$$

$$\sigma_\theta = \frac{1}{\sqrt{2\pi r}}\cos\frac{\theta}{2}\left(K_I\cos^2\frac{\theta}{2} - \frac{3}{2}K_{II}\sin\theta\right)$$

$$\tau_{r\theta} = \frac{1}{2\sqrt{2\pi r}}\cos\frac{\theta}{2}\left[K_I\sin\theta + K_{II}(3\cos\theta - 1)\right] \tag{8.2}$$

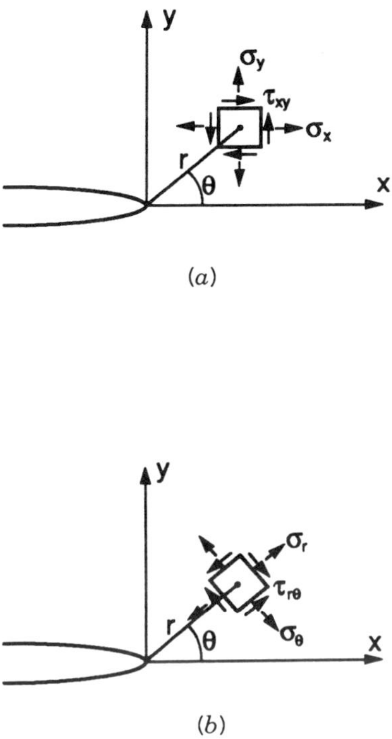

Figure 8-2 Stress notation in region of crack tip: (*a*) cartesian stress and (*b*) polar stresses.

8.2 CLASSICAL MIXED-MODE FAILURE THEORIES

Many of the early crack propagation models under mixed-mode conditions were directly related to the so-called classical failure theories. These included maximum circumferential stress, minimum strain energy density, maximum circumferential strain, shear–friction, Mohr–Coulmb, and octahedral stress failure theories. In all of these the tie-in with fracture mechanics is the use of stresses expressed in terms of K_{I} and K_{II}.

8.2.1 Maximum Circumferential Stress Criterion

This criterion was proposed by Erdogan and Sih[4] and is based on the assumption that the crack initiates from its tip in a direction normal to the maximum circumferential stress σ_θ. In Fig. 8-3 an inclined crack at angle β to the principal stress direction is subjected to the stresses applied at infinity σ_β, τ_β. By using Eqs. (8.2) and Mohr's circle analysis, one obtains

$$
\begin{aligned}
K_{\mathrm{I}} &= [\tfrac{1}{2}(\sigma_1 + \sigma_2) + \tfrac{1}{2}(\sigma_1 - \sigma_2)\cos(2\beta)]\sqrt{\pi a} \\
K_{\mathrm{II}} &= [\tfrac{1}{2}(\sigma_2 - \sigma_1)\sin(2\beta)]\sqrt{\pi a}
\end{aligned}
\tag{8.3}
$$

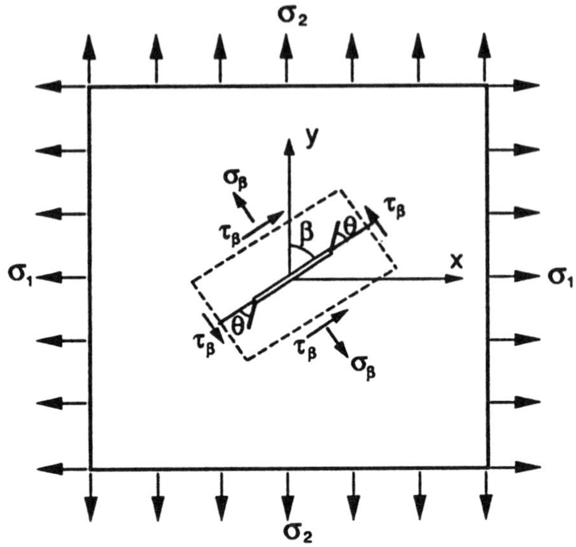

Figure 8-3 Inclined crack in biaxial stress field.[1]

If m_s is the ratio of σ_1/σ_2 of the principal stresses, then Eq. (8.3) can be rewritten as

$$\begin{aligned} K_{\mathrm{I}} &= \sigma_2\sqrt{\pi a}\,[m_s + (1 - m_s)\sin^2\beta] \\ K_{\mathrm{II}} &= \sigma_2\sqrt{\pi a}\,(1 - m_s)\sin\beta\cos\beta \end{aligned} \tag{8.4}$$

Crack propagation occurs at an angle θ (cf. Fig. 8-3) that maximizes σ_θ (i.e., σ_θ becomes a principal stress and $\tau_{r\theta} = 0$). Thus

$$\tan\frac{\theta}{2} = \frac{1}{4}\left[\frac{K_{\mathrm{I}}}{K_{\mathrm{II}}} \pm \sqrt{\left(\frac{K_{\mathrm{I}}}{K_{\mathrm{II}}}\right)^2 + 8}\right] \tag{8.5}$$

From Eqs. (8.4) and (8.5), the crack propagation angle θ and crack inclination angle β are related by

$$[m_s + (1 - m_s)\sin^2\beta]\sin\theta + [\tfrac{1}{2}(1 - m_s)\sin(2\beta)](3\cos\theta - 1) = 0 \tag{8.6}$$

Equation (8.6) can be transformed into[1]

$$\begin{aligned} &2(1 - m_s)\sin(2\beta)\,[\tan(\tfrac{1}{2}\theta)]^2 - 2[m_s + (1 - m_s)\sin^2\beta]\,[\tan(\tfrac{1}{2}\theta)] \\ &\quad - (1 - m_s)\sin(2\beta) = 0 \end{aligned} \tag{8.7}$$

If $m_s = 1$, then $\theta = 0$. In this case, since the stress field at infinity is uniform, the crack extension is collinear due to symmetry. If $m_s = 0$, there is a discontinuity for $\beta = 0$. In fact, $\theta(\beta = 0, m_s = 0) = 0$ due to the symmetry, while[1]

$$\lim_{\beta \to 0^+} \theta(\beta, m_s = 0) = 70°$$

While Eq. (8.5) determines the direction of crack propagation, an additional criterion for instability is needed. This was taken to be $K_I = K_{Ic}$. In other words, propagation is assumed to be in mode I regardless of stress state and internal crack inclination. Thus

$$\sqrt{2\pi r}\, \sigma_\theta = K_{Ic} \tag{8.8}$$

By introducing the dimensionless stress intensity factors

$$K_I^* = \frac{K_I}{K_{Ic}} \qquad K_{II}^* = \frac{K_{II}}{K_{Ic}} \tag{8.9}$$

the fracture locus may be plotted in the K_I^*-versus-K_{II}^* plane. Solving

$$\begin{aligned} K_I^* \sin\theta + K_{II}^* (3\cos\theta - 1) &= 0 \\ [K_I^* \cos^2(\tfrac{1}{2}\theta) - \tfrac{3}{2} K_{II}^* \sin\theta]\cos(\tfrac{1}{2}\theta) &= 1 \end{aligned} \tag{8.10}$$

for K_I^* and K_{II}^* by varying θ, all the points of the locus are defined. They are symmetric with respect to the K_I^* axis and valid only in the half plane $K_I^* \geq 0$, as shown in Fig. 8-4. It was proved by Nuismer[5] that the maximum stress criterion is totally equivalent to the maximum strain energy release rate criterion.

8.2.2 Minimum Strain Energy Density Criterion

Based on Sih[6] and Sih and Macdonald,[7] the general strain energy density function at the crack tip is of the form

$$\frac{dU}{dV} = \frac{1}{r}\,(a_{11}K_I^2 + 2a_{12}K_I K_{II} + a_{22}K_{II}^2 + a_{33}K_{III}^2) \tag{8.11}$$

where dU is the change of the strain energy and $dV = r\,dr\,d\theta$ is the volume of the differential element. The coefficients a_{ij} for the plane-strain conditions are

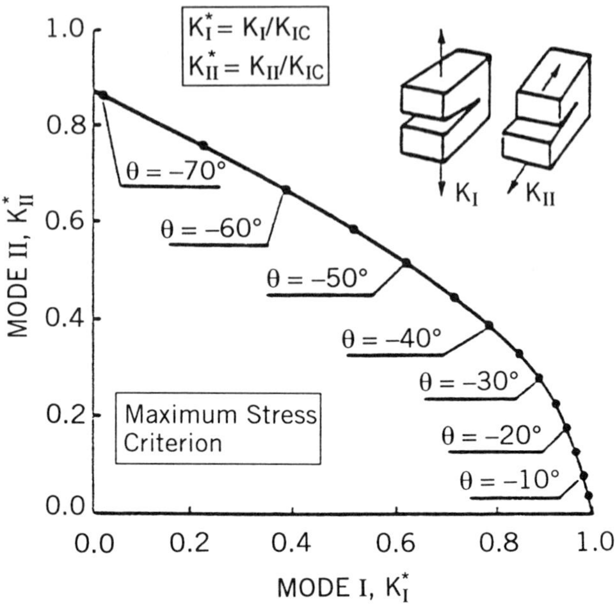

Figure 8-4 Fracture locus, maximum stress criterion.[1]

$$
\begin{aligned}
a_{11} &= \frac{1+\mu}{8\pi E}(3 - 4\mu - \cos\theta)(1 + \cos\theta) \\
a_{12} &= \frac{1+\mu}{8\pi E}(2\sin\theta)(\cos\theta - 1 + 2\mu) \\
a_{22} &= \frac{1+\mu}{8\pi E}[4(1-\mu)(1-\cos\theta) + (1+\cos\theta)(3\cos\theta - 1)] \\
a_{33} &= \frac{1+\mu}{2\pi E}
\end{aligned}
\tag{8.12}
$$

The function dU/dV thus possesses a $1/r$ singularity at the crack tip.

A strain energy density factor is defined as

$$
\frac{S_G(\theta)}{r} = \frac{dU}{dV} \tag{8.13}
$$

The following criteria were then proposed[6,7]:

1. The crack initiation direction is assumed to correspond to the minimum strain energy density factor:

$$\frac{\partial S_G(\theta)}{\partial \theta} = 0 \quad \text{and} \quad \frac{\partial^2 S_G(\theta)}{\partial \theta^2} > 0 \quad \text{for } \theta = \theta_0 \tag{8.14}$$

where θ_0 is the direction for the minimum strain energy density factor.

2. The crack starts to propagate when $S_G(\theta)$ reaches a critical value S_{Gc} at $\theta = \theta_0$.

For mode I crack extension, $K_{\rm II} = K_{\rm III} = 0$, $\theta_0 = 0$, and therefore $a_{11} = (1 - 2\mu)(1 + \mu)/2\pi E$. The critical value of the strain energy density factor, S_{Gc} can be related to the critical value of the stress intensity factor, $K_{\rm Ic}$, for the plane-strain condition,

$$S_{Gc} = \frac{(1 - 2\mu)(1 + \mu)}{2\pi E} K_{\rm Ic}^2 \tag{8.15}$$

In the case of mixed-mode in-plane loading, $K_{\rm I} \neq 0$, $K_{\rm II} \neq 0$, and $K_{\rm III} = 0$. By the application of Eqs. (8.11) and (8.12), an expression for finding the stationary values of S_G is obtained:

$$\begin{aligned}\frac{\partial S_G}{\partial \theta} = {} & K_{\rm I} \frac{1 + 2\mu}{4E} \sin\theta(\cos\theta - 1 + 2\mu) \\ & + K_{\rm I}K_{\rm II} \frac{1 + 2\mu}{2E} [2\cos^2\theta + (2\mu - 1)\cos\theta - 1] \\ & + K_{\rm II}^2 \frac{1 + 2\mu}{4E} \sin\theta(-3\cos\theta + 1 - 2\mu) = 0\end{aligned} \tag{8.16}$$

This results in a fourth-order equation:

$$\begin{aligned}& 2(1 - \mu)K_{\rm I}K_{\rm II}\left(\tan\frac{\theta}{2}\right)^4 + (3K_{\rm II}^2 - K_{\rm I}^2)\left(\tan\frac{\theta}{2}\right)^3 \\ & + [(2\mu - 1)(K_{\rm I}^2 - K_{\rm II}^2) - 6K_{\rm I}K_{\rm II}]\left(\tan\frac{\theta}{2}\right)^2 + (K_{\rm I}^2 - 3K_{\rm II}^2)\left(\tan\frac{\theta}{2}\right) \\ & + [(2\mu - 1)(K_{\rm I}^2 - K_{\rm II}^2) + 2K_{\rm I}K_{\rm II}\mu] = 0\end{aligned} \tag{8.17}$$

The two factors $K_{\rm I}$ and $K_{\rm II}$ are functions of the loading ratio m_s and crack inclination angle β.

Substituting Eqs. (8.4) into Eq. (8.17), it is possible to obtain the crack inclination angle β as a function of θ. The results are displayed graphically in Fig. 8-5 for $\mu = 0.3$.

The crack growth conditions are

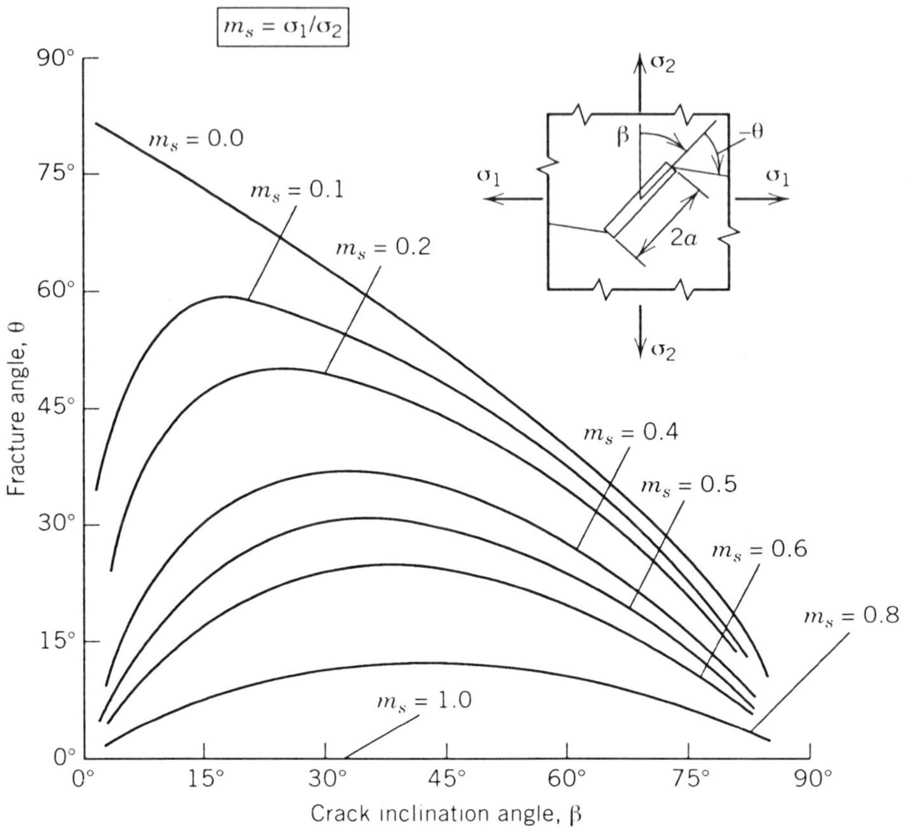

Figure 8-5 Crack branching angle θ versus crack inclination angle β, minimum strain energy density criterion.[1]

$$\frac{\partial S_G}{\partial \theta} = 0 \quad \text{and} \quad S_G(K_\mathrm{I}, K_\mathrm{II}) = S_{Gc} \tag{8.18}$$

Equations (8.18) can be restated as

$$(K_\mathrm{I}^*)^2 \sin\theta\,(\cos\theta - 1 + 2\mu) + (K_\mathrm{II}^*)^2 \sin\theta(-3\cos\theta + 1 - 2\mu)$$
$$+ 2K_\mathrm{I}^* K_\mathrm{II}^*\,[2\cos^2\theta + (2\mu - 1)\cos\theta - 1] = 0$$

$$(K_\mathrm{I}^*)^2\,[(3 - 4\mu - \cos\theta)\,(1 + \cos\theta)] + 4K_\mathrm{I}^* K_\mathrm{II}^* \sin\theta[\cos\theta - (1 - 2\mu)]$$
$$+ (K_\mathrm{II}^*)^2\,[4(1 - \mu)\,(1 - \cos\theta) + (1 + \cos\theta)\,(3\cos\theta - 1)]$$
$$- 4(1 - 2\mu) = 0$$

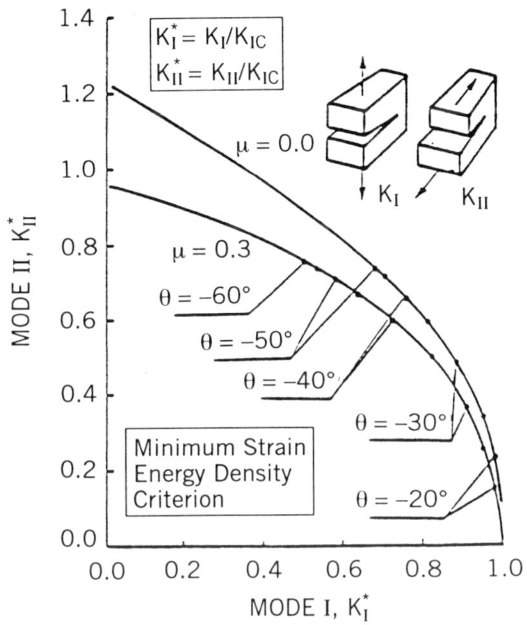

Figure 8-6 Fracture loci, minimum strain energy density criterion.[1]

The results of Eqs. (8.19) for fracture loci are plotted in Fig. 8-6 for Poisson's ratio $\mu = 0.0, 0.3$, respectively. These may be compared to the results obtained from the maximum circumferential stress criterion plotted in Fig. 8-4. It is seen that two theories agree rather well for $\mu = 0.3$ and also that the influence of Poisson's ratio is not insignificant.

8.2.3 Maximum Circumferential Strain Criterion

Maiti and Smith[8] proposed that crack propagation occurs when ϵ_θ reaches a limiting value of $\epsilon_{\theta c}$. The value of ϵ_θ is expressed as

$$\epsilon_\theta = \frac{K_I}{E\sqrt{2\pi r}}\left[\cos\frac{\theta}{2}\left(\frac{3}{4} - \frac{5}{4}\mu\right) + \frac{1}{4}\cos\left(\frac{3\theta}{2}\right)(1 + \mu)\right]$$
$$+ \frac{K_{II}}{E\sqrt{2\pi r}}\left[\sin\frac{\theta}{2}\left(-\frac{3}{4} + \frac{5}{4}\mu\right) - \frac{3}{4}\sin\left(\frac{3\theta}{2}\right)(1 + \mu)\right] \quad (8.20)$$

where the value of μ for concrete is about 0.15–0.20. The critical value $\epsilon_{\theta c}$

may be obtained by considering the condition $\sigma_r = 0$ for $\theta = 0$ and is given by

$$\epsilon_{\theta c} = \frac{K_{Ic}(1 - \mu)}{E\sqrt{2\pi r}} \tag{8.21}$$

8.2.4 Shear–Friction Criterion

Melin[9] showed that for pure mode II crack growth, the crack will be extended without directional change, but if mode I starts to dominate, the crack will be extended via a kink. A crack of length $2a$ subjected to different values of pressure p and shear stress τ (Fig. 8-7) was studied. The crack was assumed to form at the angle of θ_d shown.

The maximum value of K_{II} was determined to be associated with the following values of θ_d:

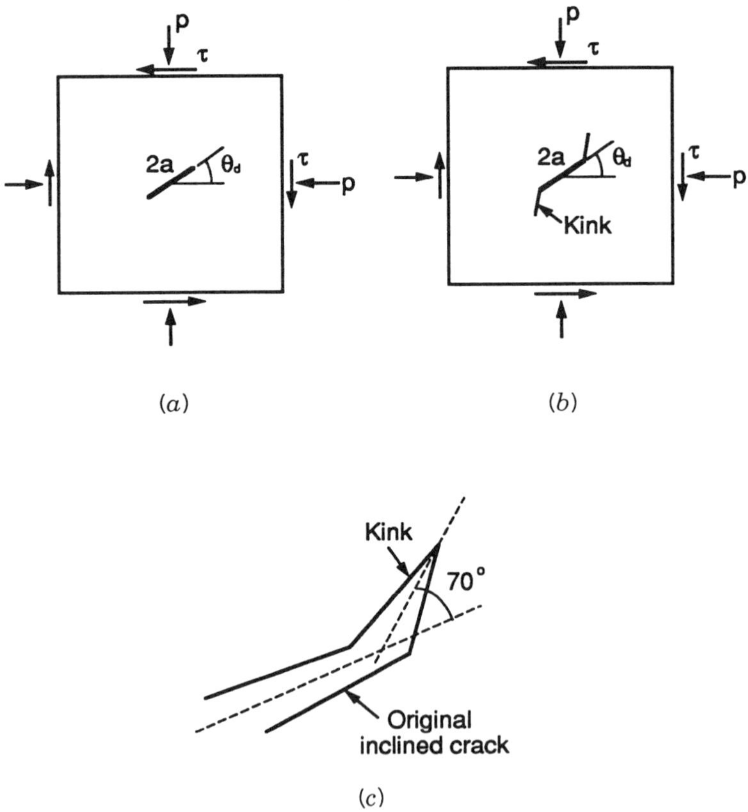

Figure 8-7 Formation of kink based on shear–friction theory: (*a*) crack configuration, (*b*) kinked crack configuration, and (*c*) angle of kink.[9]

$$\theta_d = \begin{cases} 0 & \text{for } \dfrac{p}{\tau} \leq 0 \\[2ex] \dfrac{1}{2\,\sin(p/\tau)} & \text{for } 0 < \dfrac{p}{\tau} \leq \dfrac{f}{\sqrt{1+f^2}} \\[2ex] \dfrac{1}{2\,\tan f} & \text{for } \dfrac{f}{\sqrt{1+f^2}} < \dfrac{p}{\tau} < \dfrac{\sqrt{1+f^2}}{f} \end{cases} \tag{8.22}$$

where f is the coefficient of friction along the crack surfaces. A crack originally situated in this direction will grow in mode II without change of direction with a stress intensity factor

$$\frac{K_{\mathrm{II,max}}}{\sqrt{\pi a}} = \begin{cases} \tau & \text{for } \dfrac{p}{\tau} \leq 0 \\[2ex] \sqrt{\tau^2 - p^2} & \text{for } 0 < \dfrac{p}{\tau} \leq \dfrac{f}{\sqrt{1+f^2}} \\[2ex] -fp + \tau\sqrt{1+f^2} & \text{for } \dfrac{f}{\sqrt{1+f^2}} < \dfrac{p}{\tau} < \dfrac{\sqrt{1+f^2}}{f} \end{cases} \tag{8.23}$$

provided that $K_{\mathrm{II,max}}$ reaches the critical value for growth, $K_{\mathrm{II}c}$, and that mode I growth via a kink, for some reason, is prevented.

Melin assumed a very small kink at a different angle from the original crack (Fig. 8-7*b*) and calculated the stress intensity factors at the tip of the kink. He found that K_{I} for the kink is maximum at an angle of about 70° from the original crack (Fig. 8-7*c*). He also found that if the crack continues to grow in mode I, it will follow a smooth path toward a direction perpendicular to the largest principal stress and the crack will be open so that no frictional sliding is involved. A typical contour of the separated surface of the main crack and the kink is shown in Fig. 8-7*c*.

A criterion for deciding whether the crack will grow in mode I by formation and extension of a kink or grow in mode II in the original crack direction might be established by comparing the stress intensity factors $K_{\mathrm{I,max}}$ for the small kink with $K_{\mathrm{II,max}}$ for the main crack. If $K_{\mathrm{I}c}$ and $K_{\mathrm{II}c}$ denote the critical stress intensity factors for modes I and II, respectively, mode II crack growth will be preferred to mode I if

$$k_F = \frac{K_{\mathrm{II,max}}}{K_{\mathrm{I,max}}} > \frac{K_{\mathrm{II}c}}{K_{\mathrm{I}c}} = k_{Fc} \tag{8.24}$$

For the case of $f = 0.4$ it was concluded that the crack extension will be in mode I if $p/\tau < 2.2$, it will continue in mode II if $2.2 \leq p/\tau \leq 2.4$, and no growth can occur if $p/\tau > 2.7$.

8.2.5 Other Failure Theories

Other failure theories could be used in the same way as the ones just described. It is possible to consider the Mohr–Coulomb theory and the octahedral stress theory.

For the Mohr–Coulomb theory

$$\tau_{r\theta} \geq C_1 + C_2\sigma_\theta \tag{8.25}$$

where constants C_1 and C_2 may be determined from conditions associated with K_{Ic} and K_{IIc}.

Octahedral stress is

$$\tau_0 = \tfrac{1}{3}\sqrt{2}[\tfrac{1}{2}(\sigma_r + \sigma_\theta)]^2 + 3\sqrt{[\tfrac{1}{2}(\sigma_r - \sigma_\theta)]^2 + \tau_{r\theta}^2} \tag{8.26}$$

and propagation occurs when $\tau_0 = \tau_{0c}$, or

$$\tau_0 = C_3 + C_4\sigma_0 \tag{8.27}$$

where $\sigma_0 = (\sigma_r + \sigma_\theta)/3$ and constants C_3 and C_4 may be determined from conditions associated with K_{Ic} and K_{IIc}.

8.2.6 Comments

As described above by Melin,[9] it appears unlikely that crack extension under the effects of normal and shearing stresses will occur unless for some reason K_{IIc} is smaller than K_{Ic} or a high confining pressure (biaxial) is present. Both the maximum stress criterion and minimum strain density criterion assume crack propagation is in mode I for combined loadings. The question then is, can crack propagation actually occur in mode II? If it does, then K_{IIc} must be a fracture property. How can it be measured?

8.3 OTHER FRACTURE MECHANICS MODELS

8.3.1 Models Based on Fracture Energies G_I and G_{II}

The total energy release rate for mixed-mode behavior is given by Eqs. (2.74) and (2.75):

$$G = G_I + G_{II} \tag{8.28}$$

where $G_{\mathrm{I}} = (K_{\mathrm{I}})^2/E'$ and $G_{\mathrm{II}} = (K_{\mathrm{II}})^2/E'$ and $E' = E$ for the plane-stress condition and $E' = E/(1 - \mu^2)$ for the plane-strain condition. Thus, crack propagation would occur for $G = G_c$, some critical energy release rate that should be a material constant. For this to be generally true, it must be valid for mode I deformation only and $G_c = G_{\mathrm{I}c} = (K_{\mathrm{I}})^2/E'$. For mode II deformation only, $G_c = G_{\mathrm{II}c} = (K_{\mathrm{II}})^2/E'$, or in other words, $K_{\mathrm{II}c} = K_{\mathrm{I}c}$. Since there is no reason why $K_{\mathrm{II}c}$ should be the same as $K_{\mathrm{I}c}$, evidently simply considering the sum of the energies to be constant at failure is not correct. As a result, the following variations are suggested.

Di Leonardo[10] proposed

$$K_{\mathrm{I}}^2 + \lambda_{21} K_{\mathrm{II}}^2 = K_{\mathrm{I}c}^2 \tag{8.29}$$

where λ_{21} is a constant. Zhou et al.[11] suggested

$$\lambda_{12} K_{\mathrm{I}} + K_{\mathrm{II}} = K_{\mathrm{II}c} \tag{8.30}$$

where λ_{12} is a constant. More generally, it was proposed[3] that

$$\left(\frac{K_{\mathrm{I}}}{K_{\mathrm{I}c}}\right)^{\lambda_1} + \left(\frac{K_{\mathrm{II}}}{K_{\mathrm{II}c}}\right)^{\lambda_2} = 1 \tag{8.31}$$

where λ_1 and λ_2 are material exponents.

8.3.2 Jenq and Shah Model

This model uses the mode I fracture parameters $K_{\mathrm{I}c}^s$ and CTOD_c as described in Chapter 5.[12] This model has been extended to describe mixed-mode fracture behavior. In their model, Jenq and Shah defined a mixed-mode stress intensity factor as

$$K = \sqrt{K_{\mathrm{I}}^2(\theta) + K_{\mathrm{II}}^2(\theta)} \tag{8.32}$$

where θ again represents the branch angle, or "kink," from the original crack and $K_{\mathrm{I}}(\theta)$ and $K_{\mathrm{II}}(\theta)$ are the modes I and II stress intensity factors at the kink. The crack is considered to propagate when $K = K_c$ and $\mathrm{CTD} = \mathrm{CTD}_c$, where K_c is the critical stress intensity factor, CTD is the crack tip displacement, and CTD_c is the critical crack tip displacement as shown in Fig. 8-8. The CTD is obtained as a vector sum of the crack tip opening displacement (CTOD) and the crack tip sliding displacement (CTSD). In a similar manner at the crack mouth, the CMOD and CMOS add vectorally to produce the crack mouth displacement (CMD). The fracture parameters are obtained directly from the mode I case:

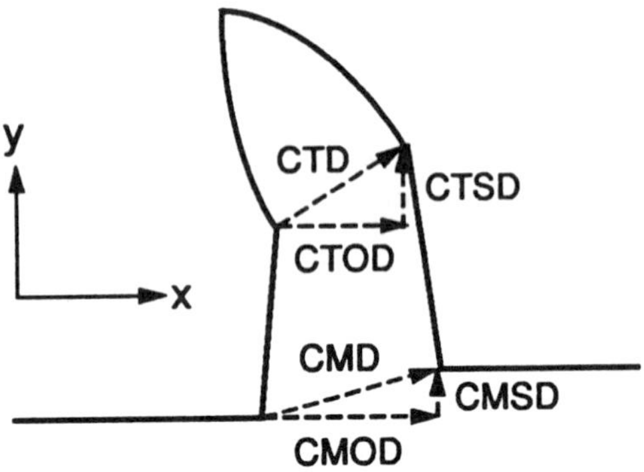

Figure 8-8 Composition of crack tip displacement and crack mouth displacement.[12]

$$K_c = K_{Ic} = K_{Ic}^s \qquad \text{CTD}_c = \text{CTOD}_c \tag{8.33}$$

8.4 PLANAR MIXED-MODE TESTING METHODS

A very extensive listing of research dealing with measurement of fracture properties of concrete and rock subjected to mixed-mode loads is given in reference.[1] Some, but not all, of the proposed testing geometries are presented in Fig. 8-9. These all have been used to determine the fracture behavior at the macrolevel. In this section the following testing geometries will be discussed in detail: three-point bend beam, four-point bend beam, quadratic double-edge notched specimen, and the tension–shear specimen.

8.4.1 Three-Point Bend Beam

Jenq and Shah[12] conducted mixed-mode experiments using beams with geometry shown in Fig. 8-10. The parameters to be varied were notch location x, notch depth a_0, and beam size. Different mix designs were used to obtain paste (P), mortar (M), and concrete (C). Only results for C-series beams are presented here. The mix design was a cement–sand–aggregate–water ratio of 1.0 : 2.6 : 2.6 : 0.55 with a maximum coarse aggregate size (A) of 0.375 in. (9.52 mm). The beams were made in two sizes of (span × depth × thickness): 24 × 6 × 2.25 in. (609.6 × 152.4 × 57.2 mm) and 12 × 3 × 1.13 in. (304.8 × 76.2 × 28.6 mm).

Because of the tortuosity of the crack line, the crack was modeled on a macroscale as shown in Fig. 8-11. Typical crack trajectories for different notch locations ($\gamma_1 = 2x/S$) for $a_0 = \frac{1}{3}$ depth are shown in Fig. 8-12. Figure 8-13 gives the theoretical predictions of crack initiation angles with the experimental results of all specimens tested. Despite the large scattering observed from the experimentally measured final crack angles, the theoretical predictions of crack initiation angles seem to be reasonable. A conventional (LEFM) critical mix mode stress intensity factor K_c^L was computed using a FE approach using the

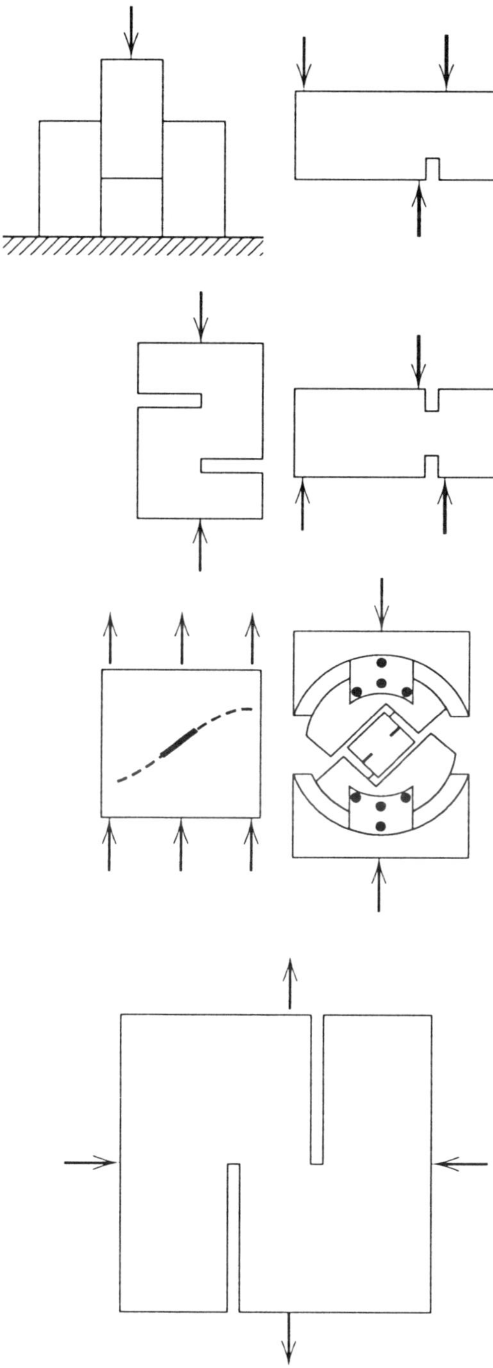

Figure 8-9 Mixed-mode test geometries.

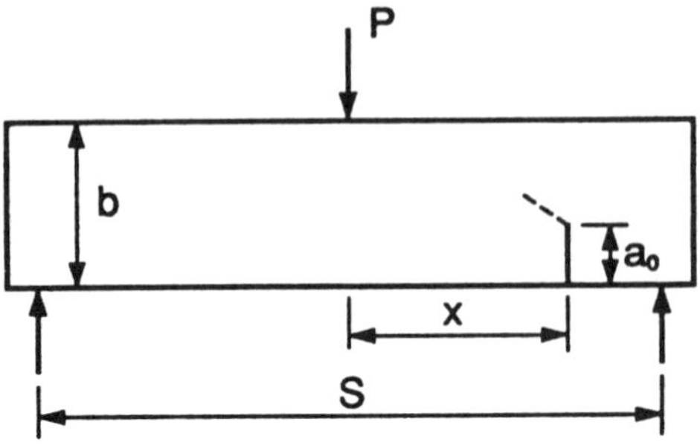

Figure 8-10 Three-point bend beam with off-center notch.[12]

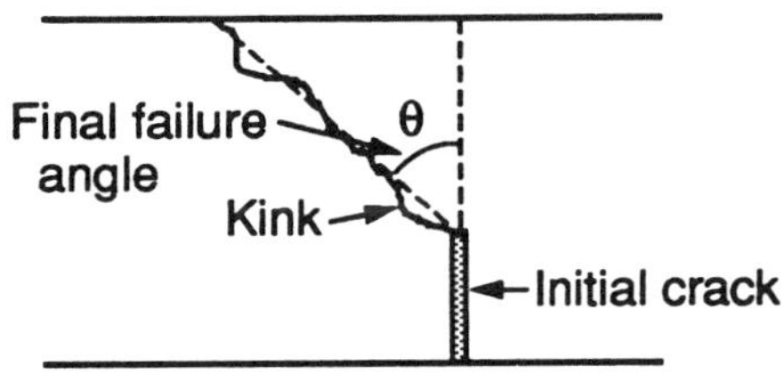

Figure 8-11 Definition of final failure angle.[12]

measured peak load, original notch length, and crack initiation angle. Values of these are plotted in Fig. 8-14 for the two beam sizes and different notch locations. It is seen that K_c^L is dependent on the specimen size. The variation in K_c^L with notch depth for $\gamma_1 = 2x/s = 0.5$ (quarter-span location) is shown in Fig. 8-15.

In order to apply the two-parameter fracture model [Eq. (8.33)], values of $K_{Ic}^s = 959$ psi · in.$^{1/2}$ (1.54 MN · m$^{-3/2}$) and CTOD$_c$ = 0.000394 in. (0.01 mm) were obtained using the procedure described in Chapter 6. The compressive strength was 4979 psi (34.3 MPa) and modulus of elasticity was 4760 ksi (32.8 GPa).

The peak load was underestimated by the conventional maximum energy release rate criterion for large beams and overestimated for small beams due to the size effect. Furthermore, using the conventional maximum energy release rate criterion, the failure location will always occur at the location of the notch and the predicted failure load will approach infinity when the off-set ratio is close to unity or the notch depth approaches zero. This prediction contradicts the experimental results presented in Figs. 8-16 and 8-17. In these it is seen that the predictions obtained from the two-parameter fracture model agree well with the experimental data. In this model, the tractions acting along the crack surfaces were assumed to be negligible. This assumption may be reasonable for the present testing configuration in which the K_{II}/K_I ratios are relatively small. However, for higher K_{II}/K_I ratios, it was speculated that the tractions

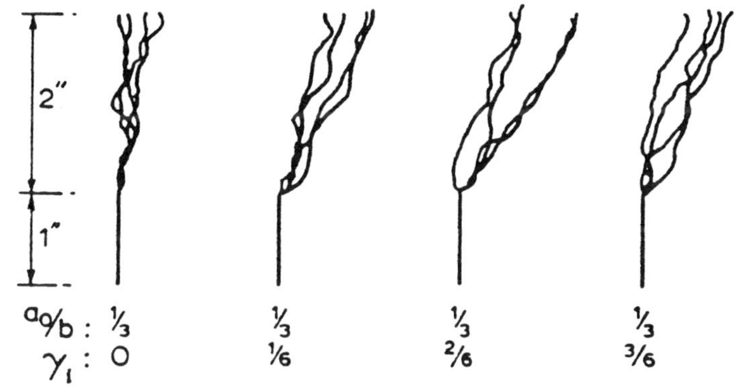

Crack trajectories of small specimens - C-series.

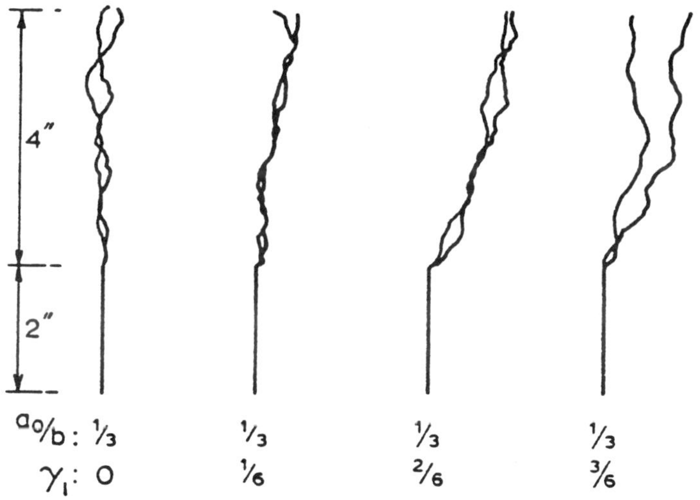

Crack trajectories for large specimens - C-series.

Figure 8-12 Crack trajectories for *C*-series specimens.[12]

that occurred due to aggregate interlock effect may not be negligible and have to be considered in the analysis.

A similar three-point bending configuration but with two notches, as shown in Fig. 8-18, was tested by Swartz et al.[13] The beams were tested in load control using a closed-loop system with a stiff frame. The time to peak load generally exceeded 30 s. The fracture energy associated with a diagonal crack was obtained by measuring the area under the load point displacement (LPD) plot and utilizing Eq. (6.6) since load and displacement were applied upward. This equation was modified to account for the crack path inclined at angle θ

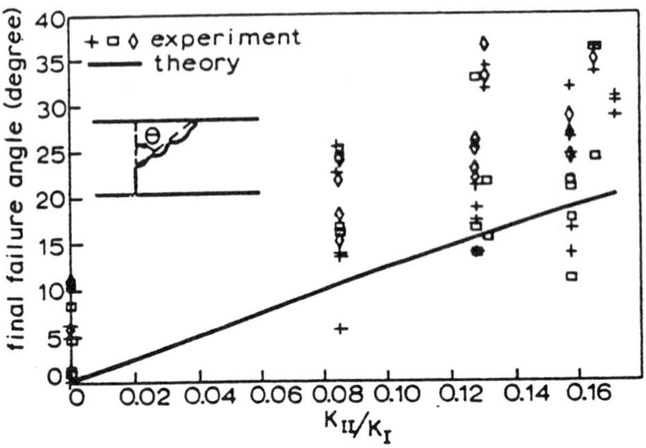

(a)

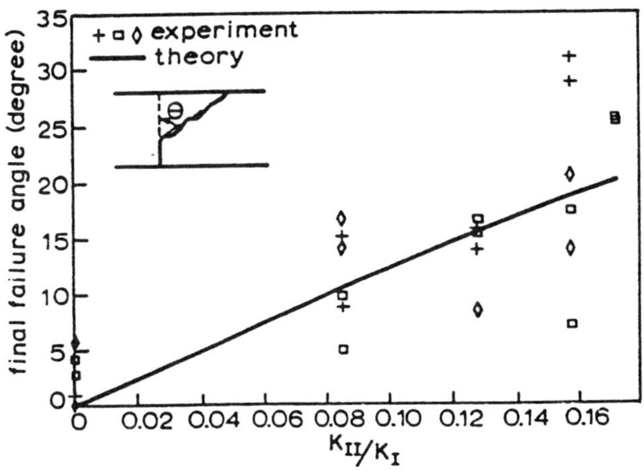

(b)

Figure 8-13 Comparison of theoretical predictions and experimental measurements: (*a*) small specimens and (*b*) large specimens.[12]

(Fig. 8-11) and a surface roughness factor of 1.15 was used. Equation (6.6) was thus modified to

$$G_F = \frac{(W_0 - P_w\delta_0)\cos\theta}{1.15(b - a_0)t} \tag{8.34}$$

The results obtained for an initial notch depth ratio of 0.67 are plotted in Fig. 8-19.[13] It is seen that fracture energy values increase substantially as the

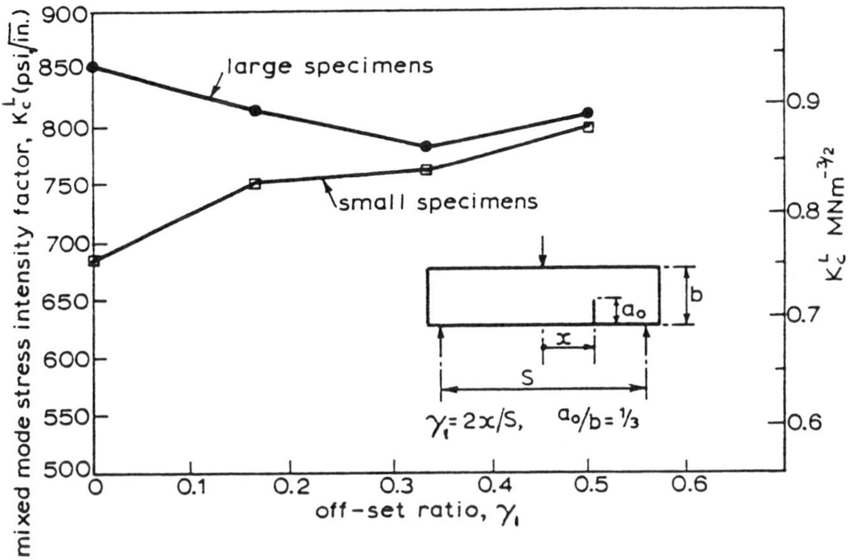

Figure 8-14 Size effect of conventional mixed-mode critical stress intensity factor for different offset ratios, *C*-series specimens.[12]

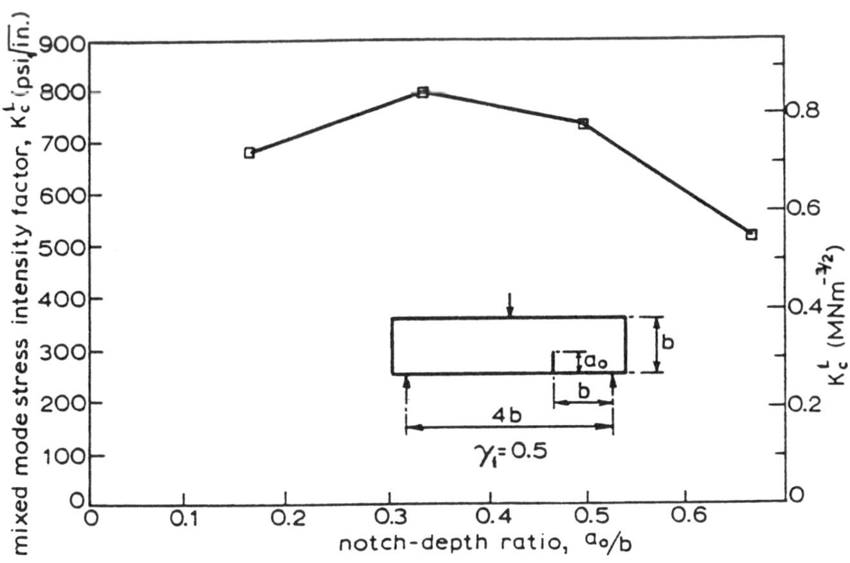

Figure 8-15 Size effect of conventional mixed-mode critical stress intensity factor for different notch depth ratios.[12]

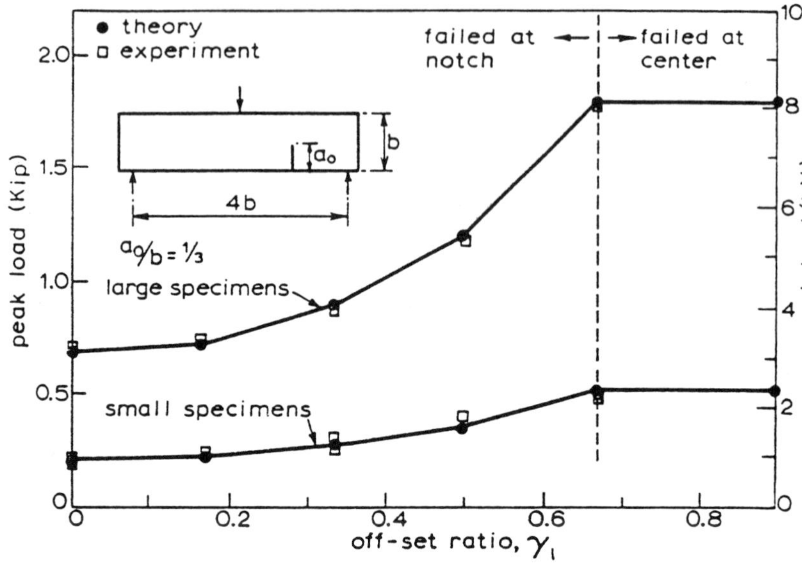

Figure 8-16 Comparison of experimental peak load and theoretical predictions using two-parameter fracture model for different offset ratios, *C*-series specimens.[12]

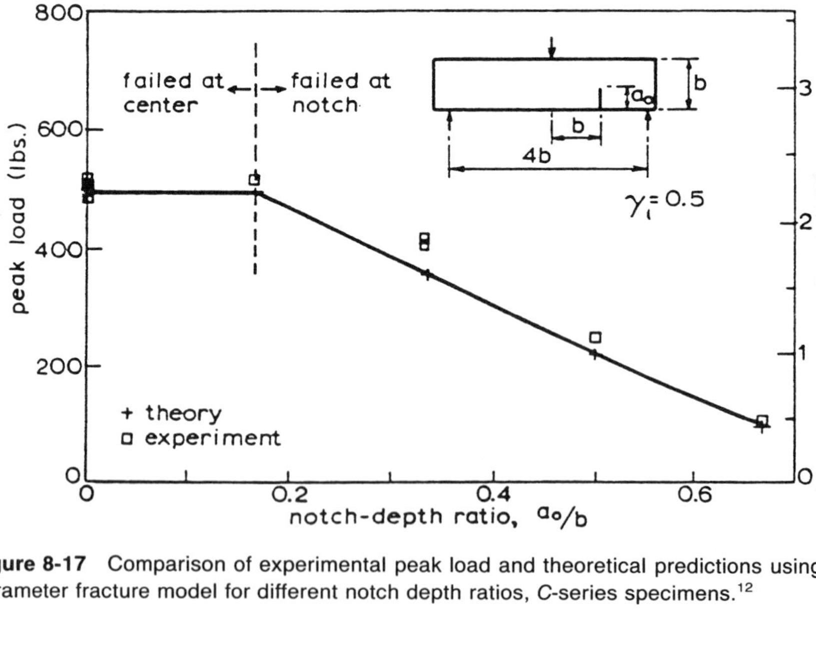

Figure 8-17 Comparison of experimental peak load and theoretical predictions using two-parameter fracture model for different notch depth ratios, *C*-series specimens.[12]

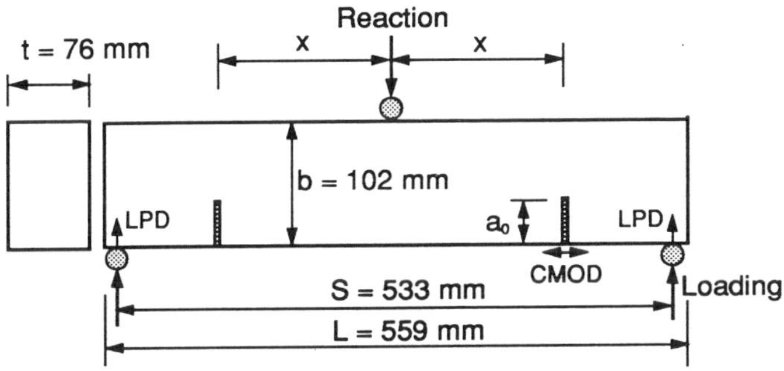

Figure 8-18 Three-point bend specimen with two notches.[13]

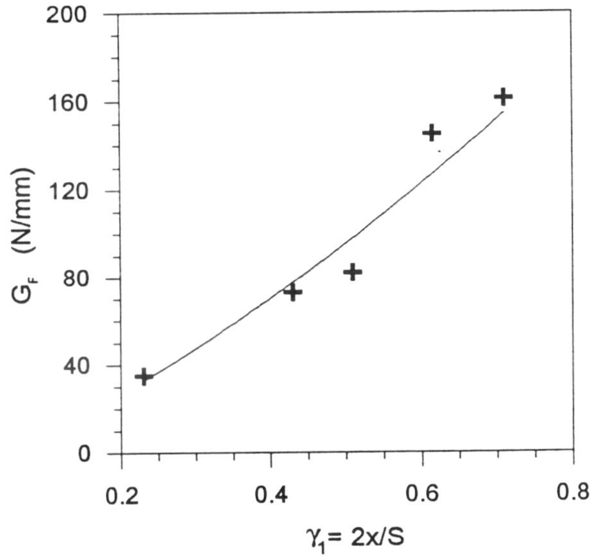

Figure 8-19 Fracture energy values measured using three-point bend beams with two notches.[13]

notch location changes from near midspan to near the support. For this testing geometry and the notch locations used, failure always proceeded from a notch. The increase in fracture energy implies a strong influence of aggregate interlock and friction on the fracture behavior.

In none of the three-point bending tests described was it possible to obtain values of fracture parameters that might be imputed to be due to mode II deformation behavior. Indeed, if friction and aggregate interlock are not considered, then on a macroscale, use of the mode I fracture parameters appears to be sufficient to predict the mixed-mode behavior of these beams. On the other hand, for reinforced concrete beams, with or without stirrups, the effects of friction and interlock cannot be ignored.

8.4.2 Four-Point Bend Beam

In the belief that mode II (shearing) fracture properties distinct from mode I should exist, researchers have attempted to define testing geometries where self-similar crack propagation occurs with only mode II deformations. Again, the geometries shown in Fig. 8-9 attest to the ingenuity of researchers on this quest. One such geometry, a beam-type specimen in four-point bending (4PB) was developed by Iosipescu.[14] This is shown schematically in Fig. 8-20 along with the normal shear and bending moment diagrams. A critical parameter in achieving the desired pure shear deformation behavior at the center of the beam is the notch angle α_p. Dasgupta and Sirkis[15] have verified experimentally, using Moiré interferometry, the analytical results of others that, for an isotropic material, the angle $\alpha_p = 102.6°$. This geometry was also analyzed by Kumoso and Hull,[16] who gave important support details, as shown in Fig. 8-21. Their test arrangement was used on composite materials.

Three salient points to be noted are (i) the geometry provides constant shear

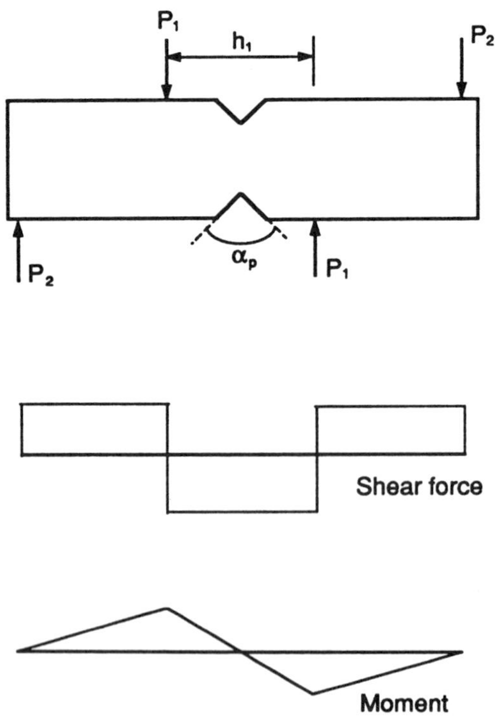

Figure 8-20 A 4PB beam proposed by Iosipescu.[14]

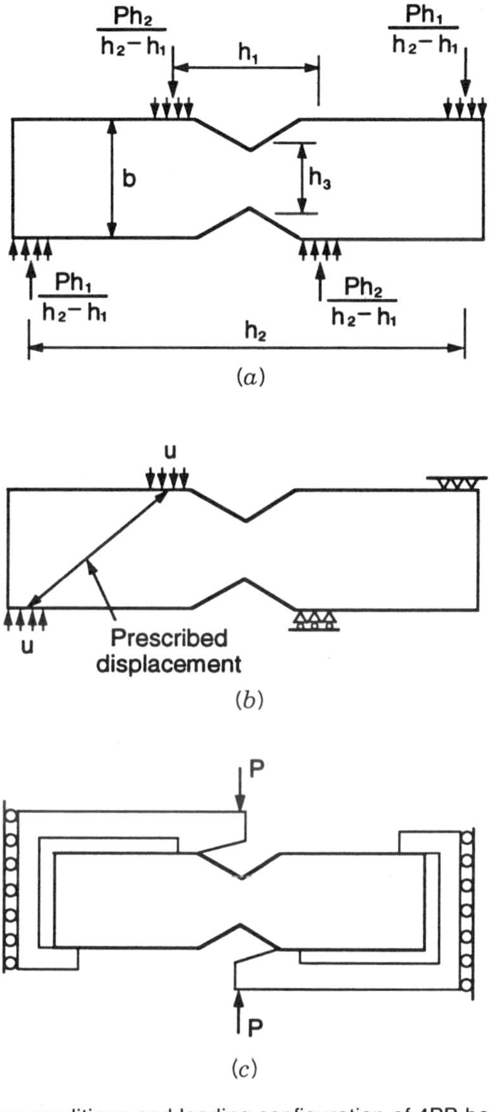

Figure 8-21 Boundary conditions and loading configuration of 4PB beam used by Kumoso and Hull.[16]

and almost zero moment at the test section, (ii) the test section is doubly notched (the original geometry had a groove around the entire test region), and (iii) the test fixture is designed to prevent rotation of the two halves of the test specimen.

No investigators of the mixed-mode fracture behavior of concrete have used this geometry precisely as shown in Fig. 8-21 with the important features noted above. Rather, simpler, and approximate, geometries have been the subject of investigations by Arrea and Ingraffea,[17] Ballatore et al.,[18] Bazant and Pfeiffer,[19,20] Biolzi,[21] Bocca et al.,[22] Ingraffea and Panthaki,[23] Schlangen and van Mier,[24,25] Swartz et al.,[26] and Swartz and Taha.[27,28]

8.4.2.1 Results by Bazant and Pfeiffer Symmetrically notched beams of concrete and mortar, as shown in Fig. 8-22, loaded near the notches by concentrated forces that produce a concentrated shear force zone were tested to failure by Bazant and Pfeiffer.[19] They arrived at the following conclusions:

1. Shear fracture (i.e., mode II fracture) of concrete exists at least on a macroscale.
2. Like the tensile (mode I) fracture, the shear fracture follows the size effect law of blunt fracture. This implies that a large fracture process zone must exist at the fracture front and that NLFM should be used, except possibly for extremely large structures.
3. The maximum aggregate size d_a appears acceptable as a characteristic length for the size effect law. This further implies that the size of the fracture process zone at the maximum load is approximately a certain fixed multiple of the maximum aggregate size.
4. The R-curve for the shear fracture may be obtained from the size effect law as the envelope of all fracture equilibrium curves for geometrically similar specimens of various sizes.
5. The shear (mode II) fracture energy for the present type of test appears to be about 25 times larger that tensile (mode I) fracture energy. This large difference may probably be explained by the fact that shear fracture energy includes, not only the energy to create inclined tensile micro-

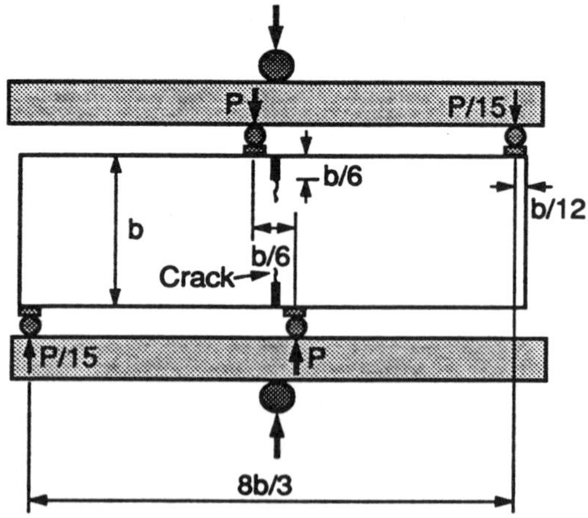

Figure 8-22 Mode II fracture specimen and testing arrangement of Bazant and Pfeiffer.[19]

cracks in the fracture process zone, but also the energy required to break the shear resistance due to interlock of aggregate and other asperities on rough crack surfaces behind the crack front. On a microlevel, such a mode II failure may arise from bands of microcracks as sketched in Fig. 8-23.

6. The direction normal to the maximum principal stress cannot be considered, in general, as the direction of crack propagation in concrete. Rather, fracture seems to propagate in the direction for which the energy release rate from the entire structure is maximized.
7. The test specimen in Fig. 8-22 does not undergo perfectly antisymmetric deformation. Nevertheless, the symmetric deformation component has apparently only a small disturbing effect because, in the sense of LEFM, the mode I stress intensity factor is not positive. According to both linear and nonlinear FE analysis, the maximum principal tensile stresses occur at the notches. So the cracks must originate also at the notches and propagate inward. This means that the fracture mode is essentially shear and that the effect of the transverse tensile strain at the middle of the ligament cannot be significant.

8.4.2.2 Results by Ingraffea et al. The beam geometry used by Arrea and Ingraffea[17] is shown in Fig. 8-24. The crack trajectory is considerably different from that shown in Fig. 8-22. Two major differences are that the beam in Fig. 8-22 is doubly notched and the interior load points are much closer together than those in Fig. 8-24. In discussing the test results of Bazant and Pfeiffer,[19] Ingraffea and Panthaki[23] showed by analysis—both classical elasticity and FEs—that the specimens tested by Bazant and Pfeiffer[19] fractured in a manner quite similar to that which occurs in a splitting tensile strength test. The following conclusions were drawn by Ingraffea and Panthaki:[23]

1. Limited tensile cracking from the notch tips, although not reported by Bazant and Pfeiffer[19] is likely to have occurred and should be observable using a crack detection technique more sensitive that the naked eye.

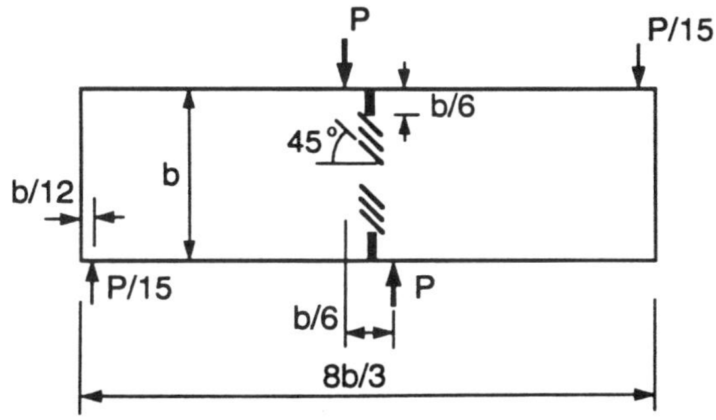

Figure 8-23 Mode II fracture arising from band of mode I microcracks.[19]

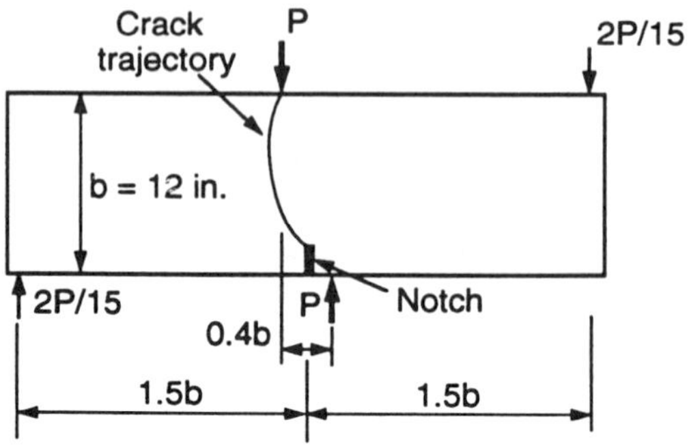

Figure 8-24 Single-notch beam geometry used by Arrea and Ingraffea.[17]

2. By moving the principal loads closer together than in the Arrea and Ingraffea[17] tests, the intensity of the shear stress in the region between notch tips is decreased rather than increased. The shear stress distribution, magnitude, and trajectory in this region cannot be predicted using elementary beam theory.
3. The major principal stress in the region between the notches is tensile, and the direction is roughly constant at about 10° above the horizontal. This means that the plane of principal shear is far from being vertical.
4. The stress state between the notch tips, despite the existence of the notches and the minor loads, is startingly similar to that in the central region of a cube or cylinder loaded by diametrically opposed point loads (the Brazilian test).
5. By using the solution for the principal tension at the center of a hypothetical Brazilian specimen inscribed within the beam specimen, one can predict the peak loads for the beam specimen with very good to good accuracy.
6. Based on the above conclusions, it is apparent that the beams suffered tensile, not shear, fracture, with a crack nucleating in the central region of the beam, roughly vertical, and propagating toward the notch tips.

Usually, shear failures are caused by tensile stresses, as shown in Fig. 8-25.[1] The trajectory of a brittle crack always follows the direction normal to the local principal tensile stress (Fig. 8-25*b*) according to the maximum circumferential stress criterion. Arrea and Ingraffea[17] argued about a transition between two different failure modes by varying the distance h_1 of the central supports. When the distance is large, crack branching is favored and a crack trajectory develops from the crack tip to the opposite force (Fig. 8-25*b*). On the contrary, when the distance is small, the ultimate tensile strength is exceeded at the center of the specimen (Fig. 8-25*c*) and a nearly vertical crack nucleates far from the crack tip and propagates upward and downward as in the Brazilian splitting

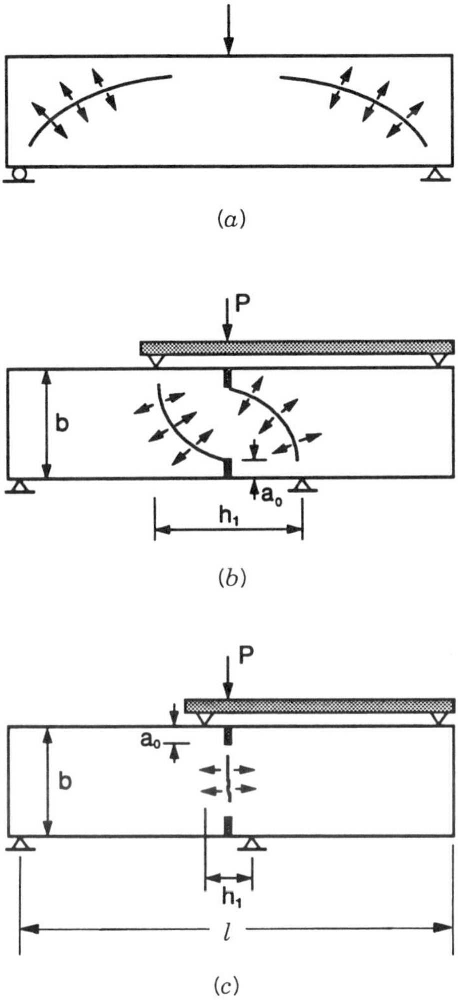

Figure 8-25 Shear failures of beam caused by tensile stress: (*a*) classical beam failure, (*b*) mixed mode crack propagation, and (*c*) tensile splitting.[1]

test. In addition, when the distance is small, St. Venant's principle is not valid between the central support and the elementary beam theory fails completely.

It is clear now that, for the Ingraffea et al. test, there are two potential failure mechanisms in competition, by varying the distance h_1 (Fig. 8-25) as well as by varying the size scale b of the specimen. Mixed-mode crack branching is favored by increasing the distance h_1 or the size scale b (the ratio of h_1/b being constant). On the contrary, ultimate tensile failure at the center is favored by decreasing the distance h_1 or size scale b.

In the Ingraffea et al. test, stress intensification is produced in both the crack tip regions and the stress intensity factors for modes I and II can be expressed respectively as (Figs. 8-25b,c)

$$K_{\mathrm{I}} = \frac{P}{t\sqrt{b}} f_{\mathrm{I}}\left(\frac{1}{b}, \frac{a}{b}, \frac{h_1}{b}\right)$$
$$K_{\mathrm{II}} = \frac{P}{t\sqrt{b}} f_{\mathrm{II}}\left(\frac{1}{b}, \frac{a}{b}, \frac{h_1}{b}\right) \tag{8.35}$$

where f_{I} and f_{II} are the respective shape functions, which can be determined numerically, and t is the specimen thickness.

When the distance h_1 tends to zero, the mode I stress intensification vanishes and

$$\lim_{h_1/b \to 0} K_{\mathrm{I}} = 0 \tag{8.36}$$

If the fracture criterion of Di Leonardo[10] Eq. (8.29) is adopted, then by combining Eqs. (8.29) and (8.36), one obtains[1]

$$\frac{P_{\max}}{t\sqrt{b}} = \frac{K_{\mathrm{I}c}}{\sqrt{f_{\mathrm{I}}^2 + \lambda_{21}^2 f_{\mathrm{II}}^2}} \tag{8.37}$$

If the geometric ratios $1/b$ and a/b are constant and h_1/b is varying, Eq. (8.37) can be transformed as follows:

$$\frac{P_{\max}}{f_t\, bt} = \frac{B_s}{f(h_1/b)} \tag{8.38}$$

where f_t is the tensile strength, $f(h_1/b)$ is the geometric function, and B_s is a brittleness number defined as

$$B_s = \frac{K_{\mathrm{I}c}}{f_t\, \sqrt{2b}} \tag{8.39}$$

On the other hand, if a diametrically loaded disk inscribed in the beam specimen as shown in Fig. 8-26 is imagined, one can obtain the central tensile stress σ by the simple Brazilian test formula:

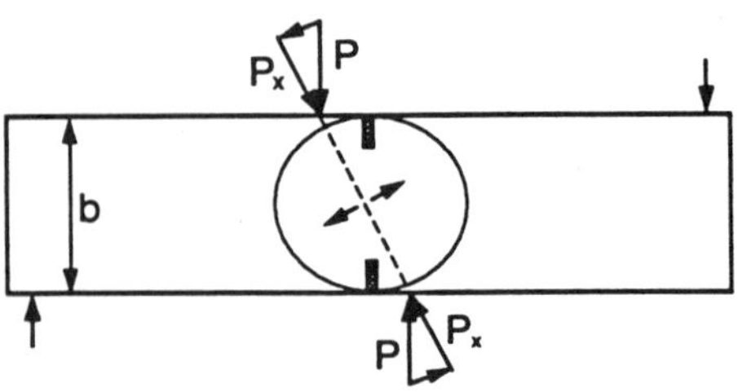

Figure 8-26 Beam inscribed with diametrically loaded disk.[1]

$$\sigma = \frac{2Pb}{\pi t(b^2 + h_1^2)} \tag{8.40}$$

The maximum load due to the achievement of the tensile strength f_t in the center of the beam is given in dimensionless form:

$$\frac{P_{max}}{f_t t b} = \frac{\pi}{2}\left[1 + \left(\frac{h_1}{b}\right)^2\right] \tag{8.41}$$

Equations (8.38) and (8.41) are in competition by varying the value of the brittleness number B_s or by varying the ratio h_1/b. If the latter is fixed, there is a size-scale transition that is described in Fig. 8-27. For small B_s values (i.e., for large sizes), LEFM crack branching precedes ultimate tensile strength failure and Eq. (8.41) overestimates the maximum load.[1] Vice versa, for large B_s values (i.e., for small sizes), ultimate tensile strength failure precedes LEFM crack branching and Eq. (8.38) overestimates the maximum load, which can be given by Eq. (8.41) with a very good accuracy.

The experimental results by Bazant and Pfeiffer[19] are reported in Fig. 8-28, where the maximum load divided by the beam area is plotted against the inverse root $b^{-1/2}$. The experimental points describe the transition between an inclined straight line passing through the origin and a horizontal asymptote. The former represents the LEFM instability, whereas the latter represents the ultimate tensile strength instability. It is important to realize that, with a mixed-mode crack configuration, larger (specimen and crack) sizes are needed than with a mode I crack configuration in order to utilize LEFM correctly. For example, mortar specimens 600 mm deep are necessary at least to apply mixed-mode LEFM in Fig. 8-28, whereas specimens 150 mm deep are usually sufficient for mode I.

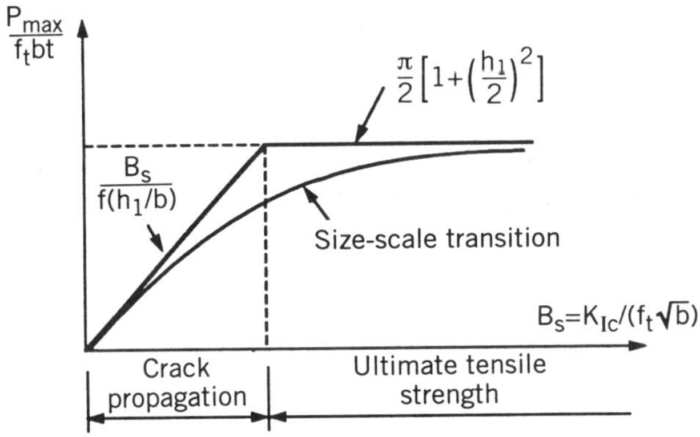

Figure 8-27 Size-scale transition between LEFM and ultimate strength.[1]

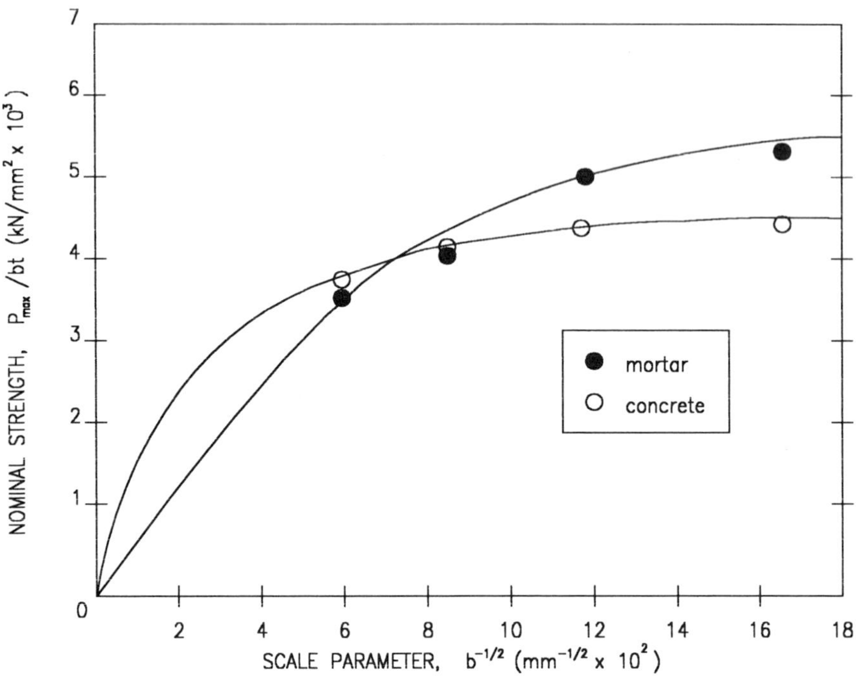

Figure 8-28 Experimental results of Bazant and Pfeiffer.[19]

8.4.2.3 Results by Swartz et al. Swartz et al.[26] presented experimental and analytical results and fracture parameters for beams loaded in four-point shear with a single-edge, starter notch located in a region of high shear and low bending moment, similar to the specimen shown in Fig. 8-24. The following conclusions were made:

1. If LEFM is assumed, the FE analysis indicates that crack propagation starts in mode II at an angle of approximately $\pm 70°$. However, in no case did this occur experimentally. The crack direction changed almost immediately after slow growth was initiated and became about 30°–40°. In this mode the FE program indicates propagation in a mode I condition. It is also noted that the crack path is far from smooth and is likely to be associated with mixed-mode behavior (rather than mode I) due to aggregate interlock and friction.
2. The influences of aggregate interlock and friction forces must be great. At the end of the fracture process the crack surfaces are subjected to compressive stresses.
3. Although the results of the energy estimates show similar variations with the notch depth a_0/b, as compared to analytical results, the values are

much different. These results become independent of notch depth at $a_0/b = 0.5$.

4. No matter which approach is used to determine K_{IIc} for concrete, the resulting value is much greater than K_{Ic}.
5. The presence of a possible process zone is not accounted for. This effect may be great for small-size beams and deserves further study. If the process zone is large, then LEFM is not valid.
6. The fracture energy values are about 8–10 times those obtained for mode I tests on the same concrete using the same beam size and similar procedures. This is similar to that reported by Bazant and Pfeiffer[19] in which they reported a 30 times difference.

Another study was conducted by Swartz and Taha[27,28] in which double notched beams were used. These were subjected to shear loads with and without additional axial, compressive loads. The setup for shear loads only is shown in Fig. 8-29. Displacements that were measured included crack mouth sliding (CMSD), LPD, and axial load displacement (LD). The overall test setup showing how axial load was applied is shown in Fig. 8-30.

The beam testing program consisted of three parts: (1) testing notched beams without axial load (but the fixture shown in Fig. 8-30 was in place), (2) 4PB tests of unnotched specimens and tensile splitting tests of portions of notched beams after their failure, and (3) testing notched 4PB beams with axial load also applied. For the beams tested without axial force, the notch depths were 0–2 in. (50 mm). For the beams with additional axial load the notch depths were 1.3 in. (33 mm). The tension rods shown in Fig. 8-30 are external to the

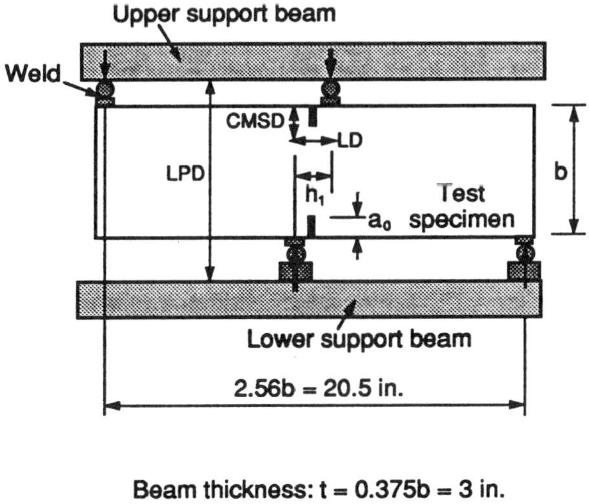

Figure 8-29 Experimental setup of Swartz and Taha.[28]

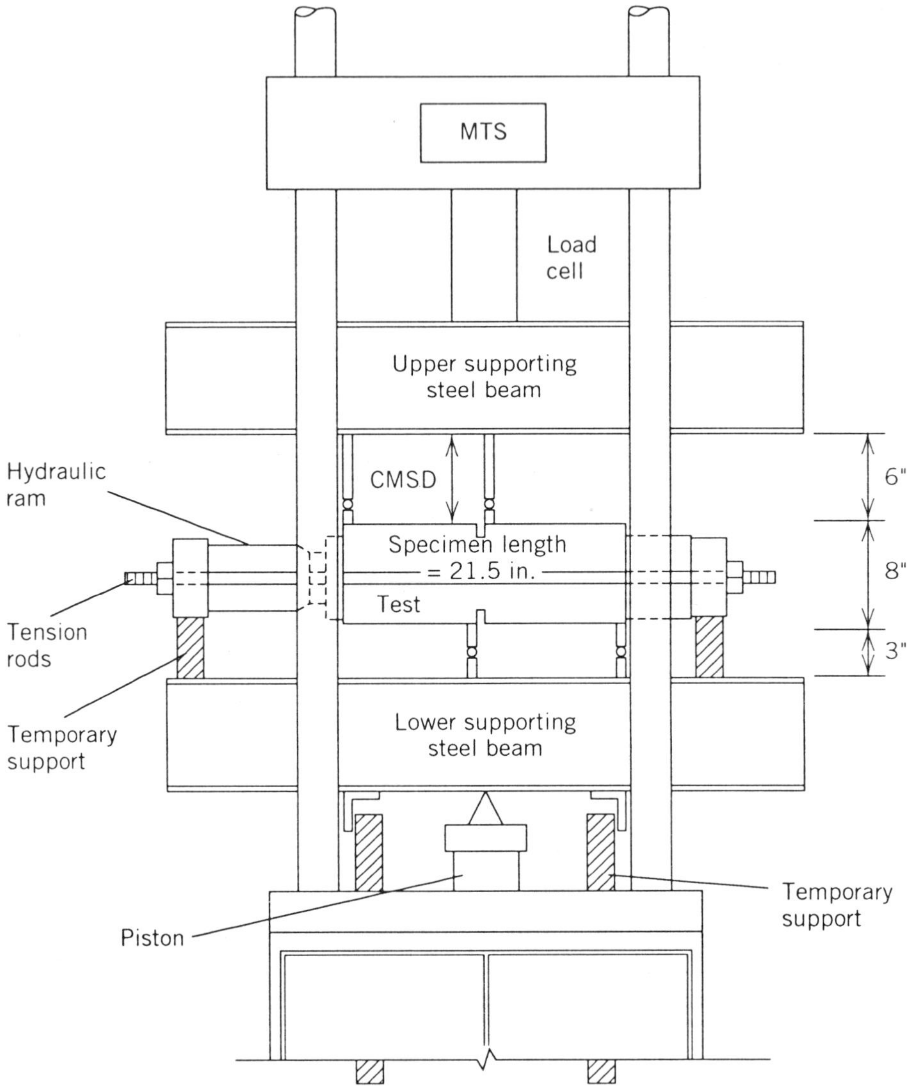

Figure 8-30 Test setup with shear loads and axial compressive load.[28]

concrete beam and carry no shear load. The left-end platen was allowed to rotate while the right was not. The axial force was monitored and maintained constant throughout the test.

Steel spreader beams were used to place the loads and the distance between interior load points was $h_1 = 1.32$ in. (33.5 mm) for beams without axial force and $h_1 = 1.91$ in. (48.5 mm) for beams with axial force. This gave beam loads of $0.9395P$ and $0.0605P$ (no axial force) and $0.9148P$ and $0.0852P$ (with axial force). For the beams without axial force all load points had 1-in.- (25-mm-) square steel bearing pads and 1-in.- (25-mm-) diameter steel rollers. The beams with axial force had interior load point steel bearing plates with a width of 1.75 in. (44.4 mm) to avoid local splitting failures. All beam tests were conducted using an electrohydraulic, closed-loop system with a stiff frame. Strain (displacement) control based on the CMSD response was used in most of the tests.

Beams without Axial Force. Despite the use of strain control, these beams failed in an unstable manner and exhibited a snapback response on the LPD measurement. A typical final crack pattern is shown in Fig. 8-31 along with analytical results. All the notched beams had similar patterns. The unnotched beams (2, 10, and 32), however, failed with a straight-line crack running be-

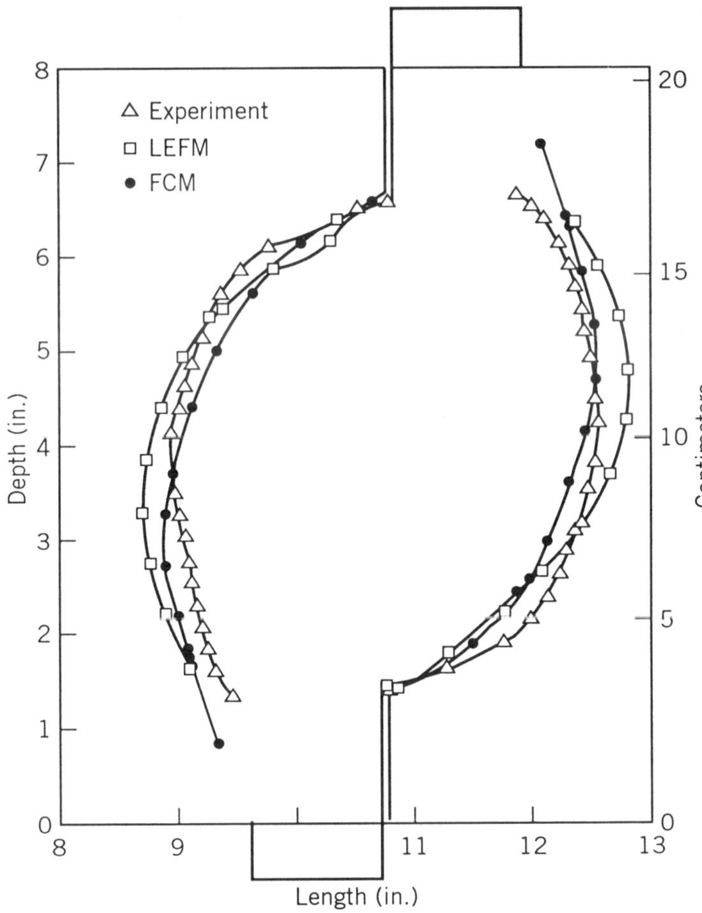

Figure 8-31 Crack pattern for beam without axial force.[28]

tween the centers of the interior load points, exactly as in a split-cylinder failure.

Beams with Axial Force. The results of these tests are summarized in Table 8-1. The loading sequence was to apply axial compression first, then zero the various displacement gages, followed by application of the lateral (4PB) load using either strain control (on the CMSD) or load control. Typical load-versus-displacement (CMSD, LD) traces are shown in Fig. 8-32. There were no unstable failures and the end of the trace indicates either stopping the test or the end of the range of the transducer. Note that positive LD indicates axial extension.

For all these beams the final crack pattern was as shown in Fig. 8-33 for beam 24. When the maximum load was reached, two cracks (denoted (1)) started simultaneously from the notch tips. The average angle of inclination was 50°. Following the peak load, these cracks stabilized and another crack (2) started at midspan almost at middepth. This crack was monitored using a low-power magnifying lens and generally had a length of about 2 in. (51 mm) when first observed. During the postpeak loading this crack eventually propagated through the beam depth. After this, the two halves of the beam continued to slide relative to each other at a load of about half the peak load.

Since the failure leading to the separation was in tension, despite the pres-

Table 8-1 Beams with Axial Force[a]

Test Number	Axial Force (kN)	Percent of f'_c	Shear Load (kN)	θ_0 (deg)	Crack, $\underline{1}$ (mm)	Type of Control	Net Center Stress (MPa)
28	2.2	0.4	102.3	50	170	Strain	3.93
21	4.4	0.8	105.0	50	147	Strain	3.87
20	11.1	1.9	109.0	51	127	Strain	3.53
26	13.4	2.3	114.4	40	132	Strain	3.58
22	15.6	2.6	107.2	49	102	Strain	3.13
13	22.2	3.8	127.3	48	91	Load	3.43
33	26.7	4.5	130.8	45	86	Strain	3.24
23	33.4	5.7	129.5	56	81	Load	2.68
27	40.0	6.8	163.8	46	79	Load	3.56
14	44.5	7.6	157.5	50	76	Load	2.97
34	44.5	7.6	155.5	50	74	Load	2.88
25	48.9	8.3	171.8	51	69	Load	3.20
24	55.6	9.5	171.8	55	53	Load	2.70
15	66.8	11.4	186.9	41	51	Load	2.47
Mean							3.22
CV							13.9%

[a]From ref. 28. Note: $a_0 = 33.0$ mm but the net area in compression was calculated using only one notch because of the nature of the crack path. CV = coefficient of variation.

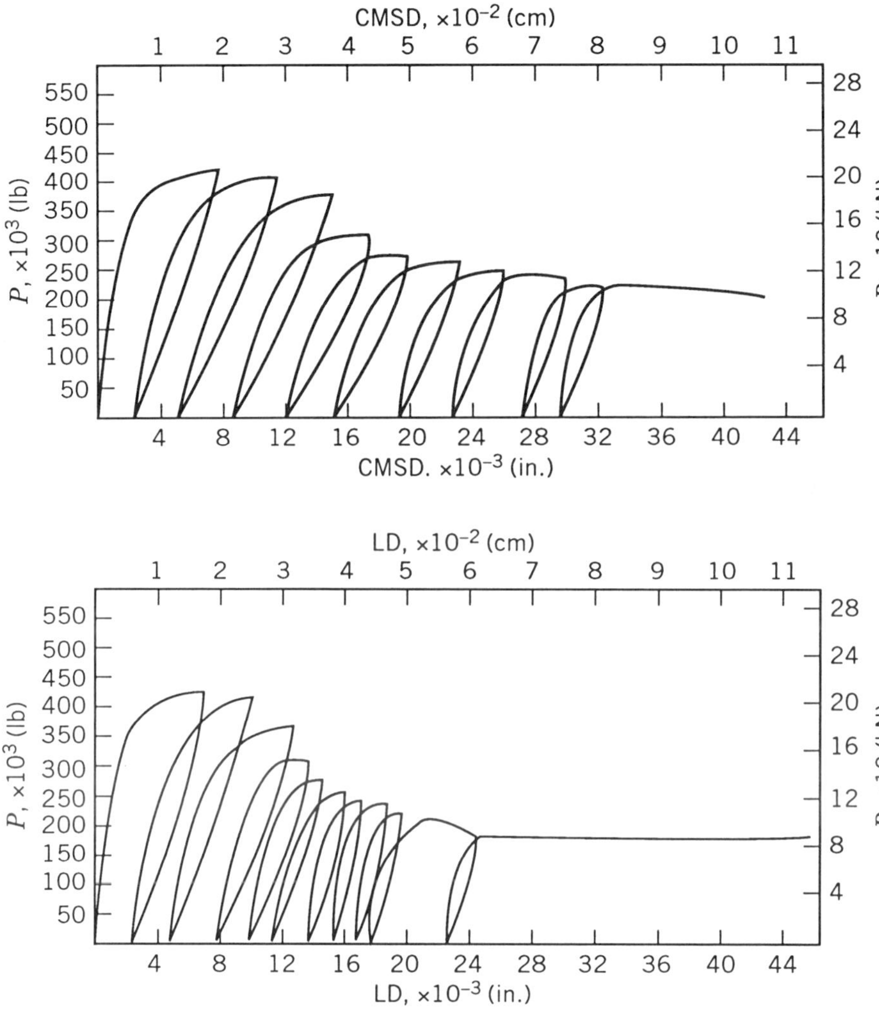

Figure 8-32 Load displacements for beam with shear loads and axial compression.[28]

ence of the axial force, the net, average stress was determined by calculating the splitting tension at peak load using the Brazilian formula and subtracting from this the compressive stress on the net area. This resulted in a net tension stress in each case. Values are presented in Table 8-1.

Conclusions for these beams with axial compression in addition to the shear load were the following:

1. Application of initial compression did not prevent the formation of cracks at the notch tips. These extended until the peak shear load was reached and then stopped when a third crack started in the center of the beam.

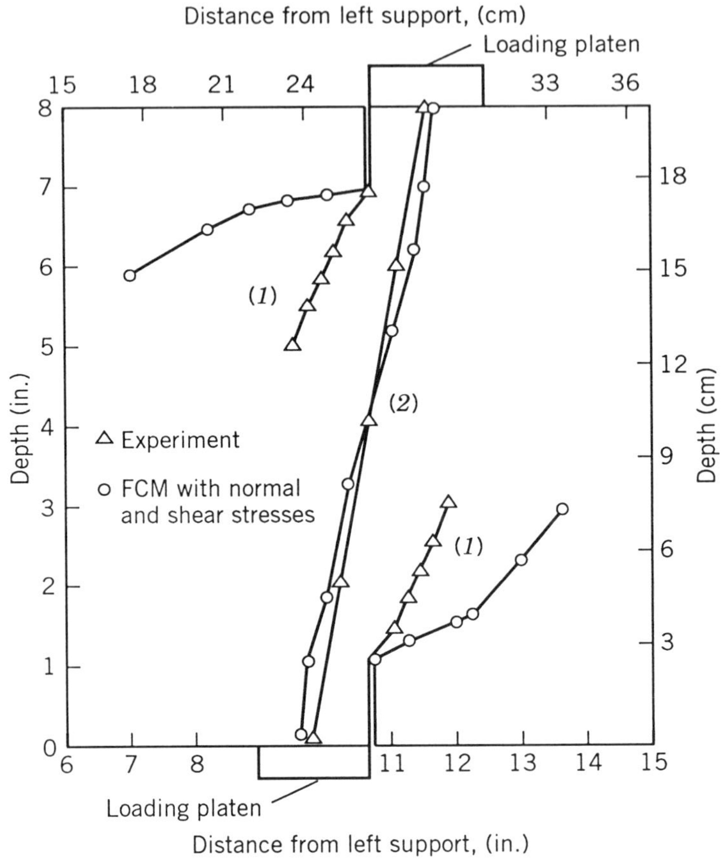

Figure 8-33 Crack pattern for beam with shear loads and axial compression.[28]

2. The beams ''failed'' due to extension of the central crack. However, because of friction and aggregate interlock, the beam still carried about 50% of their peak shear loads after complete penetration of the third crack through the beam.
3. Increasing the axial compression merely increased the peak shear load at which the third crack formed. This occurred with average, net tensile stress at middepth of 468 psi (3.22 MPa). This stress was somewhat less than the splitting strength obtained from the beams without notches and the splitting tests whose average value was 547 psi (3.77 MPa).

8.4.2.4 Results from RILEM Round-Robin Test The RILEM Committee 89-FMT proposed a round-robin study of a mixed-mode test using a geometry with a single notch, as shown in Fig. 8-34. A suggested testing arrangement is given in Fig. 8-35.[1] The specimen dimensions and maximum

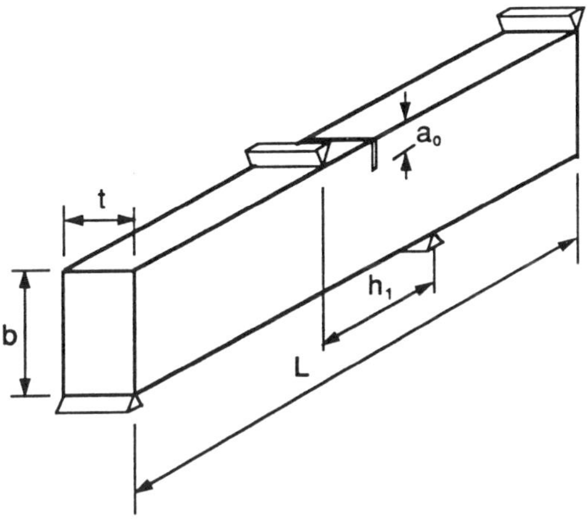

Figure 8-34 Testing geometry of RILEM round-robin test.[1]

aggregate sizes to be used are listed in Table 8-2. This geometry was analyzed and tested extensively by Schlangen and van Mier,[24,25] who considered effects of support frame stiffness, friction, one or two notches, load placement, and beam size.

Their testing arrangement shown schematically in Fig. 8-36*a* is intended to apply loads to single-edge notch (SEN) beams in the ratios indicated, which are also the ratios required in the RILEM round-robin test. They[25] pointed out that support conditions used in the RILEM test shown in Fig. 8-35 would induce friction at the loading points that would influence the stress distribution in the specimen. They devised testing arrangements, shown in Fig. 8-36, in which free rotation could occur—and friction minimized (Fig. 8-36*b*)—and in which the relative movement of all support points could be fixed (Fig. 8-36*c*). In both of these arrangements it was possible to measure all loads and reactions. As will be shown later, this arrangement resulted in the loads actually varying in distribution as intensity increased. To correct this problem, the testing device was modified as shown in Fig. 8-37. Again, free rotation of supports or relative fixing is permitted. The major change in the new setup is the use of the steel beam at the bottom to which the bottom supports are attached. This in turn is attached to the testing machine base through a central hinge. This arrangement provided identical frame support conditions at top and bottom. The results of this design change are shown in Fig. 8-38, where reaction ratios are plotted against beam displacement δ_2 (cf. Fig. 8-35). Recalling that the ratio of applied forces F_1/F_2 (or F_{1r}/F_{2r}) is 10, where meanings of F_1, F_2, F_{1r}, and F_{2r} are given in Fig. 8-36*a*, it is seen that the original test setup is essentially inadequate. This very important problem has not been investigated by any of the previous researchers mentioned in this chapter. The performance shown by the new testing frame is clearly much better, although even here the

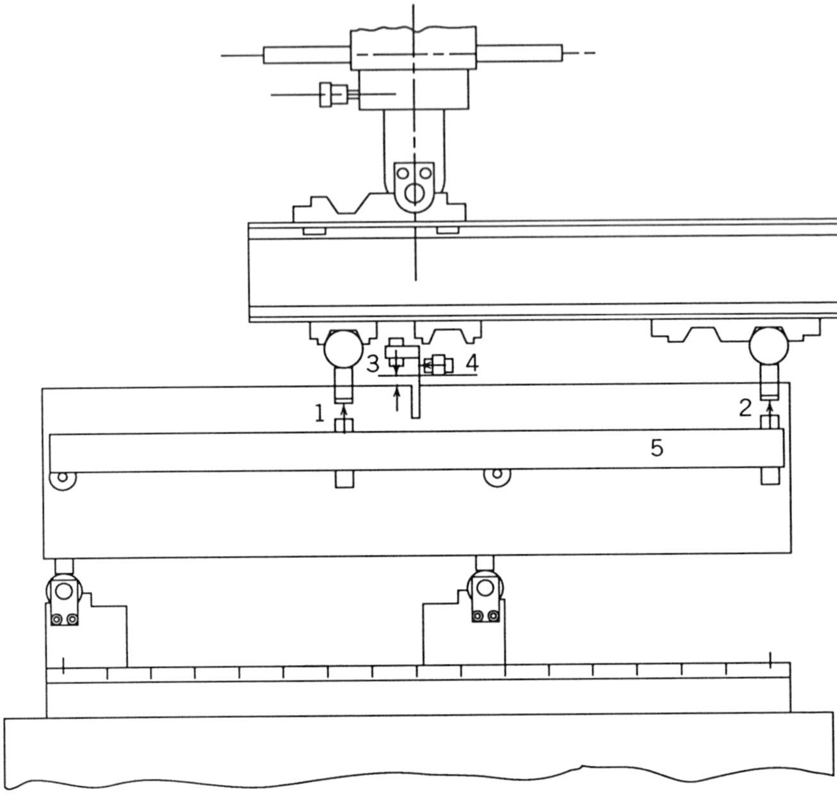

Figure 8-35 Testing arrangement for mixed mode round-robin test of RILEM[1]: LVDT for measurement of (1) δ_1 deflection, (2) δ_2 deflection, (3) CMSD, and (4) CMOD; (5) reference bar for δ_1 and δ_2 measurement.

Table 8-2 Specimen Dimensions: RILEM Mixed-Mode Test[a]

Maximum Aggregate, d_a (mm)	Thickness, t (mm)	Depth, b (mm)	Length, L (mm)	Span, S (mm)	Crack Length, a (mm)	Distance, c (mm)
10	100	100	440	400	20	40
	100	200	840	800	40	80
20	100	200	840	800	40	80

[a] From ref. 1.

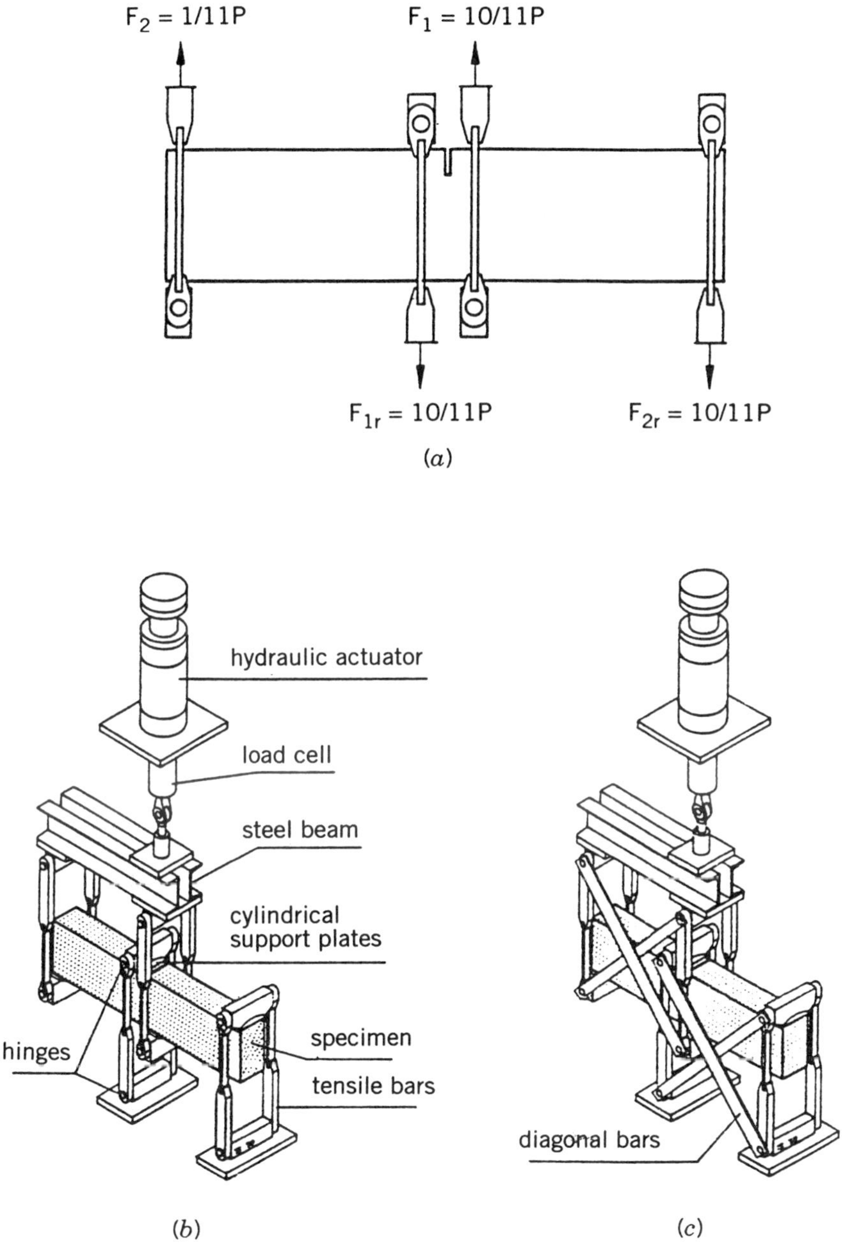

Figure 8-36 Beam in 4PB: test arrangement of Schlangen and van Mier: (*a*) schematic of desired load distribution, (*b*) asymmetric device with freely rotating supports, and (*c*) asymmetric device with fixed support.[25]

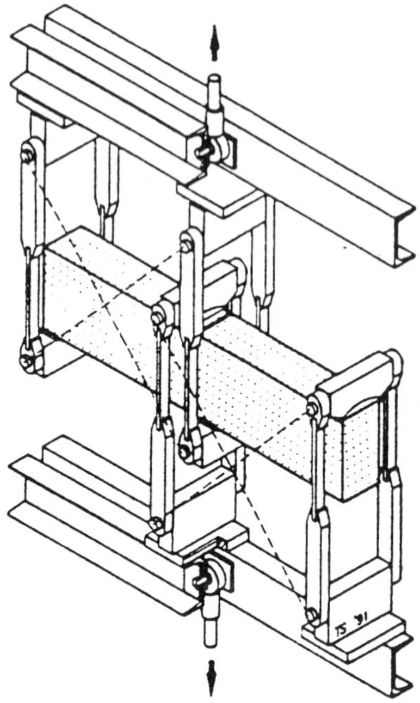

Figure 8-37 New symmetric loading device.[25]

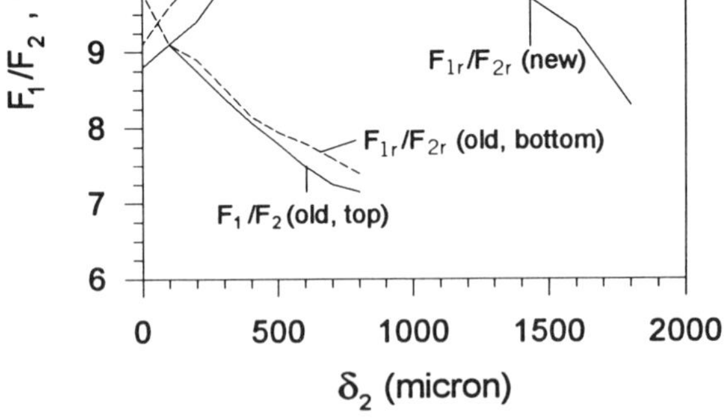

Figure 8-38 Distribution of loads versus deformation δ_2 in two tests (one with notch at top and one with notch at bottom of beam) in old asymmetric device and one test in new symmetric device.[25]

effect of misalignment or some other eccentricity has caused the F_1/F_2 ratio to be slightly different from 10.

The SEN beams were in two sizes (cf. Fig. 8-34): the small size, with $a_0 = 20$ mm, $b = 100$ mm, $h_1 = 40$ mm, $L = 400$ mm, $t = 100$ mm, and the large size, with $a_0 = 40$ mm, $b = 200$ mm, $h_1 = 80$ mm, $L = 800$ mm, $t = 100$ mm.

Double-edge notch (DEN) beams were also tested. For these, the loading frame was modified so that $F_1 = F_{1r} = 15P/16$ and $F_2 = F_{2r} = P/16$ were achieved. This resulted in a narrower shear zone. Again, two sizes were tested: depth b values of 150 and 300 mm. The other dimensions were $a_0 = b/6 = h_1$ and $L = 2.5b$.

Results for crack patterns for the SEN beams tested with freely rotating supports are presented in Fig. 8-39. It is seen that these are identical with respect to size of specimen and very similar to the results of Arrea and Ingraffea,[17] as shown in Fig. 8-24. A comparison of crack patterns to those for SEN beams using the fixed-support testing arrangement is shown in Fig. 8-40 (top). It is seen that when free movement of the supports is prevented, two failure mechanisms develop. The curved crack (1) starts as in the case of rotating supports but does not propagate completely due to the horizontal confinement of the beam. When this crack arrests, a second crack (2), which is a flexural crack, starts at an unnotched location. This propagates through the beam to failure.

For the DEN beams, the contrasting failure patterns are shown in Fig. 8-40

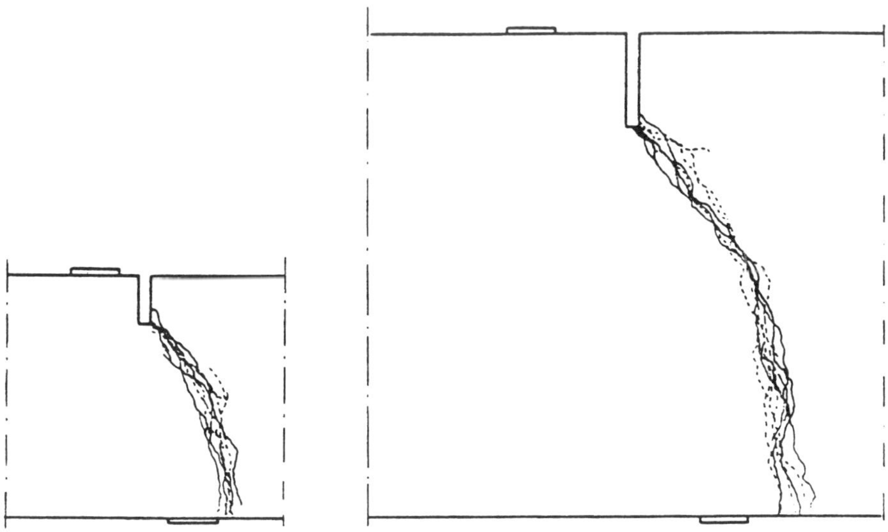

(a) (b)

Figure 8-39 Experimental crack patterns in: (a) three small and (b) three large SEN beams loaded between freely rotating supports.[25]

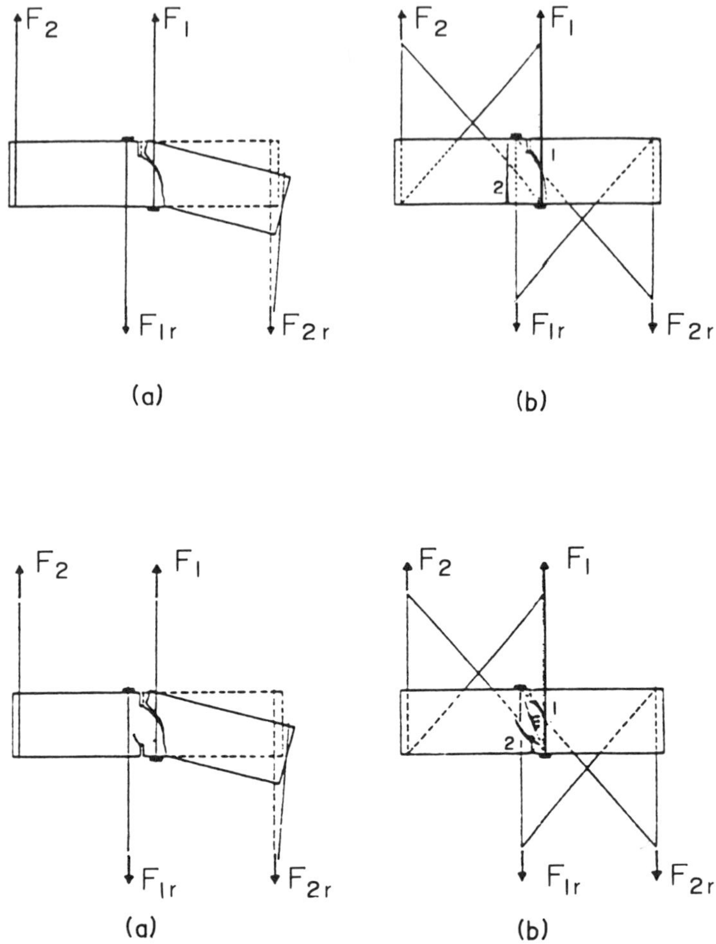

Figure 8-40 Failure mechanisms: top, SEN beams; bottom, DEN beams: (*a*) Rotating supports. (*b*) Fixed supports.[25]

(bottom). The pattern associated with rotating supports shows cracks growing from both notches. However, after peak load only one crack continues to failure. This is rather similar to the results of Swartz and Taha[28] presented in Fig. 8-31, although in their case both cracks propagated completely through the beam.

When the supports are fixed against rotation, a pattern of cracks propagating from the notches (1 and 2) is obtained. These are both arrested due to the horizontal forces and a third crack grows in-between these and propagates to the interior support points, causing failure. This pattern is identical to that obtained by Swartz and Taha[28] for this geometry and for beams loaded in axial compression in addition to shear, as shown in Fig. 8-33.

8.4.2.5 Comments on Four-Point Bend Beam From the results of the investigations presented here it is clear that a number of factors need to be considered when testing for shear fracture properties using this general geometry. These include beam geometry (single notch, double notch, ratio of notch depth to beam depth) and testing geometry (spacing of loads, support geometry including relative movement between load points). Based on these considerations at least five possible failure patterns will be obtained; four are shown in Fig. 8-40, and a fifth, the so-called Brazilian splitting failure, is shown in Fig. 8-25.

From an experimental viewpoint therefore essentially no conclusions with respect to possible mode II fracture parameters have been reached using this geometry. In a number of FE analyses performed on this geometry,[22,23,28,29] it has been possible to model cracking behavior by using only mode I fracture parameters. Two examples of these analyses were presented in Chapter 5 with results displayed in Figs. 5-21 and 5-23.

8.4.3 Biaxial Test Rigs

In principle, the setup of Swartz and Taha[28] shown in Fig. 8-30 could be considered a biaxial test arrangement. However, it is quite limited in load paths that can be applied. In practice the path chosen was always compression, held at a constant value, followed by shear of varying magnitude. As mentioned earlier, these specimens never actually failed, but rather testing was terminated when the magnitude of the shear deformation exceeded the capacity of the transducer.

More elaborate setups are those used at the Stevin Laboratory, Technical University of Delft, the Netherlands, and Lund Institute of Technology, Sweden. These two setups are presented below.

8.4.3.1 Test at Technical University of Delft At the Stevin Laboratory of Technical University of Delft a biaxial test rig was developed in which a separate control for mode I and mode II displacement in a mixed-mode test is possible.[30] In the current approach, crack growth under combined tension and shear is studied under well-known boundary conditions. It is assumed that a crack initiates as soon as the principal stress reaches the tensile strength of the material.

The biaxial test rig consists basically of two independent rigid frames that are fixed in an overall frame by means of plate springs. An exploded view of the apparatus is shown in Fig. 8-41. One of the two rigid frames consisted of two coupled frames, and the second frame can slide in between these coupled frames. The coupled frames can move in the horizontal direction and are fixed to the overall frame in the vertical direction. The middle frame can move vertically and is fixed in the horizontal direction. To both the middle frame and the coupled outer frames a load cell and a hydraulic actuator, with a capacity of 100 kN in tension or compression, are connected as indicated. Square DEN

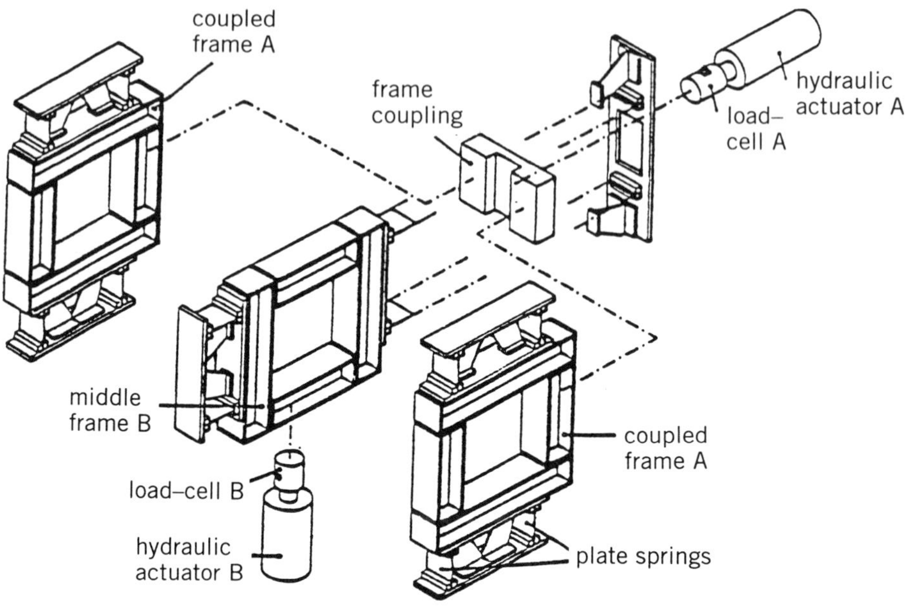

Figure 8-41 Exploded view of biaxial test rig of Stevin Laboratory.[1]

concrete plates are loaded in the apparatus. The experiments resemble the well-known shear-box experiments from rock mechanics.

The loading procedure is clarified in Fig. 8-42. In uniaxial tension, a concrete plate is glued between the upper side of the middle frame and the lower side of the coupled outer frames (Fig. 8-42*a*). When the glue has set, the specimen is loaded by moving the middle frame upward. Since the coupled outer frames are fixed in the vertical direction, a tensile stress develops in the concrete specimen, which will eventually fracture the plate (Fig. 8-42*b*). The concrete plates are double- or single-edge notched plates in order to generate crack growth at a known location. This facilitates deformation-controlled testing. The same procedure is used for applying a shear load to a precracked specimen as shown in Fig. 8-43. Cracking of the concrete specimens of different sizes was studied when different load paths were followed as reported by van Mier and Nooru-Mohamed.[31–33] For these studies, square specimens 200 × 200 × 50 mm were used with other dimensions shown in Fig. 8-43. The load P could be in tension (as shown) or compression, and the load P_s could be applied with a clockwise couple (as shown) or with a counterclockwise couple. A number of combinations and load paths/sequences are presented in Fig. 8-44. The strong influence of selected load path on the cracking response is illustrated in Figs. 8-45 and 8-46. For example, in Fig. 8-45*a* the crack opening paths δ–δ_s for the load path 4*b* and 6*b* experiments intersect at point A. At this point the loads P and P_s should be the same (since the δ_s are

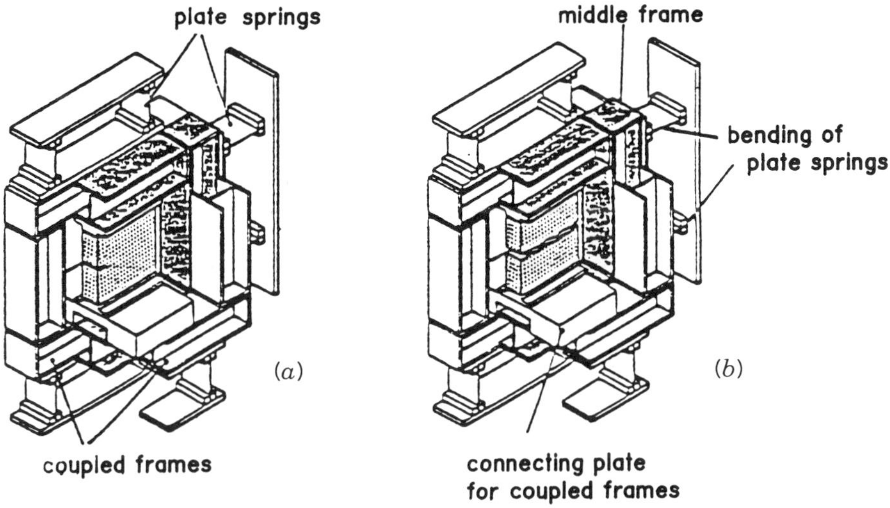

Figure 8-42 Sectional view of biaxial apparatus: (*a*) at beginning of experiment and (*b*) after cracking. Note bending of plate springs of middle frame in (*b*) as result of upward movement of this frame with respect to coupled frame.[1,33]

the same for both paths) but the loads are actually different. The related load values are plotted in Fig. 8-45*b*. The difference between the expected loads P, P_s and actual loads are denoted ΔP, ΔP_s, respectively. These would be zero if the expected and actual loads coincided. Using results from path intersections from a number of tests, ΔP and ΔP_s are plotted in Fig. 8-46. There is significant deviation—as high as about 50% of the peak value of P. The change in behavior is due to the crack geometry (which type of crack appears first) because of irregularities that do not allow complete closure, dilatency, and other effects (multiple cracking, crack branching, etc.).

8.4.3.2 Test at Lund Institute of Technology At the Lund Institute of Technology, a setup to test specimens in direct tension or compression with or without additional shear loads was developed by Hassanzadeh et al.[34,35] The setup is shown schematically in Fig. 8-47. Details are shown in Fig. 8-48. The entire setup is rather compact and can be used in a commercial testing frame with closed-loop control. The specimens are notched rectangular prisms with dimensions shown in Fig. 8-49.

In the testing program reported the tensile load is applied first until a process zone has just formed; then the shear load is imposed. The loading path is governed by relationships between the normal displacement δ_1 and shear displacement δ_2. Two types of relationship were considered:

$$\delta_1 = \delta_2 \tan \alpha_1 \qquad \delta_1 = \alpha_2 \sqrt{\delta_2} \tag{8.42}$$

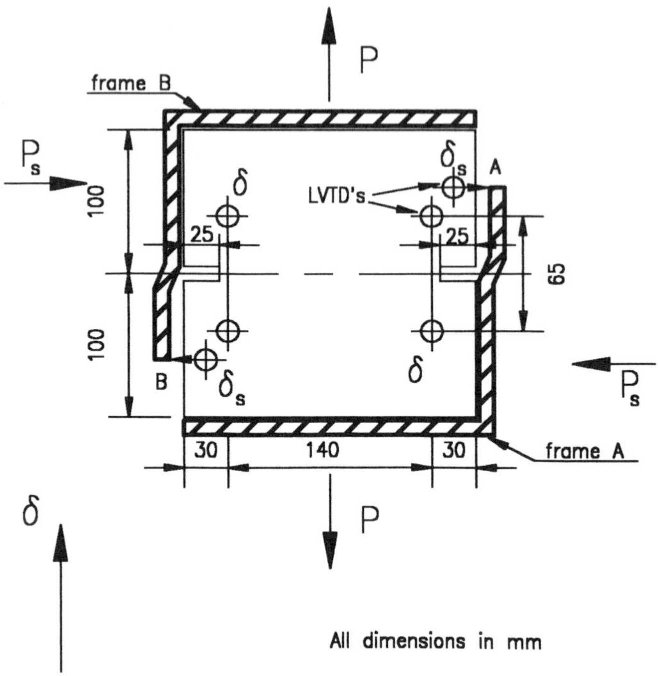

Figure 8-43 A DEN square plate for mixed-mode tests.[1,33]

In the first, linear relation $30° \le \alpha_1 \le 75°$, and in the second, parabolic relation $0.4 \le \alpha_2 \le 0.9\ \text{mm}^{1/2}$. The displacements were taken as zero at the start of the applied shear load, and thus results presented are associated with postpeak or softening response of the material. As an example of the influence of path dependency on response, a path shown in Fig. 8-50*a* was selected. This is shown as a solid line and consists of a straight line with $\alpha_1 = 60°$ (section *a*), a parabolic with $\alpha_2 = 0.4\ \text{mm}^{1/2}$ (section *b*), another straight line with $\alpha_1 = 76°$ (section *c*), and another parabolic with $\alpha_2 = 0.6\ \text{mm}^{1/2}$ (section *d*). A comparison of tests results is given in Fig. 8-50*b* for normal stresses and displacements. Five sets of results are presented: parabolic load paths for $\alpha_2 = 0.4, 0.6\ \text{mm}^{1/2}$, linear load path for $\alpha_1 = 60°$, and two tests using the mixed path in Fig. 8-50*a* with wet and dry specimens. Remembering that the starting point for these tests is an initial tension up to peak load, the normal stress response is interesting to consider first. As normal displacements continue to increase, the material continues to soften, as is normally observed. However, the presence of the shearing loads actually drives the normal stresses into compression. In looking at the shearing stresses it is seen that, for the two parabolic paths, these stresses increase continuously while the linear and mixed paths show a clear peak stress followed by a softening-type decrease.

It is also of interest to look at the intersection points of the various paths in

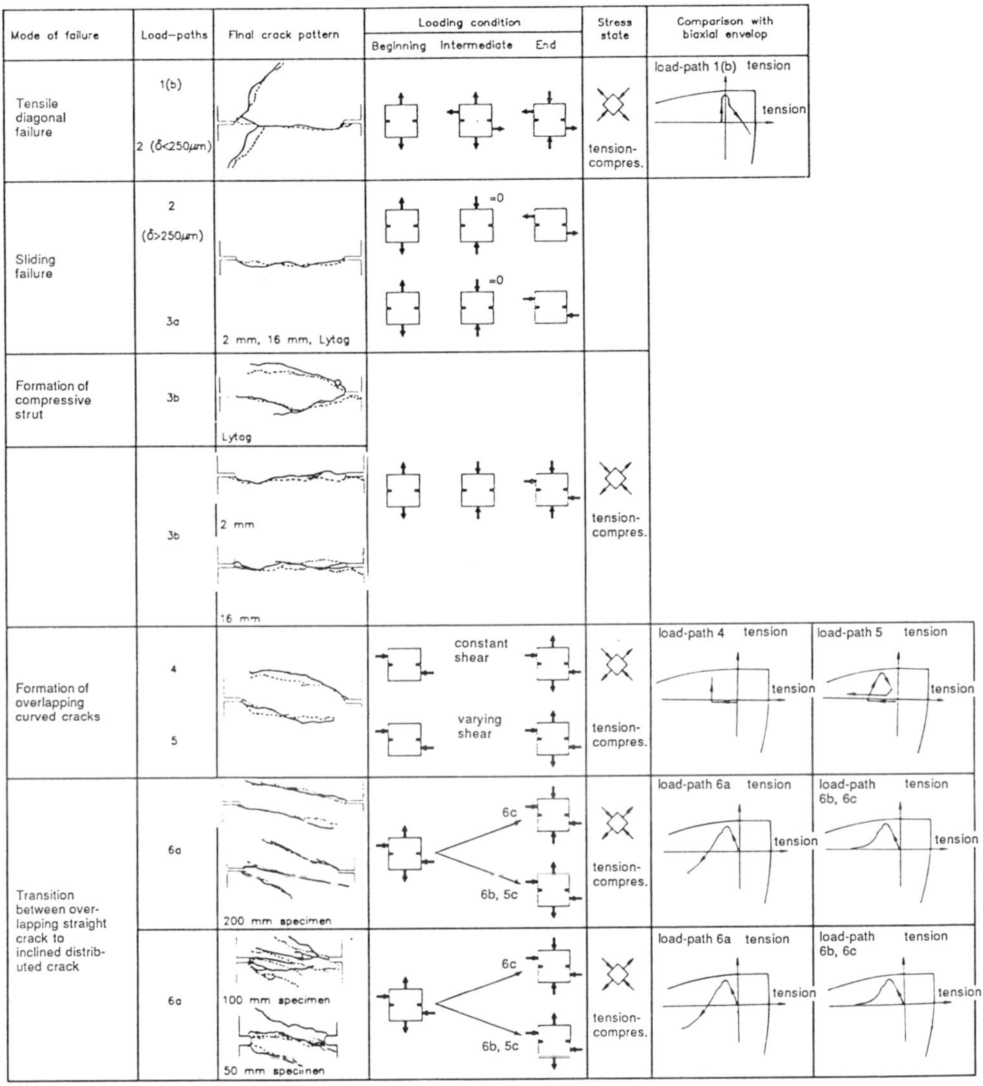

Figure 8-44 Various load paths and failure mechanisms for different tests.[33]

Fig. 8-50*a*. If the material behavior were path independent, the stresses at these intersections on the displacement diagram would be the same. Comparing the results at the second intersection point ($\alpha_1 = 60°$, $\alpha_2 = 0.6\ \text{mm}^{1/2}$, mixed) for a shearing displacement of about 0.12 mm, a large difference is seen—the shearing stress varies from about 0.5 MPa for the $\alpha_2 = 0.6\ \text{mm}^{1/2}$ path to about 2.0 MPa for the mixed dry path. Another observation concerns comparing the continuation of the $\alpha_2 = 0.6\ \text{mm}^{1/2}$ path and both mixed paths

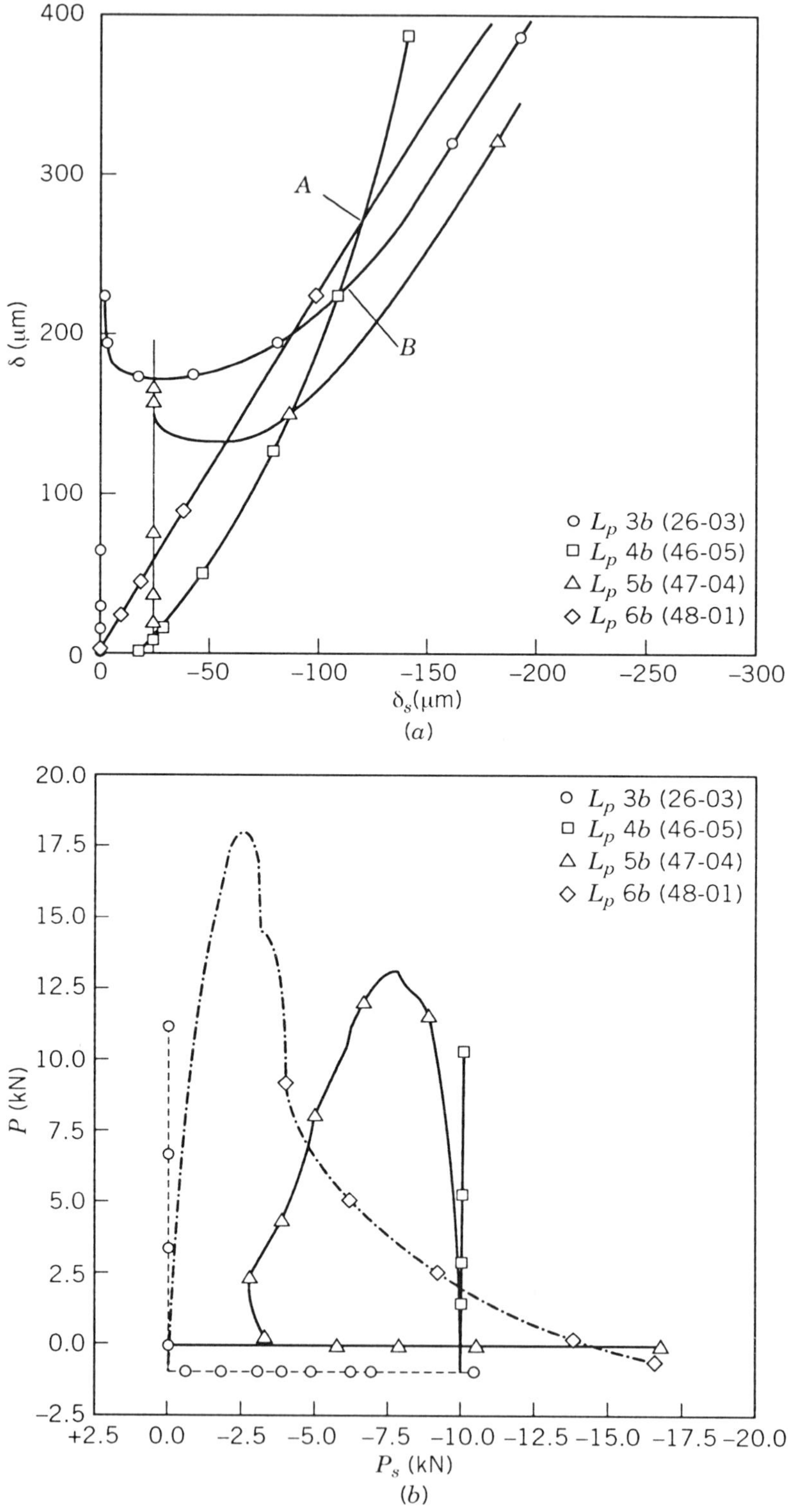

Figure 8-45 (a) Crack opening paths δ–δ_s and (b) P–P_s response.[33]

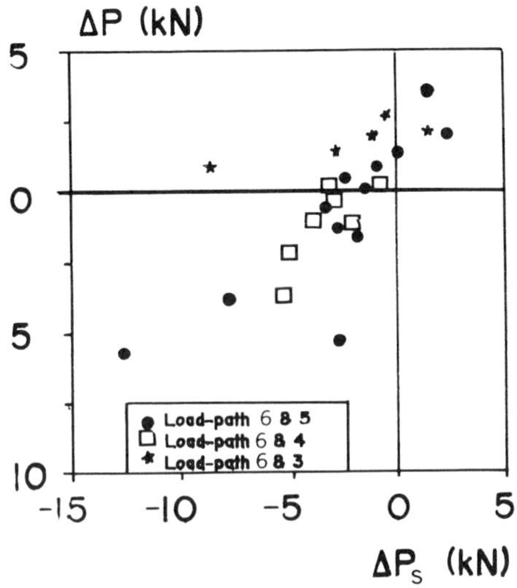

Figure 8-46 Deviation of intersection points in P–P_s plane.[33]

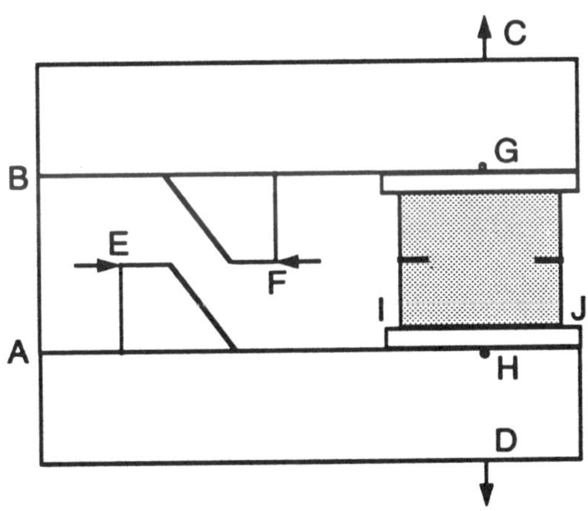

Figure 8-47 Testing schematic.[35]

after a shearing displacement of 0.12 mm. All three displacement paths are now identical, but the stress paths are quite different.

These results are in basic agreement with the tests at Stevin Laboratory[33] and show clearly the effects of initial crack formation and damage on subsequent load displacement response. Thus, one should conclude that what is being measured is really *not* a material response but a structural response where the structural type is created by the initial damage.

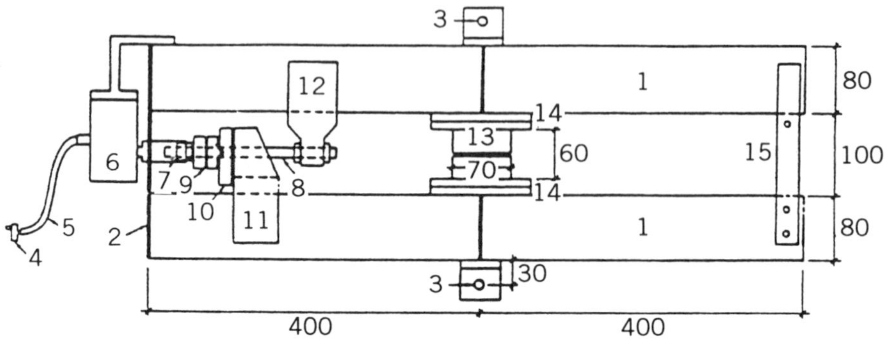

1– Box beams (height=80 mm, width=40 mm thickness-3 mm)
2– Steel sheet (height=260 mm, width=100 mm, thickness=1 mm)
3– Connection to the testing machine
4– Crank
5– Flexible axle
6– Gear box
7– Nut
8– Steel bar with strain gauges
9– Roller support
10– Freely sliding support
11– Fixed support
12– Fixed support
13– Specimen
14– Steel plates
15– Lock and freely sliding support

Figure 8-48 Mixed-mode testing arrangement at Lund Institute of Technology.[35]

8.4.4 Tests at Cardiff

Barr et al.[36] developed and applied two compact test specimen geometries suitable for studying the shear strength of concrete and rocks. The first test geometry was a symmetrically notched compact beam specimen loaded adjacent to the notches by forces that created a concentrated shear zone in the near absence of bending moments. One disadvantage of the geometry was that a large number of nonshear fractures were obtained in experimental work. A numerical analysis of the geometry showed that tensile zones are created in the area adjacent to the roots of the notches. These tensile stresses may be the cause of some of the nonshear types of failure observed in practice.

A numerical study was carried out on the various potential errors in the use of the above geometry. Errors in the positioning of the loads, errors in the notch depths, and errors in both the loads and notch depth were investigated. The numerical results showed that errors in the positioning of the applied loads were not significant. On the other hand, errors in the notch depths can result in a bending stress situation being created in the shear zone resulting in a nonshear type of failure. In the case of the combined errors, the greater influ-

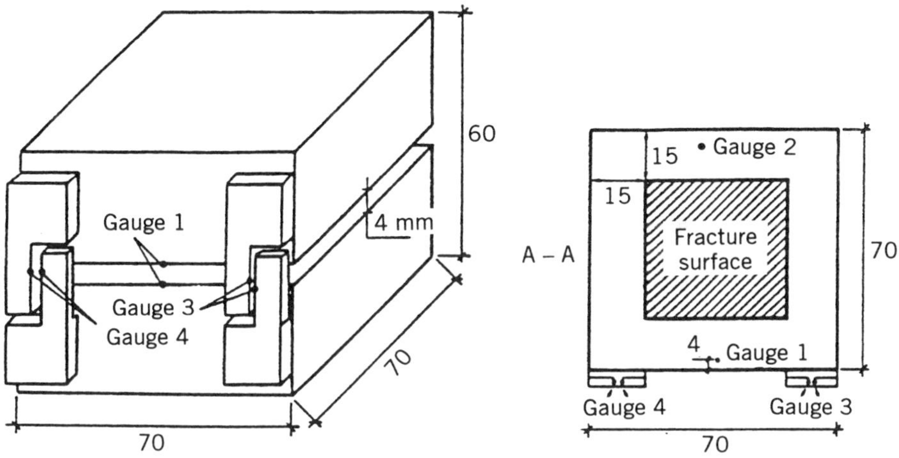

Figure 8-49 Specimen geometry and locations of clip gages (in mm). Normal displacements are recorded by gages 1 and 2. Shear displacements are recorded by gages 3 and 4.[35]

ence was the error in the loading position for shallow notches and errors in the notch depth for the deeper notches. The main conclusion of the numerical study was that all three types of errors investigated can result in a nonshear failure in the test specimens. Thus great care is essential when using this test geometry.

In view of the above results, a circumferentially notched compact shear test specimen geometry was developed (Fig. 8-51). This test specimen geometry gave reproducible results with a low coefficient of variation, generally below 10%. The shear strengths reported for various concrete mixes showed that the shear strength values were independent of the notch depth used and also of the maximum size of the coarse aggregate used in the mix. However, these results are limited in scope, and future work on size effects using larger test specimen sizes is required before firm conclusions can be drawn regarding the effects of the size of the coarse aggregates.

The two test specimen geometries discussed are suitable for additional development and can be used for further studies of the shear performance of concrete and rocks. A more compact version of the rectangular beam specimen was used to investigate size effects in plain and fiber-reinforced concrete mixes at Cardiff (Fig. 8-52). Both geometries are suitable for studies of size effects since both can be made more compact. Furthermore, the circumferentially notched cylindrical specimen is suitable for studying rocks subjected to shear. The cylindrical specimen is also suitable for use in the testing of concrete and/or rock cores.

A number of other geometries have been proposed recently by Barr and Hughes[37] and are shown in Fig. 8-53.

Another geometry investigated by Arslan et al.[38] is shown in Fig. 8-54. This

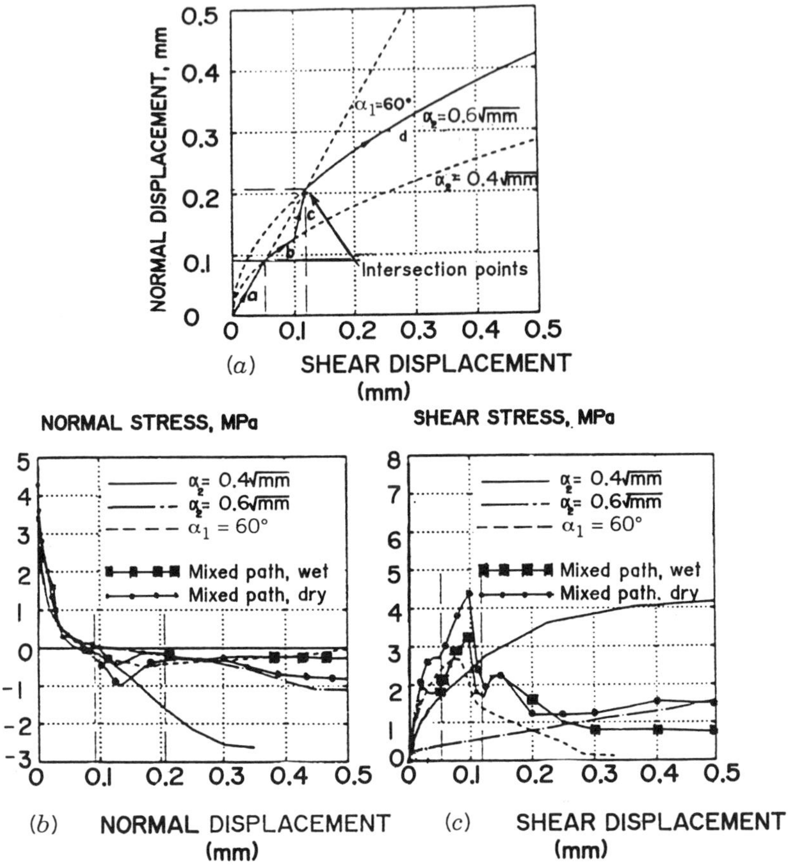

Figure 8-50 Path-dependent behavior.[35]

"punch-through" geometry follows that developed by Luong.[39] In Fig. 8-54*a*, a number of possible testing configurations is shown, whereas in Fig. 8-54*b* actual dimensions are given. The analyses presented in Fig. 8-55 along with test results indicate that tensile fracture develops between the inner surface at the bottom notch and the outer surface at the top notch. They thus conclude that this geometry has failed to remove the influence of peak tensile stress.

Finally, a recent testing geometry[40] using a standard 150-mm cube is shown in Fig. 8-56. Four separate failure planes can be created by two sawcuts of depth h as shown. These cuts are called the main notches. Four additional cuts of depth a could be used. These are called secondary notches. Depending on values of h and a, it is possible to induce mixed-mode I–II, mixed-mode I–III, or mixed-mode I–II–III behavior. By using four failure planes in different regions of the sample, influence of the inhomogeneous nature of concrete may

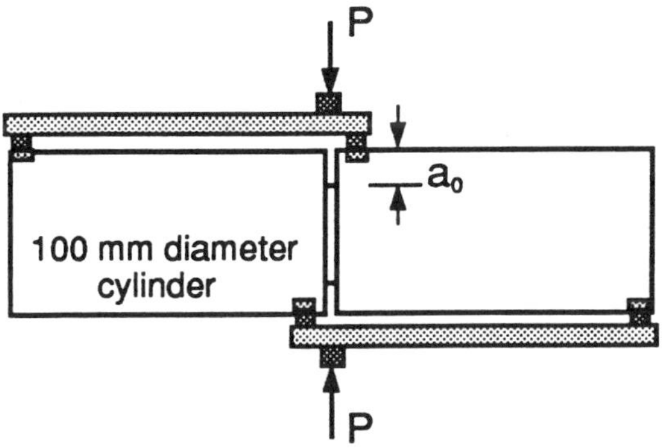

Figure 8-51 Circumferentially notched cylindrical shear test specimen.[36]

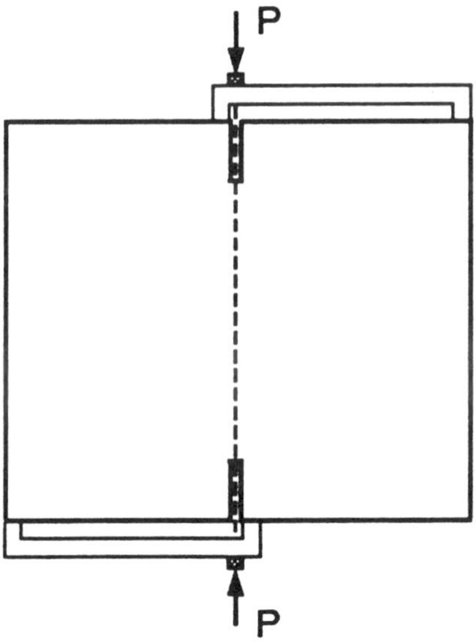

Figure 8-52 Compact cube/prism shear test specimen.[36]

be minimized. Arslan et al.[40] conclude that this specimen geometry is capable of investigating the full mixed-mode behavior of fiber-reinforced and plain concrete.

8.4.5 Arcan Specimen

The Arcan specimen and its holder are shown in Fig. 8-57. This was first proposed for use with fiber-reinforced composite materials by Arcan et al.[41] Extensive numerical and photoelastic studies were conducted on the entire system.[41-43] Some of the results are presented for the assembly shown in Fig. 8-58. Photoelastic patterns of the test section are shown in Fig. 8-59 in which

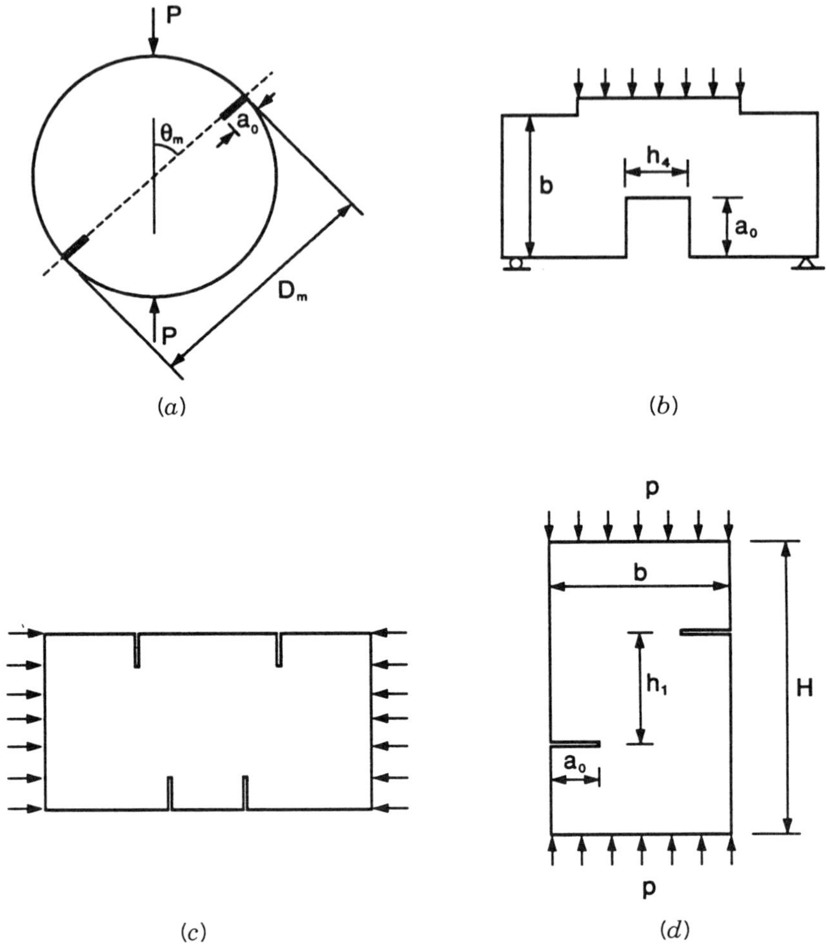

Figure 8-53 Other proposed mixed-mode testing geometries: (*a*) disk specimen, (*b*) wide notched prism, (*c*) modified loading arrangement for (*b*), and (*d*) double notched prism.[37]

it is seen that this area is in a state of almost pure shear. These results are plotted in Fig. 8-60 for three width-to-depth ratios. The versatility in this test arrangement comes from the possibility of rotating the entire setup in the testing machine to achieve different deformation modes in the test area. For the configuration shown in Fig. 8-58 almost pure mode II determination occurs. By rotating the frame counterclockwise 90°, a pure mode I deformation state is obtained. By rotating the holder, mixed-mode combinations of shear–tension or shear–compression may be obtained. This device is similar, but not identical, to one proposed by Izumi et al.[44] The Arcan device has not been used to evaluate concrete or rock to our knowledge. Arcan, in private correspondence,

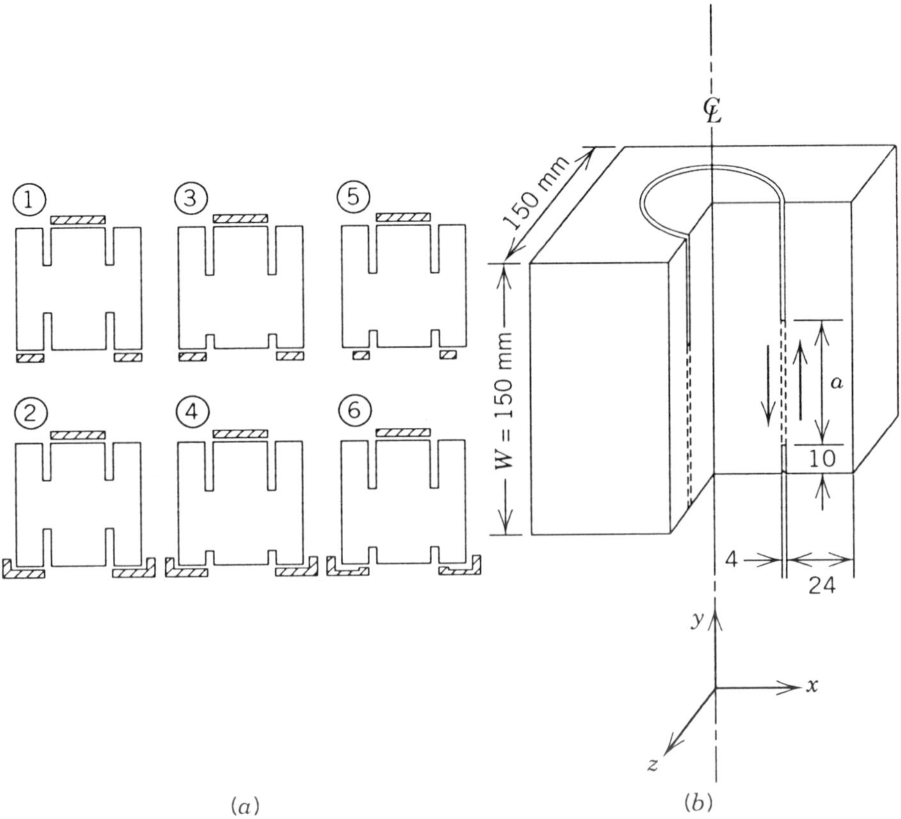

Figure 8-54 Punch-through test specimen: (*a*) different geometries and (*b*) specimen used for cases 3–6.[38]

(1988) has suggested dimensions of $a_0 = 120$ mm, $b = 720$ mm, and $t = 100$ mm for the test section if the maximum aggregate size is about 20 mm. With these dimensions, the holding device becomes very large and heavy. However, it should be no more difficult to construct this device than the other devices or systems described in this chapter. This device deserves strong consideration as a tool to investigate mixed-mode fracture in concrete and rock in a systematic and controlled manner.

8.4.6 Semicircular Bend Specimen

The semicircular bend specimen geometry, shown in Fig. 8-61, was developed by Chong and his associates[45,46] for use with rock samples. It consists of a disk cut from the core and then cut in half to form two bending test specimens.

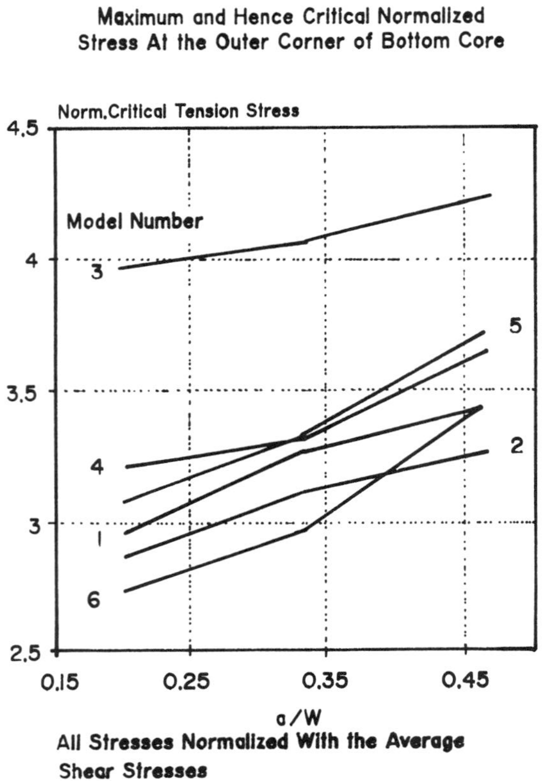

(*a*)

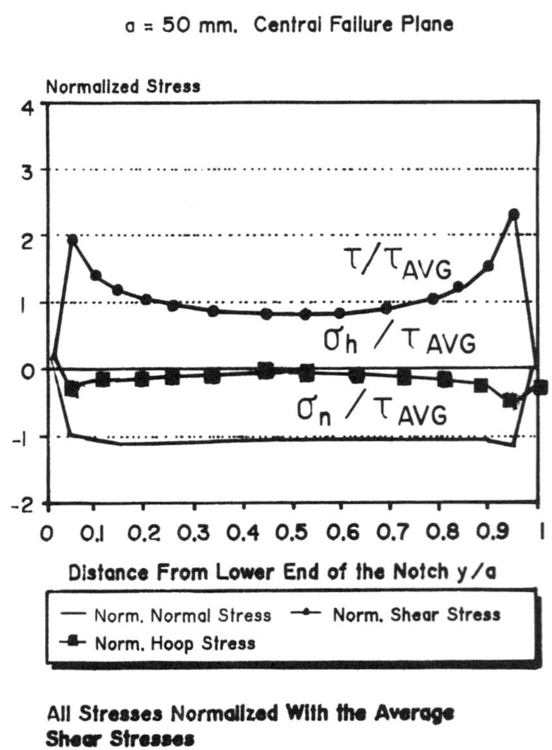

(*b*)

Figure 8-55 Analysis of test specimen geometries: (*a*) comparison of critical tension stresses and (*b*) stress distribution along proposed shear plane, geometry 6.[38]

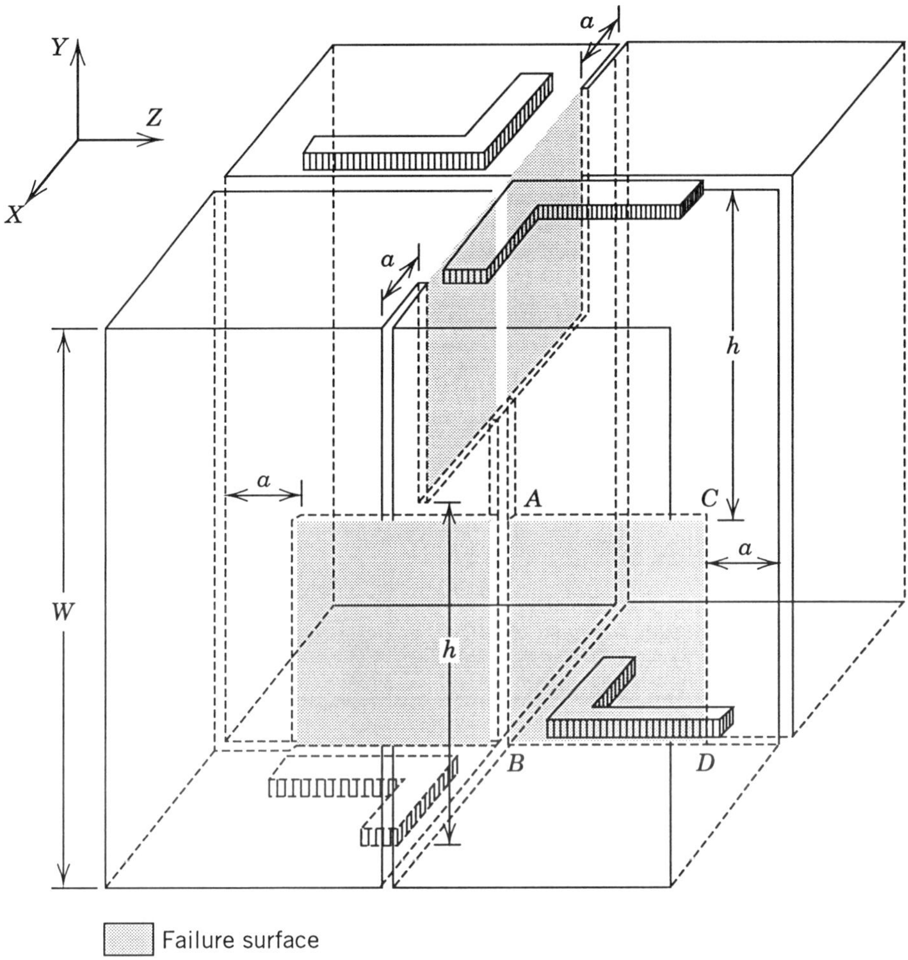

Figure 8-56 General geometry of full mixed-mode fracture specimen.[40]

The central notch may be vertical for mode I testing or inclined for mixed-mode testing. Another possible mixed-mode test geometry is with the notch off center. This geometry has been analyzed extensively and numerous tests conducted on rock samples. One major use for this would be to obtain fracture parameters for in situ concrete from cores. A detailed description of this test specimen and data evaluation for mode I testing were given in Chapter 6.

Because of the relatively small size, on the order of 100–150 mm in diameter, correlations would need to be made with fracture properties obtained from other tests on the same material. The simplicity and ease of use make this an attractive specimen.

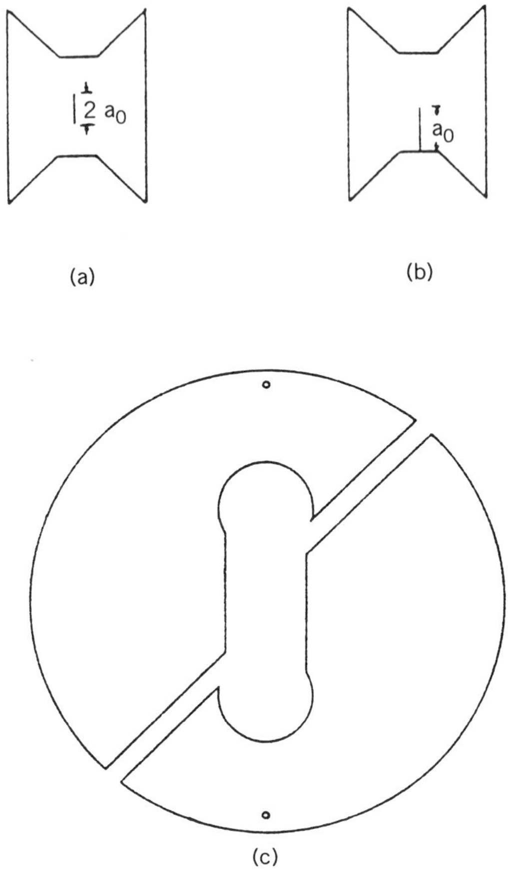

Figure 8-57 Arcan test specimen and frame: (*a*) center notched specimen, (*b*) edge notched specimen, and (*c*) holder frame.[42]

8.4.7 Applications to Rock

The following methods, described above, have been applied to rock: single-edge cracked plate in four-point bending (SC4PB),[2] single-edge cracked half-disk specimen in asymmetrical three-point bending (HDB),[2] single-edge cracked plate in asymmetrical three-point bending (SC3PB),[2] and the core-based specimen of Luong.[39] In addition to these, other methods presented in reference[2] are the central through cracked Brazilian disk (CCBD) in diametrical compression (Figs. 8-62 to 8-64), compact tension and shear (CTS) specimen (Fig. 8-65), compact double-shearing (CDS) specimen as given in Fig. 8-66, and the short beam specimen in compression (SBC) as shown in Fig. 8-67.

In all of these methods, formulas are developed for mode I and mode II

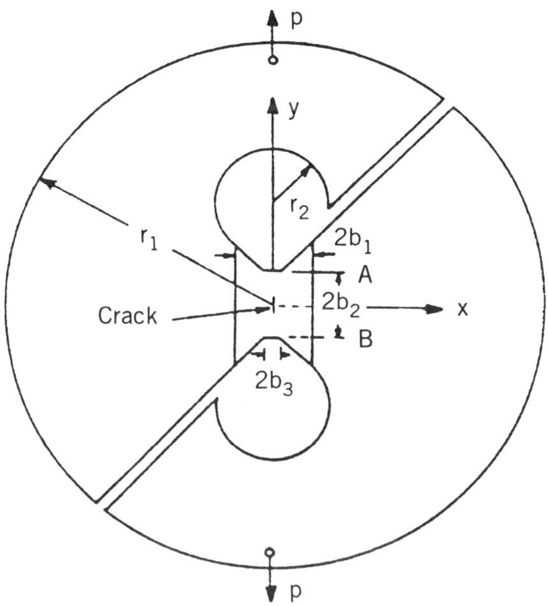

Figure 8-58 Specimen and frame assembly for photoelastic studies (r_1 = 190 mm, r_2 = 40 mm, b_1 = 36 mm, b_2 = 30 mm, and $b_3/b_2 = \frac{2}{15}, \frac{3}{15}, \frac{4}{15}$).[42]

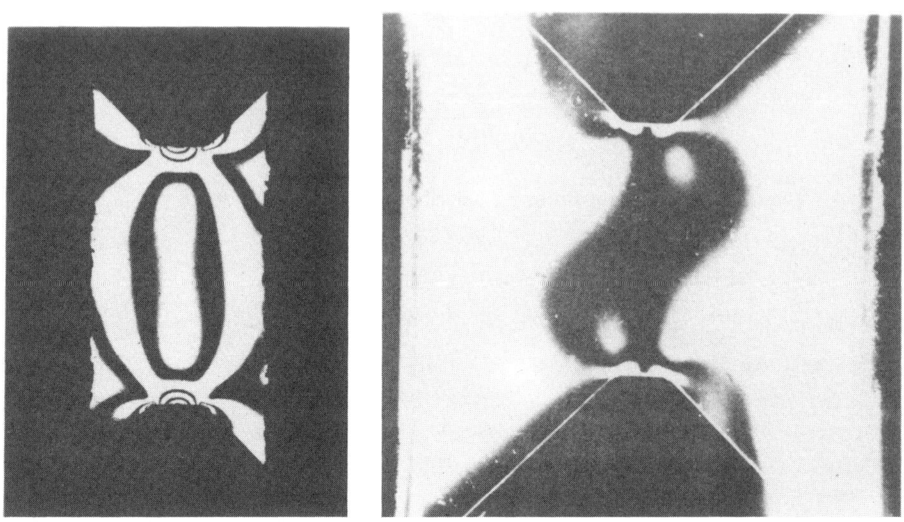

(*a*) (*b*)

Figure 8-59 Photoelastic patterns for specimen with $b_3/b_2 = \frac{3}{15}$: (*a*) isochromatics and (*b*) 45° isoclinics.[42]

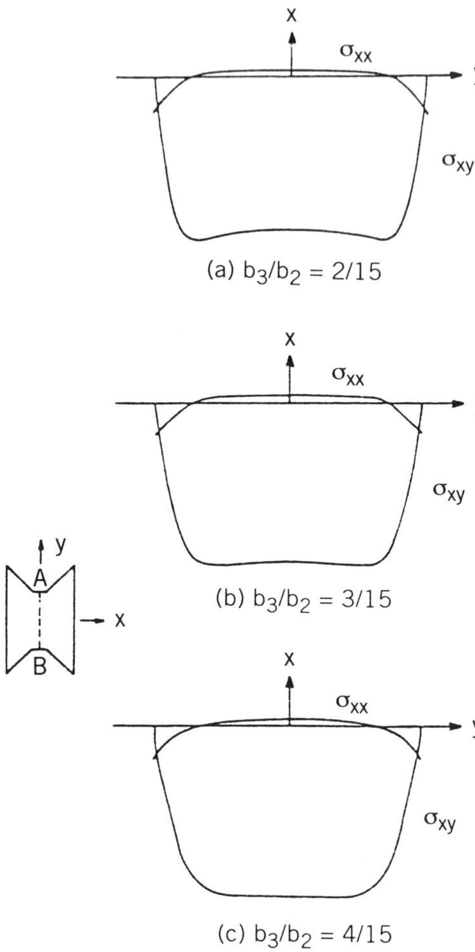

Figure 8-60 Normal stress σ_{xx} and shearing stress σ_{xx} along $x = 0$.[42]

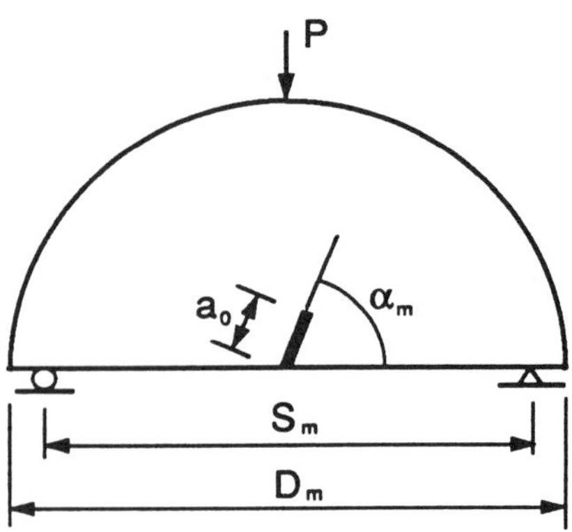

Figure 8-61 Semicircular bend mixed-mode test specimen.[45,46]

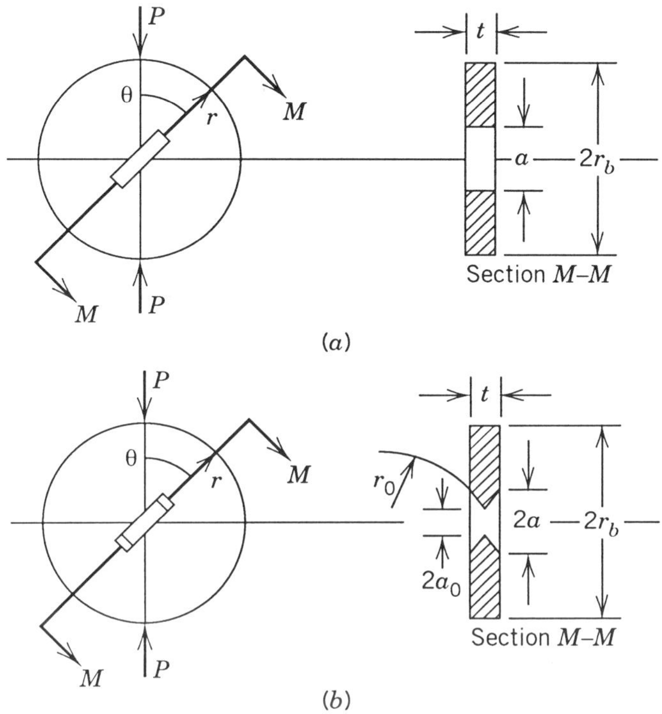

Figure 8-62 Brazilian disk specimen in diametrical compression: (*a*) internal straight-through crack and (*b*) internal chevron notched crack.[2]

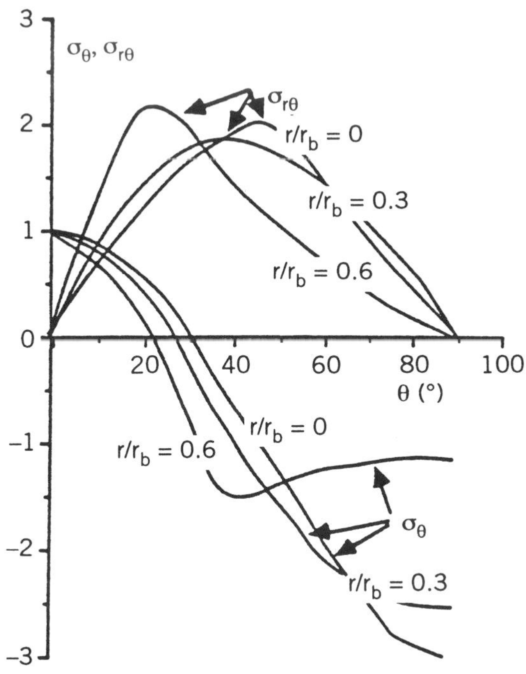

Figure 8-63 Stress distributions in uncracked Brazilian disk for r/r_b = 0, 0.3, 0.6.[2]

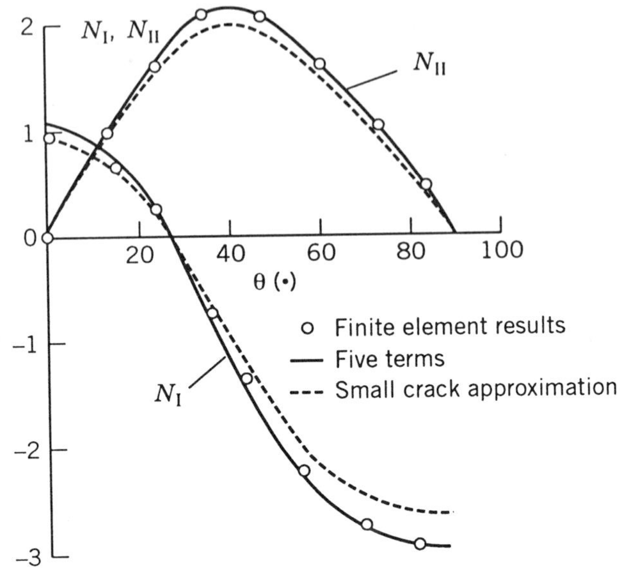

Figure 8-64 Stress intensity factors for small crack approximation at $a/r_b = 0.3$.[2]

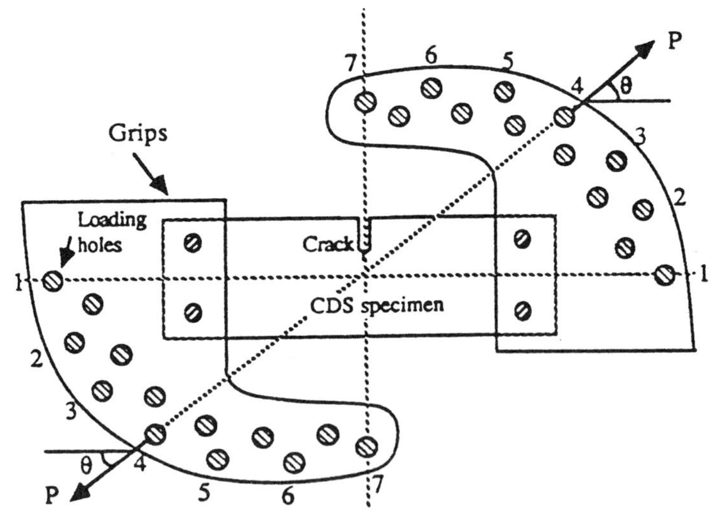

Figure 8-65 Compact tension and shear specimen and loading grips.[2]

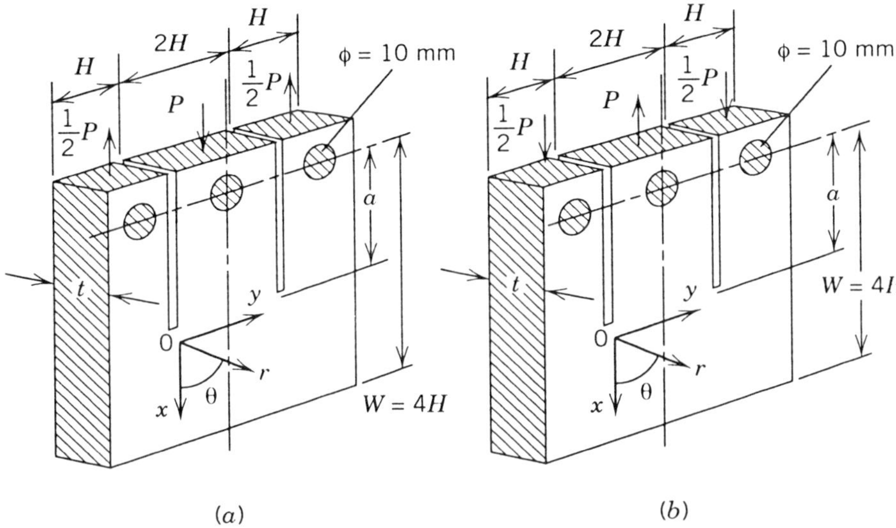

Figure 8-66 Compact double-shear specimen: (*a*) central load downward and (*b*) central load upward.[2]

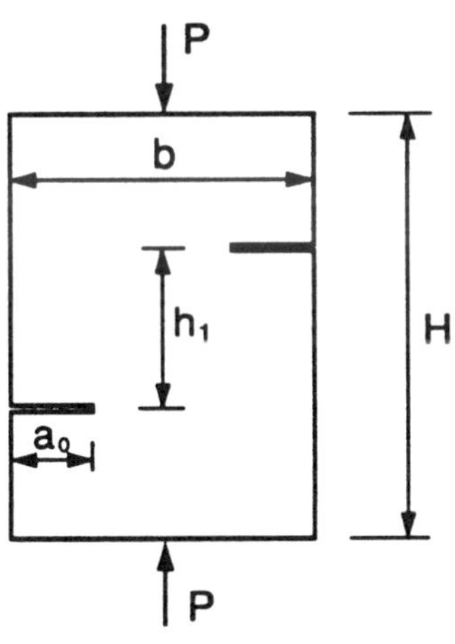

Figure 8-67 Short beam in compression.[2]

stress intensity factors. For an appropriate specimen, the mode II fracture toughness is obtained from

$$K_{IIc} = K_{II}\,(P_{max},\, a_c,\, g_F) \tag{8.43}$$

where P_{max} and a_c are the maximum load and the corresponding crack length, respectively, and g_F represents the geometry of the specimen.

Considering the CCBD specimen of Fig. 8-62, the stress distributions are plotted in Fig. 8-63 for $r/r_b = 0, 0.3, 0.6$, where r is the radial distance from the center of the crack and r_b is the disk radius. The stresses are nondimensionalized in terms of $\sigma_0 = P/(\pi r t)$. For $r/r_b = 0$ the dimensionless stress can be expressed as

$$\sigma_\theta = \cos^2\theta - 3\sin^2\theta \qquad \sigma_{r\theta} = 2\sin(2\theta) \tag{8.44}$$

On the other hand, it is seen from Fig. 8-63 that zero values of $\sigma_{r\theta}$ occur only at $\theta = 0°, 90°, 180°$, but zero values of σ_θ occur at $\theta = 30°, 27°, 21°$ for $r/r_b = 0, 0.3, 0.6$, respectively. This implies that a pure mode II loading can be obtained for θ between 21° and 30° depending on the r/r_b ratio. These observations are the basis for the consideration of the Brazilian disk as a mixed-mode test specimen.

Formulas for the stress intensity factors K_{I} and K_{II} were developed by Atkinson et al.[47] for this geometry containing a central, straight-through crack. The numerical solutions they obtained are

$$K_{\mathrm{I}} = \frac{P\sqrt{a}}{\sqrt{\pi r_b t}}N_{\mathrm{I}} \qquad K_{\mathrm{II}} = \frac{P\sqrt{a}}{\sqrt{\pi r_b t}}N_{\mathrm{II}} \tag{8.45}$$

where r_b and t are the radius and thickness of the disk, respectively. For $a/r_b \leq 0.3$, Atkinson et al. gave

$$\begin{aligned} N_{\mathrm{I}} &= 1 - 4\sin^2\theta + 4\sin^2\theta(1 - 4\cos^2\theta)\left(\frac{a}{r_b}\right)^2 \\ N_{\mathrm{II}} &= \left[2 + (-5 + 8\cos^2\theta)\left(\frac{a}{r_b}\right)^2\right]\sin(2\theta) \end{aligned} \tag{8.46}$$

For $a/r_b = 0.3$, these expressions are plotted in Fig. 8-64. These are compared with FE results and more accurate expressions for N_{I} and N_{II}. Putting $N_{\mathrm{I}} = 0$ in Eq. (8.46) results in an expression for θ in terms of a/r_b. Using these values of θ, a/r_b, and maximum load in Eq. (8.45), one obtains the value of $K_{\mathrm{II}c}$ by this method. In order to obtain crack length a—different from an initial notch—it is necessary to precrack the specimen. This is done using a mode I loading in order to grow a straight crack from the initial notch.

The CTS specimen shown in Fig. 8-65 was proposed by Richard.[48] It is similar in concept to the Arcan device.[41] It is stated that a mode II condition exists for $\theta = 90°$ and expressions are given for the stress intensity factor.[2] The advantages to this test setup are (i) large specimens can be used, particularly advantageous for nonhomogenous/discontinuous materials, and (ii) it provides uniform testing conditions for a variety of mixed-mode loading cases.

Some disadvantages are that (i) the test specimen is not core based and must be machined and (ii) tensile loading is required.[2]

The CDS specimen shown in Fig. 8-66 was introduced by Jones and Chisholm.[49] This is similar in concept to the punch-through specimen of Barr[38] and Luong.[39] If the loads are applied as in Fig. 8-66*a*, a compressive (negative) value of K_I is obtained at the notch tip, whereas the reverse is true for the loading in Fig. 8-66*b*. This test specimen has been shown to have a relatively low (but not zero) influence of mode I, that is, $K_I/K_{II} < 0.13$.[2] It is considered to be a useful specimen for rocks having well-bedded planes.

The SBC specimen shown in Fig. 8-67 was introduced first for concrete testing by Liu et al.[50] Again, the conditions of pure mode II do not exist and compressive as well as tensile normal stresses are also present in the region of the crack line. The calculations are fairly simple due to the availability of calibration formulas for specific geometries. The loading conditions are also fairly simple, but again, the specimen geometry for rock must be obtained by machining.

Reference[2] reports a large compilation of values of mode II fracture toughness data obtained through the use of these tests. It would appear, however, that intrinsically the same problems exist for rock testing as they do for concrete. In fact, are mode II fracture values obtained? Concrete is really no more heterogeneous than rock (it may be less in fact) if one considers the scale of modeling. All of these methods for assessing mode II fracture toughness for rock are based on LEFM. Again, depending on the level, or scale, of observation, it is well known that a significant process zone exists in rock fracture; see, for instance, Meridith et al.[51] Many of the methods attempt to use a "critical crack length," but its definition is not clear, especially in light of three-dimensional fracture processes. And again, no accepted testing standard or general agreement on an appropriate method exists.

8.5 NONPLANAR MIXED-MODE FRACTURE

Nonplanar mixed-mode behavior would include the combinations modes I–III, modes II–III, and modes I–II–III deformations. The most common case of pure torsion of a round member includes not only mode III deformation but also, due to Poisson's effect and dilation, mode I deformation. Only limited testing has been done to examine fracture behavior in torsion; see references 52–57 and, as mentioned earlier, reference 40 (Fig. 8-56). Practical applications would include assessment of the torsional strength of concrete spandral beams and other structure members subjected to torsional shear.

Bazant et al.[52] tested plain and reinforced microconcrete beams in three sizes to evaluate the validity of the size effect law stated in terms of torsion. If the shearing stress v_u is proportional to T/d_t^3, where T is the applied torque and d_t is a representative dimension of the beam, then the size effect law may be expressed as

$$v_u = C_1\left(1 + \frac{d_t}{\lambda_0 d_a}\right)^{-1/2} \tag{8.47}$$

for NLFM or

$$v_u = C_2 d_t^{-1/2} \tag{8.48}$$

for LEFM. The parameters λ_0, C_1, and C_2 are constants and d_a is the size of the aggregate. The test setup is shown in Fig. 8-68, and results for plain concrete are shown in Fig. 8-69. The upper part of Fig. 8-69 shows that the response is more appropriately considered to be NLFM. Eq. (8.47) can be rearranged into a linear form as

$$\frac{1}{v_u^2} = \frac{1}{C_1^2} + \frac{1}{C_1^2\lambda_0}\frac{d_t}{d_a} \tag{8.49}$$

or

$$Y = C + AX \tag{8.50}$$

where $Y = 1/v_u^2$, $C = 1/C_1^2$, $A = C/\lambda_0$, and $X = d_t/d_a$. This is plotted in Fig. 8-69, in which it is seen that good agreement is obtained for these limited data. The specimen sizes d_t were 1.5, 3, and 6 in. (38.1, 76.2, and 152.4 mm). Other results for reinforced beams are presented in Chapter 9.

Tokatly and Barr[57] tested cylinders in torsion to investigate the suitability of the size effect law. Cylinder diameters ranged from 80 to 200 mm, as shown in Table 8-3. The cylinders were grooved with presented values for groove depth–diameter ratios of $\frac{1}{4}$ and $\frac{1}{5}$. The size effect plots using four diameters are presented in Fig. 8-70. A reasonable agreement with the size effect law was obtained. They had poorer agreement when they used response from three of the four sizes and then predicted the response of the fourth diameter specimen from the calculated results of the other three. However, it should be noted that the basic 2 : 1 ratio requirement on d_t was then severely violated. It might also be observed that the size effect law really should be weaker for torsion specimens because the surface effects are probably constant for different diameters.

As a final example, the work of testing cylinders in torsion by Keuser and Walraven[56] is presented. The setup is shown in Fig. 8-71. As shown in the upper figure, the outer diameter is 290 mm and the inner diameter is 250 mm, giving a wall thickness of 20 mm. The height of the cylinder is 120 mm, but the ends were glued into the steel holder shown in the lower part of Fig. 8-71

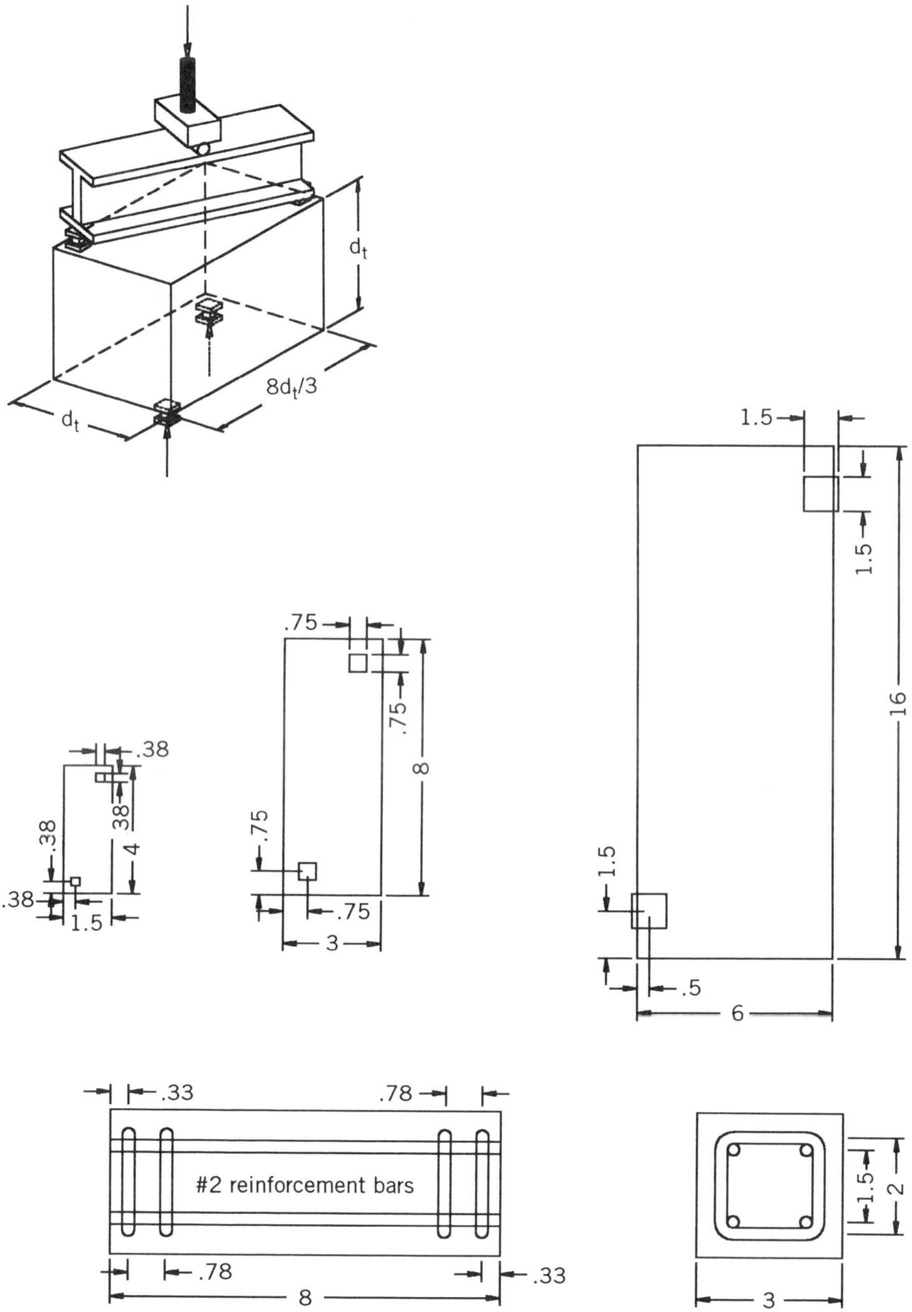

Figure 8-68 Torsion beams tested by Bazant et al.[52]

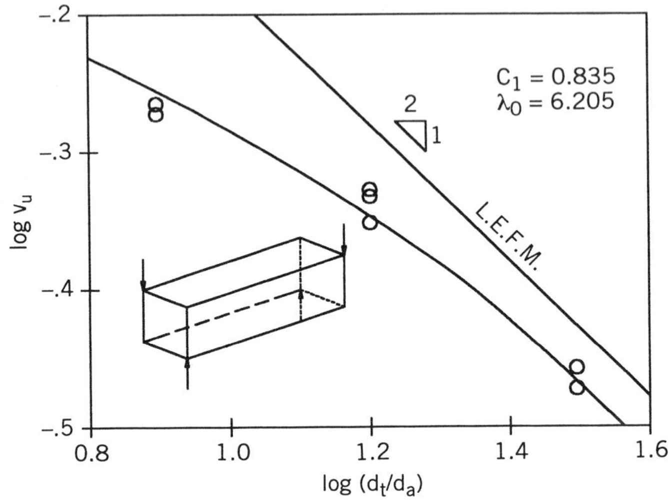

(*b*)

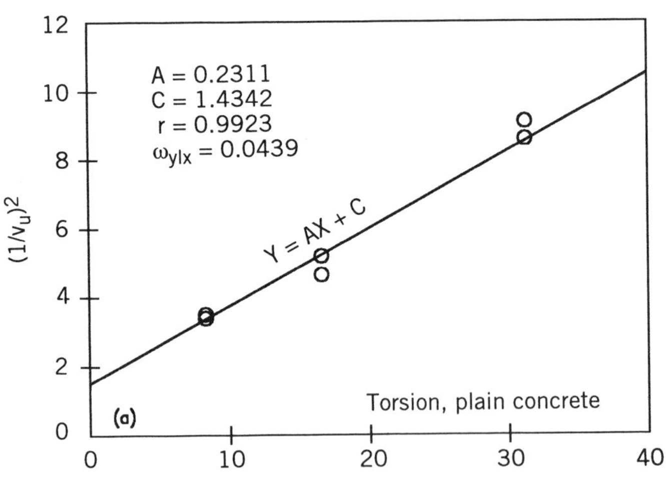

(*a*)

Figure 8-69 Observed size effects for unreinforced beams in torsion.[52]

Table 8-3 Maximum Torque Results for Notched Cylinders Tested by Tokatly and Barr[a]

Notch Depth Ratio	Cylinder Diameter, d_t (mm)	Notch Depth (mm)	Average Maximum Torque (N-m)	Coefficient of Variation (%)
$d_t/4$	80	20	143.6	8.6
	100	25	237.4	4.7
	150	37.5	896.8	2.9
	200	50	1628.0	10.8
$d_t/5$	80	16	249.2	8.2
	100	20	403.1	3.0
	150	30	1244.4	6.6
	200	40	2908.4	2.6

[a]From ref. 57.

so that only 20 mm of free surface remained. The torsion crack path was thus confined to this region and no notch was used. The crack widths were measured at four quarter points around the circumference of the specimen. In addition, it was possible to apply tensile loading along with the torsional moment. Based on their tests they concluded that for this type of mixed-mode problem the axial and shear mechanisms do not interact in a significant way.

8.6 SUMMARY

Various theories for propagation of mixed-mode and mode II cracks have been reviewed. Applications of these theories to concrete and rocks have been discussed. The numerous mixed-mode tests presented in this chapter seem to have a number of common features.

1. All methods, at least initially, were motivated by a desire to obtain mode II fracture mechanics properties. In our view, none of these succeeded in achieving this goal.
2. All the measurement methods presented have been at what might be called the macroscale. That is, there are well-defined, coherent cracks or crack zones. It is likely that different deformation modes interact in a complicated manner due to boundary effects, relative aggregate–paste strength, surface texture of aggregates, and distribution of different types and sizes of aggregates. These effects have been described in Chapter 4, but correlations to the macrolevel to predict overall response have yet to be done. Further applications using these ideas will be presented in Chapter 9.
3. There is general agreement within the community of researchers in concrete that planar mixed-mode crack propagation can be modeled using

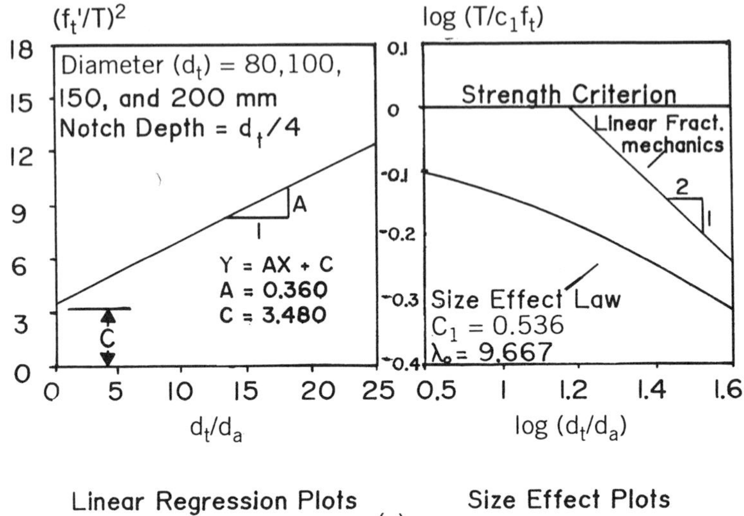

(a)

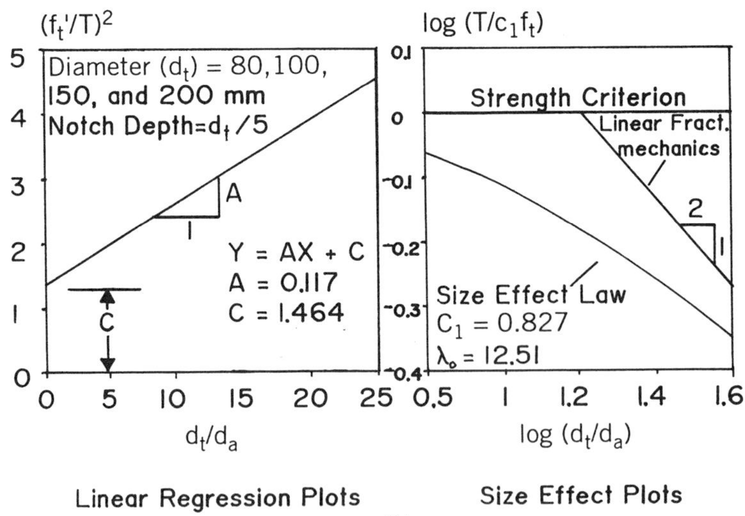

(b)

Figure 8-70 Torsion test size effect results: (*a*) notch depth = $d_t/4$ and (*b*) notch depth = $d_t/5$.[57]

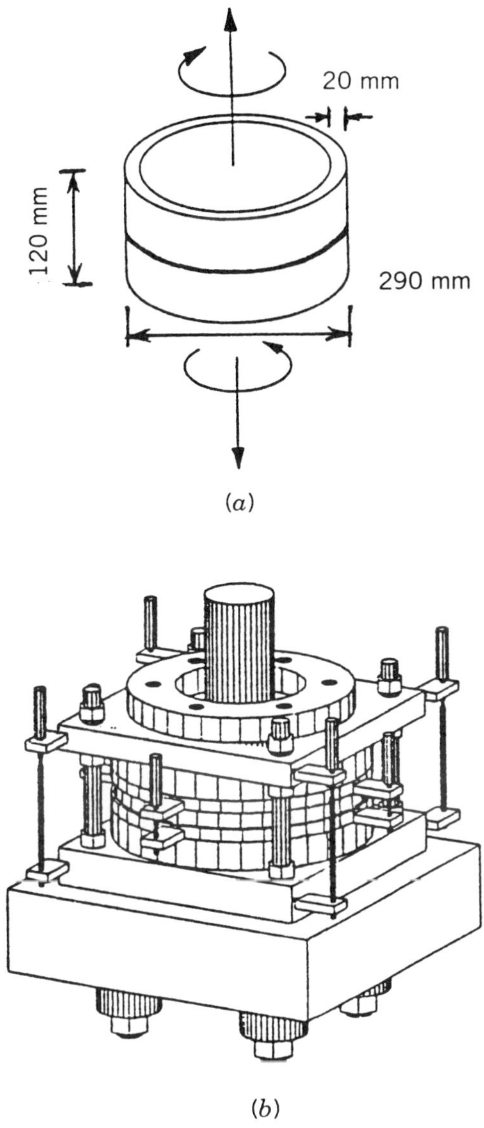

Figure 8-71 Torsion setup of Keuser and Walraven: (*a*) schematic of testing priniciple and (*b*) testing arrangement.[56]

only mode I fracture parameters. Results obtained for three-dimensional fracture problems are limited and inconclusive.

4. Finally, recent results by Landis and Shah[58] have suggested that if mode I fracture is modeled at the microlevel, this will be due to mixed-mode and mode II deformation behavior as sketched in Fig. 8-72.

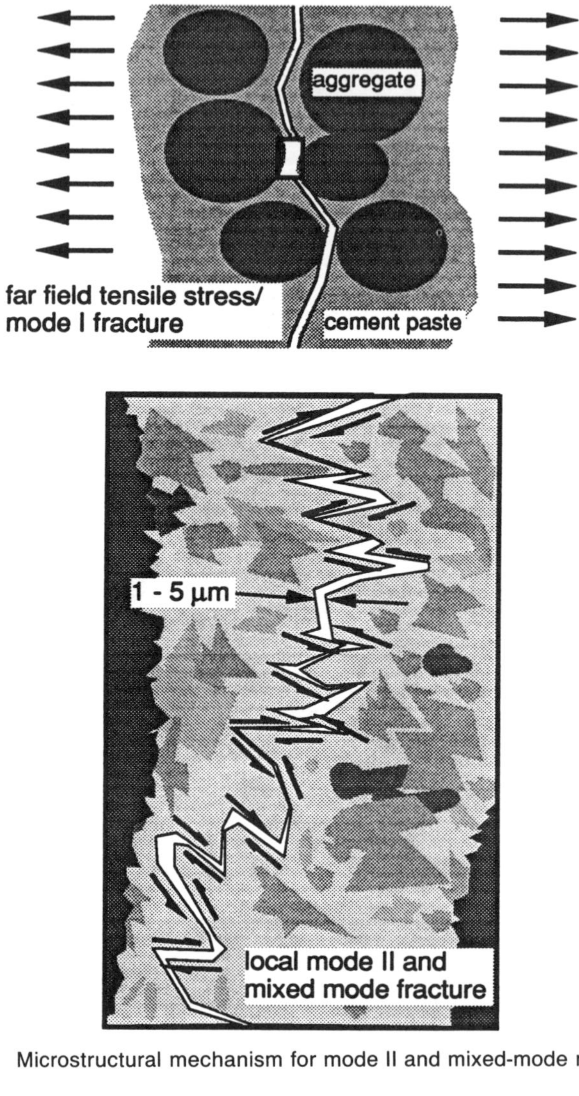

Figure 8-72 Microstructural mechanism for mode II and mixed-mode microcrack.[58]

REFERENCES

1. Carpinteri, A., and Swartz, S. E., ''Mixed-Mode Crack Propagation in Concrete,'' in Report of RILEM Committee 89-FMT, *Fracture Mechanics Test Methods for Concrete,* edited by Shah, S. P., and Carpinteri, A., Chapman & Hall, London, 1991, pp. 129–197.
2. Whittaker, B. N., Singh, R. N., and Sun, G., in *Rock Fracture Mechanics: Principles, Design and Applications,* Elsevier Science, Amsterdam, The Netherlands, 1992, pp. 311–348.

3. Broek, D., *Elementary Engineering Fracture Mechanics,* 4th rev. ed., Martinus Nijhoff, Dordrecht, The Netherlands, 1986.

4. Erdogan, F., and Sih, G. C., "On the Crack Extension in Plates under Plane Loading and Transverse Shear," *Journal of Basic Engineering,* Vol. 85, 1963, pp. 519–527.

5. Nuismer, R. J., "An Energy Release Rate Criterion for Mixed Mode Fracture," *International Journal of Fracture,* Vol. 11, 1975, pp. 245–250.

6. Sih, G. C., "Some Basic Problems in Fracture Mechanics and New Concepts," *Engineering Fracture Mechanics,* Vol. 5, 1973, pp. 365–377.

7. Sih, G. C., and Macdonald, B., "Fracture Mechanics Applied to Engineering Problems: Strain Energy Density Fracture Criterion," *Engineering Fracture Mechanics,* Vol. 6, 1974, pp. 381–386.

8. Maiti, S. K., and Smith, R. A., "Comparison of the Criteria for Mixed Mode Brittle Fracture Based on the Preinstability Stress–Strain Field, Part II, Pure Shear and Uniaxial Compressive Loading," *International Journal of Fracture,* Vol. 24, 1984, pp. 5–22.

9. Melin, S., "When Does a Crack Grow under Mode II Conditions?" *International Journal of Fracture,* Vol. 30, 1986, pp. 103–114.

10. Di Leonardo, G., "Fracture Toughness Characterization of Materials under Multiaxial Loading," *International Journal of Fracture,* Vol. 15, 1979, pp. 537–552.

11. Zhou, Q., Gong, S., and Chen, Z., "Compression Shear Fracture Mechanism for Marble," in *Proceedings of the SEM-RILEM International Conference on Fracture of Concrete and Rock,* edited by Shah, S. P., and Swartz, S. E., Houston, TX, June 17–19, 1987, pp. 580–586.

12. Jenq, Y. S., and Shah, S. P., "Mixed Mode Fracture of Concrete," *International Journal of Fracture,* Vol. 38, 1988, pp. 123–142.

13. Swartz, S. E., Lu, L. W., and Tang, L. D., "Mixed-Mode Fracture Toughness Testing of Concrete Beams in Three-Point Bending," *Materials and Structures,* Vol. 21, 1988, pp. 33–40.

14. Iosipescu, N., "New Accurate Procedure for Single Shear Testing of Metals," *Journal of Materials,* Vol. 2, No. 3, 1967, pp. 537–566.

15. Dasgupta, A., and Sirkis, J., "Experimental Validation of Iosipescu Shear Test Using Geometric Moirè," in *Developments in Theoretical and Applied Mechanics,* Vol. 15. *Proceedings of SECTAM15,* edited by Hanagud, S. V., Kamat, M. P., and Veng, C. E., Georgia Institute of Technology, Atlanta, 1990, pp. 86–93.

16. Kumoso, M., and Hull, D., "Mixed Mode Fracture of Composites Using Iosipescu Shear Test," *International Journal of Fracture,* Vol. 102, 1987, pp. 83–102.

17. Arrea, M., and Ingraffea, A. R., "Mixed-Mode Crack Propagation in Mortar and Concrete," Report 81-13, Department of Structural Engineering, Cornell University, Ithaca, NY, 1981.

18. Ballatore, E., Carpinteri, A., Ferrara, G., and Melchiorri, G., "Mixed Mode Fracture Energy of Concrete," *Engineering Fracture Mechanics,* Vol. 35, 1990, pp. 145–157.

19. Bazant, Z. P., and Pfeiffer, P. R., "Test of Shear Fracture and Strain Softening

in Concrete,'' paper presented at the 2nd Symposium on the Interaction of Non-Nuclear Munitions on Structures, Panama City Beach, FL, 1985.

20. Bazant, Z. P., and Pfeiffer, P. A., ''Shear Fracture Tests of Concrete,'' *Materials and Structures,* Vol. 19, 1986, pp. 111–121.
21. Biolzi, L., ''Mixed Mode Fracture in Concrete Beams,'' *Engineering Fracture Mechanics,* Vol. 35, 1990, pp. 187–193.
22. Bocca, P., Carpinteri, A., and Valente, S., ''Size Effects in the Mixed Mode Crack Propagation: Softening and Snap Back Analysis,'' *Engineering Fracture Mechanics,* Vol. 35, 1990, pp. 159–170.
23. Ingraffea, A. R., and Panthaki, M. J., ''Analysis of Shear Fracture Tests of Concrete Beams,'' in *Proceedings of Finite Element Analysis of Reinforced Concrete Structures,* Tokyo, Japan, 1985, pp. 151–173.
24. Schlangen, E., and van Mier, J. G. M., ''A FE-Supported Investigation of Mixed Mode Fracture in Concrete,'' in *Proceedings of the 9th International Conference on Experimental Mechanics,* Copenhagen, Denmark, August 20–24, 1990, pp. 1403–1412.
25. Schlangen, E., ''Experimental and Numerical Analysis of Fracture Processes in Concrete,'' Ph.D. Dissertation, Technical University of Delft, The Netherlands, 1993.
26. Swartz, S. E., Lu, L. W., Tang, L. D., and Refai, T. M. E., ''Mode II Fracture Parameter Estimates for Concrete from Beam Specimens,'' *Experimental Mechanics,* Vol. 28, No. 2, 1988, pp. 146–153.
27. Swartz, S. E., and Taha, N. M., ''Mixed Mode Crack Propagation and Fracture in Concrete,'' *Engineering Fracture Mechanics,* Vol. 35, 1990, pp. 137–144.
28. Swartz, S. E., and Taha, N. M., ''Crack Propagation and Fracture of Plain Concrete Beams Subjected to Shear and Compression,'' *ACI Structural Journal,* Vol. 88, No. 2, 1991, pp. 169–177.
29. Rots, J. G., and Blaauwendraad, J., ''Crack Models for Concrete: Discrete or Smeared? Fixed, Multi-Directional or Rotating?'' *Heron,* Vol. 34, No. 1, 1989.
30. Reinhardt, H. W., Cornelissen, H. A. W., and Hordijk, D. A., ''Mixed Mode Fracture Tests on Concrete,'' in *Fracture of Concrete and Rock,* edited by Shah, S. P., and Swartz, S. E., Springer-Verlag, New York, 1989, pp. 117–130.
31. van Mier, J. G. M., and Nooru-Mohamed, M. B., ''Fracture of Concrete under Tensile and Shear-like Loading,'' in *Fracture Toughness and Fracture Energy: Test Methods for Concrete and Rock,* edited by Mihashi, H., Takahashi, H., and Wittmann, F. H., Balkema, Rotterdam, The Netherlands, 1989, pp. 549–563.
32. van Mier, J. G. M., ''Mixed-Mode Fracture of Concrete under Different Boundary Conditions,'' in *Proceedings of the 1989 SEM Spring Conference on Experimental Mechanics,* Cambridge, MA, May 28–June 1, 1989, pp. 51–58.
33. Nooru-Mohamed, M. B., ''Mixed-Mode Fracture of Concrete: An Experimental Approach,'' Ph.D. Dissertation, Technical University of Delft, Delft, The Netherlands, 1992.
34. Hassanzadeh, M., Hillerborg, A., and Zhou, F. P., ''Tests of Material Properties in Mixed Mode I and II,'' in *Proceedings of the SEM-RILEM International Conference on Fracture of Concrete and Rock,* edited by Shah, S. P., and Swartz, S. E., Houston, TX, June 17–19, 1987, pp. 353–358.

35. Hassanzadeh, M., ''Behavior of Fracture Process Zones in Concrete Influenced by Simultaneously Applied Normal and Shear Displacement,'' Ph.D. Dissertation, Lund Institute of Technology, Lund, Sweden, 1992.

36. Barr, B., Hasso, A., and Khalifa, S., ''A Study of Mode II (Shear) Fracture on Notched Beams, Cylinders and Cores,'' in *Proceedings of the SEM-RILEM International Conference on Fracture of Concrete and Rock,* edited by Shah, S. P., and Swartz, S. E., Houston, TX, June 17–19, 1987, pp. 370–382.

37. Barr, B., and Hughes, T. G., ''A Comparative Study of Three Proposed Geometries Used to Investigate the Fracture Behavior of Materials under Mixed Mode Loading,'' *Proceedings of the 1989 SEM Spring Conference on Experimental Mechanics,* Cambridge, MA, May 28–June 1, 1989, pp. 59–65.

38. Arslan, A., Hughes, T. G., and Barr, B. I. G., ''Mixed Mode Fracture in Compact Shear Specimens,'' in *Fracture Behavior and Design of Materials and Structures,* Vol. II, edited by Firrao, D., EMAS, United Kingdom, 1990, pp. 676–681.

39. Luong, M. P., ''Tensile and Shear Strengths of Concrete and Rock,'' *Engineering Fracture Mechanics,* Vol. 35, 1990, pp. 127–135.

40. Arslan, A., Hughes, T. G., and Barr, B. I. G., ''Mixed Mode Fracture—Including Torsion in a New Compact Test Specimen Geometry,'' in *Fracture Processes in Concrete, Rock and Ceramics,* edited by van Mier, J. G. M., Rots, J. G., and Bakker, A., Chapman & Hall, London, 1991, pp. 737–746.

41. Arcan, M., Hashin, Z., and Volashin, A., ''A method to Produce Uniform Plane-Stress States with Applications to Fiber-Reinforced Materials,'' *Experimental Mechanics,* Vol. 18, No. 4, 1978, pp. 141–146.

42. Arcan, M., and Banks-Sills, L., ''Mode II Fracture Specimen-Photoelastic Analysis and Results,'' in *Proceedings of the Seventh International Conference on Experimental Stress Analysis,* Haifa, Israel, August 23–27, 1982, pp. 187–201.

43. Banks-Sills, L., and Bortman, Y., ''Mixed Mode Fracture Specimens: Analysis and Testing,'' *International Journal of Fracture,* Vol. 30, 1986, pp. 181–202.

44. Izumi, M., Mihashi, H., and Nomura, N., ''Fracture Toughness of Concrete for Mode II,'' in *Fracture Toughness and Fracture Energy of Concrete,* edited by Wittmann, F. W., Elsevier, Amsterdam, 1986, pp. 347–354.

45. Chong, K. P., and Kuruppu, M. D., ''New Specimen for Fracture Toughness Determination of Rock and Other Materials,'' *International Journal of Fracture,* Vol. 26, 1984, pp. R52–R62.

46. Chong, K. P., Kuruppu, M. D., and Kuszmaul, J. S., ''Fracture Toughness Determination of Rocks with Core-Based Specimens,'' in *Fracture of Concrete and Rock,* edited by Shah, S. P., and Swartz, S. E., Springer-Verlag, New York, 1989, pp. 13–25.

47. Atkinson, C., Smelser, R. E., and Sanchez, J., ''Combined Mode Fracture Via the Cracked Brazilian Test Disk,'' *International Journal of Fracture,* Vol. 18, No. 4, 1982, pp. 279–291.

48. Richard, H. A., ''Examination of Brittle Fracture Criteria for Overlapping Mode I and Mode II Loading Applied to Cracks,'' in *Application of Fracture Mechanics to Materials and Structures,* edited by Sih, G. C., et al., Martinus Nijhoff, The Hague, 1984, pp. 309–316.

49. Jones, D. J., and Chisholm, D. B., "An Investigation of the Edge-Sliding Mode in Fracture Mechanics," *Engineering Fracture Mechanics,* Vol. 7, 1975, pp. 261–270.

50. Liu, K., Barr, B. I. G., and Watkins, J., "Mode II Fracture of Fiber Reinforced Concrete Materials," *International Journal of Cement Composites and Lightweight Concrete,* Vol. 7, 1985, pp. 93–101.

51. Meridith, P. G., Ayling, M. R., Murrell, S. A. F., and Sammonds, P. R., "Cracking, Damage and Fracture in Stressed Rock: A Holistic Approach," in *Toughening Mechanisms in Quasi-Brittle Materials,* edited by Shah, S. P., Kluwer Academic, Dordrecht, 1991, pp. 67–89.

52. Bazant, Z. P., Sener, S., and Prat, P. C., "Fracture Mechanics Size Effect and Ultimate Load of Beams under Torsion," in *Fracture Mechanics Application to Concrete,* edited by Li, V. C., and Bazant, Z. P., ACI SP-118, American Concrete Institute, Detroit, MI, 1989, pp. 171–178.

53. Luong, M. P., "Fracture Behavior of Concrete and Rock under Mode II and Mode III Shear Loading," in *Fracture of Concrete and Rock: Recent Developments,* edited by Shah, S. P., Swartz, S. E., and Barr, B., Elsevier Applied Science, London, 1989, pp. 18–26.

54. Xu, D., and Reinhardt, H. W., "Softening of Concrete under Torsional Loading," in *Fracture of Concrete and Rock: Recent Developments,* edited by Shah, S. P., Swartz, S. E., and Barr, B., Elsevier Applied Science, London, 1989, pp. 39–50.

55. Sarkar, S., "The Behavior and Failure of Plain Concrete under the Action of Combined Compression and Torsion," in *Fracture of Concrete and Rock: Recent Developments,* edited by Shah, S. P., Swartz, S. E., and Barr, B., Elsevier Applied Science, London, 1989, pp. 475–493.

56. Keuser, W., and Walraven, J., "Fracture of Plain Concrete under Mixed Mode Conditions," in *Fracture of Concrete and Rock: Recent Developments,* edited by Shah, S. P., Swartz, S. E., and Barr, B., Elsevier Applied Science, London, 1989, pp. 625–634.

57. Tokatly, Z. Y., and Barr, B., "Size Effects in Mode III Fracture," in *Fracture Processes in Concrete, Rock and Ceramics,* edited by van Mier, J. G. M., Rots, J. G., and Bakker, A., E & F SPON, London, 1991, pp. 473–482.

58. Landis, E. N., and Shah, S. P., "Recovery of Microcrack Parameters in Mortar Using Quantitative Acoustic Emission," *Journal of Nondestructive Evaluation,* Vol. 12, No. 4, 1993, pp. 219–232.

9

APPLICATIONS OF FRACTURE MECHANICS TO CONCRETE STRUCTURES

9.1 BEHAVIOR OF CONCRETE STRUCTURES AND FRACTURE MECHANICS

Concrete is a tension-weak material, and its tensile strength is only 8–15% of its compressive strength. Since plain concrete structures almost always fail due to propagation of tensile cracks, the fracture mechanics concepts presented in previous chapters can be directly applied to them. For engineering applications, the tension-weak nature of concrete is usually reduced by using different types of reinforcement. For reinforced concrete structures, cracks generally initiate at locations where the principal tensile stress (or the strain energy release rate) exceeds the material tensile strength (or the fracture toughness) under service loads, and they are oriented perpendicular to the direction of this stress. Further propagation of these cracks will depend on distributions and magnitudes of both the principal tensile stress and the material fracture resistance. The presence of conventional reinforcements may have little effect on initiation of a crack, but they do provide considerable resistance to both propagation and opening of the crack. As a result, the service performance, failure mode, and loading capacity of the reinforced concrete structures depend, not only on structural geometry and loading configuration, but also on amount, location, and orientation of the reinforcement used.[1]

The influence of reinforcement on the crack propagation and failure mechanisms of the same type of structures can be illustrated by a flexural member, as shown in Fig. 9-1. An underreinforced concrete beam with a reinforcement ratio of ρ, which is smaller than the balanced reinforcement ratio ρ_b, is shown in Fig. 9-1a. When the beam is subjected to bending, some tensile cracks will initiate and propagate on the tensile (bottom) side of the beam. At a certain

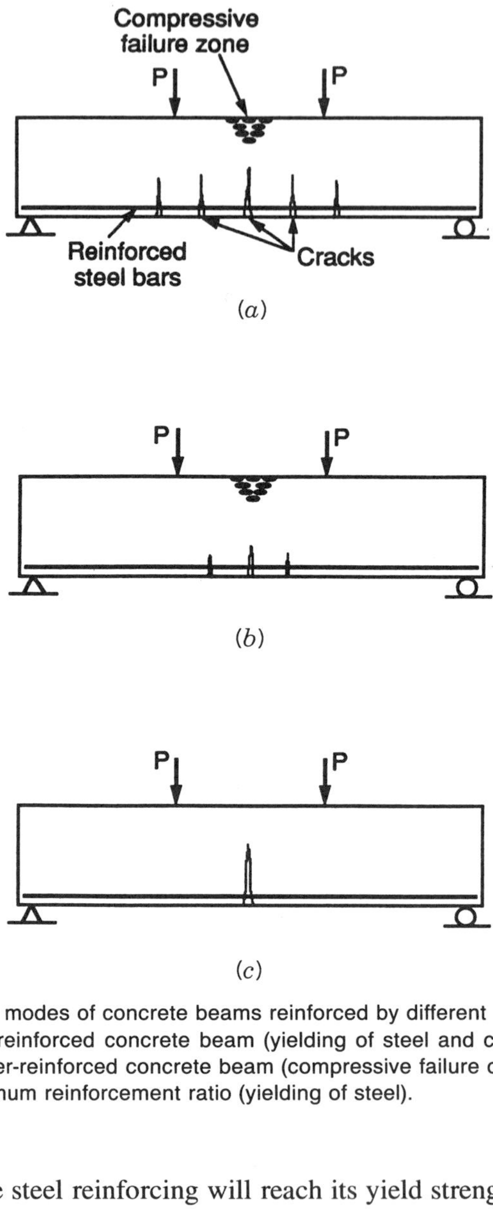

Figure 9-1 Failure modes of concrete beams reinforced by different amounts of reinforcement: (*a*) an under-reinforced concrete beam (yielding of steel and compressive failure of concrete), (*b*) an over-reinforced concrete beam (compressive failure of concrete), and (*c*) a beam with the minimum reinforcement ratio (yielding of steel).

loading level, the steel reinforcing will reach its yield strength before the concrete reaches its maximum compressive capacity. Since the tensile force in the steel basically remains constant with further loading, increase of the load primarily results from an increase of the moment arm to the center of compression in concrete on the top of the cracked sections. Whenever the concrete reaches its maximum compressive capacity, the beam collapses. Therefore, the collapse of the underreinforced concrete beam is characterized by both yielding of the steel and compressive failure of concrete at the cracked section. With the presence of a suitable amount and distribution of the steel, several tensile cracks usually develop during this "tensile–ductile" failure, and damage localization (single crack) is reduced by formation of these multiple cracks. For an overreinforced concrete beam as shown in Fig. 9-1*b*, since a larger amount of steel bars ($\rho > \rho_b$) is used, these steel bars may not yield when the beam fails. In this case, the collapse of the beam results from the compressive failure of concrete. It is noted that some tensile cracks may develop in this compressive-brittle failure of the beam. However, since the lengths of the cracks are less than the critical crack length, the failure of the beam is not controlled by propagation of these cracks. A beam with a reinforcement ratio less than the minimum reinforcement ratio, $\rho_{\min}$, is shown in Fig. 9-1*c*. Since the amount of the steel is not sufficient to arrest crack propagation, the collapse of the beam is only due to the propagation of a single crack that develops at the section of the maximum bending moment.

Conventional design of concrete and reinforced concrete structures includes the ultimate-limit analysis and the service performance analysis. The so-called ultimate-limit analysis uses strength-based failure criteria for determining the loading capacity of the structures. In terms of failure mechanisms observed from experimental studies, the ultimate-limit analysis calculates loading behavior of a structure by combining stress equilibrium, strain compatibility, and constitutive laws of materials at failure.[2] In addition, a strength criterion is usually used to access failure of the structure. On the other hand, the so-called service performance analysis is primarily based on either linear elastic approaches (for deflection) or empirical approaches (for cracking). After the dimensions of a structure and the amount and distribution of the reinforcement are determined using the ultimate-limit analysis, the service performance analysis is further conducted to assure that the structure has a satisfactory performance, regarding usually both deflection and cracking, under service loads. Even though the conventional design approaches are capable of predicting both the service performance and the loading capacity of different structures by introducing some empirical constants, fracture processes and failure mechanisms of concrete and reinforced concrete structures are not completely and explicitly taken into account by these approaches. Problems resulting from using the conventional design methods may include the following:

(i) Validity of the ultimate-limit analysis may require a structure showing some type of yield plateau before failure on its load–displacement relationship (ductile failure). This yield plateau does exist in a certain range for some structures, as shown in Fig. 9-2. However, failures of many structures are often quasi-brittle, instead of ductile, and are characterized by a softening load–displacement curve after a maximum load (Fig. 9-2). Apparently, it is unsafe for one to use the ultimate-limit analysis based on the ductile failure of structures to calculate the loading capacity of a structure with a quasi-brittle failure.

(ii) Since only stress and strain conditions at the critical section are usually

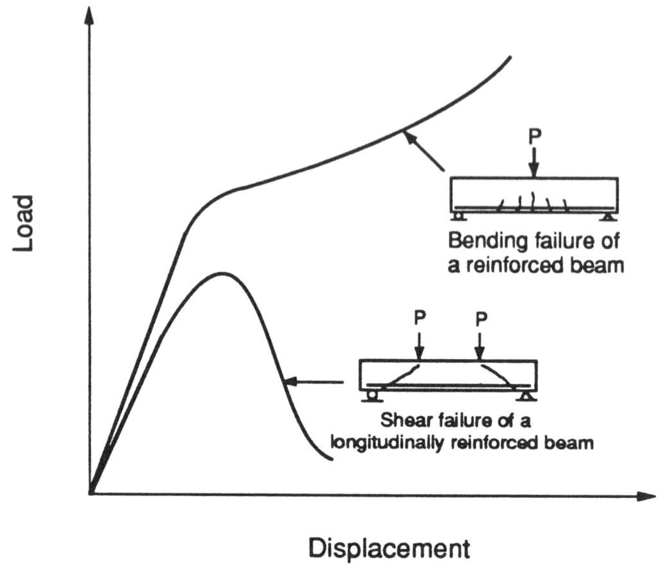

Figure 9-2 Ductile and quasi-brittle failure of reinforced concrete structures.

considered, the ultimate-limit analysis implies that failure of a structure occurs only at the critical section. However, experimental observations have revealed that the fracture process in concrete occurs in a certain range of zone (fracture process zone) instead of a single critical section. An additional empirical constant is usually necessary for the ultimate-limit analysis to account for the effect of the fracture process zone.

(iii) Since the fracture process zone has a certain size, energy criteria, which can take fracture energy within the fracture process zone into account, rather than the strength criterion for the ultimate-limit analysis should be used in conjunction with stress equilibrium, strain compatibility, and constitutive laws of materials to describe the failure of a structure.

(iv) Stress–strain curves as shown in Fig. 9-3 are usually used to determine constitutive laws for concrete under tension and compression in the ultimate-limit analysis. However, many studies have shown that softening (descending) parts of the stress–strain curves are not material properties, but depend on the dimension of the measured specimen, the loading rate, and the gage length of measurement. As a result of the strain localizations, softening behavior of concrete should be described using stress–displacement curves rather than the stress–strain curves.

(v) It is known that the nominal strength of concrete structures often decreases with increasing size of specimens tested. This size effect on the nominal strength cannot be explained by the strength-based failure criterion used in the ultimate-limit analysis. Even though this size effect may be reduced by

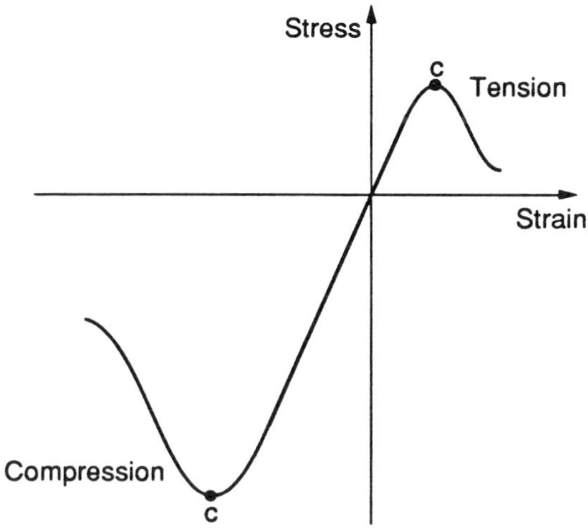

Figure 9-3 Stress–strain curves for concrete used in conventional plastic limit analysis (curves after point c depend on measurement gage length and loading rate).

using a certain amount of reinforcement, the strength-based failure criterion may be somewhat unconservative for the design of large structures.

(vi) Reinforcement can provide bridging over concrete matrix cracks. Some debonding and slipping of the steel–concrete interfacial bond usually occurs in the vicinity of the cracked matrix. These debonding and slipping effects influence the service performance and the loading capacity of a reinforced concrete structure. However, this influence is difficult to account for in the conventional design approaches.

(vii) In the conventional analysis, deflection of reinforced concrete structures is usually calculated based upon linear elastic approaches in conjunction with some empirical assumptions on mechanical properties of cracked sections. These assumptions on the cracked sections may not reasonbly include the influence of fracture process and require some additional empirical constants.

(viii) In the conventional analysis, a crack width is evaluated primarily using empirical methods that require some empirical constants. In the past, many efforts have been made to experimentally determine these constants. It is noted that previously obtained experimental data on the crack width are very scattered. This is partially because the crack width depends, not only on the geometry and loading condition of the structure, but also on the size of the structure. This size effect on the crack width cannot be accounted for by the conventional analysis unless additional empirical constants are used.

These problems associated with use of the conventional design methods for the service performance and failure of both concrete and reinforced concrete structures can be reduced or solved by using fracture mechanics. Applications of fracture mechanics to concrete structures can provide a rational basis for both service performance and failure analyses and lead to a better understanding of presently used empirical provisions in the current design code. A brief summary of applications of fracture mechanics has been given by ACI Committee 446.[3] Based on fracture mechanics, strain-softening behavior, which results from the phenomenon of strain localization, can be simulated by propagation of a crack. Failure of concrete structures can be described by energy-based criteria, rather than a strength criterion, in conjunction with stress equilibrium, strain compatibility, and constitutive laws. Influence of the fracture process zone is automatically included in the energy criteria. The influence of debonding and sliding on the loading capacity of structures can be uniformly taken into account by the energy criteria in conjunction with concrete cracks. The size effect on both nominal strength and crack width can be explained by the influence of the fracture process zone. Fracture mechanics has revealed that constitutive laws of a material are governed by material fracture parameters such as the fracture toughness and the critical crack opening displacement, which are independent of specimen geometry and size. Based on these material fracture parameters, the mechanical response of concrete and reinforced concrete structures, including both service and ultimate stages, can be predicted by the fracture mechanics approaches.

Since the mechanical response of reinforced concrete structures is influenced by the amount of reinforcement, different fracture mechanics considerations should be made in terms of various amounts of reinforcement. Since only a single major crack develops for structures with reinforcement ratios less than $\rho_{\min}$, fracture mechanics approaches, presented in previous chapters, can be easily applied to these structures. For underreinforced structures with $\rho_b > \rho > \rho_{\min}$, several cracks are usually developed under loading. Fracture mechanics in terms of a single crack may be modified to include interaction among multiple cracks. Since debonding and slipping usually occur at the interface between concrete and steel for these structures, an accurate description of crack propagation should include influence of debonding and slipping energy. Additional considerations for the compressive failure of concrete may be needed for some geometries of structures. For overreinforced concrete structures with $\rho > \rho_b$, failure of these structures is due to the concrete reaching its compressive capacity instead of propagating tensile cracks. Unless a greater understanding of the correlation between fracture mechanics and compressive failure of concrete, which will be discussed in Chapter 10, is developed, fracture mechanics methods must be modified with use of some empirical constants in order to simulate failure of these overreinforced concrete structures.

In addition, in some structures such as reinforced concrete shells and shear walls, crack propagation is influenced not only by the amount of reinforcement but also by the location and orientation of reinforcement. When cracks form in these structures, the bridging forces provided by the reinforcement can arrest the propagation of the cracks. The magnitudes of the bridging forces depend on the location and orientation of the reinforcement. Since shells and shear walls are large-scale structures, the effect of the fracture process zone may be neglected for the fracture analysis. As a result, the fracture mechanics approaches may be a powerful tool for the failure analysis of these structures.

In general, to apply fracture mechanics to reinforced concrete structures, one needs to be concerned not only with the structural geometry and loading conditions but also with the amount, location, and orientation of reinforcement used. Fracture mechanics approaches for different plain and reinforced concrete structures will be presented in the following sections.

9.2 SIZE EFFECT ON NOMINAL STRENGTH OF PLAIN CONCRETE SPECIMEN

As discussed earlier, experimental results have indicated that the nominal strength of concrete usually decreases with increasing size of structure and then remains constant. Traditionally, the size effect on concrete strength was explained by Weibull's weakest link statistical theory.[4,5] Weibull's theory determines strength of a structure based on the probability of failure associated with the largest defect in the material used. The lower strength for larger size structures is thus due to the fact that the larger the structure, the higher the probability to encounter the largest defect in the material. This explanation is certainly valid for one-dimensional structures such as a long chain and may be extended to some multidimensional structures after introducing some simplified hypotheses. However, since a stable crack propagation prior to failure exists for quasi-brittle materials such as concrete, this size effect on strength can be better explained by the presence of cracks with a fracture process zone rather than Weibull's weakest link in the material.[6] When a concrete structure is loaded, the strain energy produced by the applied load is converted to the energy consumed to create new fracture surfaces and the energy absorbed in the fracture process zone. For a large-size structure the latter is negligible compared to the former, whereas for a small-size structure these can be comparable. Therefore, the larger the structural size, the lower the nominal strength. However, the concrete strength approaches a constant when the size of the concrete structure becomes very large.

It has been proven that the size effect on concrete strength can be predicted by using fracture mechanics approaches presented in previous chapters.[7–10] Based on the fracture criterion of LEFM ($K = K_{Ic}$), the nominal strength of a structure, σ_{Nc}, can be expressed as

$$\sigma_{Nc} = \frac{K_{Ic}}{g_1(a_c/D)\sqrt{\pi a_c}} = \frac{K_{Ic}}{g_1(a_c/D)\sqrt{\Delta a_c(1 + a_0/\Delta a_c)\pi}} \tag{9.1}$$

where K_{Ic} is the critical stress intensity factor of a material used, $a_c = a_0 +$

Δa_c is the critical crack length, a_0 is the initial crack length, Δa_c is the critical crack extension, D is the characteristic size of the structure, and $g_1(a/D)$ is the geometric function of the structure.

To simplify Eq. (9.1), a series of geometrically similar structures with a constant value of a_0/D is considered. The geometric function $g_1(a/D)$ can be expanded into a Taylor series at point $a/D = a_0/D$. If only up to linear terms are considered, the following equation is obtained:

$$g_1\left(\frac{a_0}{D}\right) \approx g_{10} + \frac{\Delta a_c}{D} g'_{10} \tag{9.2}$$

where $g_{10} = g_1(a_0/D)$ and $g'_{10} = D[\partial g_1(a/D)/\partial a]$ for $a = a_0$. Substituting Eq. (9.2) into Eq. (9.1) results in

$$\sigma_{Nc} = \frac{K_{Ic}}{(g_{10} + g'_{10}\ \Delta a_0/D\sqrt{\Delta a_c(1 + a_0/\Delta a_c)\pi}} \tag{9.3}$$

By introducing the notation

$$Bf_t = \frac{K_{Ic}}{(g_{10} + g'_{10}\ \Delta a_c/D)\sqrt{\pi \Delta a_c}} \tag{9.4}$$

and

$$D_0 = \frac{D\ \Delta a_c}{a_0} \tag{9.5}$$

Eq. (9.3) can be rewritten as

$$\sigma_{Nc} = \frac{Bf_t}{\sqrt{1 + D/D_0}} \tag{9.6}$$

This is identical in form to the size effect model proposed by Bazant.[10]

As mentioned before, the size effect on concrete strength is an experimental phenomenon that primarily results from the presence of cracks with the fracture process zone. Equation (9.6) describes the size effect based on a series of geometrically similar structures with a constant value of a_0/D. For real materials, however, the value of the initial flaw, a_0, depends on the microstructure of the material and may not linearly increase with an increasing dimension of the structure. It is noted that although the use of the concept of geometrically similar structures can greatly simplify the size effect model, it is not a necessary condition to explain the experimentally observed size effect phenomonen.

Tang et al.[11] used the two-parameter fracture model to determine values of

K_{Ic} and a_c in Eq. (9.1). They found that the size effect on the nominal strength of concrete can be directly predicted by Eq. (9.1), provided that the value of a_c was determined from the condition of CTOD = CTOD_c. The results that they obtained for three-point bend beams are shown in Fig. 9-4. They predicted that the nominal strength decreases with increasing beam depth and approaches a constant value when the beam depth is sufficiently large. This prediction shows a trend that is similar to experimental observations.

Since the direct tension test is usually difficult to perform, the split-tension test shown in Fig. 9-5 is often used to measure the tensile strength of concrete. The measured trend in the nominal strength change with size for split-tension cylinders is shown in Fig. 9-6.[12, 13] The nominal strength decreases at first and then increases before approaching a limit value for this type specimen. Tang et al.[11] found that this trend on the nominal strength of the cylinder may also be explained by the two-parameter fracture model. They applied a finite element analysis to the cylinder, and conditions $K_I = K_{Ic}^s$ and CTOD = CTOD_c from the two-parameter fracture model were used as criteria for failure of the cylinder. Since the actual load on the cylinder is distributed on a certain width as shown in Fig. 9-5, analyses were performed for the cylinder subjected to a concentrated load and a distributed load, respectively. The obtained nominal strengths for cylinders subjected to the distributed load with the 12-mm load distribution width (t_1 = 6 mm) are compared with these for the concentrated load in Fig. 9-7. It is seen that the predicted critical nominal stress for the distributed load shows the same trend as the experimental data given in Fig.

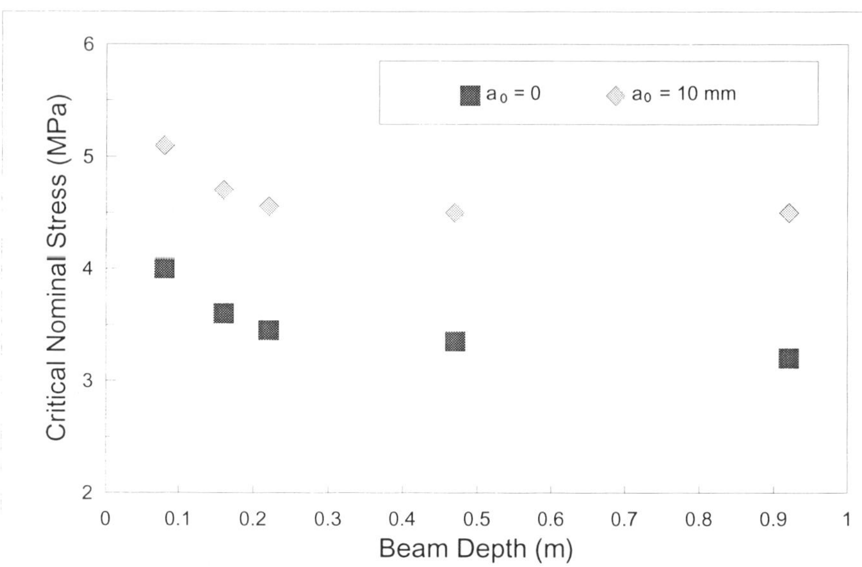

Figure 9-4 Predicted nominal strength of three-point bend beams with different depths.[11]

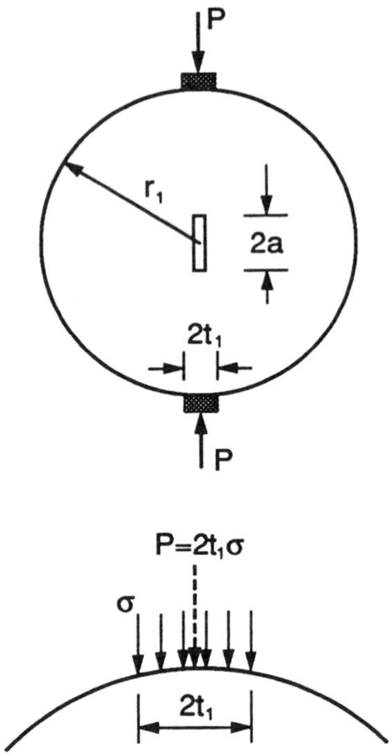

Figure 9-5 Split-tension test for measuring tensile strength of concrete.

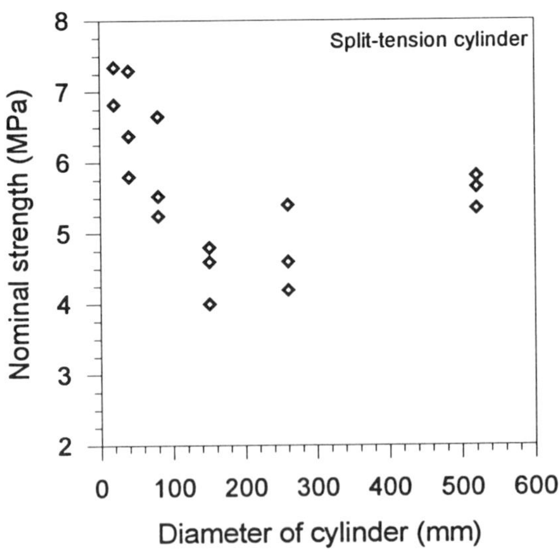

Figure 9-6 Measured size dependence of split-tension strength of concrete.[13]

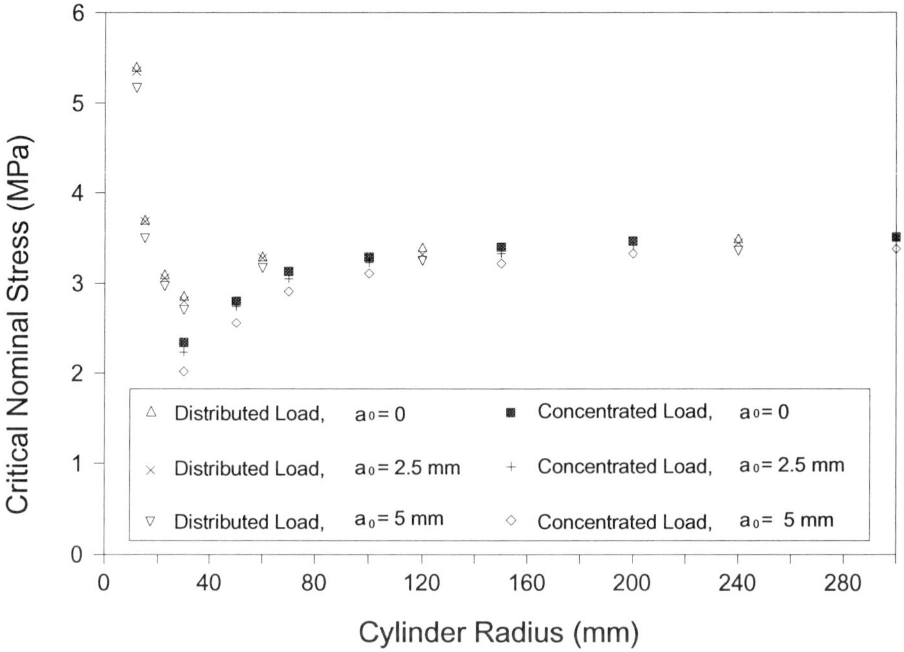

Figure 9-7 Size effect in split-tension cylinder under distributed and concentrated load predicted by two-parameter fracture model.[11]

9-6. The results obtained here suggest that the load distribution width is an important parameter for the split-tension cylinder used to evaluate the tensile strength of concrete. Idealizing the actual distributed load as the concentrated load in many previous studies may result in a gross misinterpretation of the experimental results. Bazant et al.[14] explained this observation on the size effect on splitting-tensile strength by the size effect model.

9.3 TENSION OF REINFORCED CONCRETE MEMBERS

The loading capacity of a concrete tensile member can be greatly increased by using steel reinforcement as shown in Fig. 9-8. The member usually cracks under service loads due to low tensile strength of concrete. However, the presence of an appropriate amount of reinforced steel bars prevents the member from failing after cracking. If the interface bond between the steel and the concrete is adequate, more cracks will be developed as the applied load increases. Both debonding and slipping may play an important role in failure of the member. Crack spacing decreases and crack width increases with increasing load. Failure of the member is characterized by yielding of the steel. Since

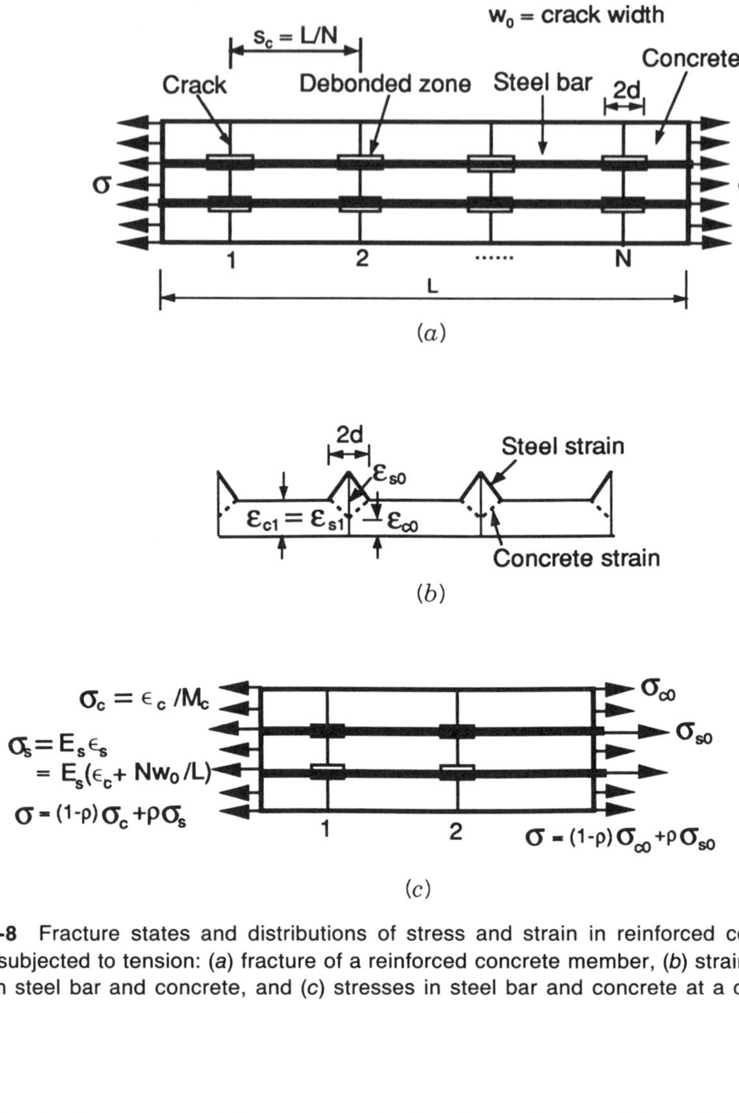

Figure 9-8 Fracture states and distributions of stress and strain in reinforced concrete member subjected to tension: (*a*) fracture of a reinforced concrete member, (*b*) strain distributions in steel bar and concrete, and (*c*) stresses in steel bar and concrete at a cracked section.

the mechanical response of the member is dominated by cracking of the concrete, fracture mechanics is a suitable approach to describe failure of the member.

Bazant and Oh[15] used an energy criterion to describe cracking of reinforced concrete structures. They assumed that all cracks formed simultaneously; that is, the interaction among cracks was neglected. Both the loading capacity and the minimum crack spacing were then calculated from the energy criterion by assuming zones of energy release during cracking. Ouyang and Shah[16] developed a fracture mechanics approach to predict cracking of reinforced concrete members subjected to tension. By balancing rates of change of the strain energy, the debonding energy, and the sliding energy, equations describing the cracking behavior of reinforced concrete tensile members are derived. Their approach takes interaction among multiple cracks into account. Change of strain, crack spacing, and crack width with respect to whole loading history can be predicted from the model. Their model is presented below.

9.3.1 Energy Criterion for Crack Propagation

Mechanical response of a reinforced concrete tensile member can be modeled based on Fig. 9-8*a*, where s_c represents the crack spacing, L stands for the length of the specimen, b and t are the width and the thickness of the specimen, ρ represents the reinforcement ratio of the steel bars, and E_s and E are the modulus of elasticity of the steel and the concrete, respectively. A debonded length, $2d$, is associated with each steel bar at the cracked sections. The crack spacing s_c is related to the number of cracks N by

$$s_c = \frac{L}{N} \tag{9.7}$$

To evaluate energy dissipations during the multiple cracking, strain distributions in reinforcing bars and concrete should be known. Since slip usually occurs on a debonded interface, values of steel (or concrete) strain at a cracked and an uncracked section are different. As shown in Fig. 9-8*b*, the steel strain ϵ_{s1} is equal to the concrete strain ϵ_{c1} in perfectly bonded sections. At a cracked section, strains in the steel bar and the concrete are denoted as ϵ_{s0} and ϵ_{c0}, respectively. It is assumed that strains in both the steel bar and the concrete vary linearly from the perfectly bonded section to the cracked section. The crack width is denoted as w_0. The average concrete and steel strains at the loading ends are ϵ_c and ϵ_s, respectively, where ϵ_c does not include the effect of w_0, and ϵ_s does include the effect of w_0. The relationship between ϵ_c and ϵ_s is

$$\epsilon_s = \epsilon_c + \frac{Nw_0}{L} \tag{9.8}$$

It is noted that due to the presence of the fracture process zone in the concrete the value of ϵ_{c0} is not equal to zero.[17] The steel is assumed to be a linearly elastic material.

Propagation of transverse cracks in a reinforced concrete tensile member is usually accompanied by debonding and sliding at the interface of reinforcing bars and concrete. Based on fracture mechanics, energy equilibrium during cracking requires that the strain energy release rate should be equal to the sum of the rates of the debonding energy, the sliding energy, and the fracture resistance of the concrete, which results in

$$-\frac{1}{bt}\frac{\partial \varphi_c}{\partial N} = R_{\mathrm{I}cf} + \frac{1}{bt}\frac{\partial(\varphi_d + \varphi_s)}{\partial N} \tag{9.9}$$

where φ_c is the strain energy of concrete containing N cracks without debonding and sliding, φ_d is the total bebonding energy on all debonded interfaces associated with N cracks, φ_s is the total sliding energy on all debonded interfaces, and $R_{\mathrm{I}cf}$ is the fracture resistance of the plain concrete member, which has the same dimensions and material properties with the corresponding reinforced concrete member. To use Eq. (9.9) to describe crack propagation in reinforced concrete tensile members, values of $R_{\mathrm{I}cf}$, φ_c, φ_d, and φ_s should be determined. It is noted that Eq. (9.9) is only valid for the steady-state propagation of multiple cracks when concrete is reinforced with an adequate amount of steel bars. The R-curve approach proposed by Ouyang and Shah,[18] presented in Chapter 7, is used to determine fracture resistance of plain concrete members $R_{\mathrm{I}cf}$ [see Eqs. (7.34), (7.35), (7.38), (7.39), and (7.44)]. Determinations of φ_c, φ_d, and φ_s will be discussed below.

9.3.2 Determination of Strain, Debonding, and Sliding Energies

The strain energy of the concrete containing N cracks without debonding or sliding, φ_c, can be written as

$$\varphi_c = \frac{Lbt(1-\rho)\sigma_c\epsilon_c}{2} = \frac{Lbt(1-\rho)\epsilon_c^2}{2M_c} \tag{9.10}$$

where M_c is the compliance of the concrete with N cracks and relates stress to strain.

To evaluate the value of M_c, the differential scheme method[19] is used to account for interaction among cracks. The strain energy of the member with N cracks (Fig. 9-8*a*) is

$$\phi_N = \tfrac{1}{2}M_c Lbt\sigma_c^2 \tag{9.11}$$

On the other hand, the strain energy of the member with $N + \Delta N$ cracks is

$$\phi_{N+\Delta N} = \tfrac{1}{2}(M_c + \Delta M_c)Lbt\sigma_c^2 \tag{9.12}$$

It is assumed here that the load σ_c remains constant when the number of cracks increases from N to $N + \Delta N$. Based on LEFM, the strain energy loss due to a crack propagating over $\Delta a = b\,\Delta N$ is

$$\Delta\phi = t\,\Delta N\left[\int_0^{a_{cf}} \frac{K_{\mathrm{I}}^2(a)}{E^*}\,da + \frac{K_{\mathrm{I}}^2(a_{cf})(b - a_{cf})}{E^*}\right]$$
$$= t\,\Delta N\left[\int_0^{a_{cf}} M^*\sigma_c^2 a\left[g_1\left(\frac{a}{b}\right)\right]^2 da + M^*\sigma_c^2 a_{cf}\left[g_1\left(\frac{a_{cf}}{b}\right)\right]^2 (b - a_{cf})\right] \quad (9.13)$$

where E^* and M^* are the elastic modulus and the compliance of the material around the crack tip, $g_1(a/b)$ is the geometry function for the stress intensity factor, and a_c is the critical crack length. Since experimental observations indicate that cracks are always initiated from one side of the member, a single-edge notch tensile member as shown in Fig. 9-9 is used to calculate the stress

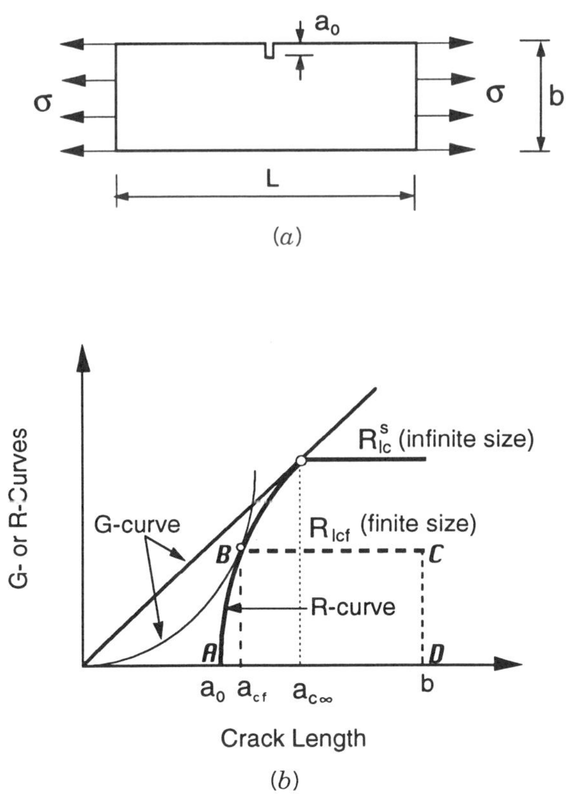

Figure 9-9 Single-edge notch specimen and proposed *R*-curve: (*a*) a single edge notch specimen and (*b*) the proposed *R*-curve.

intensity factor in this study. For such a member, $g_1(a/b)$ is given by (see Chapter 2)

$$g_1\left(\frac{a}{b}\right) = \sqrt{\pi}\left[1.12 - 0.231\,\frac{a}{b} + 10.55\left(\frac{a}{b}\right)^2 - 21.72\left(\frac{a}{b}\right)^3 + 30.39\left(\frac{a}{b}\right)^4\right] \tag{9.14}$$

The value of the expression in the bracket in Eq. (9.13) is equal to the area under the R-curve, that is, the area $ABCD$ in Fig. 9-9 when $E^* = E$. Using the self-consistent approximation, the local compliance M^* can be replaced by the global compliance M_c. Therefore, Eq. (9.13) can be rewritten as

$$\Delta\phi = \tfrac{1}{2}LbtM_c\xi\sigma_c^2\,\Delta N \tag{9.15}$$

where

$$\xi = \frac{2}{Lb}\left[\int_0^{a_{cf}}\left[g_1\left(\frac{a}{b}\right)\right]^2 a\,da + a_{cf}(b - a_{cf})\left[g_1\left(\frac{a_{cf}}{b}\right)\right]^2\right] \tag{9.16}$$

Using the energy balance for the unreinforced material ($\phi_{N+\Delta N} - \phi_N = \Delta\phi$), the following equation governing compliance of the concrete member with N cracks is derived:

$$\frac{\Delta M_c}{\Delta N} = \xi M_c \tag{9.17}$$

When the value of N is adequately large, Eq. (9.17) can be solved as a differential equation. By solving Eq. (9.17) and using the boundary condition $M_c = 1/E$ for $N = 0$, where E is the modulus of elasticity of concrete, the compliance of the concrete with N cracks is obtained,

$$M_c = \frac{e^{\xi N}}{E} \tag{9.18}$$

It is noted that when the value of N is small, Eq. (9.18) only gives an approximate estimation for the value of M_c. Substituting Eq. (9.18) into Eq. (9.10), the value of φ_c can be calculated.

Noting that the debonded surface of a steel bar for each of the cracks is $4\pi r_s d$ and the number of steel bars crossing each crack is $\rho/\pi r_s^2$, the total debonded energy of N cracks can be calculated as

$$\varphi_d = \frac{4N\rho\ d\gamma_a\ bt}{r_s} \tag{9.19}$$

where γ_d is the debonding energy on the unit area of interface between steel and concrete, which is regarded as a material property in the present model, and r_s is the radius of the steel bar.

The debonded length $2d$ is assumed to be a constant here. A similar assumption has been used by different researchers. According to Gilbert,[20] for a reinforced concrete member containing deformed bars, the value of d may be determined as follows:

$$d = \frac{r_s}{20\rho} \tag{9.20}$$

In order to calculate the sliding energy, the crack width should be first evaluated. The load transfer capacity of cracked concrete members can be described by a stress–separation curve.[21] For simplicity, a linear softening stress–separation curve is used as below:[22]

$$\sigma_{c0} = kf_{cf}\left(1 - \frac{w_0}{w_c}\right) \tag{9.21}$$

where σ_{c0} is the cohesive stress after cracking, w_c is the critical crack separation, f_{tf} is the tensile strength for a finite-size specimen, and k is a constant that is smaller than unity. The values of f_{tf} as well as R_{Icf} can be determined together by using the R-curve approach proposed by Ouyang and Shah.[18]

The crack width w_0 is the sum of the steel elongation u_{s0} and the concrete recovery u_{c0} after debonding, as shown in Fig. 9-10*a*. Since the values of u_{c0} and u_{s0} can be evaluated from the area *abc* and the area *def* in Fig. 9-10*b*, the following result is obtained:

$$w_0 = 2(u_{c0} + u_{s0}) = d(\epsilon_{s0} - \epsilon_{c0}) = d\left(\frac{\sigma_{s0}}{E_s} - \frac{\sigma_{c0}}{E}\right) \tag{9.22}$$

where $\epsilon_{c0} = \sigma_{c0}/E$ is approximately used because ϵ_{c0} is a local quantity defined at the crack face.

Stresses and strains in reinforcing bars and concrete at a loading end and a cracked section are shown in Fig. 9-8*c*, respectively. Substituting these relationships into Eq. (9.22) and using Eqs. (9.8) and (9.21) result in

$$\begin{aligned}\frac{w_0}{d} &= \frac{\sigma - (1 - \rho)\sigma_{c0}}{E_s\rho} - \frac{\sigma_{c0}}{E} \\ &= \left(\frac{1 - \rho}{E_s\rho M_c} + 1\right)\epsilon_c + \frac{Nw_0}{L} - \eta\left(1 - \frac{w_0}{w_c}\right)\end{aligned} \tag{9.23}$$

where

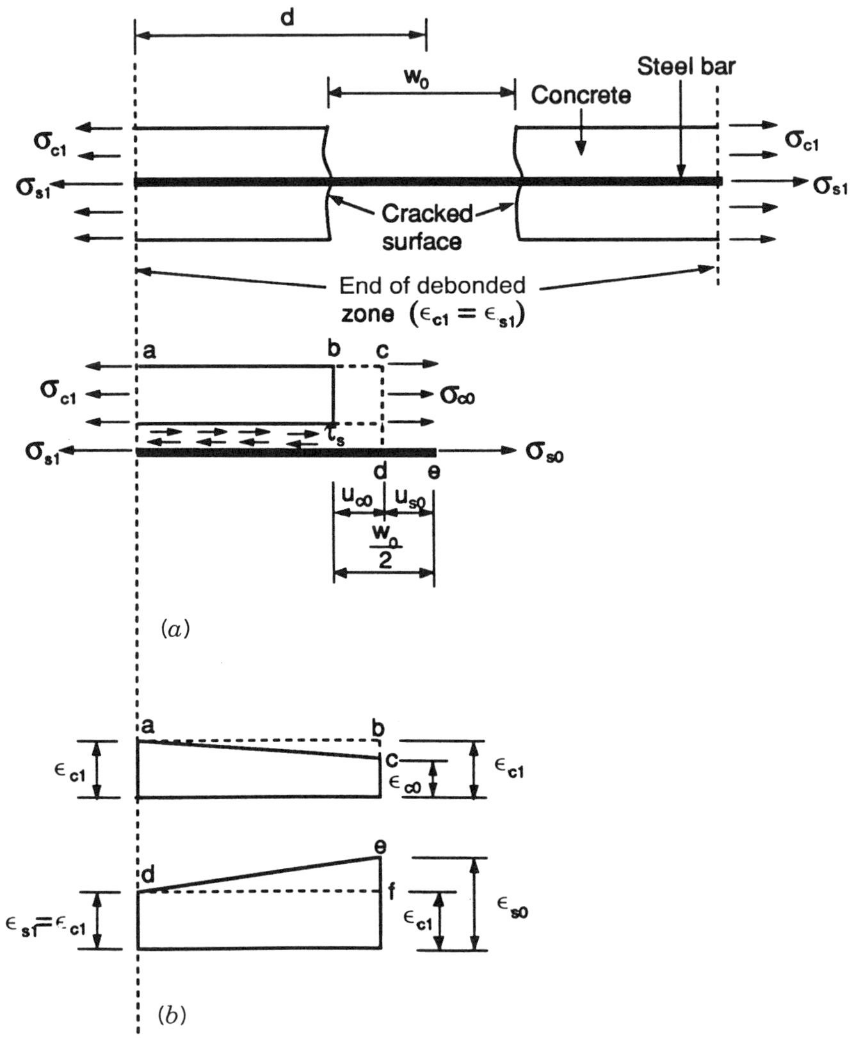

Figure 9-10 Deformation, strain, and stress distributions at cracked section and associated debonded zone: (*a*) deformation and stress distributions at a cracked section and associated debonded zone and (*b*) strain distributions in steel bar and concrete.

$$\eta = \frac{(1-\rho)f_{tf}}{\rho E_s} + \frac{f_{tf}}{E} \tag{9.24}$$

As a result, the crack width w_0 can be solved from Eq. (9.23),

$$\frac{w_0}{d} = \frac{[(1-\rho)/(\rho E_s M_c) + 1]\epsilon_c - \eta}{1 - Nd/L - \eta d/w_c} \tag{9.25}$$

Equation (9.25) expresses the crack width w_0 as a function of the average concrete strain ϵ_c. It is noted that the value of w_0 predicted by Eq. (9.25) is the crack width on the surface of the reinforcing steel bar instead of the concrete.

The concrete will slide back after debonding. This sliding energy can be calculated based on the crack width obtained above. The energy dissipated due to sliding is

$$\varphi_s = \frac{4N\rho bt}{r_s}\int_0^d \tau_s(u_s - u_c)\,dx \tag{9.26}$$

where τ_s is the frictional stress resulting from sliding, which is assumed to be a constant in the present study, the terms u_s and u_c are the displacements of the steel bar and the concrete, respectively. Based on the strain distributions of the steel bar and the concrete within the debonded zone indicated in Fig. 9-8*a* and using Eqs. (9.22) and (9.25), Eq. (9.26) can be rewritten as

$$\begin{aligned}\varphi_s &= \frac{4\rho\tau_s Nbt}{r_s}\,\frac{d^2(\epsilon_{s0} - \epsilon_{c0})}{6} \\ &= \frac{2\rho\tau_s Nd^2bt}{3r_s}\,\frac{[(1-\rho)/(\rho E_s M_c) + 1]\epsilon_c - \eta}{(1 - Nd/L - \eta d/w_c)}\end{aligned} \tag{9.27}$$

9.3.3 Relationships among Crack Spacing, Strains, and Stresses

Substituting Eqs. (9.10), (9.18), (9.19), and (9.27) into Eq. (9.9), the following equation governing the relationship between N and ϵ_c can be derived:

$$\begin{aligned}&\frac{L\xi(1-\rho)E\epsilon_c^2}{2e^{\xi N}} - \frac{\begin{array}{c}2d^2\rho\tau_s\{[(1-\rho)(1-N\xi)E/(E_s\rho e^{\xi N}) + 1] \\ (1-\eta d/w_c) + (1-\rho)\,dN^2\,E\xi/(LE_s\rho e^{\xi N})\}\epsilon_c\end{array}}{3r_s(1 - Nd/L - \eta d/w_c)^2} \\ &+ \frac{2\rho\tau_s\,d^2\eta(1-\eta d/w_c)}{3r_s(1 - Nd/L - \eta d/w_c)^2} - \frac{4\rho\,d\gamma_d}{r_s} - R_{\mathrm{I}cf} = 0\end{aligned} \tag{9.28}$$

Equation (9.28) is a quadratic function of ϵ_c. For a given value of N, the corresponding value of ϵ_c can be solved. It is noted that Eq. (9.28) is valid under the condition

$$w_0 \le w_c \qquad 2d \le s_c$$

$$\frac{Nd}{L} + \frac{\eta d}{w_c} < 1 \tag{9.29}$$

For a given number of cracks N, which relates to the crack spacing through Eq. (9.7), the corresponding matrix strain ϵ_c is calculated from Eq. (9.28). After knowing the values of N and ϵ_c, the average concrete stress σ_c and the average steel stress σ_s can be computed from

$$\sigma_c = \frac{\epsilon_c}{M_c} = E\epsilon_c e^{-\xi N}$$

$$\sigma_s = E_s \epsilon_s = E_s\left(\epsilon_c + \frac{Nw_0}{L}\right) \tag{9.30}$$

For the theoretical prediction, fracture properties of concrete ($K_{\mathrm{I}c}$, CTOD_c, and w_c), the fracture energy for interface bond between reinforcing steel bars and concrete (γ_d), and the frictional sliding stress (τ_s) should be known. In addition, the value of the material characteristic defect a_0 is assumed. The computational scheme is as below:

1. For a given reinforced concrete member, values of a_{cf}, f_{tf}, and $R_{\mathrm{I}cf}$ for the same-size plain concrete member are first determined based on the R-curve approach presented in Chapter 7 [Eqs. (7.34), (7.35), (7.38), (7.39), and (7.44)]. It is noted that these values of a_{cf}, f_{tf}, and $R_{\mathrm{I}cf}$ depend on the dimensions of the member.
2. The value of the debonded length d may be determined from Eq. (9.20).
3. The number of cracks N is increased step by step. For each step of the value of N, the corresponding concrete strain ϵ_c is determined using Eq. (9.28), which is a quadratic equation of ϵ_c.
4. The corresponding crack width w_0 is evaluated from Eq. (9.25). The values of s_c, ϵ_s, σ_c, and σ_s can be computed from Eqs. (9.7), (9.8), and (9.30), respectively.

9.3.4 Comparison and Discussion

The proposed model is compared with experimental results reported by different researchers. The values $K_{\mathrm{I}c}^s = 30.7\ \mathrm{N/mm}^{1/2}$, $\mathrm{CTOD}_c = 0.016$ mm, and $E = 33$ GPa, which were reported by Jenq and Shah[11] for concrete, were used for the theoretical prediction. The value of a_0 depends on the microstructure of a material. However, it was shown that the theoretical prediction of the model is not sensitive to the selection of the value of a_0.[16] Therefore, the value of a_0 was assumed to be 5 mm. The interfacial properties for the deformed steel bar were assumed to be $\gamma_d = 0.004$ N/mm and $\tau_s = 2.5$ MPa. The values $E_s = 210$ GPa and $k = 0.25$ were used.

The predicted stress–strain curve is compared with the experimental data (specimen 2) reported by Rizkalla et al.[23] in Fig. 9-11*a*. The specimens tested

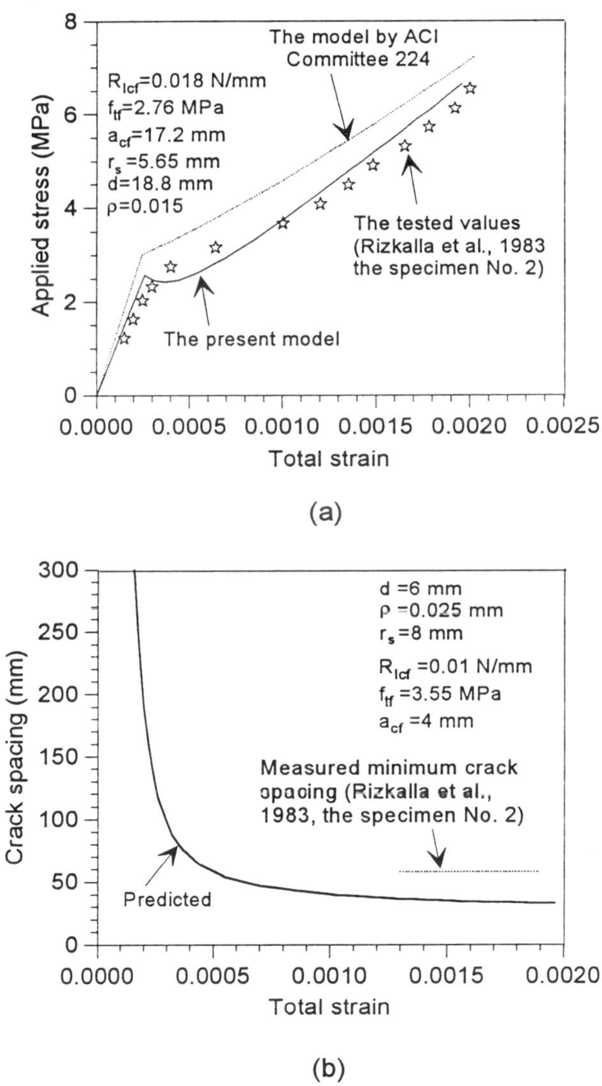

Figure 9-11 Theoretical and experimental comparison for results reported by Rizkalla et al.[16]

were 763 × 305 × 178 mm. Eight deformed steel bars with 11.3 mm in diameter were used ($\rho = 0.015$). The value of d obtained is 18.8 mm. The obtained values $R_{Icf} = 0.018$ N/mm, $a_{cf} = 17.2$ mm, and $f_{tf} = 2.76$ MPa were used for the calculation.

The tension-stiffening curve predicted by the model given by ACI Committee 224[2] for the same specimen is also shown in Fig. 9-11*a*. It is seen that the proposed theory predicts well the stress–strain curve compared to the experimental values, whereas the model by ACI Committee 224 seems to predict upper bound values for this specimen.

The predicted crack spacing is compared with the measured minimum crack spacing on the concrete surface in Fig. 9-11*b*. Note that there was one measured value of crack spacing that corresponds to the minimum value. The predicted crack spacing is a lower bound of the measured value, because the crack spacing predicted is on the reinforcing bar surface, which includes contributions from the external and the internal cracks.

The predicted crack width on the surface of the reinforcing bar is compared with the experimental data reported by Yannopoulos,[24] as shown in Fig. 9-12. The specimens tested were 100 long and 76 mm in diameter. One deformed steel bar with 16 mm in diameter was used as the reinforcement ($\rho = 0.025$). For the theoretical prediction, the cylindrical concrete specimen was replaced by an equivalent specimen with the same area of the square cross section. The obtained values $R_{Icf} = 0.0079$ N/mm, $a_{cf} = 10.3$ mm, and $f_{tf} = 2.2$ MPa were used to calculate the crack width. As indicated in Fig. 9-12, the predicted crack widths on the surface of the steel bars are close to the measured values.

It has been realized that concrete strength decreases with increasing speci-

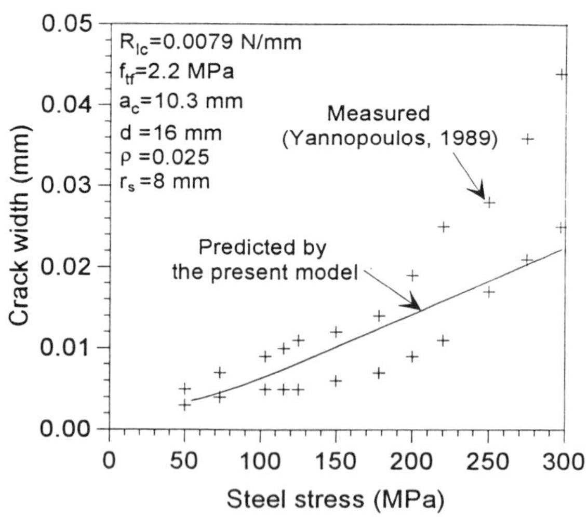

Figure 9-12 Theoretical and experimental comparison for results reported by Yannopoulos.[16]

men size. This size effect on concrete strength primarily results from the strain localization in concrete at failure. The presence of steel reinforcement can reduce this strain localization and make multiple cracks possible. Therefore, a more gentle size effect may be expected in reinforced concrete members. It is noted that this size effect on the cracking behavior of reinforced concrete members is usually ignored in many previous studies.

Since different sizes of members result in different values of R_{Icf}, a_{cf}, and f_{tf} as calculated from the R-curve approach, different cracking responses will be predicted from these obtained values using the proposed model. Therefore, the proposed model is capable of predicting the size effect on the cracking behavior of reinforced concrete members.

To demonstrate the size effect, tensile members with varying square cross-sectional dimensions and a constant length of 1000 mm were analyzed. The members were reinforced by the 16-mm-diameter deformed steel bars. The same concrete and interfacial properties as mentioned earlier were used here. The ratio of a_0/b was kept equal to 0.1 for all members. Two reinforcement ratios, $\rho = 0.015$ and $\rho = 0.03$, were studied.

The effect of the member size on the tensile strength of plain concrete, the total applied stress, and the crack width for the reinforced concrete members ($\rho = 0.015$) at the average strain value of 0.001 is plotted in Fig. 9-13. All values were divided by the results of the member with $b = 116$ mm for easier comparison. Although all quantities decrease with increasing member sizes, the decay slopes of the reinforced concrete members are smaller than that for the plain concrete. Within the studied range of member sizes, the size effect curve for the tensile strength of plain concrete is concave downward, whereas

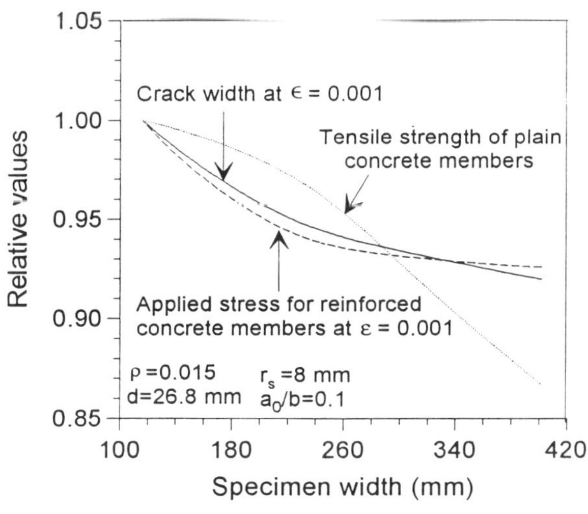

Figure 9-13 Effect of specimen size on mechanical behavior of concrete and reinforced concrete members.[16]

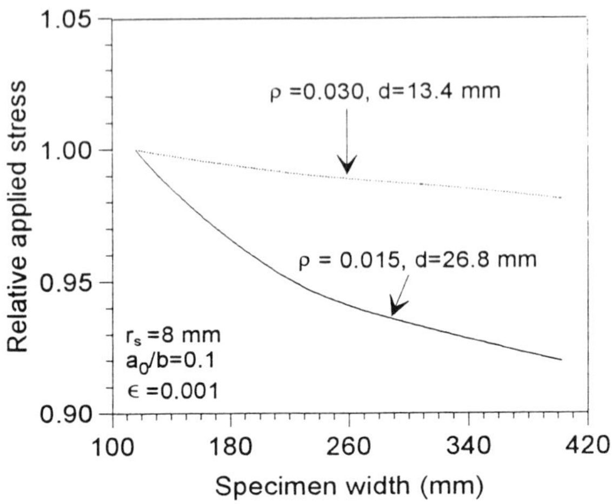

Figure 9-14 Size effect on applied stress for different reinforcement ratios.[16]

the curves for the reinforced concrete member are concave upward and their slopes approach zero. These different size effect curves may be explained by the fact that the curve for the tensile strength of the plain concrete corresponds to the single crack, whereas the curves for the reinforced concrete member are based on the multiple cracks.

The size effect curves for the applied stress value for two different reinforcement ratios $\rho = 0.015$ and $\rho = 0.03$ are shown in Fig. 9-14. It is seen that the influence of the size decreases with increasing reinforcement ratio.

9.4 BENDING OF REINFORCED CONCRETE BEAMS

As mentioned earlier, applications of fracture mechanics to reinforced concrete beams are influenced by their failure modes. Depending on the amount of reinforcement used, collapse of beams may be due to yielding of steel, a compressive failure of concrete, or a combination of both. Some fracture mechanics models for underreinforced concrete beams will be presented in this section.

9.4.1 Hillerborg Model

By combining the concept of his fictitious crack model[25] and conventional bending theory, Hillerborg[26] modified the conventional bending theory of reinforced concrete beams by taking account of the effect of the compressive strain localization. Similar to the conventional bending theory, it is assumed that plane cross sections remain plane after deformation. An elastic–plastic stress–strain curve is used for steel. Based on his fictitious crack model, Hillerborg characterized the compressive response of concrete by a linear stress–strain curve prior to the peak stress and a linear stress–displacement curve after the peak stress, as shown in Fig. 9-15. Since compressive failure of concrete usually happens within a certain characteristic length of the beam, l, a formal complete compressive stress–strain curve can be constructed by means of a summation of the influence of the two curves according to the formula

$$\epsilon_1 = \epsilon + \frac{w}{l} = \epsilon + \frac{w}{k_c \xi_1 b_1} \tag{9.31}$$

where ϵ is taken from the prepeak stress–strain curve and w from the postpeak stress–deformation curve. The compressive characteristic length $l = k_c \xi_1 b_1$ is assumed, where b_1 is the distance from the centroid of the steel to the top of the beam, $\xi_1 b_1$ is the depth of the compressive zone (Fig. 9-16), and k_c is a

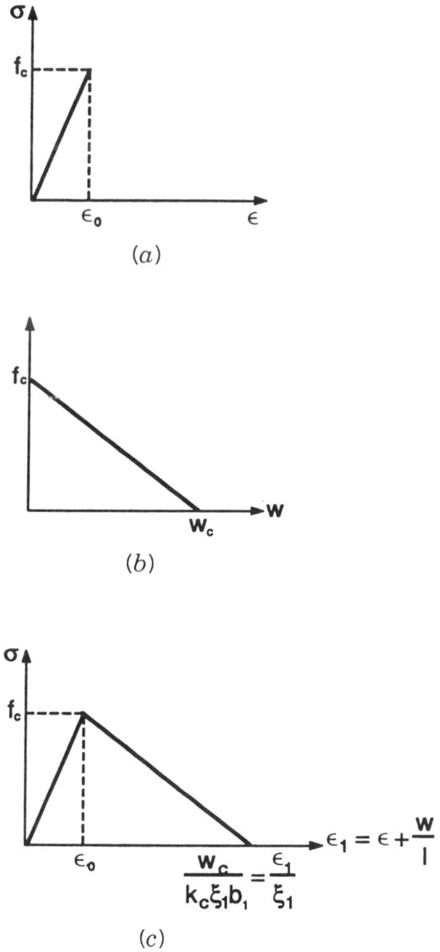

Figure 9-15 Simplified stress–strain curve for concrete under compression.[26]

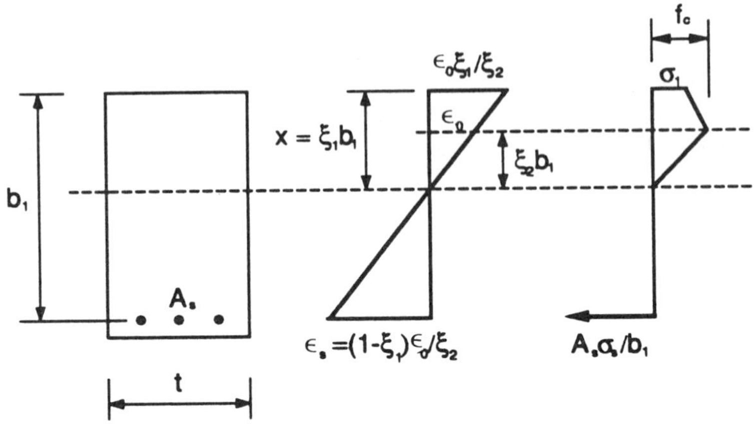

Figure 9-16 Strain–stress distributions in rectangular cross section.

proportionality factor that is approximately 0.8. To simplify further the analysis, the following definition is introduced according to Fig. 9-15:

$$\epsilon_1 = \frac{w_c}{k_c b_1} \tag{9.32}$$

When the compressive strain is smaller than the value of ϵ_0 given in Fig. 9-16, the present analysis is the same as the conventional approach. Therefore, the following analysis is limited to the case of $\epsilon > \epsilon_0$. The stresses and strains in the section of maximum bending moment as shown in Fig. 9-16 directly lead to

$$\epsilon_s = \frac{1 - \xi_1}{\xi_2} \epsilon_0 \tag{9.33}$$

and

$$\frac{\sigma_1}{f_c} = \frac{\epsilon_1/\xi_1 - \epsilon_0 \xi_1/\xi_2}{\epsilon_1/\xi_1 - \epsilon_0} \tag{9.34}$$

where ϵ_s is the strain at the centroid of steel bars, f_c is the compressive strength of concrete, σ_1 is the compressive stress of concrete at the top surface, and ξ_1 is shown in Fig. 9-16.

Equilibrium of forces on the section gives

$$\frac{A_s \sigma_s}{b_1 f_c t} = \frac{1}{2}\left[\xi_1 + \frac{\sigma_1}{f_c}(\xi_1 - \xi_2)\right] \tag{9.35}$$

where σ_s is the steel stress. By combining Eqs. (9.33)–(9.35), the bending moment M can be expressed as

$$\frac{6M}{b_1^2 f_c t} = \xi_1(3 - 2\xi_1 + \xi_2) + \frac{\sigma_1}{f_c}(\xi_1 - \xi_2)(3 - \xi_1 + \xi_2) \tag{9.36}$$

The curvature $1/r_c$ is given by

$$\frac{1}{r_c} = \frac{\epsilon_0}{\xi_2 b_1} \tag{9.37}$$

It is possible to predict the bending behavior of a beam using the above formulas. Theoretical moment–curvature diagrams predicted are indicated in Fig. 9-17, where $f_c = 40$ MPa, $f_s = 400$ MPa, $E_s = 200$ GPa, $\epsilon_0 = 0.002$, and $\epsilon_1 = 0.002, 0.008$, respectively, were used.

The stress–deformation curve used by Hillerborg is capable of modeling the compressive strain localization after the peak stress. However, the effect of the compressive strain localization has not yet been considered in many design codes such as the CEB-FIP Model Code[27] and the ACI 318-89 Code.[28] For example, the CEB-FIP Model Code gives a simplified compressive stress–strain curve as shown in Fig. 9-18*a*, where the end of the horizontal line part is defined as $\epsilon_u = 0.0035$. However, to include the strain localization as indicated in the above analysis, it is proposed[26] to define the end as $\epsilon_u = k_1/x$, as shown in Fig. 9-18*b*, where x is the depth of the compressive zone, and k_1 is a factor that may depend on material property and specimen geometry.

The above analysis uses the stress–strain curve prior to the peak stress and the stress–displacement curve after the peak stress to characterize compressive

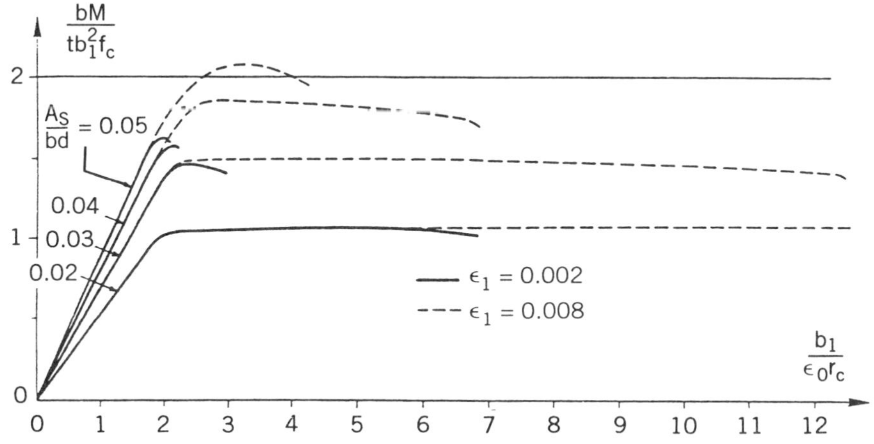

Figure 9-17 Theoretical moment–curvature diagrams.[26]

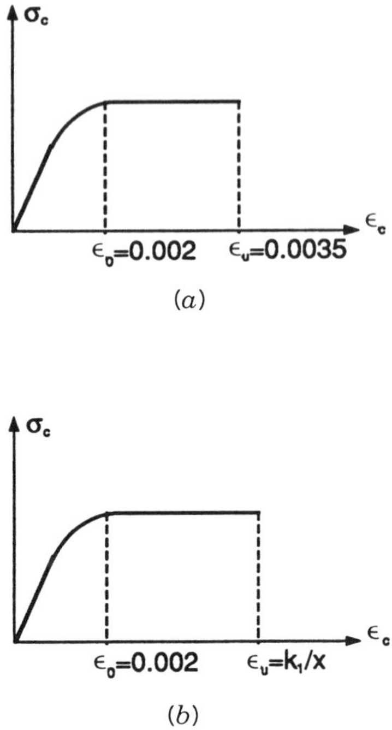

Figure 9-18 Stress–strain curves for concrete in compression: (*a*) the *CEB-FIP* Model Code and (*b*) the proposed curve.[26]

response of concrete and takes into account the compressive strain localization. However, since the influence of many factors, such as material fracture parameters G_F, f_t, and w_c, on the compressive behavior of concrete is not clear, the above analysis is very preliminary. More detailed discussions relating fracture mechanics principles to compressive failure of concrete will be discussed in Chapter 11.

9.4.2 Carpinteri and Baluch Model

Carpinteri et al.[29,30] proposed a fracture mechanics model for bending of reinforced concrete beams. In their model concrete is assumed to be a linear elastic material and steel is considered to be linear elastic–plastic. The effect of reinforced steel bars is simulated by a closing force whose magnitude is determined by a compatibility condition. Baluch et al.[31] modified this model to include the nonlinear behavior of concrete under compression. Their model will be briefly presented below.

A cracked section of a reinforced concrete beam is shown in Fig. 9-19. The section is subjected to a moment M, and the effect of the steel bar is simulated by a force F_s acting at the centroid of the steel bar. Based on LEFM,[32,33] the stress intensity factors of the beam subjected to M and F_s are, respectively,

$$K_{IM} = \frac{MY_M(a/b)}{b^{3/2}t} \tag{9.38}$$

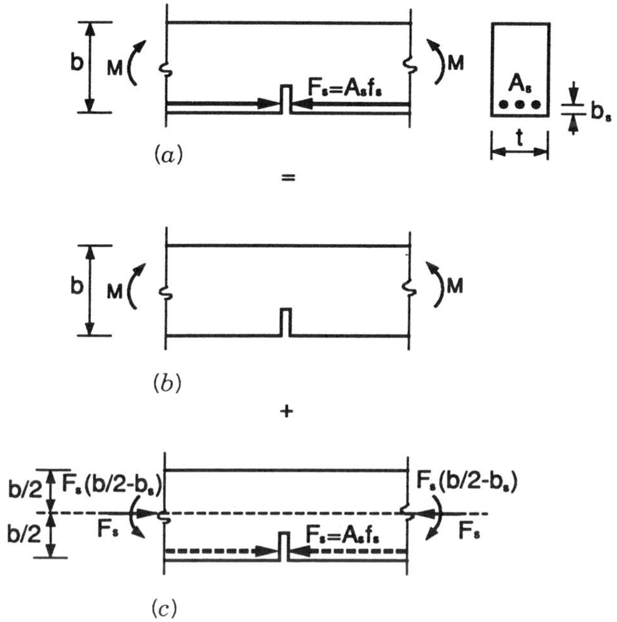

Figure 9-19 Cracked beam element.[29]

and

$$K_{IF} = -\frac{F_s Y_A(a/b)}{b^{1/2} t} - F_s\left(\frac{b}{2} - d_s\right)\frac{Y_M(a/b)}{b^{3/2} t} \tag{9.39}$$

where K_{IM} and K_{IF} are the stress intensity factors due to M and F_s, respectively, b is the depth of the beam, d_s is the distance between the bottom surface of concrete and the centroid of the steel bar, and Y_M and Y_A are geometric factors given by

$$Y_M\left(\frac{a}{b}\right) = 6\left[1.99\left(\frac{a}{b}\right)^{1/2} - 2.47\left(\frac{a}{b}\right)^{3/2} + 12.97\left(\frac{a}{b}\right)^{5/2} - 23.17\left(\frac{a}{b}\right)^{7/2} + 24.80\left(\frac{a}{b}\right)^{9/2}\right] \tag{9.40}$$

and

$$Y_A\left(\frac{a}{b}\right) = 1.99\left(\frac{a}{b}\right)^{1/2} - 0.41\left(\frac{a}{b}\right)^{3/2} + 18.70\left(\frac{a}{b}\right)^{5/2} - 38.48\left(\frac{a}{b}\right)^{7/2} + 53.85\left(\frac{a}{b}\right)^{9/2} \tag{9.41}$$

Note that Eqs. (9.38)–(9.41) are only valid for $a/b < 0.6$.

The total stress intensity factor K_I can be obtained by superposing both K_{IM} and K_{IF}. The criterion of crack propagation is $M = M_f$ when $K_I = K_{Ic}$, where M_f is the fracture moment and K_{Ic} is the critical stress intensity factor of concrete. Substituting Eqs. (9.38) and (9.39) into this criterion leads to

$$M_f = \frac{K_{Ic} b^{3/2} t}{Y_M(a/b)} + \frac{F_s b}{Y_M(a/b)} \left[Y_A\left(\frac{a}{b}\right) + Y_M\left(\frac{a}{b}\right)\left(\frac{1}{2} - \frac{d_s}{b}\right) \right] \tag{9.42}$$

Equation (9.42) describes a relationship between M_f, F_s, and a. If the values of F_s and the corresponding crack length a are known, the value of M_f can be calculated. The value of F_s may be determined by equilibrium consideration of forces acting on the section.

To account for concrete behavior in the compressive zone of the beam, a compressive stress–strain relationship is necessary. Since no explicit stress–strain curve for concrete under compression based on fracture mechanics was available, an empirical parabolic curve proposed by Hognestad was used as below (Fig. 9-20*a*):

$$\sigma = f_c \left[2\frac{\epsilon}{\epsilon_0} - \left(\frac{\epsilon}{\epsilon_0}\right)^2 \right] \tag{9.43}$$

where f_c and ϵ_0 are the maximum compressive stress and the corresponding strain of concrete, respectively. The tensile behavior of concrete is divided into two parts. A linear elastic stress–strain relationship was used up to the peak stress f_r, where f_r is the modulus of rupture of concrete. A linear softening stress–separation curve was used for the postpeak response.

The cross section of the beam with the maximum moment is shown in Fig.

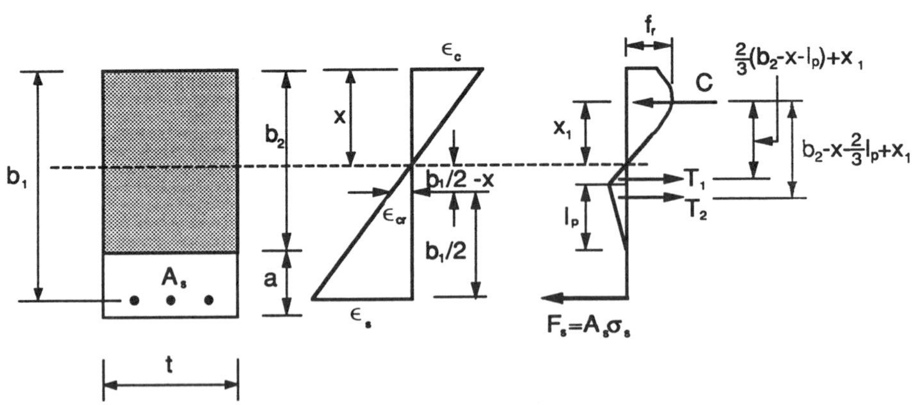

Figure 9-20 Distributions of strain and stress across beam section proposed by Baluch et al.[31]

9-20. By assuming that plane sections remain plane after deformation, a linear distribution of strain over the beam depth was obtained. Combining this strain distribution and the constitutive relations for both concrete and steel, distributions of stresses of the concrete and the steel can be derived.

The tensile stress resultants sustained by the concrete were divided into two parts, T_1 and T_2, as shown in Fig. 9-20 for convenience. Values of T_1 and T_2 are given by

$$T_1 = \tfrac{1}{2} f_r (b_2 - x - l_p) t \qquad T_2 = \tfrac{1}{2} f_r l_p t \tag{9.44}$$

where l_p is the length of the fracture process zone and notations b_2 and x are indicated in Fig. 9-20.

Summing moments about the centroid of the concrete compressive stress block results in the following expression for the fracture moment M_f:

$$\begin{aligned} M_f = {} & T_1 [\tfrac{2}{3}(b_2 - x - l_p) + x_1] \\ & + T_2 (b_2 - x - \tfrac{2}{3} l_p + x_1) + F_s (b_1 - x + x_1) \end{aligned} \tag{9.45}$$

where the notation x_1 is defined in Fig. 9-20. The fracture behavior of a reinforced concrete beam can be predicted by combining Eqs. (9.42) and (9.45).

It is noted $F_s = \sigma_s A_s$ before steel yielding, where σ_s is the tensile stress in the steel bar and A_s is the area of the steel bar. To determine the value of M_f in this case, the computational scheme is as below:

1. Select a crack length a.
2. Assume a value of the steel stress $\sigma_s < f_y$ where f_y is the yield strength of the steel, and calculate the value of $\epsilon_s = \sigma_s / E_s$.
3. Assume the depth of the neutral axis, x.
4. Use $\epsilon_c = \epsilon_s x/(d - x)$ and Eq. (9.43) to numerically calculate the concrete compressive force resultant C and its location x_1.
5. Calculate the total tensile force in concrete, $T = T_1 + T_2 = (b_1 - x - a) f_r / 2$.
6. Check the equilibrium condition of forces, $C = T + \sigma_s A_s$. If satisfied, go to step 7. Otherwise go back to step 3 and change the value of x until the equilibrium condition of forces is satisfied.
7. Determine the value of l_p at which $\epsilon = \epsilon_t$, that is, $l_p = (b_2 - x) - (b_1 - x)\epsilon_t / \epsilon$.
8. Calculate the values of T_1 and T_2 using Eq. (9.44).
9. Calculate the values of M_f based on Eqs. (9.42) and (9.45), respectively.
10. If both values of M_f are not equal, then assume a new steel stress σ_s and go back to step 2 and repeat until convergence is attained. If both

values of M_f are equal, this means that the crack will propagate and the corresponding load P_f can be found from overall equilibrium of the beam. The crack length can then increase and steps 1–10 can be repeated.

For the case of steel yielding ($\sigma_s = f_y$), the computational scheme is as follows:

1. Select a value of the crack length a.
2. Assign the steel force $F_s = F_p = f_y A_s$ and assume a value of the steel strain $\epsilon_s = \epsilon_p > \epsilon_y$.
3. Follow steps 3–10 of the presteel yield algorithm, replacing ϵ_s by ϵ_p and F_s by F_p in the relevant equations.

Baluch et al.[31] have compared predictions of the above model with the experimental data of three-point bend reinforced concrete beams tested by them, as shown in Fig. 9-21. Values of $b = 380$ mm, $t = 102$ mm and $a_0 = 38$ mm were used for these two beams. Beam B1 had values of $A_s = 50.3\ \text{mm}^2$ and $b_s = 25$ mm, whereas beam B2 had values of $A_s = 78.3\ \text{mm}^2$ and $b_s = 10$ mm. The values of $f_c = 23\ \text{N/mm}^2$, $K_{Ic} = 31\ \text{N/mm}^{3/2}$, $E_s = 205$ GPa, and $f_y = 400$ MPa were used for the theoretical prediction. According to ACI 318-89,[28] values of $E = 430 f_c$ and $f_r = 0.62 f_c$ (f_c in newtons per square millimeter) were used. The above model predicts the experimental results quite well.

It is noted that the above analysis is based on a single crack. The effect of interaction of multiple cracks, which has been observed in concrete beams reinforced with appropriate amounts of steel bars, is neglected. Values of K_{Ic}, f_r, and f_c are assumed not to change with different sizes of the beams. Both debonding and slipping energies are not taken into account in the crack propagation criterion.

A similar analysis based on the fictitious crack model has been proposed by Gerstle et al.[34] They assumed concrete as a linear elastic material in compression. A linear softening tensile stress and separation curve was applied on the cracked surface of a beam. A relationship between normalized moment and crack length can be obtained from stress balance at the cracked section. They reported that their analytical results compare favorably with those obtained from a finite element analysis.

9.5 MINIMUM REINFORCEMENT RATIOS OF CONCRETE MEMBERS

To prevent a sudden failure of a reinforced concrete structure subsequent to concrete cracking, a minimum amount of reinforcement is required. Since this

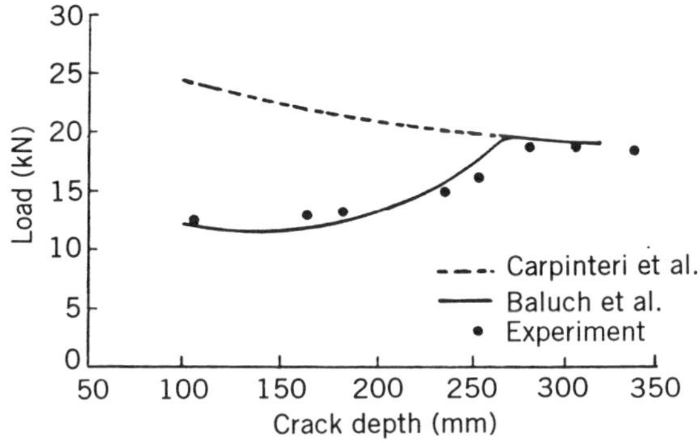

(a)

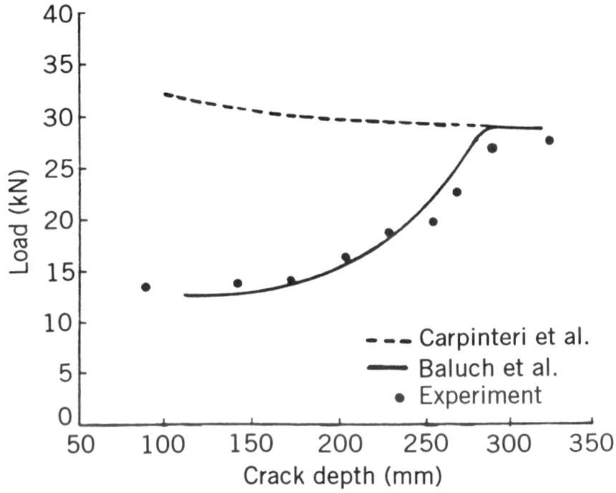

(b)

Figure 9-21 Theoretical and experimental comparisons of relationships between load and crack depth: (*a*) beam B1 and (*b*) beam B2.[31]

sudden failure occurs when a single crack develops, fracture mechanics approaches based on a single crack have been used to evaluate the minimum reinforcement ratio. Minimum reinforcement ratios for both tensile and flexural members will be discussed in this section.

9.5.1 Minimum Reinforcement Ratios for Tensile Members

Based on their analysis of reinforced concrete tensile members that has been presented in detail in Section 9.3, Ouyang and Shah[16] have derived a minimum reinforcement ratio for tensile members. Since the minimum reinforcement is required to prevent sudden failure of a member immediately after development of a single crack, the condition for the minimum reinforcement ratio is that the sum of the tensile stresses in concrete and reinforcing steel bars should be equal to the yield strength of the steel bars immediately after the first crack forms.

This leads to

$$(1 - \rho_{\min})\sigma_{cr} + \rho_{\min}\sigma_{sr} = \rho_{\min} f_y \tag{9.46}$$

where σ_{cr} and σ_{sr} are the tensile stresses in the concrete and the steel bars immediately after the first crack forms, respectively, and $\rho_{\min}$ is the minimum reinforcement ratio. To obtain the value of σ_{cr}, the corresponding strain ϵ_{cr} should be first calculated. For simplicity, both debonding and crack separation are neglected at the loading stage immediately after the first crack forms. The strain values in steel and concrete are equal at this stage due to no debonding. Thus by setting $N = 1$ and $d = 0$ in Eq. (9.28), the following is obtained:

$$\epsilon_{cr} = \epsilon_{sr} = \left[\frac{2e^{\xi}R_{Icf}}{L\xi(1 - \rho_{\min})E}\right]^{1/2} \tag{9.47}$$

Combining Eqs. (9.30), (9.46), and (9.47) results in

$$\frac{\sqrt{1 - \rho_{\min}}}{\rho_{\min}}\left(\frac{2ER_{Icf}}{L\xi e^{\xi}}\right)^{1/2} + \frac{E_s}{\sqrt{1 - \rho_{\min}}}\left(\frac{2e^{\xi}R_{Icf}}{L\xi E}\right)^{1/2} = f_y \tag{9.48}$$

Since the value of $1 - \rho_{\min}$ is approximately equal to unity, Eq. (9.48) can be further simplified as

$$\rho_{\min} = \frac{Ee^{-\xi}}{f_y\sqrt{L\xi E/2e^{\xi}R_{Icf}} - E_s} \tag{9.49}$$

It is seen that the predicted $\rho_{\min}$ depends on the moduli of elasticity of the concrete and steel, the yield strength of steel, and the values of L, R_{Icf}, and ξ. Since the values of R_{Icf} and ξ relate to sizes of members, the value of $\rho_{\min}$ is also member size dependent.

The values of $\rho_{\min}$ for concrete members with different sizes of square cross sections are shown in Fig. 9-22, where $L = 1000$ mm and the values of $K_{Ic}^{s} = 30.7\ \text{N/mm}^{1/2}$, $\text{CTOD}_c = 0.016$ mm, and $E = 33$ GPa for concrete were used for the theoretical prediction. The value of a_0 was assumed to be 5 mm. A grade 40 steel with $f_y = 276$ MPa and $E_s = 210$ GPa was used. It is seen that the predicted minimum reinforcement ratio decreases with increasing member size. ACI 318-89 Code[28] does not specify the minimum reinforcement ratio for tensile members. However, ACI 318-89 requires $\rho_{\min} = 200/f_y$ (f_y in pounds per square inch) for flexural members. This requirement is also plotted in Fig. 9-22. The predicted values approach the ACI 318-89 requirement when the member size increases. As expected, the predicted minimum reinforcement ratio for tension is larger than that for flexure as given by ACI 318-89.

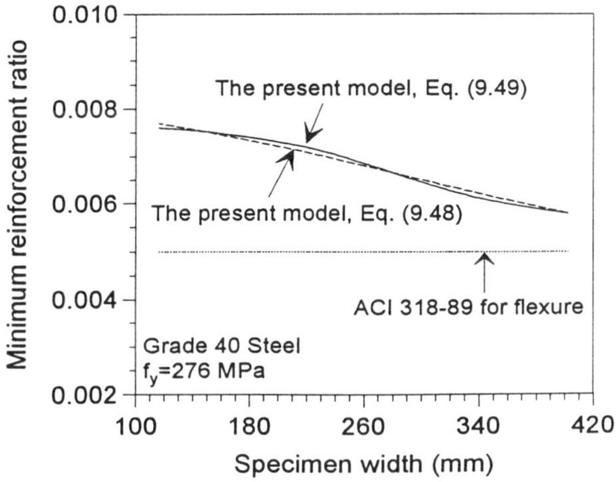

Figure 9-22 Minimum reinforcement ratios for different-size members under tension.[16]

9.5.2 Minimum Reinforcement Ratios for Flexural Members

Bosco et al.[30] have proposed a minimum reinforcement ratio for concrete beams. Physically, the minimum reinforcement ratio can be obtained from the condition that steel immediately yields when a beam cracks. Using this condition ($F_s = f_y A_s$), Eq. (9.42) can be rearranged as

$$\frac{M_f}{K_{Ic} b^{3/2} t} = \frac{1}{Y_M(a/b)} + \frac{N_P}{Y_M(a/b)} \left[Y_A\left(\frac{a}{b}\right) + Y_M\left(\frac{a}{b}\right)\left(\frac{1}{2} - \frac{d_s}{b}\right)\right] \tag{9.50}$$

where the brittleness number N_P is defined as

$$N_P = \frac{f_y b^{1/2}}{K_{Ic}} \frac{A_s}{bt} = \frac{f_y b^{1/2}}{K_{Ic}} \rho \tag{9.51}$$

and $\rho = A_s/bt$. Based on their experimental observations, Bosco and Carpinteri[35] proposed a brittleness number corresponding to a transition between ductile ($\rho > \rho_{\min}$) and brittle ($\rho < \rho_{\min}$) failures that may be expressed as a function of compressive strength of concrete,

$$(N_P)_{\rho = \rho_{\min}} = 0.1 + 0.0023 f_c \tag{9.52}$$

This transitional brittleness number is plotted against concrete strength in Fig. 9-23*a*. As a result, the minimum reinforcement ratio can be obtained by combining Eqs. (9.51) and (9.52),

$$\rho_{\min} = (0.1 + 0.0023 f_c) \frac{K_{Ic}}{f_y b^{1/2}} \tag{9.53}$$

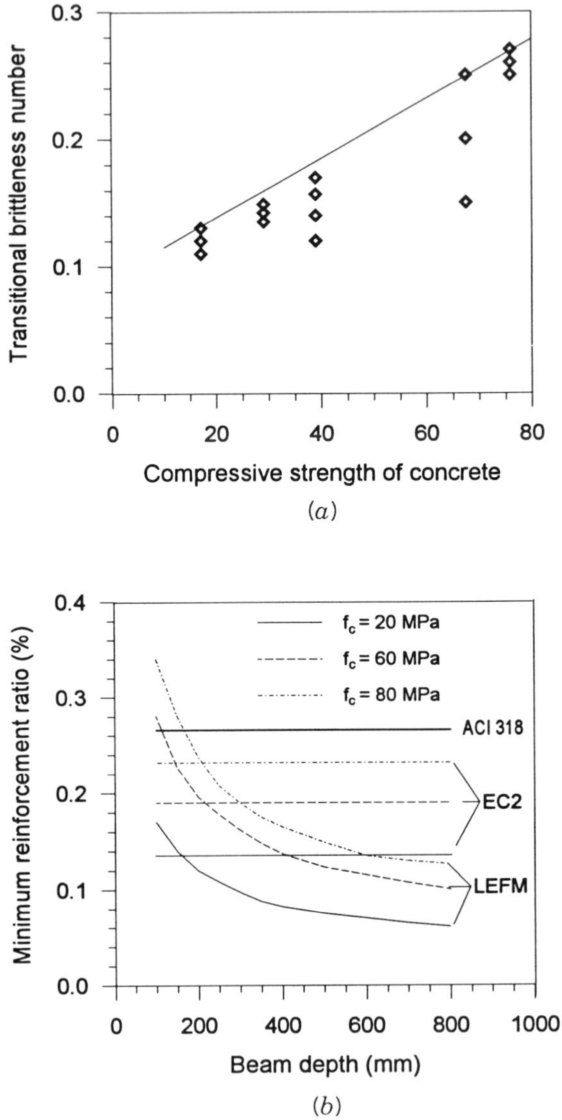

Figure 9-23 Transitional brittleness number and minimum reinforcement ratios proposed by Bosco et al: (*a*) transitional brittleness number and (*b*) minimum reinforcement ratio.[35]

The predicted values of ρ_{min} for different types of concrete are compared to the values given by ACI 318-89[28] and EC2[36] in Fig. 9-23*b*. It is seen that Eq. (9.53) predicts that the minimum reinforcement ratio decreases with increasing size of the beam in a manner similar to that of the tensile members discussed earlier.

Hawkins and Hjorteset[37] have analyzed a series of reinforced concrete beams using the finite element method in conjunction with the fictitious crack approach. According to the results obtained from their numerical model that were validated with experimental data from other studies, they proposed a minimum reinforcement ratio for reinforced concrete beams as

$$\rho_{min} = 0.175 \frac{f_f}{f_y} \frac{b}{t} \tag{9.54}$$

Note that the value of f_f is the fracture strength of the beam, which is a function of material fracture parameters and size of the beam. Using the discrete crack approach in conjunction to the finite element analysis, Hawkins and Hjorteset obtained values of f_f as shown in Fig. 9-24.

Gerstle et al.[34] interpreted a minimum reinforcement ratio as the quantity that guarantees a monotonic increase of moment–crack length curves. From the numerical study using their model based on the fictitious crack model, they proposed the minimum reinforcement ratio as

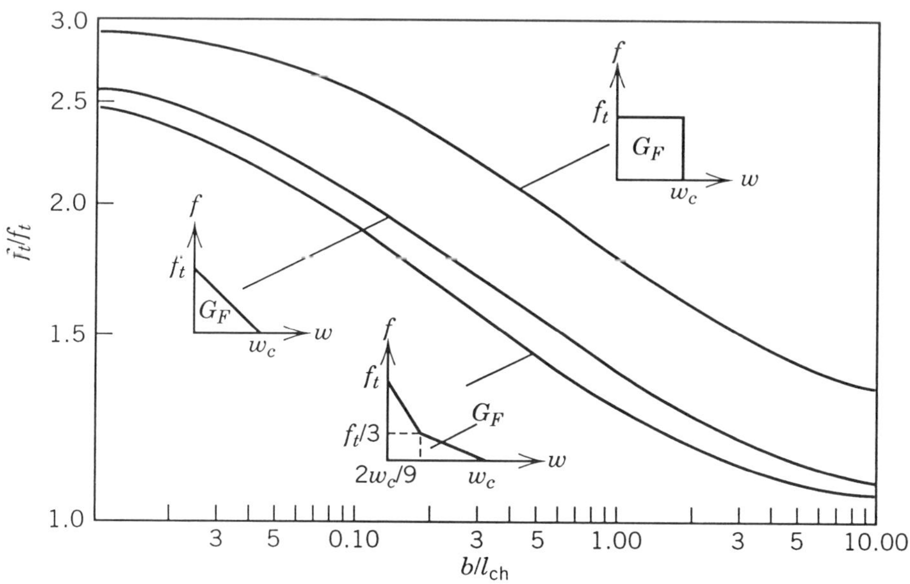

Figure 9-24 Theoretical variation in flexural strength f_t.[37]

$$\rho_{\min} = \frac{E}{E_s}\left[\left(0.0081 + 0.0148\,\frac{f_t b}{E w_c}\right)^{1/2} - 0.0900\right]^{1/2} \tag{9.55}$$

Note that the value of $\rho_{\min}$ given by Eq. (9.55) is independent of the yield strength of the steel.

Equations (9.53), (9.54), and (9.55) are based on the fictitious crack approach, and values of $\rho_{\min}$ predicted by these three equations are compared in Fig. 9-25. The minimum reinforcement ratio given by ACI 318-89 is also shown in Fig. 9-25. Values of f_t = 4 MPa, w_c = 0.08 mm, G_F = 0.16 N/mm, E = 25 GPa, and E_s = 210 GPa were used for the calculation. A grade 40 steel with f_y = 276 MPa was used. Relationships of $f_t = 0.556(f_c)^{1/2}$ (f_c in megapascals) and $G_F = (K_{Ic})^2/E$ were used for Eq. (9.53). A ratio of $b/t = 2$ was assumed for all sizes of the beams. The values of f_f in Eq. (9.54) were determined from Fig. 9-24 based on a linear softening stress–separation curve. Equations (9.53) and (9.54) predict that the values of $\rho_{\min}$ decrease with increasing beam size, whereas Eq. (9.55) predicts an opposite trend. Note that the values of $\rho_{\min}$ predicted by Eq. (9.55) are much higher than those predicted by Eqs. (9.53) and (9.54). Thus it may be that the monotonic increase of moment–crack length curves used by Gerstle et al. is an inappropriate condition for obtaining the minimum reinforcement ratio. Although the principle of fracture mechanics was applied, unlike Eq. (9.49), some statistical relationships based on experimental data or numerical analysis were also used to obtaining Eqs. (9.53)–(9.55). Also, if one thinks about it, if the nominal tensile strength decreases with increasing beam size, it follows that the minimum steel ratio must decrease as well.

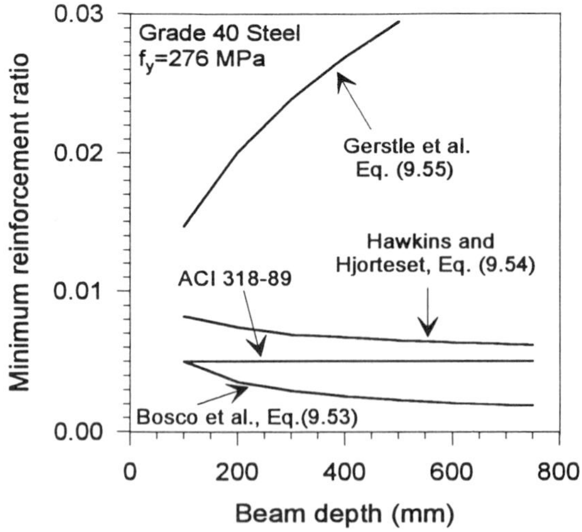

Figure 9-25 Minimum reinforcement ratios predicted by different models.

9.6 PULLOUT FAILURE OF BAR

In reinforced concrete structures, concrete and steel work together through interfacial bond. The stress in the steel is transferred to the concrete through this bond mechanism. Some bond stress will develop at the interface because of differences of stresses in concrete and steel. To completely transfer the steel stress to the concrete, a certain length of bond, which is referred to as a development length, is necessary. This development length is usually evaluated using a test of a steel bar pullout from the concrete block. Since some cracks are usually developed at the interface when the steel bar is pulled out, fracture mechanics may be a useful tool to describe this pullout slip response. It will be discussed in this section how to evaluate the pullout slip response based on fracture mechanics.

The failure mechanism for the pullout of steel is shown in Fig. 9-26. It has been concluded that two types of cracks, longitudinal interface cracks and radial cracks as shown in Fig. 9-26, are primarily responsible for the pullout slip relationship. A discrete crack approach in conjunction with a finite element analysis has been used to describe the pullout and slip response by Ingraffea et al.[38] They used the fictitious crack model to study the pullout of a deformed bar based on an axisymmetric finite element analysis. Since directions of propagation of cracks were usually not normal to the pullout force, mixed-mode cracks were used in the analysis. This means that both normal and shear stresses acted on crack surfaces. Tensile softening at each bond crack was modeled by the fictitious crack approach. The finite element analysis based on the discrete crack approach was performed for this problem. Unlike most conventional methods, it is possible to account for the influence of the radial cracks on the pullout behavior using the fracture mechanics approach in conjunction with the finite element method. The initial undeformed finite element mesh is shown in Fig. 9-27*a*, whereas an undeformed mesh corresponding to the load level of $P = 5200$ MPa (P is distributed on the cross section of the steel bar) is given in Fig. 9-27*b*. They concluded, based on the analytical results that the radial cracks, which generally remain confined to locations where the steel exits the concrete, significantly contribute to the slip when a deformed bar is pulled out from the concrete.

A recent study by Brown et al.[39] has used the fictitious fracture model in

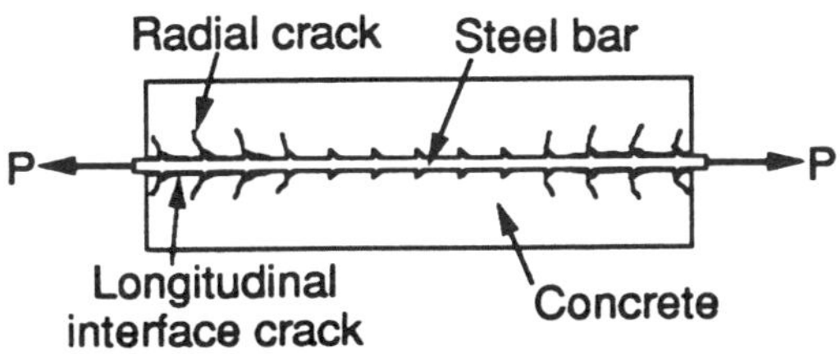

Figure 9-26 Failure mechanism for pullout of steel bar from concrete.

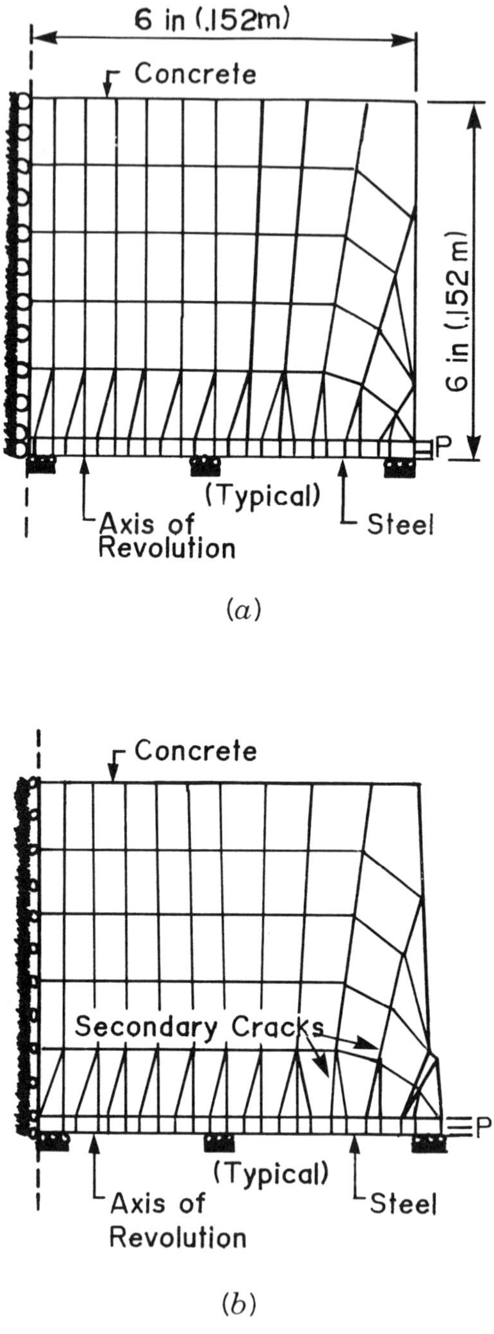

Figure 9-27 Deformed mesh from an axisymmetric FE analysis for bar pullout specimen: (*a*) undeformed shape and (*b*) deformed mesh at stage of applied stress = 520 MPa (75 ksi) with amplification factor = 44.[38]

conjunction with finite element analysis to study failure of steel–concrete bond through beam-end specimens as shown in Fig. 9-28*a*. Three-dimensional, eight-node, linear, isoparametric elements, as shown in Fig. 9-28*b*, were used to model the concrete and steel. The concrete fracture behavior was simulated by a linear softening stress–separation curve. The steel–concrete interface was

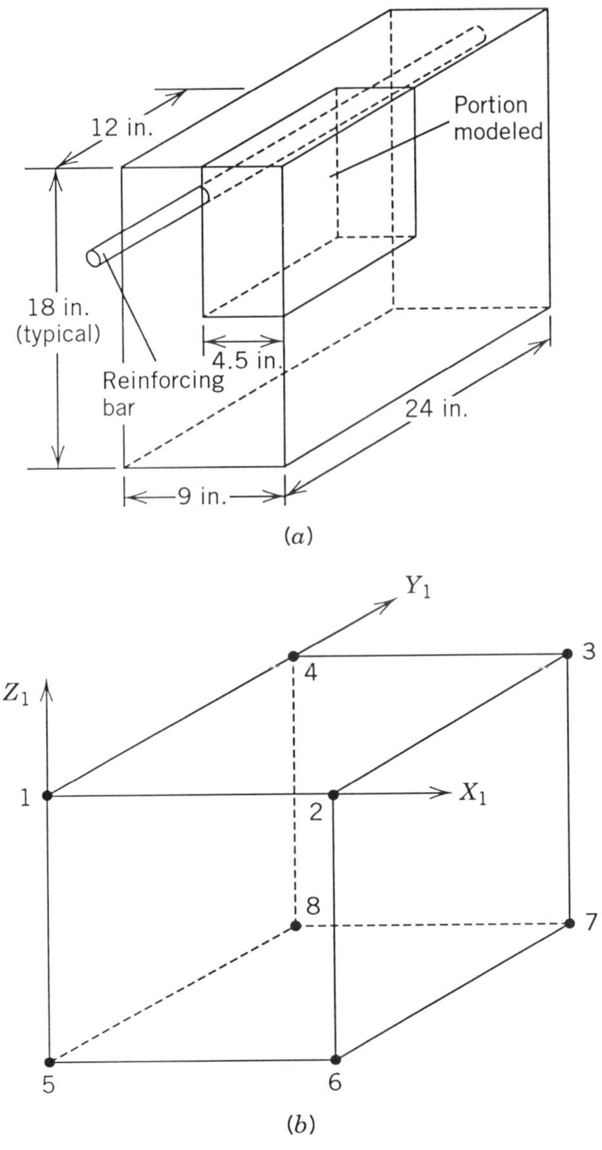

Figure 9-28 Finite element analysis of portion of a beam-end specimen: (*a*) portion of beam-end specimen.[39] (*b*) three-dimensional, eight-node isoparameter element.

modeled using special link elements that follow a Mohr–Coulomb failure law. In this study, the steel–concrete bond behavior as a function of bar size, interface conditions, and specimen configuration was studied. In Figs. 9-29*a*,*b*, the pullout forces obtained are plotted against the thickness of concrete cover and the embedment length of the steel bar, respectively. These obtained results, based on fracture mechanics, explain the experimentally observed fact that pullout force increases with increasing cover thickness and embedment length.

9.7 SHEAR IN REINFORCED CONCRETE BEAMS

When a reinforced concrete beam is subjected to transverse loading, bending moment and shear force will develop in the beam. Both moment and shear combine to create a biaxial state of stress. As shown in Fig. 9-30, opening cracks may initiate from the bottom at the midsection of the beam where the shear stress is zero. As the load increases, more and more vertical cracks initiate, and these vertical cracks propagate. Some of these vertical cracks may turn into diagonal cracks due to influence of the shear stress (i.e., diagonal tension). As a result, the beam may fail due to propagation of these diagonal tension shear cracks. It has been recognized that the diagonal shear resistance of longitudinally reinforced concrete beams consists of fracture of the concrete, aggregate interlock between cracked surfaces, and dowel action of longitudinal reinforced and steel bars between the cracked surfaces. Different fracture mechanics models have been proposed to describe this shear failure of longitudinally reinforced concrete beams (without shear reinforcement). Since so many mechanisms are involved in the diagonal shear failure of beams, all these models need to make some simplifications. These models have been summarized by Karihaloo.[40] Especially, the Bazant and Kim model[41] has been extended to beams with stirrups. All these models will be briefly presented in this section.

9.7.1 Bazant et al. Model

Bazant and Kim[41] have combined the existing code formulas for diagonal failure of longitudinally reinforced concrete beams with the size effect model,[42] and established the several required empirical constants statistically so as to achieve the best fit with the large body of available test data. The influence of the steel ratio and the shear span was considered together with size effect.

Consider the end segment of a beam shown in Fig. 9-31*a*, in which a constant shear force V acts throughout the shear span $s_v = M/V$, where V is the shear force and M is bending moment at the same cross section. The bending moment at any distance x from the support may be expressed as $M = Tjb$ where b is the depth of the beam, $T = T(x)$ is the tensile force resultant acting at the

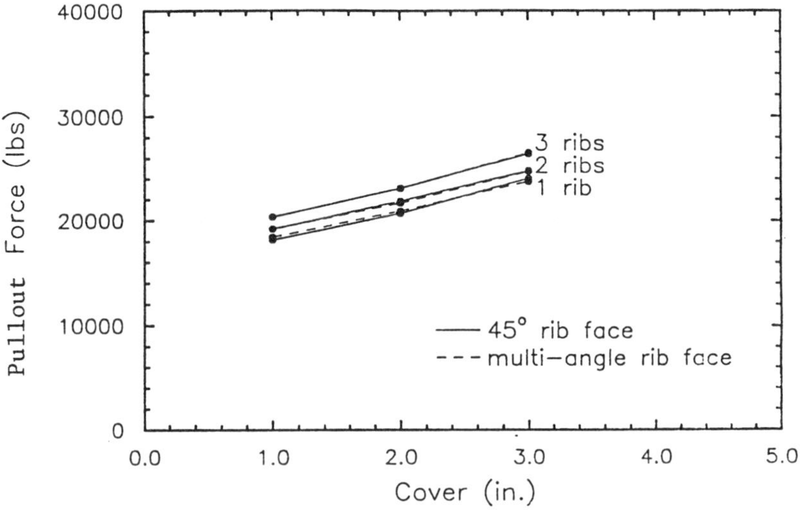

(a)

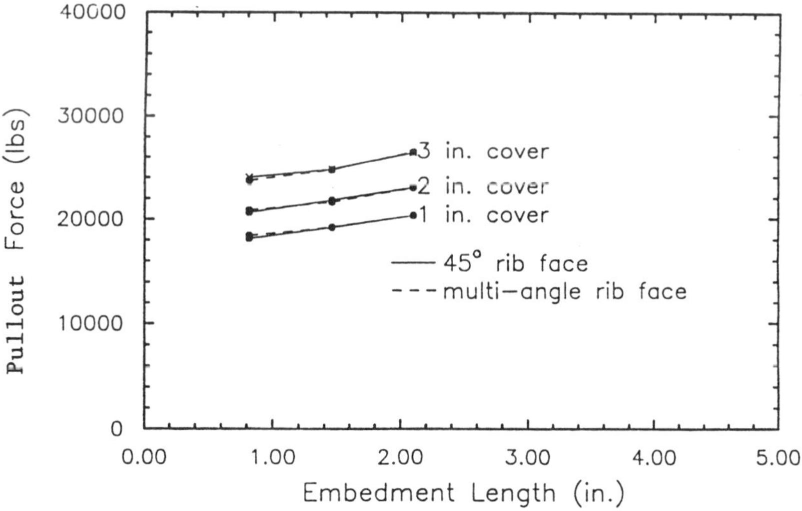

(b)

Figure 9-29 Predicted relationship among pullout force, thickness of concrete cover, and embedment length of steel bar: (*a*) pullout load versus thickness of concrete cover and (*b*) pullout versus embedment length of steel bar.[39]

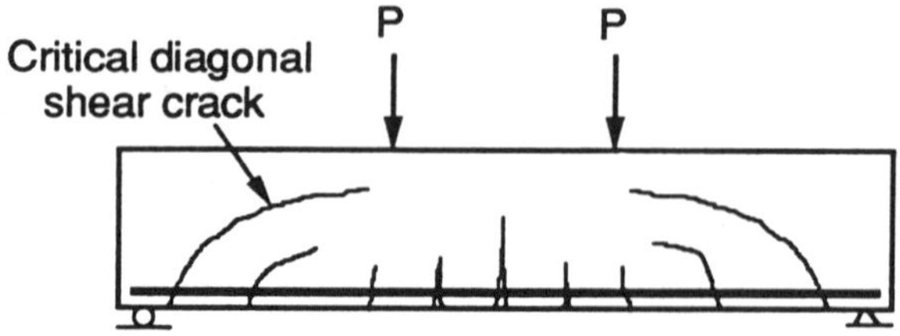

Figure 9-30 Diagonal shear failure of longitudinally reinforced concrete beam.

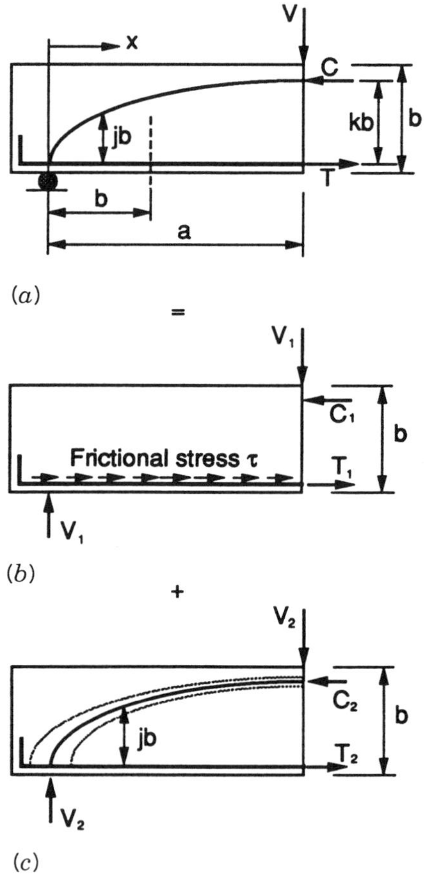

Figure 9-31 (*a*) Decomposition of shear loading capacity into (*b*) composite action and (*c*) arch action.[41]

centroid of longitudinal reinforcement, and $j = j(x)$ is a variable coefficient as shown in Fig. 9-31*a*. The shear force may be expressed as $V = dM/dx$ and may be decomposed into two components: a component V_1 resulting from the composite action of concrete and steel through the mechanism of bond slip and frictional pullout as shown in Fig. 9-31*b* and a component V_2 due to the so-called arch action shown in Fig. 9-31*c*. This leads to

$$V = V_1 + V_2 \qquad V_1 = \frac{dT}{dx} jb \qquad V_2 = \frac{dj}{dx} Tb \tag{9.56}$$

By empirically selecting values of T and j, Bazant and Kim proposed the following:

$$V_1 = k_1 \rho^{0.5 - c_1} (f'_c)^{q_1} bt \qquad V_2 = c_2 \frac{\rho^{1 - c_1}}{(s_v/b)^r} bt \tag{9.57}$$

where k_1, q_1, c_1, and c_2 are empirical constants. As a result, the total shear stress $v_s = V/bt$ is expressed as

$$v_s = k_1 \rho^{q_2} \left[(f'_c)^{q_1} + k_2 \frac{\sqrt{\rho}}{(s_v/b)^r} \right] \tag{9.58}$$

where $k_2 = c_2/k_1$ and $q_2 = 0.5 - c_1$. Equation (9.58) is similar to that used in the ACI 318-89, but with more empirical constants. To account for the size effect on concrete strength, one may replace Bf_t in Eq. (9.6) by v_s given in Eq. (9.58). This results in

$$v_s = k_1 \rho^{q_2} \left[(f'_c)^{q_1} + k_2 \frac{\sqrt{\rho}}{(s_v/b)^r} \right] \left(1 + \frac{D}{\lambda_0 d_a} \right) \tag{9.59}$$

where $D_0 = \lambda_0 d_a$, d_a is the maximum size of aggregate, and λ_0 is an empirical constant.

From a statistical analysis of existing test data, the following formula for the mean ultimate nominal shear strength resulted from Eq. (9.59):

$$v_s = \frac{10\rho^{1/3}}{\sqrt{1 + b/(25d_0)}} \left[\sqrt{f'_c} + 3000 \sqrt{\rho/(s_v/b)^5} \right] \tag{9.60}$$

where f'_c is in pounds per square inch. By considering a safety margin in design, Bazant and Kim[41] have recommended the following formula to be used in design:

$$v_u = \frac{8\rho^{1/3}}{\sqrt{1 + b/(25d_a)}} \left[\sqrt{f'_c} + 3000 \sqrt{\rho/(s_v/b)^5} \right] \tag{9.61}$$

where v_u is the ultimate shear stress.

Bazant and Kazemi[43] have tested a series of geometrically similar concrete beams that are longitudinally reinforced (without stirrups) and fail due to diagonal shear cracks. The test results indicate a significant size effect on nominal failure strength. Based on the test results, they proposed the following expressions by combining the size effect model and current ACI 318-89 formula for shear design:

$$v_u = \frac{Bv_c}{\sqrt{1 - D/D_0}} \tag{9.62}$$

where v_c is the nominal strength of concrete in diagonal shear given by ACI 318-89

$$v_c = 1.9\sqrt{f_c'} + 2500\rho \frac{V_u b}{M_u} \tag{9.63}$$

where f_c' and v_c are in pounds per square inch (1 psi = 6895 Pa) and M_u is the moment occurring simultaneously with shear force V_u at the cross section considered. It is noted that the value of v_c should not exceed $3.5(f_c')^{1/2}$.

Bazant and Kazemi[43] also found that prevention of bond slip of steel bars by providing anchorage at the ends increases the brittleness of the beams.

Bazant and Sun[44] extended the Bazant and Kim[41] model to reinforced concrete beams with stirrups. By analyzing over 400 test series from the literature, Bazant and Sun[44] proposed the following formula for diagonal failure of the beams including effects of stirrups and aggregate size:

$$v_c = k_1 \rho^{1/3}\left(\sqrt{f_c'} + 3000\sqrt{\frac{\rho}{(a/b)^5}}\right)\left(1 + \sqrt{\frac{c_0}{d_a}}\right)\left(1 + \frac{b}{\lambda_0 d_a}\right)^{-1/2} \tag{9.64}$$

where k_1 and c_0 are empirical constants.

It is noted that the Bazant and Kim[41] model is basically empirical. However, the size effect on nominal strength of concrete structures was taken into account.

9.7.2 Jenq and Shah Model

Jenq and Shah[45] have proposed a model based on their two-parameter fracture model. In their model, a diagonal failure mechanism of a longitudinally reinforced beam is assumed in Fig. 9-32, where a single diagonal linear crack is assumed to be responsible for shear failure of the beam and the value $\theta \approx 45°$ was used. Both the multiple tensile cracks in concrete and the dowel action of longitudinal reinforcement were neglected in this model. It was assumed that any crack growth in the concrete takes place when the stress intensity factor at the crack tip attains the mode I fracture toughness K_{Ic}^s. In other words, the effect of shear and other in-plane stresses on the fracture criterion is ignored. Note that it is difficult to evaluate the steel force along the beam due to the absence of precise information on longitudinal cracking, internal cracking, and

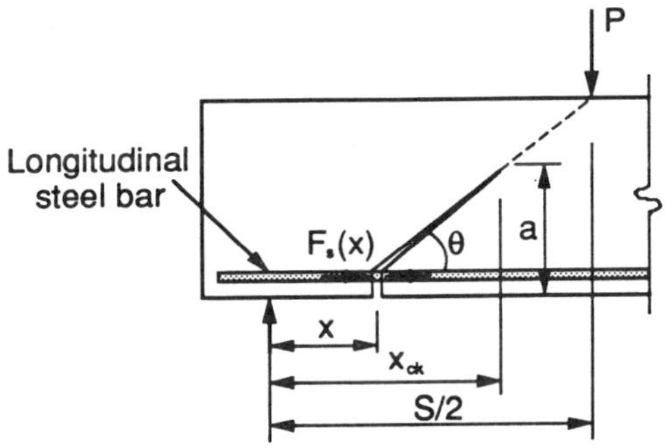

Figure 9-32 Theoretical failure mode for modeling diagonal shear failure of reinforced concrete beam.[45]

dowel action. It is assumed that the steel force along the beam, $F_s(x)$, is simply represented by a power law

$$F_s(x) = F_{s,\max}\left(\frac{2x}{S}\right)^{N_1} \leq f_y A_s \tag{9.65}$$

where $F_{s,\max}$ is the maximum steel force at the midspan ($x = S/2$), and the value of N_1 is a constant whose value will be adopted after numerical experimentation based on a bilinear approximation to the bond slip and frictional pullout behavior of the reinforcing bars. The value of $F_{s,\max}$ is regarded as a function of the average ultimate bond strength, which is assumed to be proportional to $1/b$. Based on the test data reported by Ferguson and Thompson,[46,47] Jenq and Shah[45] proposed that

$$F_{s,\max} = 2.509\, Sf_t\left(\frac{\rho}{b}\right)^{1/2} \tag{9.66}$$

where the tensile strength $f_t = 1.4705\ (K_{Ic}^s)^2/(E\ \text{CTOD}_c)$ based on the two-parameter fracture model.[8] Note that in Eq. (9.66) S is the clear span in millimeters and f_t is in megapascals. Variations of $F_s(x)$ along the shear span for different values of H are shown in Fig. 9-33. It was recommended to use $N_1 = 2.5$.[45]

The total applied force P may be divided into the force resisted by the concrete, P_m, and the force resisted by the longitudinal steel, P_s. The value of P_m may be calculated on the assumption that the crack has propagated to the current length a at a section x such that at its tip the stress intensity factor is equal to K_{Ic}^s. Based on LEFM, this leads to

$$K_{Ic}^s = K_I = \sigma_N \sqrt{\pi a}\, g_1\left(\frac{a}{b}\right) \tag{9.67}$$

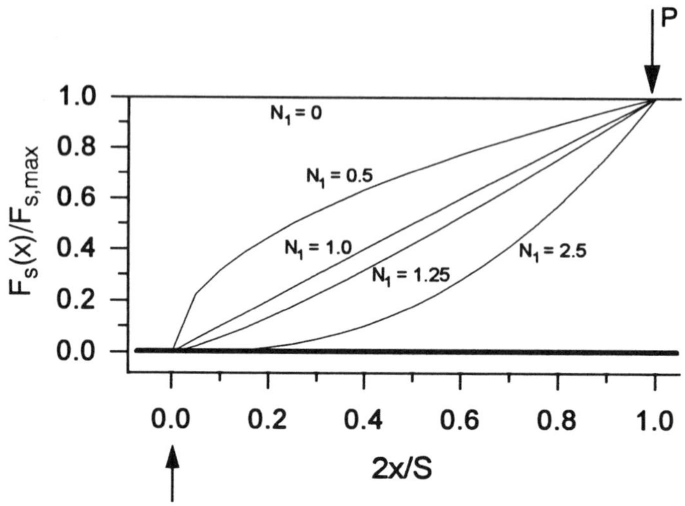

Figure 9-33 Power law variation of steel stress along shear span.[45]

where σ_N is the nominal stress at the section x and $g_1(a/b)$ is the geometric function. For three-point bending beams with $S/b = 4$, $\sigma_N = (3P_m x)/(b^2 t)$, and $g_1(a/b)$ is given by

$$g_1\left(\frac{a}{b}\right) = \frac{1.99 - a/b(1 - a/b)[2.15 - 3.93(a/b) + 2.70(a/b)^2]}{\sqrt{\pi}\,(1 + a/b)(1 - a/b)} \tag{9.68}$$

The value of P_m can be solved from Eq. (9.67):

$$P_m = \frac{K_{Ic}^s b^2 t}{3x\sqrt{\pi a}\, g_1(a/b)} \tag{9.69}$$

The value of P_s can be evaluated by the conventional strength theory, in which the dowel action, or direct shear resistance of the steel bar, is neglected. By assuming the neutral axis is located at the bottom edge of the uncracked ligament as shown in Fig. 9-32, one can obtain the following expression for P_s from equilibrium condition of forces:

$$P_s = \frac{2F_s(x)}{X_{ck}}\left(\frac{2b}{3} + \frac{a}{3} - b_s\right) \tag{9.70}$$

where b_s is the distance between the concrete surface and the centroid of the steel bars, as shown in Fig. 9-34.

As a result, the total shear resistance of the beam, v_u, is given as

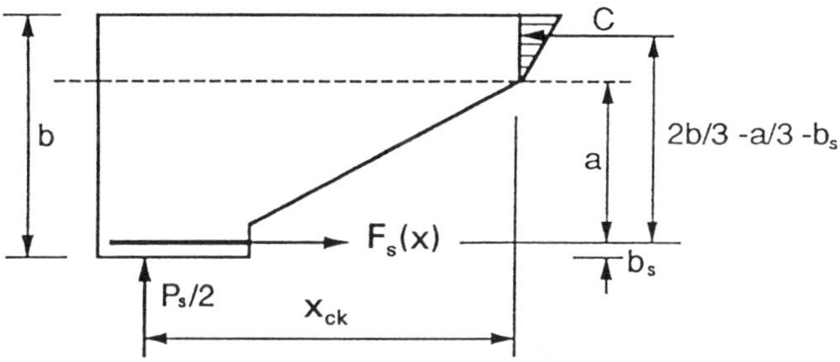

Figure 9-34 Balance condition for calculating value of P_s.

$$v_u = \frac{P}{bt} = \frac{P_m + P_s}{bt} \tag{9.71}$$

It is noted that although some features associated with the diagonal shear failure of beams can be predicted by Eq. (9.71), this model is quite simple.

9.7.3 Gustafsson and Hillerborg Model

Gustafsson and Hillerborg[48] have proposed a finite element analysis for diagonal failure of reinforced concrete beams based on the fictitious crack approach. Fracture of concrete is described by the fracture toughness G_F and the characteristic length parameter $l_{\text{ch}} = EG_F/f_t^2$.

The basic methodology of the Gustafsson and Hillerborg model is summarized in Fig. 9-35. Concrete and steel are assumed to be linear elastic materials. A single diagonal crack is assumed, and the length of the crack increases with applied force. Based on the value of the crack opening displacement w,

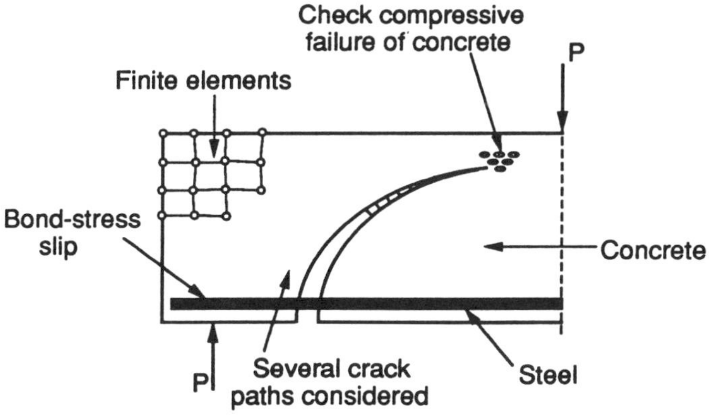

Figure 9-35 Summary of various stages in shear failure analysis according to Gustafsson and Hillerborg.[48]

the crack is divided into two parts. Two surfaces of the crack are considered to be completely separated when $w > w_c$, where w_c is the critical crack separation displacement, whereas the crack surfaces are regarded to be in contact when $w < w_c$. According to the fictitious crack model,[49] a linear softening stress–strain separation curve is applied to the cracked surfaces where $w < w_c$. This softening stress–separation curve accounts for stress transferred in the fracture process zone. A linear elastic bond slip curve is assumed. Aggregate interlock between fracture surfaces and shear force carried by the dowel action of reinforcing steel bars are neglected. Several potential crack paths shown in Fig. 9-36 should be chosen for numerical analysis, and the ultimate shear stress corresponds to the path with the minimum shear resistance.

It has been shown that the shear strength of longitudinally reinforced concrete beams is approximately proportional to $f_t(w/l_{ch})^{-1/4}$. The following result can thus be obtained due to $l_{ch} = EG_F/f_t^2$):

$$v_u = k_3\left(\frac{EG_F f_t^2}{b}\right)^{1/4} \tag{9.72}$$

where k_3 is a proportional constant. The influence of beam depth on the ultimate shear strength qualitatively as predicted by Eq. (9.72) is claimed to be consistent with the tested results reported by many investigations.[50]

9.7.4 Discussions

Based on his test results, Karihaloo[40] has systematically compared the three above models for diagonal shear failure of longitudinally reinforced concrete beams. He concluded that the Bazant and Kim[41] model is the simplest model to be used in design, because they only scale the existing code formulas to

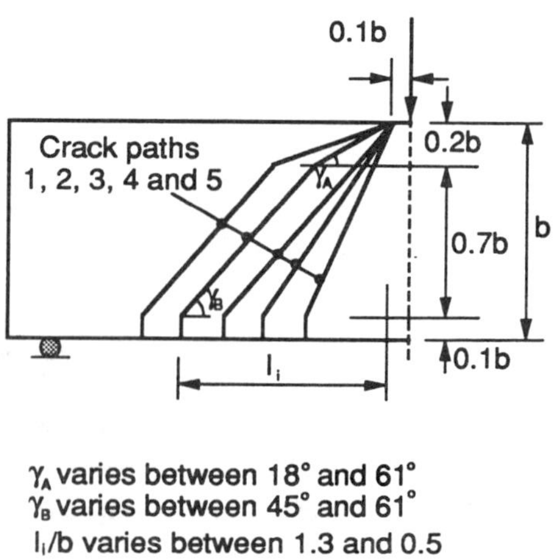

Figure 9-36 Various crack propagation paths chosen for shear strength analysis.[48]

account for the size effect appropriate for the crack growth in concrete. Their model, moreover, assumes that the steel stress is constant over the entire shear span. This would be a somewhat doubtful assumption due to ignoring any bond slip between the steel and concrete and due to considering only the frictional part of the overall pullout behavior of the reinforcing bars. On the other hand, although the Jenq and Shah[45] model can predict diagonal shear failure of concrete beams, their model seems to underestimate the diagonal shear strength of concrete beams because the influence of aggregate interlock and steel dowel action are neglected. The Jenq and Shah model predicts that v_u scales to $b^{-1/2}$, which agrees with the prediction of the Bazant and Kim[41] model and also with the implied size effect in the concept of the brittleness number N_p previously discussed [Eq. (9.51)]. The Gustafsson and Hillerborg[48] model is similar in principle to the Jenq and Shah[45] model for diagonal shear failure of concrete beams, but it can only be implemented numerically in a finite element program. The Gustafsson and Hillerborg[48] model predicts the ultimate shear strength is proportional to $b^{-1/4}$, whereas all other available models would suggest this variation to be $b^{-1/2}$. Since the scatter of the existing test data is so large,[50] both scaling laws could be claimed to fit the data reasonably well, leaving the question of the scaling law still unsolved.

After reevaluating the test results reported by Ferguson and Thompson,[46,47] So and Karihaloo[51] further proposed that the maximum steel force given by Eq. (9.66) for the Jenq and Shah model[45] may be replaced by

$$F_{s,\max} = k_4(4\pi r_s)L_s k_5 \tau_u \tag{9.73}$$

where r_s (in millimeters) is the radius of the reinforcing bar, L_s (in millimeters) is the embedment length of the bar, and τ_u (in megapascals) is the ultimate bond stress, given by

$$\tau_u = 0.4684\sqrt{f'_c}\left(\frac{\rho L_s}{2r_s}\right)^{N_2} + 0.0271(t_c - 3.0r_s) \tag{9.74}$$

where t_c (in millimeters) is the clear cover of concrete and $N_2 = -0.8205(2r_s)^{-0.2933}$ is a constant. Equation (9.73) also includes two coefficients, k_4 and k_5. The coefficient k_4 is the reduction factor on the value of $F_{s,\max}$ due to two reinforcing bars and is given by

$$k_4 = \frac{93 + 135(t/4r_s - 1) - 7(t/4r_s - 1)^2}{93 + 135(t/2r_s - 1) - 7(t/2r_s - 1)^2} \tag{9.75}$$

where t is the thickness of the beam. The coefficient k_5 is the bond stress distribution factor. For $S/(d - d_s) \leq 17.5$ the value of k_5 is given by

$$k_5 = 0.3889 + 0.0592\left(\frac{S}{b - b_s}\right) - 0.0017\left(\frac{S}{b - b_s}\right)^2 \tag{9.76}$$

For $S/(d - d_s) > 17.5$, $k_5 = 0.90$. So and Karihaloo[51] indicated that the influence of dowel action and aggregate interlock is included by the above empirical expressions. Equations (9.73)–(9.76) cover nearly all factors that influence the shear capacity of beams; the concrete strength, bar diameter and embedment length, clear concrete cover, steel ratio, beam geometry, and number of reinforcing bars. The deficiencies are that these equations are limited to beams containing no more than two longitudinal bars and no stirrups. So and Karihaloo[51] also proposed an alternative approach to estimate the maximum steel force based on behavior of pullout of a steel bar from the concrete.

So and Karihaloo[51] reported that if Eq. (9.73) instead of Eq. (9.66) is used, the Jenq and Shah[45] model can reasonably predict the shear capacity of reinforced concrete beams.

9.8 TORSION IN REINFORCED CONCRETE BEAMS

The application of fracture mechanics to torsion failure of plain concrete beams has been discussed in Chapter 8. Only reinforced concrete beams are discussed here. Based on their own experimental results of a series of geometrically similar beams (1 : 2 : 4 of $t:b:S$) as well as those of a few other investigators, Bazant and Şener[52] found that the torsional failure of longitudinally reinforced concrete beams exhibits a strong size effect on the nominal failure stress, which is close to the LEFM asymptote slope of $-1/2$, as shown in Fig. 9-37. However, the scatter in the limited experimental data for the reinforced concrete beams was quite large, and more extensive tests are needed. This scatter may be attributed to the fact that bond failure must accompany the torsional crack, and bond failure is a phenomenon of high scatter. The size effect in the torsional failure shown in Fig. 9-37 was also confirmed by nonlocal finite element analysis by Ozbolt and Eligehausen.[53]

9.9 COMBINED LOADINGS AND INTERACTION DIAGRAMS

A structural member is often subjected to combined loadings. Since the mechanical response of concrete structures depends on the loading path, it is very important to understand the influence of interaction of different loads on the failure behavior of the structures. Fracture mechanics has been applied to study interaction failure diagrams for different structures. However, since failure mechanisms under combined loading are so complicated, such an application is very preliminary.

Hawkins[54] has applied fracture mechanics to diagonal tension cracking in

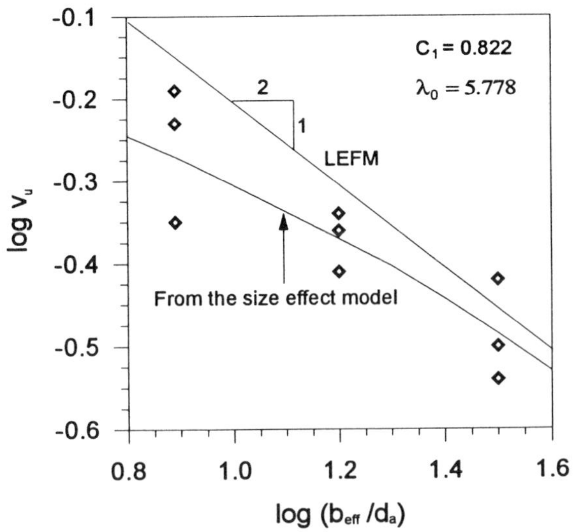

Figure 9-37 Comparisons of experimental strength with size effect model for torsional failure of reinforced concrete beams.[52]

combined torsion, bending, and shear. The combination of torsion and shear loading causes a nonsymmetric shear stress distribution. On one side of the beam the effects of the two loadings are additive, whereas on the opposite side the two loadings counteract one another. Shear–torsion interaction failure curves based on a stress criterion are normally a straight line, as shown in Fig. 9-38, because only the stress state on the critical side of the beam is included. Using the energy criterion of fracture mechanics, which allows one to consider the stress state of the whole cross section, Hawkins proposed an approximate interaction diagram given by

$$\left(\frac{T_u}{T_0}\right)^2 + \left(\frac{V_u}{V_c}\right)^2 = 1 \tag{9.77}$$

where T_u and V_u are the ultimate torque and shear force and T_0 and V_c are loading capacities of the beam under the pure torsion and the pure shear, respectively. Equation (9.74) is indicated in Fig. 9-38. It is seen that since the beneficial effects of the stress state for the less highly stressed side are included, the shear–torsion interaction curve is circular and consistent with experimental results. Hawkins also indicated that Eq. (9.77) is corroborated by the circular interaction formula used in the ACI 318-89, which also results from the lower bound approximation of plasticity.

Bazant and Kwon[55] have reported results from a series of tests on geometrically similar, tied reinforced concrete columns of different sizes, with size ratio of 1:2:4 and slenderness of 16, 32, and 48, as shown in Fig. 9-39. The

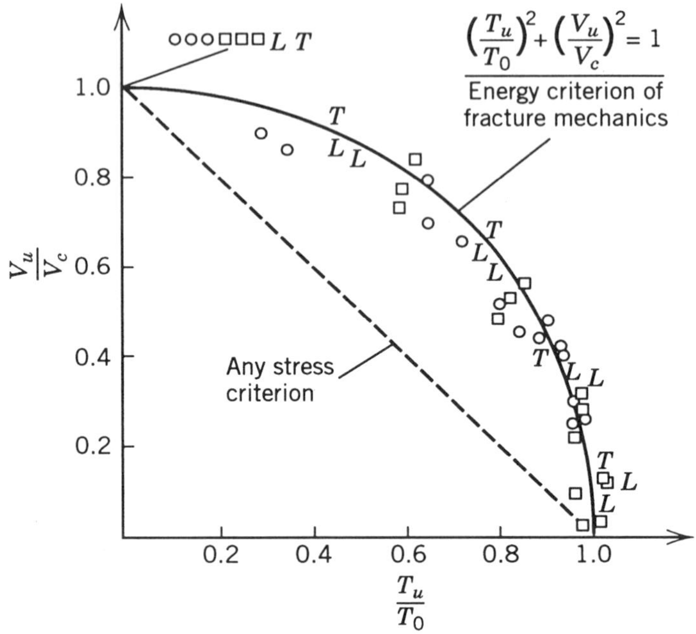

Figure 9-38 Interaction diagram for combined shear and torsion based on energy criterion.[54]

model concrete columns were made of reduced-size aggregate with maximum aggregate size of $\frac{1}{8}$ in. and reduced-size steel bars. They reported that the failure loads obtained experimentally for tied reinforced concrete columns exhibit a strong size effect, which is in agreement with the size effect model. It was also observed that with increasing slenderness of the column, the size effect became stronger and the brittleness number became higher.

9.10 ANCHOR BOLTS

Anchor bolts are used in concrete structures for local load transfer. Although it has been indicated that "anchor bolts and mechanical connections shall be designed to reach their design strength prior to anchorage failure or failure of surrounding concrete," no guidelines are given by ACI 318-89.[28] On the other hand, based on a strength criterion, ACI Committee 349[56] recommended that the nominal pullout strength of anchor bolts is expressed as $P_c = k_6 d_{\text{em}}^2 (f'_c)^{1/2}$, where k_6 is an empirical constant, d_{em} is the embedded length of an anchor bolt, and f'_c is the compressive strength of concrete. A summary of research efforts to evaluate the performance of anchor bolts has been given by Eligehausen and Sawade.[57]

The failure mechanism of pullout of an anchor bolt is shown in Fig. 9-40.

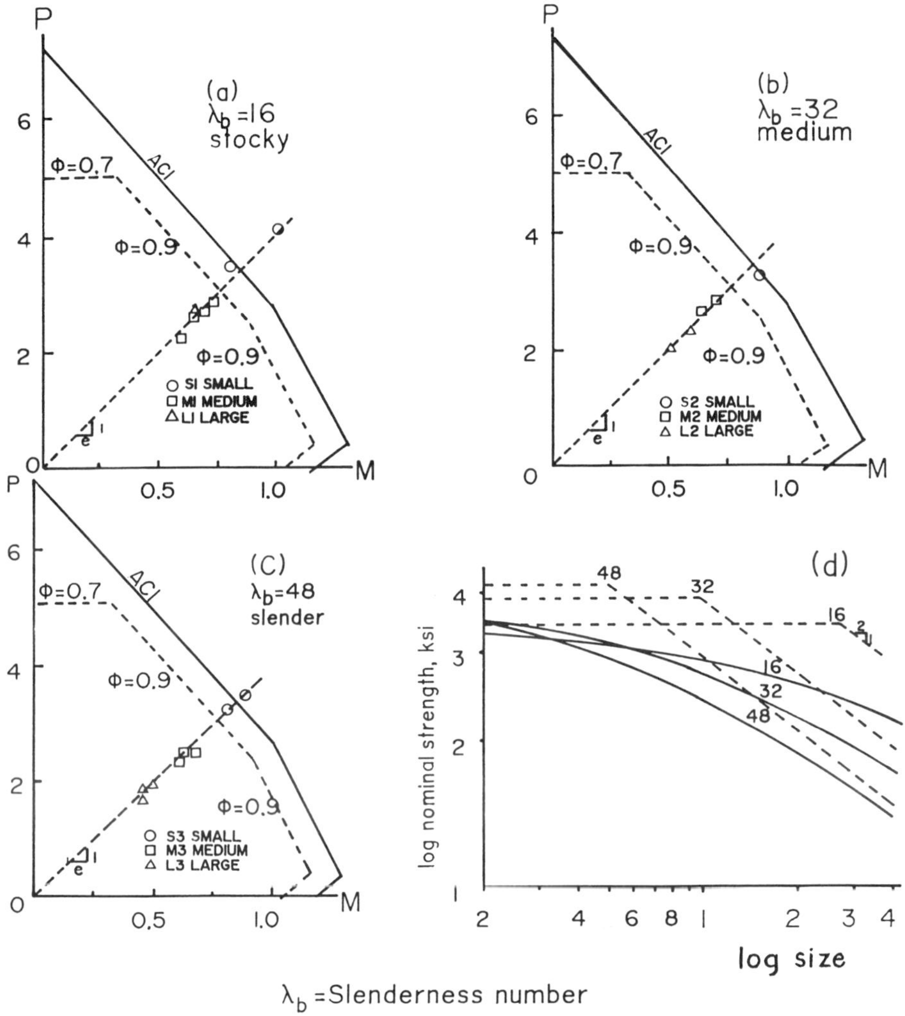

Figure 9-39 Failure of beam columns: (*a–c*) interaction diagrams for different slenderness and (*d*) size effect plot.[55]

Two types of cracks, a primary crack that tends to generate a conical failure surface of concrete and a series of radial secondary cracks that occur when an anchor bolt is made of a deformed bar, have been observed.[58] Fracture mechanics approaches have been applied to failure of the pullout of anchor bolts. The viability of fracture mechanics approaches to this problem includes the following: (i) an energy criterion can be used to describe initiation and propagation of cracks, (ii) objectivity of calculations with respect to the finite element mesh is allowed, and (iii) observed size effect on strength may be explained.

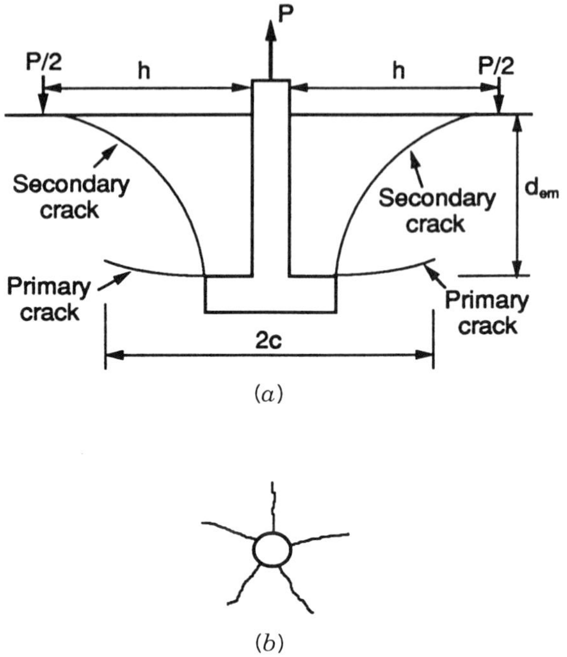

Figure 9-40 Possible primary and secondary cracks for pullout of anchor: (*a*) section and (*b*) plan.[59]

Ballarini et al.[55] used a LEFM-based two-dimensional mixed-mode approach to model crack propagation in the pullout of rigid anchor bolts. Since pullout of an anchor bolt made of a smooth bar was studied, only the primary crack is considered. Using a Green's function for the concentrated force and a dislocation in an elastic half space, they reduced the elasticity problem to solving a set of coupled singular integral equations. The mixed-mode stress intensity factor was obtained by numerically solving the integral equations. Crack profiles, growth, and stability were obtained from the model and compared with results from anchor pullout experiments. They found that when reaction forces are sufficiently far away from the location of the anchor ($h \gg c$ in Fig. 9-40), crack propagation becomes unstable. In this case, the failure surface is a by-product of the fracture process and does not contribute to the tensile capacity of the anchor as current design procedures assume. For short reaction distances and deep embedments, crack propagation is stable.

Elfgren and Ohlsson[60] have used a discrete crack approach based on the mixed-mode fictitious crack model combined with a finite element method to analyze the pullout of an anchor bolt. Only the primary crack was included in their analysis. The crack surface was assumed as a straight line from the upper edge of the anchor lead to support on the top surface. Bocca et al.[61] have performed a similar analysis but with an unknown crack path that was deter-

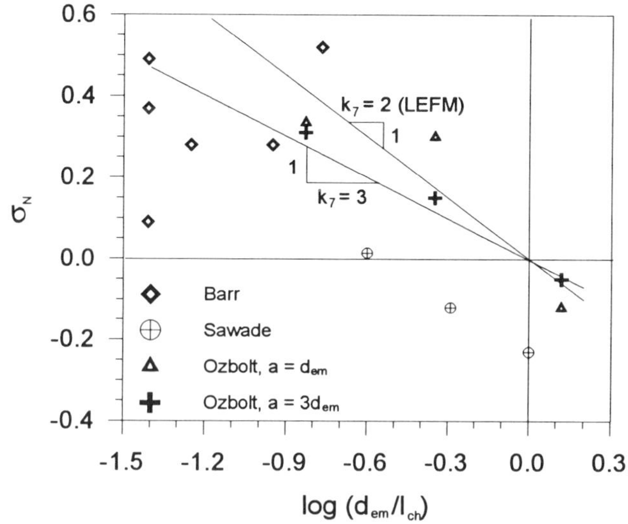

Figure 9-41 Change of pullout load with embedment length of anchor.[63]

mined based on the direction of the maximum principal tensile stress obtained during calculation.

A round-robin analysis of pullout of anchor bolts has been recently organized by RILEM Committee TC 90-FMA.[62] Preliminary results of the round-robin analysis have been summarized by Elfgren and Swartz.[63] They introduced a brittleness number B_n as a ratio of elastic energy U_e to fracture energy U_F. By assuming that U_e is related to the stress σ and fracture volume d_{em}^3, the values of U_e may be regarded to be proportional to $\sigma^2 d_{\mathrm{em}}^3/E$. On the other hand, by assuming U_F is related to the fracture energy G_F and the cracked surface d_{em}^2, the value of U_F is regarded to be proportional to $G_F d_{\mathrm{em}}^2$. By noting $\sigma = f_t$ at failure, the value of $B_n = d_{\mathrm{em}}/l_{\mathrm{ch}}$ can be obtained, where $l_{\mathrm{ch}} = EG_F/f_t^2$ is used. By plotting the nominal strength $\sigma_N = P_c/\pi d_{\mathrm{em}}^2 f_t$ against the value of B_n, they observed the size effect on the nominal strength as shown in Fig. 9-41 ($k_7 = 3$). Bazant and Şener[64] and Eligehausen and Ozbolt[65] have also indicated the size effect on the nominal pullout strength. However, they reported that this size effect is very close to LEFM ($k_7 = 2$ in Fig. 9-41).

9.11 CONCRETE PIPES

Gustafsson and Hillerborg[66] have analyzed two primary types of failure, ring failure and beam failure, as shown in Fig. 9-42, in unreinforced concrete pipes using the fictitious crack model. The ring failure as shown in Fig. 9-42*a* involves the formation of four bending fracture zones where the fictitious crack model was applied. These zones behave like softening pseudoplastic hinges.

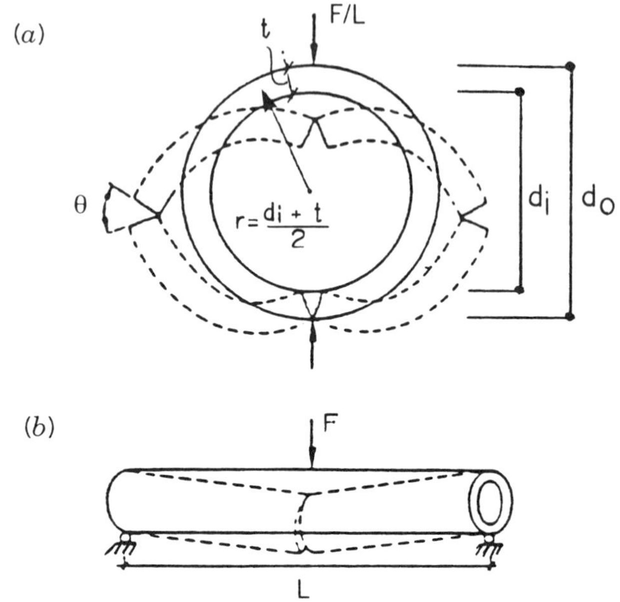

Figure 9-42 Failure mechanisms of unreinforced concrete pipe.[66]

The relation between the moment M and the rotation θ in such a hinge can be calculated by means of the fictitious crack model, and the obtained results are shown in Fig. 9-43*a*. After these properties of the hinges have been determined, analysis of ring failure can be achieved by treating the parts between the hinges as elastic elements and taking the hinge properties into account. For the beam failure mode as shown in Fig. 9-42*b*, the softening pseudohinge formation at the critical section (midspan) was similarly analyzed using the fictitious crack model. The results obtained are indicated in Fig. 9-43*b*, where the ratio between the nominal flexural strength due to the ring failure or the bending failure, f_{cr}, to the material tensile strength f_t is plotted against the relative inner diameter of the ring. The f_{cr}/f_t ratio decreases with increasing relative inner diameter of the ring. Smaller pipes are observed to be stronger and more ductile than larger pipes, as shown in Fig. 9-43b (d_i/l_{ch} is a nondimensional measure of pipe size). They also reported that the ring-type failure exhibits a greater size sensitivity than the beam failure. A size-independent MOR used currently for design purposes (ACI 318-89) is unconservative for sufficiently large sizes.

Bazant and Cao[67] have also studied the size effect in the brittle failure of unreinforced concrete pipes. They defined two nominal strengths corresponding to different types of failures. For the ring failure the nominal strength of $\sigma_{Nr} = 3P_r d_{mn}/\pi t^2$ was used, where P_r is the load for the ring failure, d_{mn} is the pipe diameter defined by the middle surface of the wall, and t is the wall

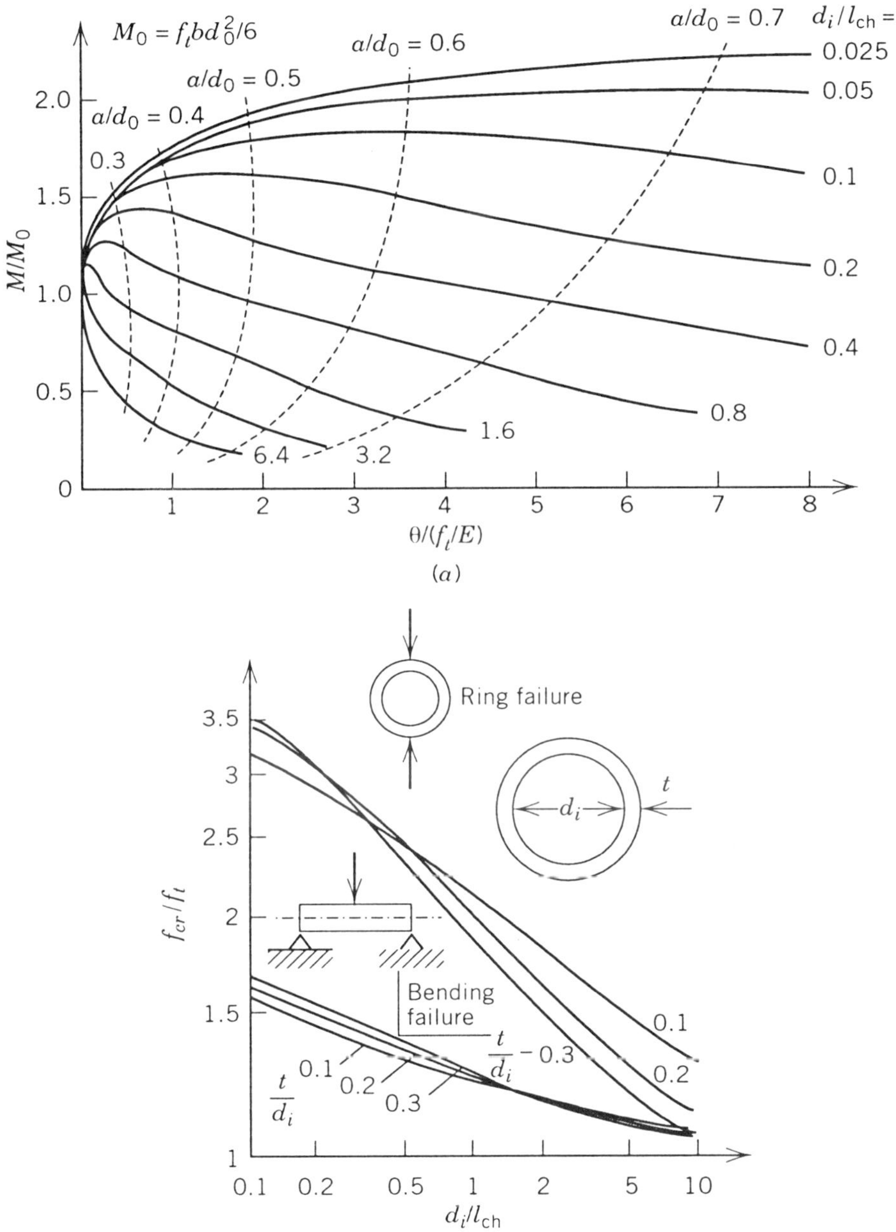

Figure 9-43 Applications of fracture mechanics to failure of unreinforced concrete pipes: (*a*) relationship between moment and rotation for ring failure and (*b*) size effect on nominal strength for ring failure and bend failure.[66]

thickness. For the beam failure the nominal strength of $\sigma_{Nc} = d_0 M_b / 2I$ was introduced, where I is centroidal moment of inertia of the ring cross section, M_b is the maximum bending moment corresponding to the beam failure, and d_0 is the outside diameter of the pipe. They found these two nominal strengths decrease with increasing size of pipes.

Both studies have shown that the design procedures for the prediction of allowable stresses in unreinforced pipes can be significantly improved with the incorporation of size effect due to NLFM.

9.12 CONCRETE DAMS

There is an opinion that the application of fracture mechanics to dams may be avoided by the so-called no-tension design, which is equivalent to elastic–plastic analysis with a yield criterion in which the tensile yield limit of concrete is regarded as zero (either Rankine criterion or Mohr–Coulomb criterion). However, it has been demonstrated[68] that such a design is not guaranteed to be safe. The stress intensity factor at the tip of a crack that satisfies the no-tension criterion according to the elastic–plastic analysis may be nonzero based on LEFM and can be either positive or negative. If the size of a dam is larger than a certain critical size, the positive value of the stress intensity factor can cause cracking of the dam when its value exceeds the fracture toughness of the material. Nonlinear fracture mechanics is necessary for most concrete structures because the size of the fracture process zone is comparable to the size of their other dimensions. Concrete dams are so large that their global failure may, in most of cases, be analyzed by LEFM.[69,70] Large unreinforced structures such as concrete gravity dams may be prone to cracking due to (1) significant thermal and shrinkage stresses, (2) poor construction joints, (3) differential movements in the foundations and abutments, and (4) large inertial loads due to seismic loadings. Most of the conventional methods based on the strength failure criterion do not allow assessment of the consequences of cracking and the effectiveness of repair methods. Use of fracture mechanics (LEFM) makes it possible to study (i) initiation of initial flaws, (ii) the performance of dams that are cracked, (iii) the chances for crack growth and stability under possible extreme loadings, and (iv) the effectiveness of repair methods.

Linsbauer et al.[69,70] have successfully used LEFM mixed-mode fracture analyses to simulate the history of crack growth and profile in the upstream and downstream sides of a doubly curved concrete arch dam (the Kölnbrein arch dam in Austria) as described in the following case study.

9.12.1 Case Study: Kölnbrein Dam

A case study in two parts is presented here to illustrate the value of the fracture mechanics method to problems of this type.

9.12.1.1 *Introduction of Dam* The dam is in Austria and has a crown cantilever height of 200 m (656 ft). The dam longitudinal section and cross section are shown in Fig. 9-44. This is a doubly curved arch dam, somewhat similar in structural action to a thin shell. These dams are considered to be highly sensitive to material and geometric imperfections in the foundation and structure.

Excessive drainage was observed in 1978 when filling the reservoir to a level of about 1860 m (or a water level from the base of about 160 m, or 525 ft). Upon emptying the reservoir an open crack on the *upstream* face was revealed (fracture '78 in Fig. 9-44*c*). The crack surface trace shown in (*a*) is about 160 m long and the crack extended into the dam–foundation interface as sketched in Fig. 9-44*c*. Repairs concentrated on the upstream foundation region and included a new grout curtain. When the reservoir was filled again following these repairs in 1982, at a level of 1890 m (190 m depth, or 10 m from the top of the dam), excessive leakage was again observed.

The new fracture system (fracture '83 in Fig. 9-44*c*) was apparently not associated with the earlier fracture but is 2–5 m (6.6–16.4 ft) above it. This fracture system also extended to the dam–foundation interface. In addition to the upstream cracks, cracks were discovered on the downstream side of the dam as well. These were centered at a level of about 14 m (45.9 ft) above the foundation level. It was believed that these cracks may have been induced into the system during construction by the grouting pressure in the top section of the arch dam. It was also suggested that the upstream crack system was generated as a result of the downstream crack system.

The cracking systems were analyzed essentially in two major steps using modeling. In the first step, stress distributions were obtained for different load combinations, assumed boundary conditions, and with and without crack systems. Based upon these results, different starter cracks were introduced. This was necessary because LEFM was used and this requires a priori the presence of one or more cracks. Using the starter cracks, propagation was allowed to take place using the *circumferential tensile stress* criterion (see Chapter 8) and $K_I \geq K_{Ic}$.

9.12.1.2. *Downstream Crack System of Dam*[69] In Fig. 9-45 are displayed results of the *stress analysis* for the downstream face using a load combination of self-weight plus grouting pressure. Parts (*a*), (*b*), and (*c*) are results when no crack is present. It is seen that a peak tensile stress of about 1.77 MPa (257 psi) is located at 10.5 m (34.4 ft) above the base. This stress is stated to be approaching the MOR of the dam material. At this location a 3-m 9.8-ft-long crack was introduced and another stress analysis conducted. The results shown in Fig. 9-45*d* reveal a new local maximum tensile stress of 1.6 MPa (232 psi) at a level 16.5 m (54.1 ft) above the base, or 6 m above the original peak stress/crack location.

The next analysis was based on LEFM and utilized two short cracks located at 10.5 m (crack 1) and 16.5 m (crack 2) above the base. A fracture toughness

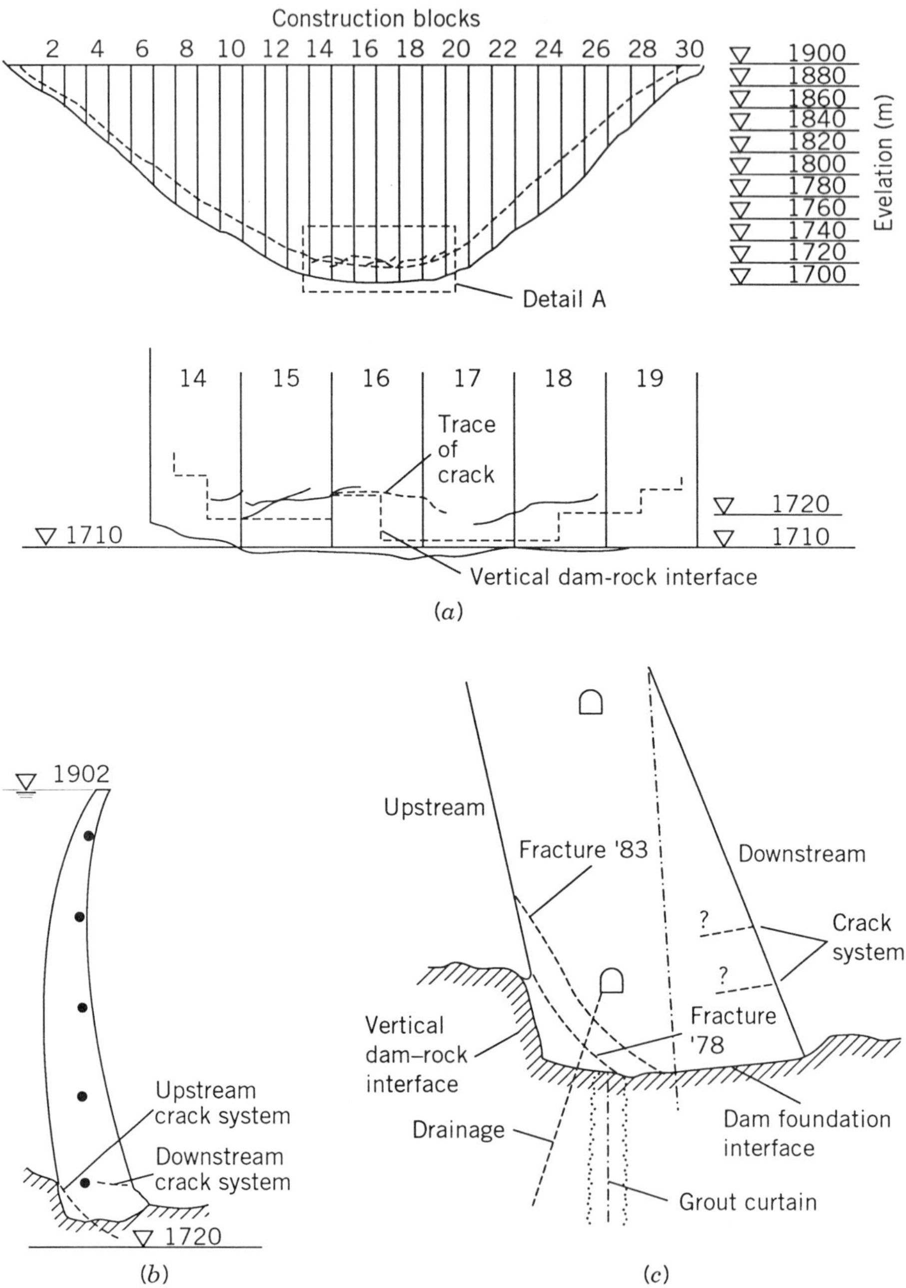

Figure 9-44 Fracture pattern development in Kölnbrein dam: (*a*) longitudinal section (upstream) showing fracture traces near heel, (*b*) cross section of dam showing relative locations of upstream and downstream crack systems, and (*c*) foundation region of dam.[69,70]

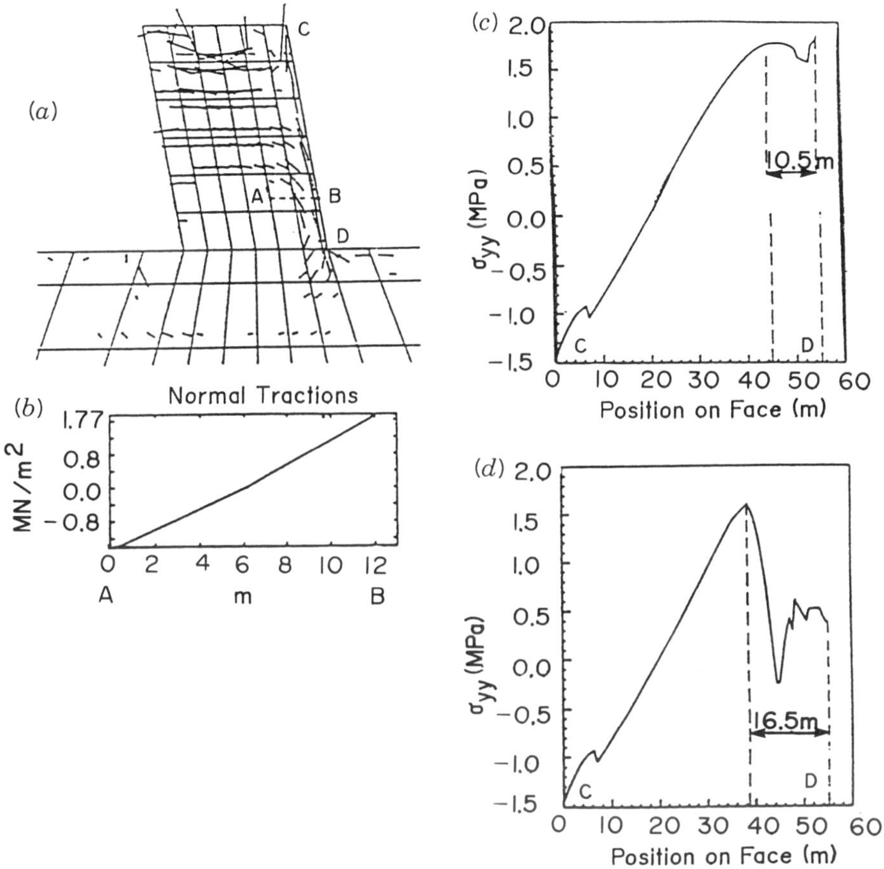

Figure 9-45 Stresses on downstream face of dam due to dead load plus grouting pressure: (*a*) vectors of tensile stress, uncracked dam; (*b*) normal stress distribution along *A–B*, uncracked dam; (*c*) stress distribution along downstream face, uncracked dam; and (*d*) stress distribution along downstream face, single crack at 10.5 m above dam base.[69,70]

value of $K_{Ic} = 2.0$ MPa m$^{1/2}$ (1820 psi in.$^{1/2}$) for the dam concrete was assumed. For these initial cracks $K_I^{(1)} = 3.18$ MPa m$^{1/2}$ and $K_I^{(2)} = 3.27$ MPa m$^{1/2}$ were obtained from the analysis. Here $K_I^{(1)}$ and $K_I^{(2)}$ are the stress intensity factors at the location of cracks 1 and 2, respectively. As a result, the cracks should propagate. The cracks were allowed to propagate with direction governed by the circumferential tensile stress criterion and discrete increments were taken. The results are shown in Fig. 9-46. In part (*a*) the crack paths are shown and in part (*b*) the values of K_I for each crack corresponding to a new crack tip location are shown. It is seen that by the time crack 1 has grown to about 12 m (39 ft), it has arrested due to $K_I^{(1)} < K_{Ic}$, whereas crack 2 continues to propagate up to about 14 m (46 ft). This is about one-third the thickness of the dam.

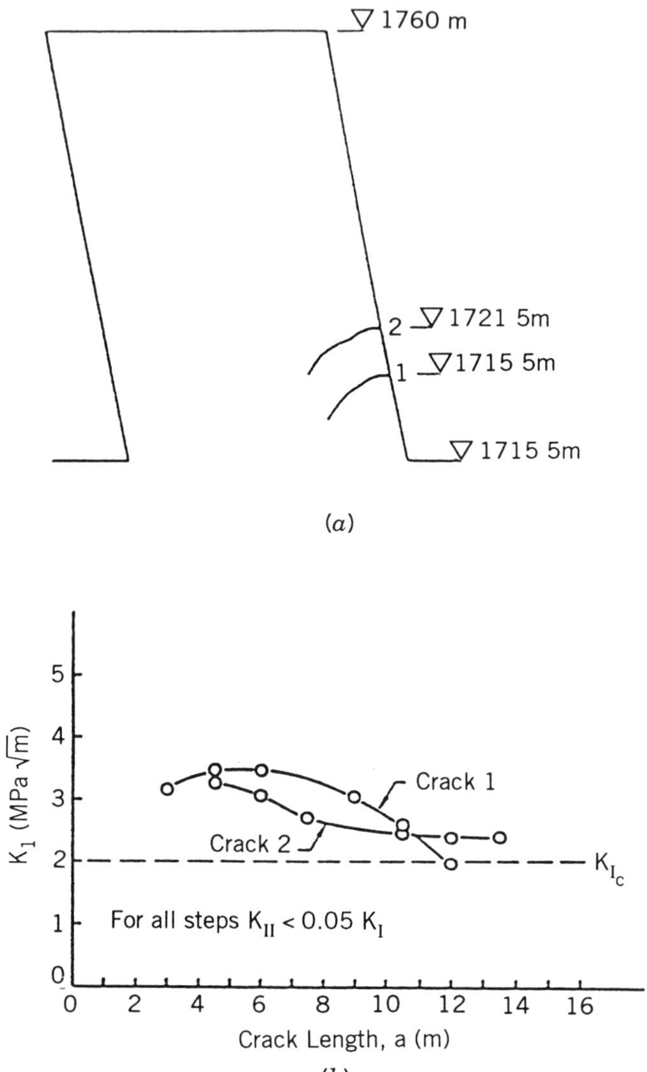

(b)

Figure 9-46 (*a*) Predicted crack trajectories after six increments of propagation and (*b*) K_I values for two cracks as function of crack length during six increments of propagation.[69,70]

The conclusions obtained from this phase of the study included the following:

1. Dead load plus grouting load are sufficient to generate tensile stresses greater than the estimated MOR of the dam material. Thus, this cracking probably occurred during construction (but evidently was not noticed).

2. The predicted locations of the downstream cracks (10.5 and 16.5 m above the toe) are in good agreement with those observed at 14 m above the toe.
3. The two cracks initiated at these locations propagated in an unstable manner, in curvilinear trajectories to about one-third the thickness, and then stabilized.
4. Filling the reservoir closed these cracks so they could not be the cause of the cracking system observed on the upstream face.
5. A predictive capability incorporating fracture mechanics could be used to control certain construction procedures, in this case grouting of construction joints, to preclude the incidence of cracking.

9.12.1.3 Upstream Crack System of Dam[70] The cracking on the upstream side involved additional complexities and more uncertainties when compared to the downstream side. These resulted in the following questions:

1. Where did the crack initiate?
2. What caused it to initiate?
3. What was its resulting trajectory?

Additional considerations concerned the formation of a second crack after initial repairs; again see Fig. 9-44*c* for details of the cracks. What were possible effects of water pressure intruding the crack faces on the crack propagation? The influence of the rock foundation on interface cracking and possible crack initiation must be considered.

To explore these questions in a systematic way, four possible cracking models for the upstream side were considered. These are shown in Fig. 9-47. The FE mesh used was typically that shown in Fig. 9-48 except for variations related to interface boundary conditions, crack initiation locations, and other reasons. The loadings shown represent the effects of a full reservoir and are an idealization of that which is felt by the one-third height, two-dimensional cantilever used in the analysis. It is further assumed that the downstream cracks do not in any way influence the upstream cracks.

Two of the models will be discussed here in some detail: the corner crack model and the foundation joint model.

The corner crack model is based on the assumption that the 1978 crack (cf. Fig. 9-44*c*) initiated at a reentrant corner at the heel of the dam point P and line PQ of Fig. 9-47*a*. The dam–rock interface PQ is assumed to be perfectly bonded. Because of the singularity at point P, it was assumed to be justifiable to insert a small crack at this location. The direction was assumed to be normal to the maximum stress obtained by an FE solution without a crack. The crack was then allowed to propagate under the same fracture conditions described earlier for the downstream cracks and the influence of water pressure was considered as an additional load on the crack faces. Figure 9-49 shows the mesh

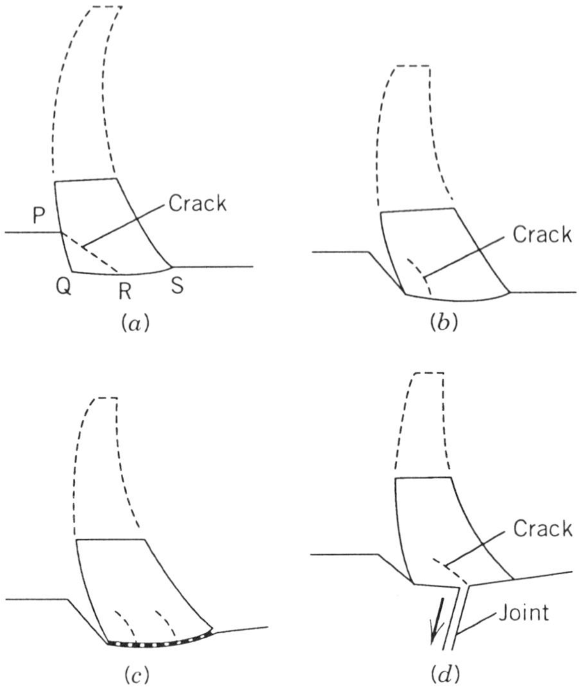

Figure 9-47 Four analyses for upstream cracking: (*a*) corner crack model, (*b*) open gap model, (*c*) variable base stiffness model, and (*d*) foundation joint model.[70]

after nine increments of propagation. It was found that $K_{\mathrm{I}} > K_{\mathrm{I}c}$ at the end of each increment and so the crack continued to extend. The K_{I} values are shown for each increment in Fig. 9-49*b*. These are extremely high as $K_{\mathrm{I}c} = 2$ MPa $\mathrm{m}^{1/2}$ and continue to increase as the crack grows. If the open gap model is used, the dam–rock interface line *PQ* is assumed to be completely unbonded so cracking starts at point *Q*. This causes separation along the base interface between rock and concrete. This separation induced high tension in the concrete base and so a new crack, shown in Fig. 9-47*b*, was initiated and allowed to propagate. It curves upward as shown but actually tends to intersect line *PQ* below point *P*, which is lower that the observed crack.

The foundation joint model is based on the idea that possibly the upstream side cracking system actually originated in the rock mass directly under the dam due to a motion of 10 mm along a vertical joint, line *AB* in Fig. 9-50*a*, triggered by filling the reservoir. It was also assumed that sufficient bond strength existed between the base of the dam and the rock foundation that this motion would initiate the crack at the position shown and confine debonding to a region of only a few meters on either side of the crack. In Fig. 9-50*b* the crack propagation after 12 increments shows the crack is about to intersect line *PQ* at about 14 m (46 ft) above point *Q*. In this model line *PQ* is 7 m in length.

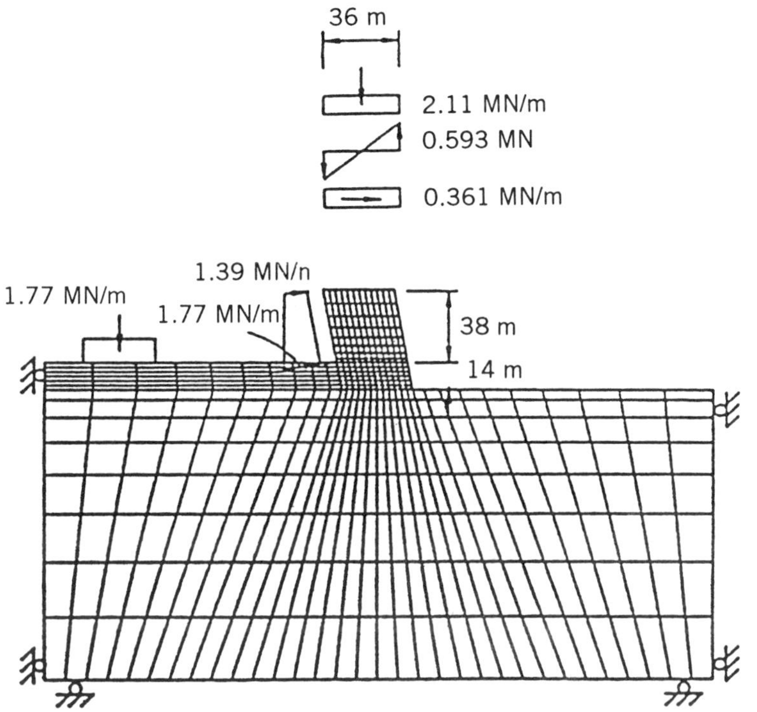

Figure 9-48 Typical overall initial FE mesh on loading used in four analyses.[70]

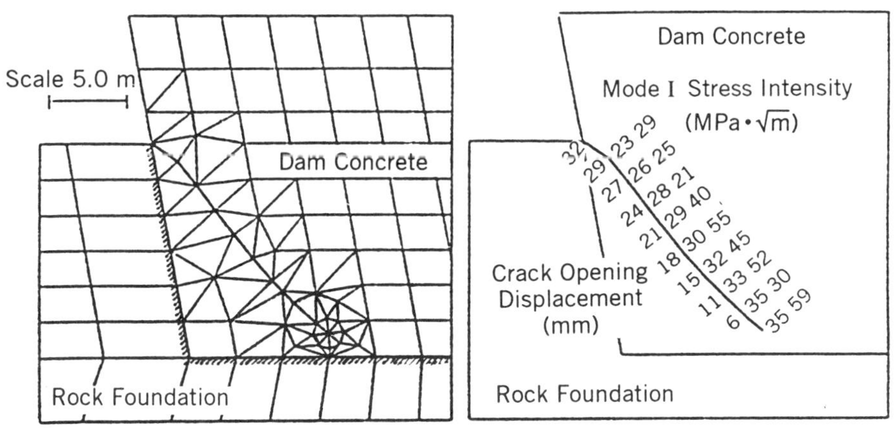

(*a*) (*b*)

Figure 9-49 Corner crack model[70]: (*a*) detail of mesh after nine increments of crack propagation and (*b*) predicted trajectory, K_I history.

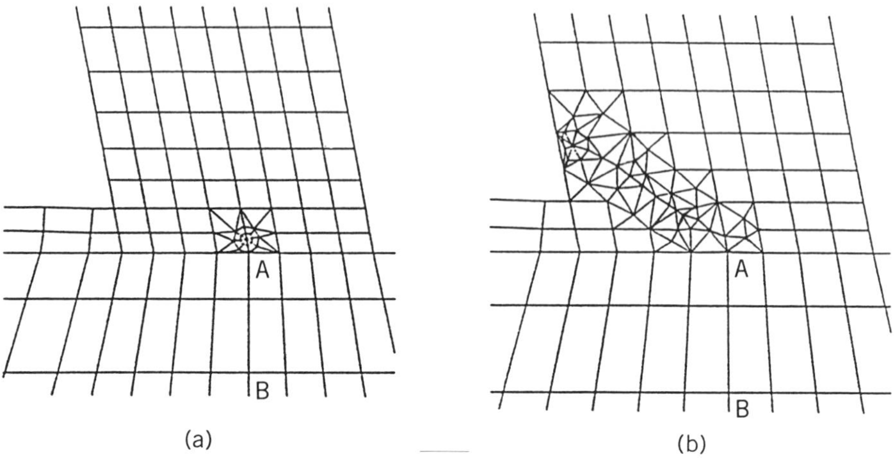

Figure 9-50 Foundation joint model: (*a*) short crack initiated at end of foundation joint, line *A–B*, and (*b*) mesh after 12 increments of cracking.[70]

The K_I values after each increment were all substantially higher than K_{Ic} and the crack propagated in an unstable manner.

The conclusions obtained from the analyses of the upstream-side cracking were the following:

1. The mechanisms investiated all yielded plausible cracks in the heel region of the dam under nearly full reservoir loading. The trajectories, at least qualitatively, agreed with the available observations.
2. The different models can cause the cracks to propagate in different directions. By comparing Figs. 9-49 and 9-50, it is seen that the corner crack model predicts crack growth in the downstream direction, whereas the foundation joint model predicts crack growth in the opposite direction.
3. Finally, it should be noted that a two-dimensional analysis was performed to analyze a very complex three-dimensional problem. As shown in Fig. 9-44, the cracking is truly three-dimensional and nonplanar. At the time of this writing (1988), a true three-dimensional approach was not available. It is still an area of current research.

9.12.2 Other Studies of Dams

Saouma et al.[71] have studied fracture of a concrete gravity dam made of roller compacted concrete using their anisotropic mixed-mode fracture analysis based on a discrete crack approach. They concluded that the nucleation and extension of a few discrete cracks on the face of the dam would result in a large decrease of the tensile stress along the faces, and the classical method of analysis is normally much more conservative than the fracture mechanics approach. The U.S. Army Corps of Engineers has provided guidelines[72] on applying the principles of fracture mechanics to assess the safety and serviceability of existing structures based on the work of Saouma et al.[73]

Zhang and Karihaloo[74] investigated the stability of a vertical central crack in the upstream face of a buttressed concrete dam. They regarded the materials as viscoelastic and allowed for tension softening. They showed that fracture mechanics can serve as a useful forensic tool in determining the possible cause for delayed instability of a previously stable, noncritical crack.

For fracture analysis of dams, it is very important to determine the proper value of fracture toughness as well as other fracture parameters such as $CTOD_c$ and c_f for a concrete with very large aggregates. Since the size of specimens used to determine fracture parameters is proportional to the aggregate size, three-point bend beams may be inconvenient for this purpose. Brühwiler and Wittmann[75] proposed the use of a wedge-splitting test, which has been presented in Chapter 6. Saouma et al.[76] used these wedge-splitting specimens to evaluate fracture properties of concrete with different sizes and shapes of aggregates. They reported that values of fracture toughness G_F are independent of aggregate size and specimen size if certain minimum dimensions are exceeded but depend on the type of aggregate. Bazant et al.[77] tested a series of geometrically similar wedge-splitting specimens of maximum cross-sectional dimension of 6 ft and applied the size effect method to obtain the fracture toughness corresponding to extrapolation to still much larger sizes.

9.13 FAILURE OF DISSIMILAR INTERFACES AND JOINTS

Fracture mechanics approaches have been used to characterize the shear failure of dissimilar interfaces and joints. Although shear friction in ACI 318-89 deals with failure of dissimilar interfaces in building members (e.g., construction joints with concrete of different ages), results discussed in this section may be also applied to a microstructural level such as the failure of aggregate–paste interfaces. Although like a mode I crack (opening type), a joint may be regarded as a material and/or geometric discontinuity, several differences between the crack and the joint should be recognized. These include that (i) the crack is subjected to a tensile stress whereas the joint may be regarded as a mode II crack (sliding type) or a mixed-mode crack (opening and sliding) subjected to a combined normal and shear loading, (ii) the path of the crack depends on the variation of the principal tensile stress whereas the direction of the joint is fixed; and (iii) debonding and sliding, instead of opening, are major failure mechanics for failure of the joint.

Reinhardt[78] treated a joint as a discontinuity subjected to a remote shear load. The shear strength was assumed to be due to cohesion of the material. Joint strength was averaged over the length of the joint. Reinhardt tested specimens consisting of two pieces of the prefabricated concrete, with a compressive strength of 70 MPa, which are adhered together at the joint using mortars with compressive strengths of 2.1, 40.6, and 55.2 Mpa, as shown in Fig. 9-51. For each type of filling mortar, specimens with different lengths of the joint were tested. He reported that the averaged joint strength t_u for the mortars with the compressive strengths of 40.6 and 55.2 MPa depends on the length of the joint, whereas that for the compressive strength of 2.1 Mpa it did not. The LEFM solutions for different arrangements of joint cracks located between two semi-infinite media have indicated that the value of τ_u is proportional to $L_s^{-1/2}$, where L_s is the length of the joint. By plotting the normalized average joint strength τ_u/τ_{u600} against l as shown in Fig. 9-51, where τ_{u600} is the average joint strength for the 600-mm-long specimen, Reinhardt reported that only specimens adhered by the mortar with the compressive strength of 55.2 MPa follow the LEFM solution. On the other hand, the values of τ_u/τ_{u600} of specimens with the compressive strength of 40.6 MPa vary with $L_s^{-1/4}$, whereas the very low strength mortar demonstrated no length effect. Reinhardt concluded that these performances for joints with different mortar may be attributed to development of a fracture process zone at the ends of the joints.

Buyukozturk et al.[79,80] have studied fracture of interfaces between dissimilar materials using a linear elastic mixed-mode fracture approach. Interface frac-

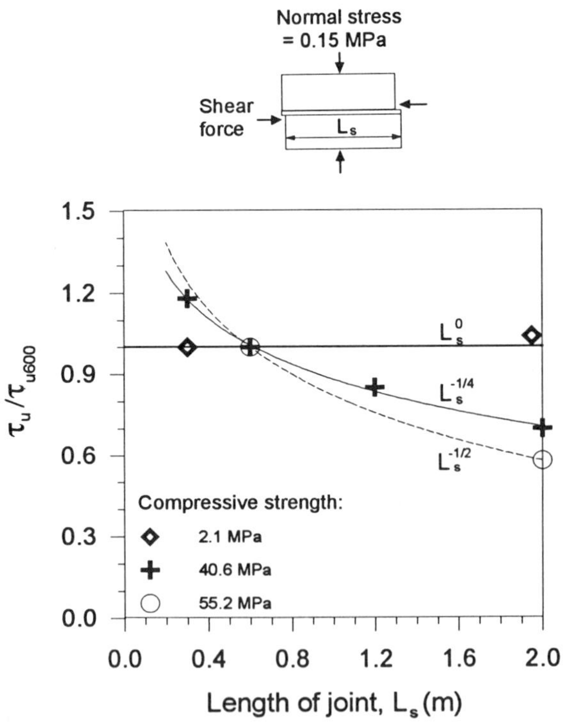

Figure 9-51 Influence of joint length on normalized shear strength.[78]

ture specimens including a thin layer of aggregate sandwiches in a mortar matrix have been used to assess the fracture toughness of the mortar–aggregate interfaces in concrete, as shown in Fig. 9-52*a*. Since the thickness of the sandwich layer is small compared to the crack length, an asymptotic relationship between the interface intensity factors and the stress intensity factors for the homogeneous problem can be obtained. Interface fracture toughness has been observed to increase with increasing effects of the shear loading relative to that of the tensile loading. The LEFM mixed mode analysis was reported to be in agreement with results from the experiments on the aggregate–mortar sandwich specimens, as shown in Fig. 9-52*b*, where ψ_1 is the phase angle that measures the contribution of shear stress to the opening experience by the interface crack.

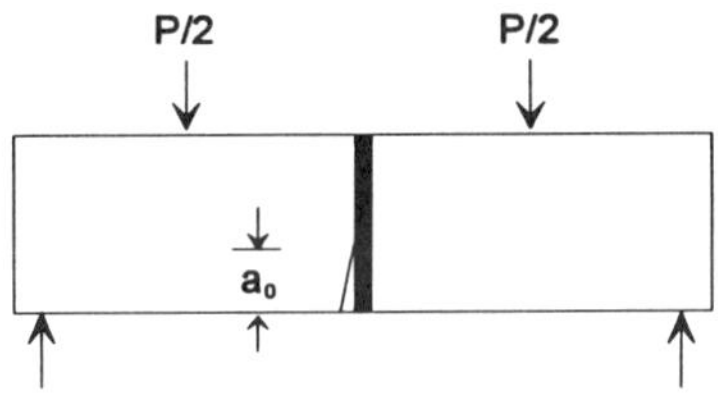

(*a*)

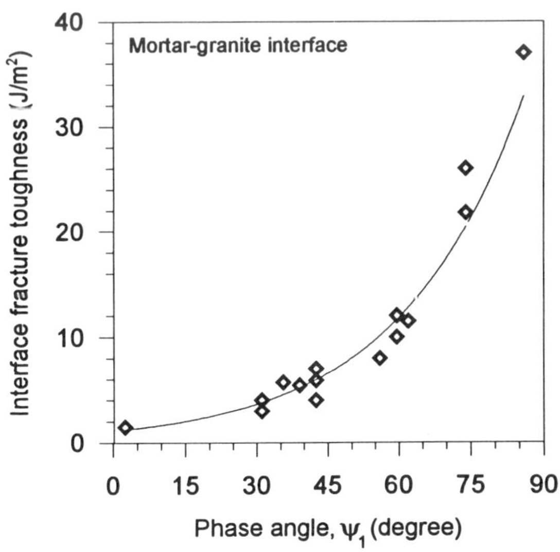

(*b*)

Figure 9-52 Fracture of mortar–granite interface: (*a*) sandwich specimen and (*b*) fracture toughness.[80]

Mitsui et al.[81] have studied fracture toughness of the paste–aggregate interface. They developed a pushout test to evaluate interfacial properties of cemetitious paste–aggregate bond, by which a cylindrical shape aggregate was pushed out from a cementitious matrix under a constant slip rate, as shown in Fig. 9-53. Interfacial properties were evaluated by applying an analytical shear-lag model[82] to the obtained pushout load slip curves, as shown in Fig. 9-53. Both backscattered electron imaging and energy dispersion analysis of x-rays were also used to characterize the microstructure of the interface. They reported that interfacial properties that include fracture toughness, stiffness, and frictional stress are related to the microstructure of the interface. Reduction of the water-to-cement ratio, addition of silica fume, and some coating of aggregate can significantly increase the interfacial properties.

Wang and Maji[83] have conducted uniaxial tension tests on mortar–limestone interface specimen to obtain the bridging stress versus crack opening displace-

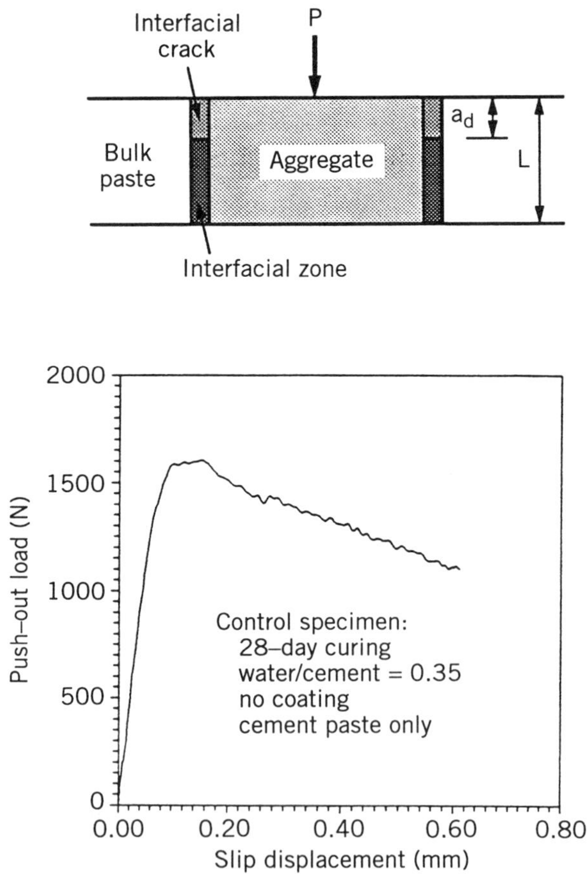

Figure 9-53 Typical pushout load and slip curve.[82]

ment curve. A scanning electron microscope was used to study the microstructure of the interface. They reported that both mechanical interlocks and chemical interaction contribute to the interface bonding between mortar and limestone, whereas only mechanical interlocking is likely responsible for the bond between mortar and rock salt. They also measured the interfacial toughness using compact tension (CT) specimens. An interfacial cohesive zone model in conjunction with the FE method was used to analyze the failure of the CT specimen. They demonstrated the similarity between interface fracture mechanics and conventional fracture mechanics.

9.14 SUMMARY

Applications of fracture mechanics to various concrete and reinforced concrete structures have been discussed in this chapter. Although applications of fracture mechanics depend on the type of structure as well as the amount, location, and orientation of reinforcement, they provide a unified and rational basis to describe failure of these structures. Many experimentally observed phenomena such as strain localization and size effect in nominal strength can be uniquely predicted by fracture mechanics. Application of results obtained to the current design code and specifications will result in better descriptions of failure of different structures and explanations of the empirical constants used. Since current understanding of the material fracture process and structural failure mechanism are limited, applications of fracture mechanics to some structures are preliminary. Intensive studies are certainly needed in this area.

REFERENCES

1. Park, R., and Pauly, T., *Reinforced Concrete Structures*, Wiley, New York, 1975.
2. Hsu, T. T. C., *Unified Theory of Reinforced Concrete*, CRC, Boca Raton, FL, 1993.
3. ACI Committee 446, ''Fracture Mechanics: Applications to Concrete Structures and Implications with Regard to the Code,'' American Concrete Institute, Detroit, MI, 1992.
4. Weibull, W., ''A Statistical Theory of the Strength of Materials,'' *Royal Swedish Academy of Engineering Society Proceedings*, Vol. 151, 1939, pp. 1–45.
5. Weibull, W., ''A Statistical Distribution Function of Wide Applicability,'' *Journal of Applied Mechanics*, Vol. 18, 1951, pp. 293–297.
6. Bazant, Z. P., Xi, Y., and Reid, S. G., ''Statistical Size Effect in Quasi-Brittle Structures: I. Is Weibull Theory Applicable?'' *Journal of Engineering Mechanics, ASCE*, Vol. 117, No. 11, 1991, pp. 2609–2622.
7. Hillerborg, A., Modeer, M., and Petersson, P.-E., ''Analysis of Crack Formation and Crack Growth in Concrete by Means of Fracture Mechanics and Finite Elements,'' *Cement and Concrete Research*, Vol. 6, No. 6, 1976, pp. 773–782.

8. Jenq, Y. S., and Shah, S. P., "A Two Parameter Fracture Model for Concrete," *Journal of Engineering Mechanics*, Vol. III, No. 4, 1985, pp. 1227–1241.
9. Bazant, Z. P., and Kazemi, M. T., "Determination of Fracture Energy, Process Zone Length and Brittleness Number from Size Effect, with Application to Rock and Concrete," *International Journal of Fracture*, Vol. 44, 1990, pp. 111–131.
10. Bazant, Z. P., "Size Effect in Blunt Fracture: Concrete, Rock, Metal," *Journal of Engineering Mechanics, ASCE*, Vol. 110, No. 4, 1984, pp. 518–535.
11. Tang, T., Shah, S. P., and Ouyang, C., "Fracture Mechanics and Size Effect of Concrete in Tension," *Journal of Structural Engineering, ASCE*, Vol. 118, No. 11, 1992, pp. 3169–3185.
12. Hasegawa, T., Shioy, T., and Okada, T., "Size Effect on Splitting Tensile Strength of Concrete," *Proceedings of Japan Concrete Institute, 7th Conference*, Japan Concrete Institute, Tokyo, Japan, 1985, pp. 309–312.
13. Bazant, Z. P., Kazemi, M. T., Hasegawa, T., and Mazars, J., "Size Effect in Brazilian Split-Cylinder Tests: Measurements and Fracture Analysis," *ACI Materials Journal*, Vol. 88, No. 3, 1991, pp. 325–332.
14. Bazant, Z. P., Kazemi, M. T., Hasegawa, T., and Mazars, J., "Size Effect in Brazilian Split-Cylinder Tests: Measurements and Fracture Analysis," *ACI Materials Journal*, Vol. 88, No. 3, 1991, pp. 325–332.
15. Bazant, Z. P., and Oh, B. H., "Spacing of Cracking in Reinforced Concrete," *Journal of Structural Engineering, ASCE*, Vol. 109, No. 9, 1983, pp. 2066–2086.
16. Ouyang, C., and Shah, S. P., "A Fracture Energy Approach for Predicting Cracking of Reinforced Concrete Tensile Members," *ACI Structural Journal*, Vol. 91, No. 1, 1994, pp. 69–78.
17. Jenq, Y.-S., and Shah, S. P., "Features of Mechanics of Quasi-Brittle Crack Propagation in Concrete," *International Journal of Fracture*, Vol. 51, pp. 1991, pp. 103–120.
18. Ouyang, C., and Shah, S. P., "Geometry-Dependent *R*-Curve for Quasi-Brittle Materials," *Journal of the American Ceramic Society*, Vol. 74, No. 11, 1991, pp. 2831–2836.
19. Kemeny, J., and Cook, N. G., "Effective Moduli, Non-Linear Deformation and Strength of a Cracked Elastic Solid," *International Journal of Rock Mechanics, Mining Science & Geomechanics Abstract*, Vol. 23, 1986, pp. 107–118.
20. Gilbert, R. I., "Shrinkage Cracking in Fully Restrained Concrete Members," *ACI Structural Journal*, Vol. 89, No. 2, 1992, pp. 141–149.
21. Gopalaratnam, V. S., and Shah, S. P., "Softening Response of Plain Concrete in Direct Tension," *ACI Journal*, Vol. 82, No. 3, 1985, pp. 310–323.
22. Owens, D. R. J., and Figueras, J. A., "Load Analysis of Reinforced Concrete Plates and Shells Including Geometrical Nonlinear Effect," *Finite Element Software for Plates and Shells*, Pineridge, Swansen, United Kingdom, 1984.
23. Rizkalla, S. H., Hwang, L. S., and Shahawi, M. El, "Transverse Reinforcement Effect on Cracking Behavior of R. C. Members," *Canadian Journal of Civil Engineering*, Vol. 10, 1983, pp. 566–581.
24. Yannopoulos, P. J., "Variation of Concrete Crack Widths through the Concrete Cover to Reinforcement," *Magazine of Concrete Research*, Vol. 147, 1989, pp. 63–68.

25. Hillerborg, A., Modeer, M., and Petersson, P.-E., "Analysis of Crack Formation and Crack Growth in Concrete by Means of Fracture Mechanics and Finite Elements," *Cement and Concrete Research*, Vol. 6, No. 6, 1976, pp. 773–782.
26. Hillerborg, A., "Fracture Mechanics Concepts Applied to Moment Capacity and Rotational Capacity of Reinforced Concrete Beams," *Engineering Fracture Mechanics*, Vol. 35, No. 1/2/34, 1990, pp. 233–240.
27. CEB-FIP Model Code 1990, First Predraft 1988, Bulletin d'Information No. 190a, 190b, Comite Euro-International du Beton (CEB), Lausanne.
28. ACI Committee 318, "Building Code Requirements for Reinforced Concrete (ACI 318-89) and Commentary (ACI 318R-89)," American Concrete Institute, Detroit, MI, 1989.
29. Carpinteri, A., "Stability of Fracturing Process in *RC* Beams," *Journal of Structural Engineering, ASCE*, Vol. 110, 1984, pp. 544–558.
30. Bosco, C., Carpinteri, A., and Debernardi, P. G., "Minimum Reinforcement in High-Strength Concrete," *Journal of Structural Engineering, ASCE*, Vol. 116, 1990, pp. 427–437.
31. Baluch, M. H., Azad, A. K., and Ashmawi, W., "Fracture Mechanics Application to Reinforced Concrete Members in Flexure," in *Applications of Fracture Mechanics to Reinforced Concrete*, edited by Carpinteri, A., Elsevier Applied Science, London, 1992, pp. 413–436.
32. Okamura, H., Watanabe, K., and Takano, T., "Application of the Compliance Concept in Fracture Mechanics," ASTM STP 536, American Society for Testing and Materials, Philadelphia, PA, 1973, pp. 423–438.
33. Okamura, H., and Watanabe, K., and Takano, T., "Deformation and Strength of Cracked Members under Bending Moment and Axial Force," *Engineering Fracture Mechanics*, Vol. 7, 1975, pp. 531–539.
34. Gerstle, W. H., Dey, P. P., Prasad, N. N. V., Rahulkumar, P., and Xie, M., "Crack Growth in Flexural Members—A Fracture Mechanics Approach," *ACI Structural Journal*, Vol. 89, No. 6, 1992, pp. 617–625.
35. Bosco, C., and Carpinteri, A., "Fracture Mechanics Evaluation of Minimum Reinforcement in Concrete Structures," in *Applications of Fracture Mechanics to Reinforced Concrete*, edited by Carpinteri, A., Elsevier Applied Science, London, 1992, pp. 347–377.
36. Commission of the European Communities, Industrial Process—Building and Civil Engineering, Eurocode No. 2, Design of Concrete Structures, Part 1: General Rules and Rules for Buildings, Commission of the European Communities, Brussels, October 1989.
37. Hawkins, N. M., and Hjorteset, K., "Minimum Reinforcement Requirements for Concrete Flexural Members," in *Applications of Fracture Mechanics to Reinforced Concrete*, edited by Carpinteri, A., Elsevier Applied Science, London, 1992, pp. 379–412.
38. Ingraffea, A. R., Gerstle, W. H., Gergely, P., and Saouma, V., "Fracture Mechanics of Bond in Reinforced Concrete," *Journal of Structural Engineering, ASCE*, Vol. 110, No. 4, 1984, pp. 871–890.
39. Brown, C. J., Darwin, D., and McCabe, S. L., "Finite Element Fracture Analysis of Steel–Concrete Bond," S. M. Report 36, University of Kansas Center for Research, Lawrence, KS, November 1993.

40. Karihaloo, B. L., ''Failure Modes of Longitudinally Reinforced Beams,'' in *Application of Fracture Mechanics to Reinforced Concrete*, edited by Carpinteri, A., Elsevier Applied Science, London, 1992, pp. 523–546.
41. Bazant, Z. P., and Kim, J.-K., ''Size Effect in Shear Failure of Longitudinally Reinforced Beams,'' *ACI Journal*, Vol. 81, 1984, pp. 456–468.
42. Bazant, Z. P., ''Size Effect in Blunt Fracture: Concrete, Rock, Metal,'' *Journal of Engineering Mechanics, ASCE*, Vol. 110, 1984, pp. 518–535.
43. Bazant, Z. P., and Kazemi, M. T., ''Size Effect on Diagonal Shear Failure of Beams without Stirrups,'' *ACI Structural Journal*, Vol. 88, No. 3, 1991, pp. 268–276.
44. Bazant, Z. P., and Sun, H. H., ''Size Effect in Diagonal Shear Failure: Influence of Aggregate Size and Stirrups,'' *ACI Materials Journal*, Vol. 83, No. 4, 1987, pp. 259–272.
45. Jenq, Y.-S., and Shah, S. P., ''Shear Resistance of Reinforced Concrete Beams—A Fracture Mechanics Approach,'' in *Fracture Mechanics: Application to Concrete*, edited by Li, V. C., and Bazant, Z. P., SP-118, American Concrete Institute, Detroit, Detroit, MI, 1989, pp. 237–258.
46. Ferguson, P. M., and Thompson, J. N., ''Development Length of High Strength Reinforcing Bars in Bond,'' *ACI Journal*, Vol. 59, 1962, pp. 887–922.
47. Ferguson, P. M., and Thompson, J. N., ''Development Length for Large High Strength Reinforcing Bars,'' *ACI Journal*, Vol. 62, 1965, pp. 71–93.
48. Gustafsson, P. J., and Hillerborg, A., ''Sensitivity in Shear Strength of Longitudinally Reinforced Concrete Beams to Fracture Energy of Concrete,'' *ACI Structural Journal*, Vol. 85, 1988, pp. 286–294.
49. Hillerborg, A., Modeer, M., and Petersson, P. E., ''Analysis of Crack Formation and Crack Growth in Concrete by Means of Fracture Mechanics and Finite Elements,'' *Cement and Concrete Research*, Vol. 6, 1976, pp. 773–782.
50. Shioya, T., Iguro, M., Nojiri, Y., Akiyama, H., and Okada, T., ''Shear Strength of Large Reinforced Concrete Beams,'' in *Fracture Mechanics: Applications to Concrete*, edited by Li, V. C., and Bazant, Z. P., ACI SP-118, American Concrete Institute, Detroit, MI, 1989, pp. 259–280.
51. So, K. O., and Karihaloo, B. L., ''Shear Capacity of Longitudinally Reinforced Beams—A Fracture Mechanics Approach,'' *ACI Structural Journal*, Vol. 90, No. 6, 1993, pp. 591–600.
52. Bazant, Z. P., and Şener, S., ''Size Effect in Torsional Failure of Longitudinally Reinforced Concrete Beams,'' *Journal of Structural Engineering, ASCE*, Vol. 113, No. 10, 1987, pp. 2125–2136.
53. Ozbolt, J., and Eligehausen, R., ''Analysis of Reinforced Concrete Beams without Shear Reinforcement Using Nonlocal Microplane Model,'' in *Fracture Processes in Concrete, Rock, and Ceramics*, edited by van Mier, J. G. M., Rots, J. G., and Bakker, A., E & F. N. Spon, London, 1991, pp. 919–930.
54. Hawkins, N., ''The Role for Fracture Mechanics in Conventional Reinforced Design,'' in *Application of Fracture Mechanics to Cementitious Composites, NATO-ARW Proceedings*, edited by Shah, S. P., Martinus Nijhoff, The Hague, 1985, pp. 639–666.
55. Bazant, Z. P., and Kwon, Y. W., ''Size Effect in Failure of Short or Slender

Reinforced Concrete Columns,'' Report No. 91-12/457s, NSF Center for Advanced Cement Based Materials, Northwestern University, Evanston, IL, 1992.

56. ACI Committee 349, ''Code Requirements for Nuclear Safety Related Concrete Structures (ACI 349-85),'' in *ACI Manual of Concrete Practice*, Part 4, American Concrete Institute, Detroit, MI, 1985, Appendix B.

57. Eligehausen, R., and Sawade, G., ''Analysis of Anchorage Behavior (Literature Review),'' in *Fracture Mechanics of Concrete Structures: From Theory to Applications*, edited by Elfgren, L., Chapman & Hall, London, 1989, pp. 263–280.

58. Elfgren, L., Ohlsson, U., and Gylltoft, K., ''Anchor Bolts Analyzed With Fracture Mechanics,'' in *Fracture of Concrete and Rock*, edited by Shah, S. P., and Swartz, S. E., Springer-Verlag, New York, 1989, pp. 269–275.

59. Ballarini, R., Shah, S. P., and Keer, L. M., ''Failure Characteristics of Short Anchor Bolts Embedded in a Brittle Material,'' *Proceedings of the Royal Society of London*, Vol. A 404, 1986, pp. 35–54.

60. Elfgren, L., and Ohlsson, U., ''Anchor Bolts Modeled with Fracture Mechanics,'' in *Application of Fracture Mechanics to Reinforced Concrete*, edited by Carpinteri, A., Elsevier Applied Science, London, 1992, pp. 267–283.

61. Bocca, P., Carpinteri, A., and Valente, S., ''Fracture Mechanics Evaluation of Anchorage Bearing Capacity in Concrete,'' in *Application of Fracture Mechanics to Reinforced Concrete*, edited by Carpinteri, A., Elsevier Applied Science, London, 1992, pp. 231–265.

62. RILEM TC 90-FMA, ''Round Robin Analysis of Anchor Bolts—Invitation,'' *Materials and Structures*, Vol. 23, No. 133, 1990, p. 78.

63. Elfgren, L., and Swartz, S. E., ''Finite Mechanics Approaches to Modeling the Pullout of Anchor Bolts,'' in *Design Based on Fracture Mechanics*, ACI SP-134, edited by Gerstle, W., and Bazant, Z. P., American Concrete Institute, Detroit, MI, 1992, pp. 63–77.

64. Bazant, Z. P., and Şener, S., ''Size Effects in Pull-out Tests,'' *ACI Materials Journal*, Vol. 85, No. 5, 1988, pp. 347–351.

65. Eligehausen, R., and Ozbolt, J., ''Size Effect in Anchorage Behavior,'' in *Fracture Behavior and Design of Materials and Structures*, edited by Firrao, D., Engineering Materials Advisory Services, Cradley Heath, UK, 1990, pp. 721–727.

66. Gustafsson, P. J., and Hillerborg, A., ''Improvements in Concrete Design Achieved through the Application of Fracture Mechanics,'' in *Application of Fracture Mechanics to Cementitious Composites, NATO-ARW Proceedings*, edited by Shah, S. P., Martinus Nijhoff, The Hague, 1985, pp. 667–680.

67. Bazant, Z. P., and Cao, S., ''Size Effect in Brittle Failure of Unreinforced Pipes,'' *ACI Journal*, Proceedings, Vol. 83, No. 3, 1986, pp. 369–373.

68. Bazant, Z. P., ''A Critical Appraisal of 'No-Tension' Dam Design: A Fracture Mechanics Viewpoint,'' *Dam Engineering*, Vol. 1, Issue 4, 1990, pp. 237–248.

69. Linsbauer, H., Ingraffea, A. R., Rossmanith, H., and Wawrzynek, P. A., ''Simulation of Cracking in Large Arch Dam: Part I,'' *Journal of Structural Engineering, ASCE*, Vol. 115, No. 7, 1988, pp. 1599–1615.

70. Linsbauer, H., Ingraffea, A. R., Rossmanith, H., and Wawrzynek, P. A., ''Simulation of Cracking in Large Arch Dam: Part II,'' *Journal of Structural Engineering, ASCE*, Vol. 115, No. 7, 1988, pp. 1615–1630.

71. Saouma, V. E., Ayari, M. L., and Boggs, H., ''Fracture Mechanics of Concrete Gravity Dams,'' in *Fracture of Concrete and Rock*, edited by Shah, S. P., and Swartz, S. E., Springer-Verlag, New York, 1989, pp. 311–333.

72. U.S. Army Corps of Engineers, ''Fracture Mechanics Analysis of Concrete Hydraulic Structures,'' Engineering Technical Letter, No. 1110-8-16 (FR), Department of the Army, Washington, DC, 1991.

73. Saouma, V. E., Broz, J. J., Brühwiler, E., and Ayari, M. L., ''Fracture Mechanics of Concrete Dams,'' Vols. I, II, and III, Reports submitted to the Electric Power Research Institute by the Department of Civil Engineering, University of Colorado, Boulder, CO, 1990.

74. Zhang, C.-Y., and Karihaloo, B. L., ''Stability of a Crack in a Linear Viscoelastic Tension-Softening Material,'' in *Fracture Mechanics of Concrete Structures, Proceedings of FraMCoS1*, edited by Bazant, Z. P., Elsevier Applied Science, New York, 1992, pp. 155–162.

75. Brühwiler, E., and Wittmann, F. H., ''The Wedge-Splitting Test, a New Method of Performing Stable Fracture Mechanics Tests,'' *Engineering Fracture Mechanics*, Vol. 35, No. 1/2/3, 1990, pp. 117–125.

76. Saouma, V. E., Broz, J. J., Brühwiler, E., and Boggs, H., ''Effect of Aggregate and Specimen Size on Fracture Properties of Dam Concrete,'' *Journal of Materials in Civil Engineering, ASCE*, Vol. 3, No. 3, 1991, pp. 204–218.

77. Bazant, Z. P., He, S., Plesha, M. E., Gettu, R., and Rowlands, R. E., ''Rate and Size Effects in Concrete Fracture: Implications for Dams,'' in *Proceedings of International Conference on Dam Fracture*, edited by Saouma, V. E., University of Colorado, Boulder, CO, 1991, pp. 413–425.

78. Reinhardt, H. W., ''Length Influence on Bond Shear Strength of Joints in Composites Precast Concrete Slabs,'' *International Journal of Cement Composites and Lightweight Concrete*, Vol. 4, No. 3, 1982, pp. 139–143.

79. Buyukozturk, O., and Lee, K. M., ''Interface Fracture Mechanics of Concrete Composites,'' in *Fracture Mechanics of Concrete Structures*, *Proceedings of FraMCoS1*, edited by Bazant, Z. P., Elsevier Applied Science, London, 1992, pp. 163–168.

80. Lee, K. M., Buyukozturk, O., and Oumera, A., ''Fracture Analysis of Mortar–Aggregate Interfaces in Concrete,'' *Journal of Engineering Mechanics*, Vol. 118, No. 10, 1992, pp. 2031–2047.

81. Mitsui, K., Li, Z., Lange, D. A., and Shah, S. P., ''Relationship between Microstructure and Mechanical Properties of the Paste–Aggregate Interface,'' *ACI Materials Journal*, Vol. 91, No. 1, 1994, pp. 30–39.

82. Stang, H., Li, Z., and Shah, S. P., ''Pullout Problem: Stress versus Fracture Mechanical Approach,'' *Journal of Engineering Mechanics*, Vol. 116, No. 10, 1990, pp. 2136–2150.

83. Wang, J., and Maji, A. K., ''Experimental Studies and Modeling of the Concrete/Rock Interface,'' paper presented at ACI Spring Convention, Vancouver, 1993.

10

FRACTURE MECHANICS AND COMPRESSIVE FAILURE

10.1 EXPERIMENTAL OBSERVATIONS OF CRACK PROPAGATION AND DAMAGE

Concrete has much higher strength in compression than in tension. As discussed in Chapter 4, many initial flaws exist in a cementitious matrix as well as at the interface between coarse aggregate and the matrix due to segregation during casting and shrinkage and thermal cracking. Propagation of these cracks is responsible for the so-called compressive failure of concrete.

Many studies have been previously conducted to observe propagation of cracks for concrete under compression. Hsu et al.[1] and Shah and Chandra[2] have observed internal cracks of compressive specimens with a microscope and x-ray photographs. Shah and Sankar[3] tested a series of concrete cylinders of 75 mm (3 in.) diameter and 150 mm (6 in.) height under uniaxial compression using a closed-loop testing machine. Axial displacement and diametric deformations at different levels of specimen height (locations *A*, *B*, and *C* in Fig. 10-1) were measured during testing. A typical compressive stress–axial displacement relationship obtained is shown in Fig. 10-1*a*, and corresponding diametric deformations at various loading levels along the specimen height are indicated in Fig. 10-1*b*. It is seen that the lateral displacement is basically uniform along the specimen height up to about 80% of the peak load (point 2). After point 2, the lateral deformation at location *B* increases much faster than those at locations *A* and *C*. This indicates that to some extent strain localization appears in the vicinity of location *B*. Shah and Sankar[3] also monitored crack patterns in both transverse and longitudinal directions using petrography techniques. Specimens loaded to each deformation level were sectioned, one transversely and one longitudinally. Slices $1\frac{1}{2}$ in. (12 mm) thick

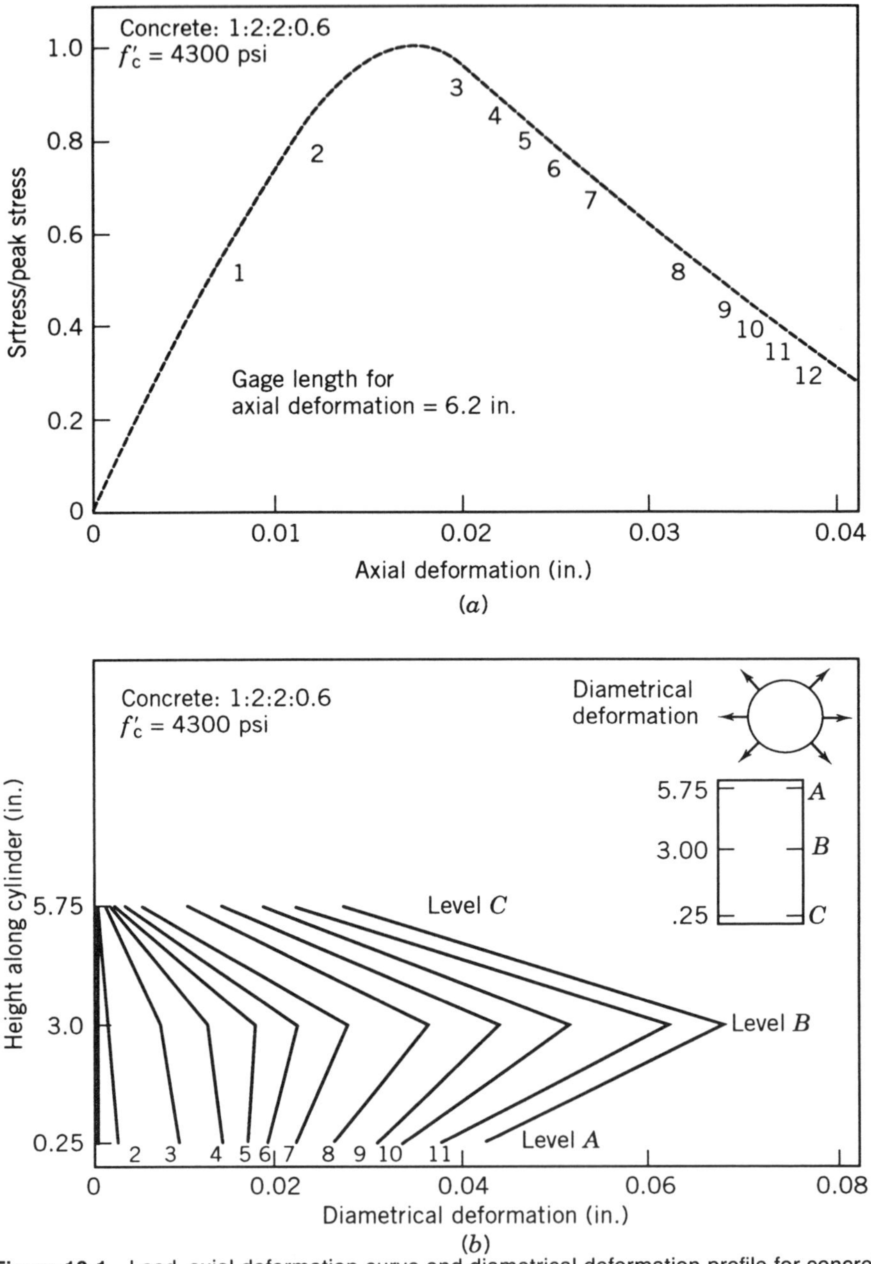

Figure 10-1 Load–axial deformation curve and diametrical deformation profile for concrete cylinder under uniaxial compression.[3]

were cut from the center of the specimen using a diamond saw. The obtained crack patterns for different loading stages are shown in Fig. 10-2. It was reported based on these observations that the extent of cracking up to about 83% of peak load is primarily limited to bond cracking between coarse aggregates and mortar matrix (Fig. 10-2*a*). These bond cracks are uniformly distributed in both transverse and longitudinal directions. When the applied stress continuously increases, these bond cracks start to coalesce into continuous cracks that can be seen at the two edges of the longitudinal section, as indicated in Figs. 10-2*b*, *c*. Cracking is uniform and extensive in both longitudinal and transverse sections further into the postpeak region (Fig. 10-2*d*). Shah and Chandra[2] observed a significant increase of specimen volume after about 80% of the peak load. A similar strain localization was also observed by Torrenti et al.[4,5] using stereophotogrammetry and speckle laser.

Maji and Shah[6] used laser holography to observe initiation and propagation of cracks in model concrete specimens. Results obtained are illustrated in Fig. 10-3. Cracks initiate at the interface between aggregates and cementitious matrix and then propagate into the matrix as applied load increases. Crack paths near the peak stress are basically in the same direction as the applied load.

Based on these experimental observations, the failure process of a uniaxial compressive concrete specimen may be summarized as in Fig. 10-4. The compressive stress–strain relationship may be divided into several stages. Stress initially linearly increases with strain. The microflaws start to open at point A, which is approximately 30% of peak stress. The interfacial flaws propagate at point B (about 50% of the peak stress). These bond cracks, which are isolated and uniformly distributed over the whole specimen, propagate in different directions based on the shapes of the aggregates. These bond cracks propagate into the mortar matrix and coalesce as a single crack or several major cracks that are in the direction of the applied compressive load at about 80% of the peak stress (point C). This is strain localization in compression. The beginning of the strain localization accelerates failure of the specimen. These major cracks stably propagate, and one of them reaches its critical length at point D. As a result, the specimen exhibits its peak compressive stress. After the peak stress, the major cracks automatically propagate even though the applied load decreases. A softening type of stress–strain curve is usually observed for the compressive failure of concrete. Approximately, the specimen continuously contracts up to point C. However, the volume of the compressed concrete starts to increase rather than contract after pont C.[2,7] Although Poisson's ratio slightly increases as the load is applied, its value suddenly greatly increases approximately after point C. It is noted that for high-strength concrete, since the interfacial bond is improved by the use of cementitious materials such as silica fume, cracks may not initiate from the interfacial bond and some cracks may penetrate through aggregates.

When a uniaxial compressive specimen is tested, some lateral restraint stress may develop at the machine–specimen boundary due to different deformations between the steel loading platen and concrete.[8] The crack pattern (failure mode)

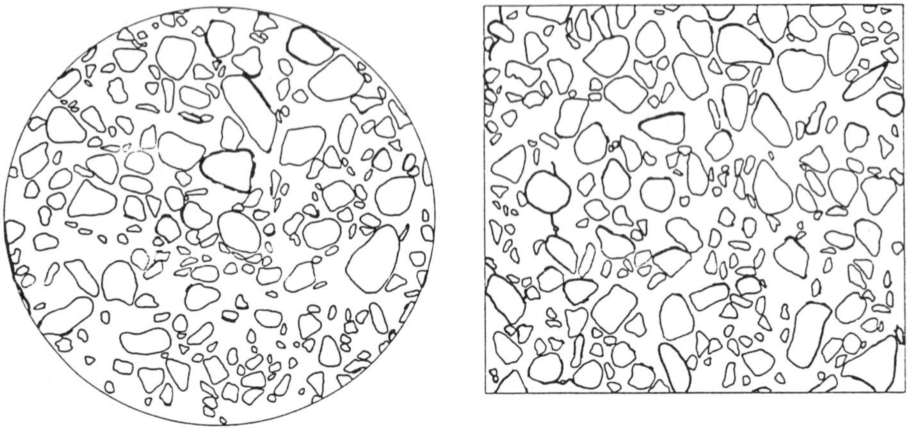

Transverse slice *Longitudinal slice*

(a)

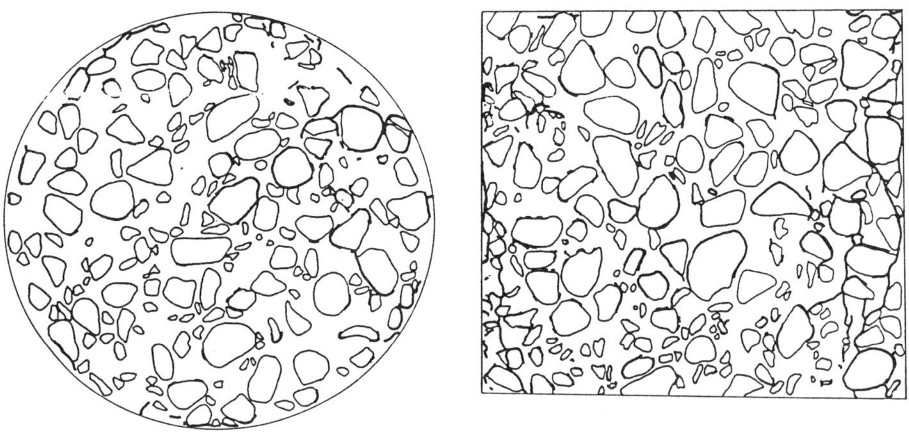

Transverse slice *Longitudinal slice*

(b)

Figure 10-2 Transverse and longitudinal crack patterns at different loading stages: (*a*) up to 83% of the peak load, (*b*) up to the peak load, (*c*) up to 93% of the peak stress in the post-peak region, and (*d*) up to 81% of the peak stress in the post-peak region.[3]

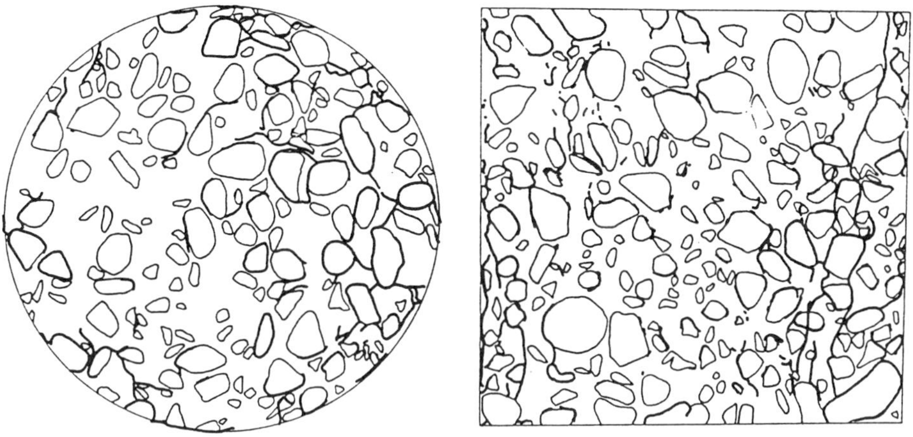

Transverse slice *Longitudinal slice*

(*c*)

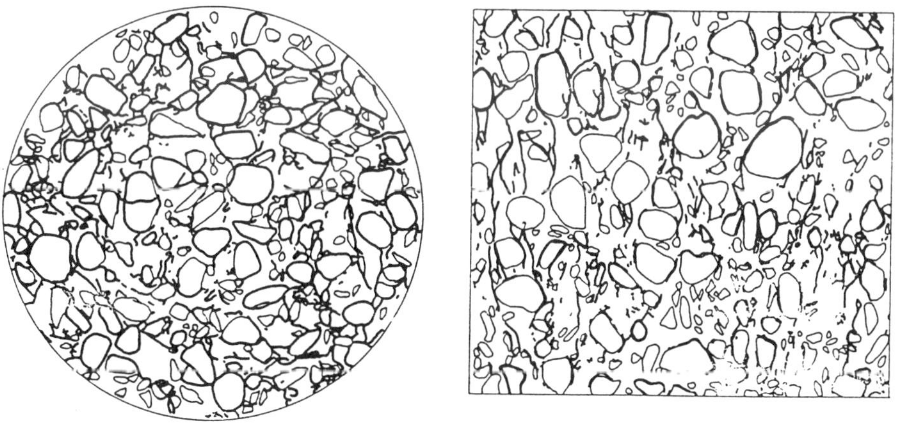

Transverse slice *Longitudinal slice*

(*d*)

Figure 10-2 (*Continued*)

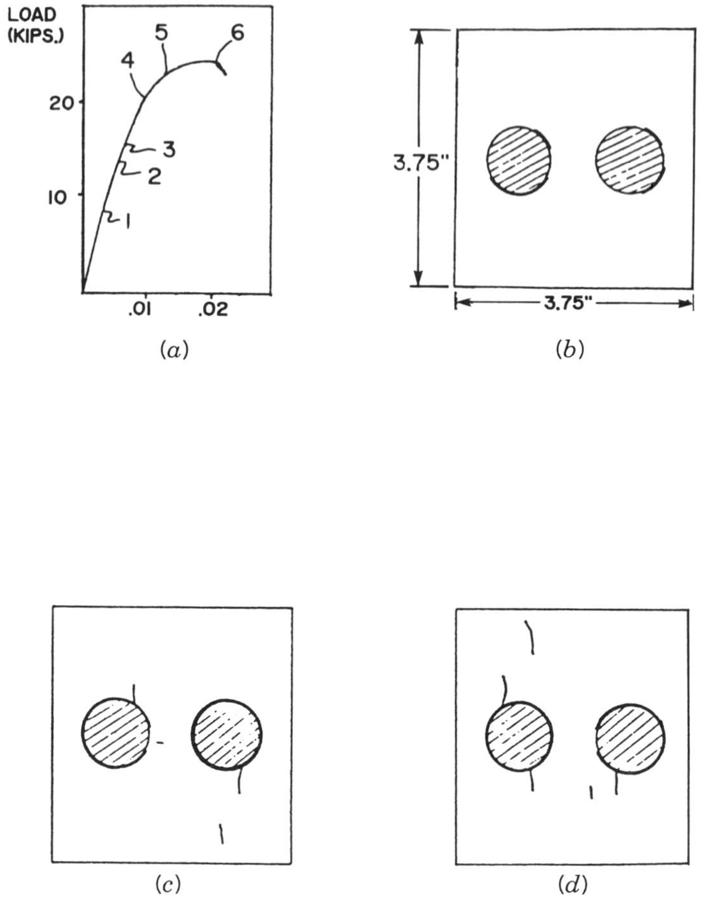

Figure 10-3 Propagation of cracks in model concrete plate under uniaxial compression observed by laser holographic interferometry: (*a*) load displacement curve, (*b*) crack pattern at point 1, (*c*) crack pattern at point 2, (*d*) crack pattern at point 3, (*e*) crack pattern at point 4, (*f*) crack pattern at point 5 and (*g*) crack pattern at point 6.[5]

of a compressive specimen is influenced by this lateral restraint stress because it results in triaxial compressive stress states at the end regions of the specimen. These triaxial stress states arrest cracking in the end regions, which may delay or prevent strain localization and formation of macrocracks. Therefore, when plain steel loading platens are used for testing of a short specimen, a hourglass-type failure mode as shown in Fig. 10-5*a* is usually obtained. This type of failure mode cannot be used to characterize compressive failure of concrete. The influence of the lateral restraint stress can be minimized by using various mechanisms such as a brush-type loading device or lubricants. In this case, a significant amount of prominent vertical splitting cracks accompanied by some

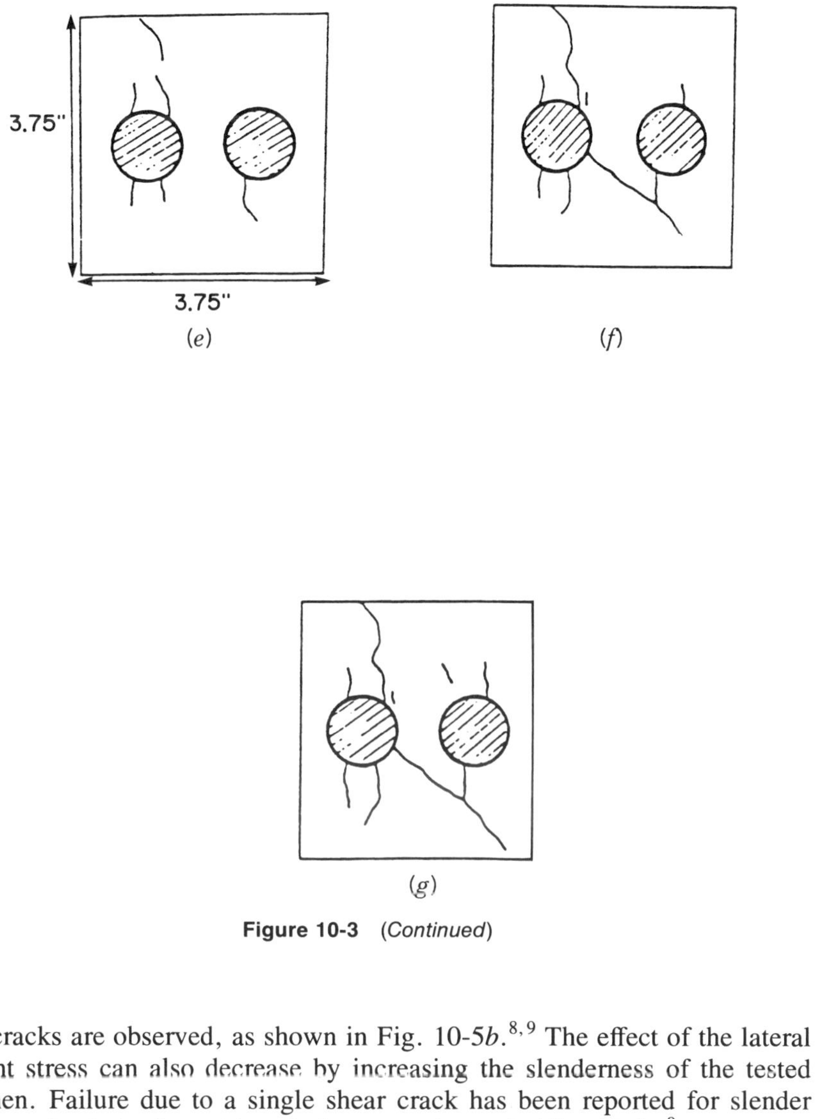

Figure 10-3 *(Continued)*

shear cracks are observed, as shown in Fig. 10-5*b*.[8,9] The effect of the lateral restraint stress can also decrease by increasing the slenderness of the tested specimen. Failure due to a single shear crack has been reported for slender specimens loaded using a hinged platen, as shown in Fig. 10-5*c*.[9]

Based on the above discussion, in a manner similar to tensile failure, the compressive failure of concrete is characterized by a strain localization evidenced by coalescence of one or several major cracks accompanied by sudden increase of Poisson's ratio. The major cracks formed after the strain localization are in directions of local maximum principal compressive stress, which globally coincide with the applied compressive load, rather than in a direction perpendicular to the applied force in tension. Since a crack results from tensile stresses developed primarily due to the effect of Poisson's ratio, several major cracks, instead of a single crack in the case of tension, are often observed after

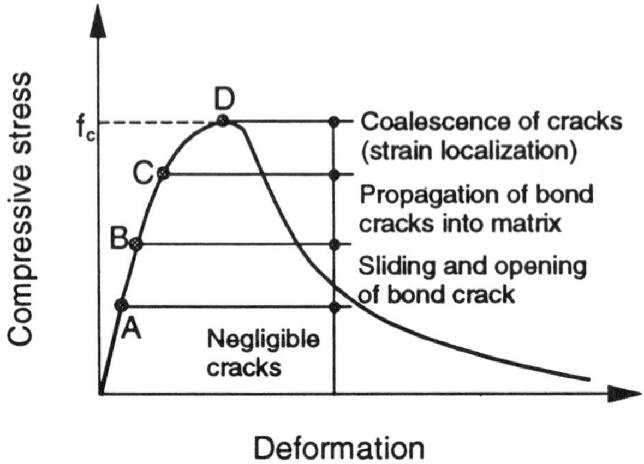

Figure 10-4 Stages of crack propagation in concrete under uniaxial compression.

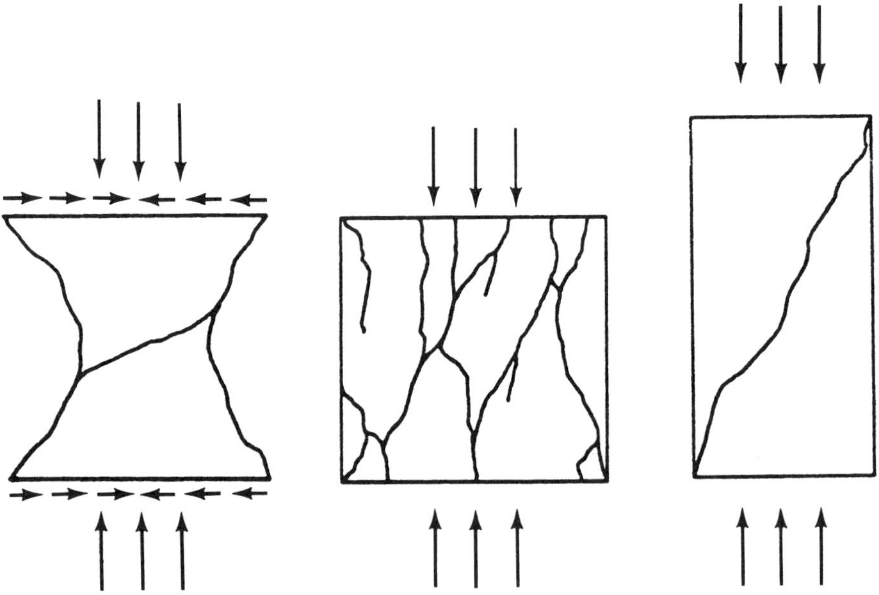

(*a*) (*b*) (*c*)

Figure 10-5 Failure modes of concrete specimens under uniaxial compression.[8,9]

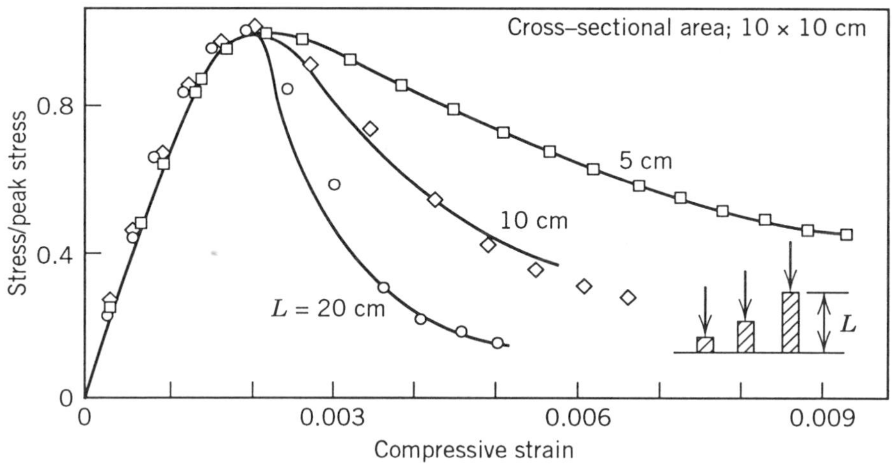

Figure 10-6 Influence of specimen height on uniaxial compressive stress–strain curves.[9]

the strain localization. The presence of strain localization allows one to use fracture mechanics to describe the failure behavior of concrete under compression.

The fact that fracture mechanics is a suitable approach to describe compressive failure of concrete is confirmed by experimental results reported by van Mier,[9] as shown in Fig. 10-6. His results indicate that postpeak compressive stress–strain curves depend on the height of prismatic specimens. Specimens with different heights have almost identical stress–strain behavior up to peak stress. However, longer specimens exhibit less strain after the peak stress compared to shorter specimens. This experimental observation indicates that many previous constitutive models[10,11] based on softening types of stress–strain relationships may not be appropriate for concrete. On the other hand, fracture mechanics can be used to reasonably describe the observed strain localization. An extensive review of experimental and theoretical studies on compressive failure of concrete has been given by Vonk.[12] Various approaches for compressive failure of concrete based on fracture mechanics will be discussed in this chapter.

10.2 FRACTURE MODELS FOR FAILURE RESULTING FROM INTERFACIAL BOND CRACKS

Both interfacial cracks and matrix voids exist in concrete before loading. A major crack usually initiates from an interfacial crack. This may be due to the facts that (i) the size of an interfacial crack is usually much larger than that of the matrix voids and (ii) the interfacial layer (transition zone) has a poor microstructure that results in a weaker fracture resistance.[13] The presence of the interfacial transition zone greatly influences the fracture behavior of concrete. Concrete may be regarded as a three-phase material consisting of aggregate, cementitious matrix, and an interfacial transition zone.[14]

Since the weakest link in concrete is the interfacial bond layer, many investigations have been conducted to study the microstructure and mechanical properties of the interfacial transition zone. Hsu and Slate[15] studied the tensile bond strength between aggregate and paste or mortar. They reported that for different types of aggregates the tensile bond strength ranges approximately from 40 to 70% of that for the paste or mortar. Shah and Winter[16] used a unit consisting of a circular-cylindrical piece of aggregate, an interfacial thin layer, and a cementitious matrix, as shown in Fig. 10-7, to simulate structural behavior of concrete under uniaxial compression. In this figure the ratio of d_1/r_a depends primarily on the relative amounts of aggregate and mortar in a given mix, and its value ranges from 0.2 to 0.6 for conventional concrete.[17] A theoretical model to predict compressive stress–strain curves of concrete was derived based on such a unit. In Shah and Winter's model[16] both the aggregate and the matrix were assumed to be linear elastic materials, whereas the interfacial bond was assumed to follow an elastic-plastic stress–strain relationship. Initially, the interfacial bond between the aggregate and the matrix is intact,

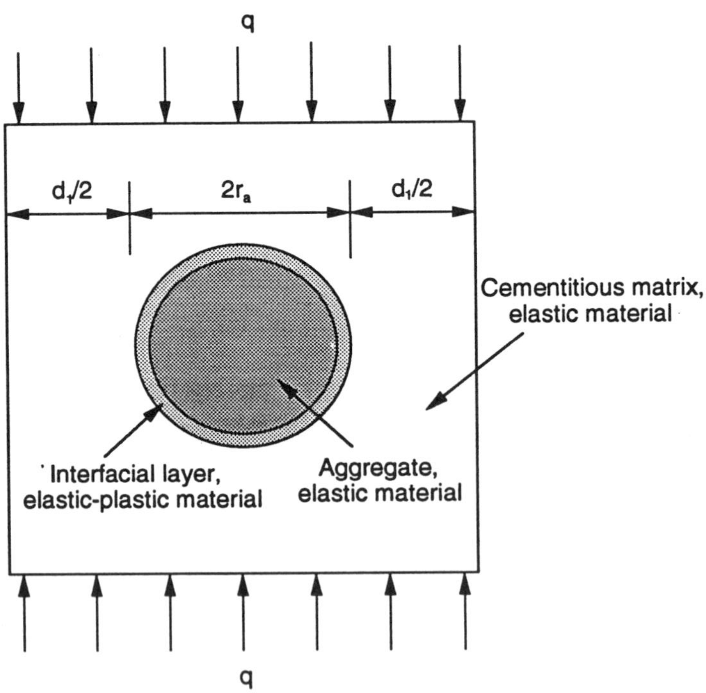

Figure 10-7 Structural unit to model compressive response of concrete proposed by Shah and Winter.[16]

and the unit behaves linearly elastically. Since the bond layer has a lower strength, some part of the bond layer reaches its yield stress at a certain load level. This yielded zone in the bond layer, which can be regarded as a bond crack, propagates with increasing load. Failure of the bond layer is described by Mohr's failure theory. After the bond layer starts to yield, the global relationship stress–strain becomes nonlinear. The above procedure was analytically implemented by Shah and Winter.[16] Their model is capable of predicting the stress at which the bond cracks initiate and the matrix cracking begins. Nonlinear stress–strain curves of concrete under uniaxial compression result from the progressive bond cracking. Note that in this model only the interfacial debonding was simulated, and the effect of matrix cracking on the stress–strain response was neglected. Some of the bond cracks propagate into the cementitious matrix when strain localization begins, and isolated bond cracks at interfaces of different aggregates coalesce into one or several major cracks. Since the matrix cracking has a significant effect on the stress–strain behavior after the strain localization, this model may only be valid up to the beginning of strain localization (at about 80% of the peak stress).

Shah and Winter's model was significantly improved by Buyukozturk et al.[18] They proposed a unit including nine cylindrical aggregate pieces placed in three symmetric rows about horizontal and vertical axes, as shown in Fig. 10-8. A plane-stress FE analysis was applied to the three different phases of the materials to obtain stress–strain curves of the unit. Since nine aggregate pieces were included, their model can be used to study the coalescence of isolated bond cracks at interfaces of different aggregates after strain localization. Results from the FE analysis indicated that cracks first initiate at weak bond layers surrounding different aggregates, and at about 70% of the peak load, these bond cracks start to propagate into the cementitious matrix. As the applied load further increases, several major cracks appear, and the unit fails at the peak stress. They reported that predicted crack patterns coincided with experimental observations and that the predicted and observed axial stress–strain curves up to the peak stress were in good correlation. Palaniswamy and Shah[19] further applied this nine-aggregate-piece model to concrete under cylindrical triaxial stresses. It is noted that the interfacial layer may not be weak for high-strength concrete due to the use of silica fume. Therefore, this approach may not be applied effectively to high-strength concrete.

Vonk[12] has used a more complicated FE model to study failure of compressive specimens due to interfacial cracks.

10.3 MULTILEVEL FRACTURE MODEL OF ZAITSEV AND WITTMANN

10.3.1 Concrete as Multilevel Hierarchy System

Zaitsev and Wittmann[20] and Zaitsev[21] proposed a multilevel fracture model to describe compressive failure of concrete. Based on its microstructure, a ce-

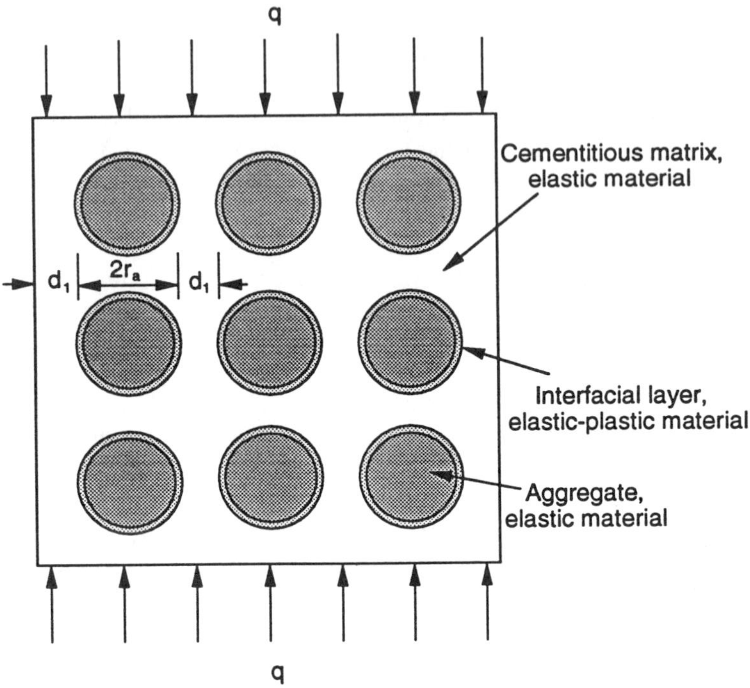

Figure 10-8 Structural unit to model compressive response of concrete proposed by Buyukozturk et al.[18]

mentitious material may be divided into macrolevel, mesolevel, and microlevel for predicting its fracture behavior. Inhomogeneity on each level was represented by a type of inclusion. Linear elastic fracture mechanics was used to describe fracture at these different material levels. To assure validity of LEFM, the minimum dimension of the representative volume element (RVE) was required to be at least four times the maximum grain size. The inhomogeneity on the macrolevel was given by ''big'' inclusions such as coarse aggregates. The typical linear dimension of the RVE for this level was assumed to be on the order of magnitude of 100 mm. On the mesolevel, the mortar volume between aggregates was considered. Inclusions on this level were average grains of fine aggregates. The typical linear dimension of the RVE for the mesolevel was an order of magnitude of 10 mm. On the microlevel, hardened cement paste volumes between fine aggregates were considered. A main inhomogeneity was represented by big capillary pores. The maximum size of these pores may have an order of magnitude of 0.01 mm, and the typical linear dimension of the RVE for the microlevel may be approximately 0.5 mm.

10.3.2 Crack Propagation in Hardened Cement Paste

Crack propagation in a hardened cement paste is based on the microlevel consideration and may be described by the simplified model shown in Fig. 10-9*a*. The big pores on the microlevel were modeled by a set of circular holes with a radius of r_h, and two line cracks were assumed to associate with each hole. It was simply assumed that all the cracks are coplanar and are in the most dangerous orientation (the direction of applied compressive load). When the load increases, cracks start to propagate in a stable way that required $K_{\mathrm{I}} = K_{\mathrm{I}c}$, where K_{I} is the stress intensity factor and $K_{\mathrm{I}c}$ is the critical stress intensity factor. This condition can be described by the expression

$$q = -K_{\mathrm{I}c}\sqrt{\frac{\pi}{4r_h}\frac{(1 + \lambda_h)^7}{(1 + \lambda_h)^2 - 1}} \tag{10.1}$$

where $\lambda_h = l_h r_h$ is the relative crack length and q is the applied compressive load, which is negative for compression. The detailed derivation of Eq. (10.1) and a more general condition with different crack lengths at both sides of a

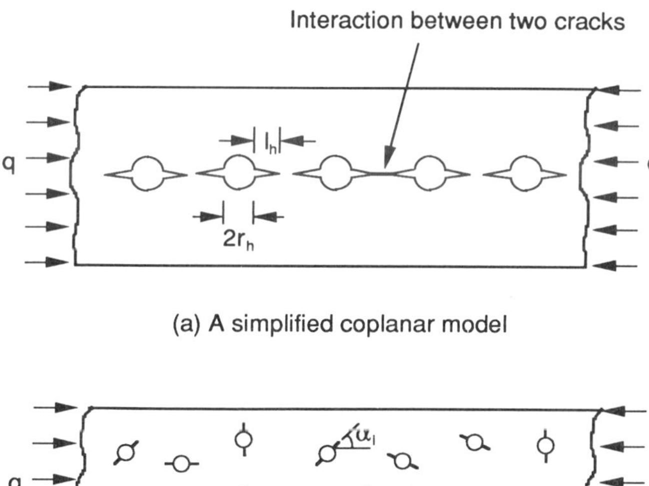

(a) A simplified coplanar model

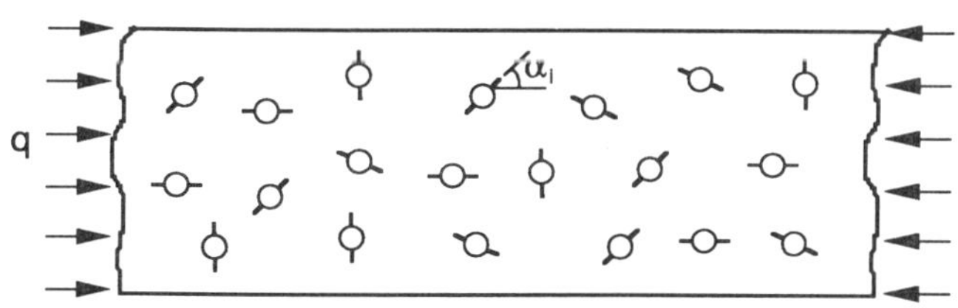

(b) A two-dimensional model with randomly distributed cracks

Figure 10-9 Models for structure of hardened cement paste.[20]

hole are also given by Zaitsev.[21] When the load further increases, cracks begin to interact. Beyond a certain load the crack propagation becomes unstable and the specimen fails.

In the above simplified analysis, a series of coplanar cracks is assumed. However, in a real hardened cement paste, cracks are oriented at random. It may be assumed that all pores are statistically uniformly distributed, as shown in Fig. 10-9*b*. Each pore may be assumed to have two preexisting cracks, and the length of the cracks is uniformly distributed within the range of $0 < l_h < 2r_h$. The angle of crack orientation with respect to the applied load, α_i, is also uniformly distributed within the limits between 0 and 2π. The interaction of cracks was taken into account by assuming that two cracks will interact and coalesce if α_1 and α_2 are both below $\pi/6$ or α_1 is below $\pi/6$ while α_2 is above $5\pi/6$. The random structure was created by a computer program. If crack propagation is studied by the Monte Carlo method, the results are similar to those obtained by using the one-dimensional model shown in Fig. 10-9*a*.

For both model structures (the coplanar cracks and the two-dimensional crack arrangement shown in Fig. 10-9), the equations describing crack propagation for stable crack propagation can be written in a general form as

$$q = -\frac{K_{Ic}}{\sqrt{2r_h}}\, g(\lambda_h) \tag{10.2}$$

where $g(\lambda_h)$ is a function of the related crack length λ_h.

Simulation of crack propagation as described above makes it possible to estimate the effect of pore size distribution on the fracture mechanism, mechanical strength, and strain behavior of hardened cement paste. It has been found that by increasing the mean size of pores while keeping the quantity constant, the ultimate load will decrease.[20] It is noted that a Griffith-type fracture criterion was used for both Eqs. (10.1) and (10.2) because only the critical stress intensity factor K_{Ic} was involved.

10.3.3 Crack Propagation in Concrete

Fracture of concrete may be treated on the mesolevel of the hierarchical system of the structure of materials. The main defects in concrete result from bleeding, shrinkage, and other processes in fresh and hardening concrete. Most of these cracks are interfacial bond cracks between hardened cement paste and coarse aggregates. The actual structure of concrete may be modeled by a unit thickness elastic plate containing inclusions (coarse aggregate particles). Aggregates with random polygonal shape are randomly distributed in a homogeneous matrix.

First consider a randomly inclined crack in a homogeneous plate loaded at infinity, as shown in Fig. 10-10. It can be shown that as the compressive load increases, two branching cracks of mode I may be created at the ends of the

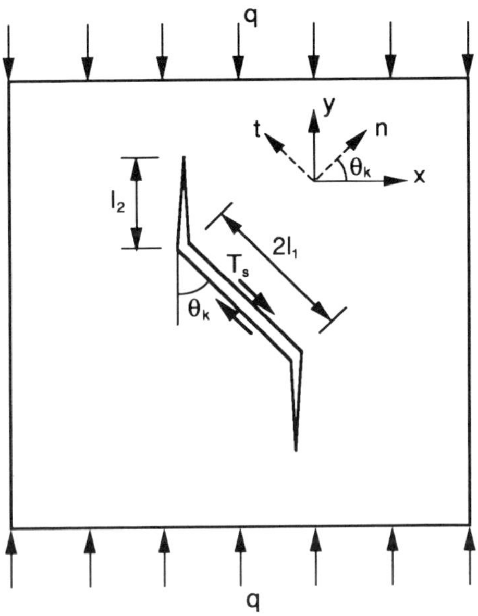

Figure 10-10 Development of branching cracks from inclined crack.[20]

initial inclined crack with a length of $2l_1$.[22,23] It is assumed here that the path of the two mode I cracks is in the same direction as the applied load. In this case, the following relationship for a stable crack growth can be obtained:

$$\frac{p}{\sqrt{\pi l_2}} = K_{Ic} \tag{10.3}$$

where l_2 is the length of the branching crack (Fig. 10-10),

$$p = T_s \sin \theta_k \tag{10.4}$$

and T_s is the resulting force of shear stress σ_s, which causes sliding of the two opposite sides of the inclined crack. By taking the coefficient of friction f into consideration, T_s can be expressed as

$$T_s = 2l_1\sigma_s = -2l_1 q\,(\sin \theta_k \cos \theta_k - f \sin^2 \theta_k) \tag{10.5}$$

The value of q is negative for compression based on the coordinate system given in Fig. 10-10. Substituting Eqs. (10.4) and (10.5) into (10.3) leads to

$$q = -\frac{\sqrt{\pi l_2}}{2l_1}\frac{K_{Ic}}{A(\theta_k, f)} \tag{10.6}$$

where $A(\theta_k, f) = (\sin \theta_k \cos \theta_k - f \sin^2 \theta_k) \sin \theta_k$. It is seen from Eq. (10.6) that a crack (the value of l_2) will steadily propagate when the applied load increases.

Now consider a homogeneous matrix with one polygonal inclusion, representing an aggregate particle in an infinite matrix, as shown in Fig. 10-11. An initial interfacial crack with length $2l_1$ is assumed to be located along one side AB. This problem can be treated in a similar manner as one with an inclined crack in a homogeneous matrix. But concentrations of shear and normal stresses have to be taken into account. This can be done by introducing coefficients of normal and shear stress concentration, k_n and k_s, respectively. Since the length of the initial inclined crack, $2l_1$, is smaller than the length of the side AB, $2L_1$, the inclined crack initially propagates in mode II (shear mode), as shown in Fig. 10-11*a*. Propagation of the shear crack is governed by

$$q_{\mathrm{II}}^{\mathrm{IF}} = -\frac{K_{\mathrm{II}c}}{\sqrt{\pi l_1}\, D^{\mathrm{IF}}(\theta_k, f)} \tag{10.7}$$

where

$$D^{\mathrm{IF}} = k_s \sin\theta_k \cos\theta_k - f\, k_n \sin^2\theta_k \tag{10.8}$$

Superscript IF is used to denote the interface hereafter.

The shear crack reaches the length of $2L_1$ and stops, because further crack

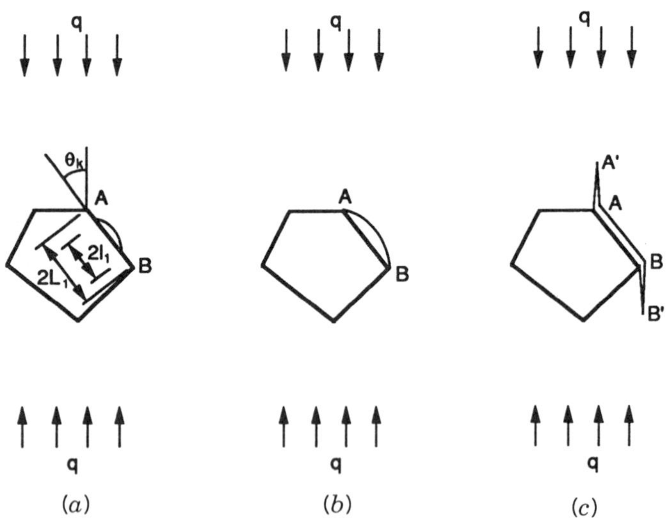

Figure 10-11 Crack propagation in matrix with aggregate: (*a*) an initial crack, (*b*) crack propagation along interface, and (*c*) crack branching.[20]

propagation in the same inclined direction would take place through the matrix. The mode II critical stress intensity factor of the matrix, $K_{\text{II}c}^{M}$, is thought to be greater than that of the interface. On the other hand, the value of $K_{\text{II}c}^{M}$ is also thought to be greater than the mode I critical stress intensity factor of the matrix, $K_{\text{I}c}^{M}$. Therefore, when the inclined crack propagates into the matrix as the load increases, it becomes a mode I crack. The relationship between the compressive load and $K_{\text{I}c}^{M}$ is given by

$$q = -\frac{\sqrt{\pi l_2}}{2L_1} \frac{K_{\text{I}c}^{M}}{A^{\text{IF}}(\theta_k, f)} \tag{10.9}$$

where $A^{\text{IF}}(\theta_k, f) = D^{\text{IF}}(\theta_k, f)\sin\theta_k$, and l_2 is the distance AA' shown in Fig. 10-11c. According to Eq. (10.9), crack length l_2 will steadily increase with increasing load.

For a matrix with two or several aggregates as shown in Fig. 10-12, the same procedure with Fig. 10-11 is repeated. But it is assumed that the branching crack AA' meets a second aggregate as it propagates. Since aggregates usually have higher fracture resistance than the interface in normal concrete, further crack growth will follow the interface MN as a crack of either mode I

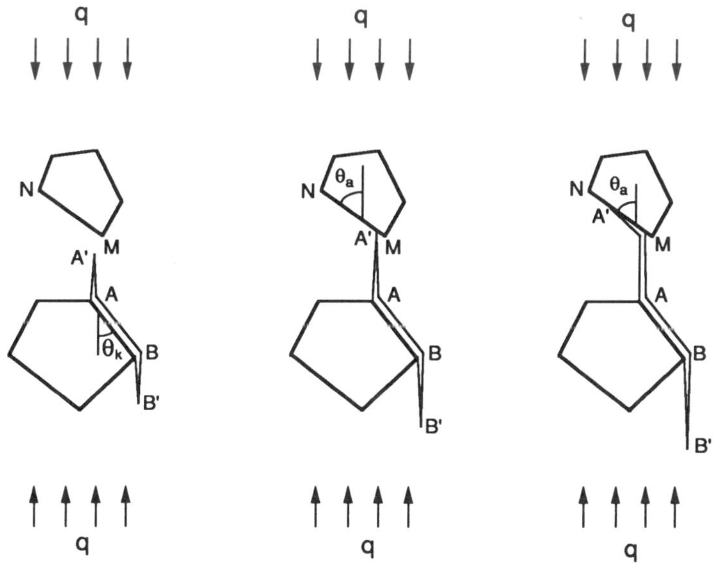

Figure 10-12 Crack propagation in matrix with several aggregates: (*a*) a crack path shown in Fig. 10-11, (*b*) crack meets a second aggregate, and (*c*) crack propagates along interface *MN* with $\theta_a > 0$.[20]

or mode II. When the crack on the interface *MN* is mode I, the relationship between the compressive load $q_{\mathrm{I}}^{\mathrm{IF}}$ and the mode I interfacial fracture toughness $K_{\mathrm{I}c}^{\mathrm{IF}}$ is given by

$$q_{\mathrm{I}}^{\mathrm{IF}} = -\frac{2K_{\mathrm{I}c}^{\mathrm{IF}}\sqrt{\pi l_2}/L_1}{A^{\mathrm{IF}}(\theta_k, f)[3\cos(\theta_a/2) + \cos(3\theta_a/2)] - 3B^{\mathrm{IF}}(\theta_k, f)[\sin(\theta_a/2) + \sin(3\theta_a/2)]} \tag{10.10}$$

When the crack on the interface *MN* is mode II, the relationship between the compressive load $q_{\mathrm{II}}^{\mathrm{IF}}$ and the mode I interfacial fracture toughness $K_{\mathrm{II}c}^{\mathrm{IF}}$ is

$$q_{\mathrm{II}}^{\mathrm{IF}} = -\frac{2K_{\mathrm{II}c}^{\mathrm{IF}}\sqrt{\pi l_2}/L_1}{A^{\mathrm{IF}}(\theta_k, f)[\sin(\theta_a/2) + \sin(3\theta_a/2)] - B^{\mathrm{IF}}(\theta_k, f)[\cos(\theta_a/2) + 3\cos(3\theta_a/2)]} \tag{10.11}$$

where $q_{\mathrm{I}}^{\mathrm{IF}}$ and $q_{\mathrm{II}}^{\mathrm{IF}}$ are the critical loads to propagate the crack along the path *MN* as mode I and as mode II, respectively, and $B^{\mathrm{IF}}(\theta_k, f) = D^{\mathrm{IF}}(\theta_k, f)\cos\theta_k$. Both values of $q_{\mathrm{I}}^{\mathrm{IF}}$ and $q_{\mathrm{II}}^{\mathrm{IF}}$ depend on values of fracture resistances $K_{\mathrm{I}c}^{\mathrm{IF}}$ and $K_{\mathrm{II}c}^{\mathrm{IF}}$ and are negative for compression.

Zaitsev[21] proposed that whether crack propagation is mode I or mode II depends significantly on the sign of θ_a, as shown in Fig. 10-12. For $\theta_a < 0$ a crack of mode I is to be expected. For $\theta_a > 0$ as given in Fig. 10-12, a crack of mode II is more probable. Zaitsev[21] further indicated that since mode II cracks are facilitated by shear components of applied pressure, the presence of normal confining components of the applied compressive load makes the formation of opening cracks in the interface less likely as compared to the formation of shear cracks. As a consequence, in a material with randomly distributed inclusions (where the probabilities of occurrence of positive and negative values of θ_a are equal), new interfacial cracks will propagate primarily in mode II. This means that a resulting crack running through the whole specimen will contain some interfacial parts. Thus, the resulting crack will be slightly inclined and not exactly parallel to the direction of the applied compressive load, as shown in Fig. 10-13*b*.

More complicated situations including that the crack may propagate into aggregates, which may happen for high-strength concrete and lightweight concrete, have also been discussed by Zaitsev and Wittmann.[20,21]

Zaitsev and Wittmann[20,21] have applied the above procedure to compressive specimens made of hardened cement paste and normal concrete. The obtained cracked patterns for the two types of materials are shown in Fig. 10-13. Zaitsev[21] also applied the proposed method to concrete subjected to a biaxial state of stress and reported that the predicted biaxial failure envelope matches with experimental results quite well.

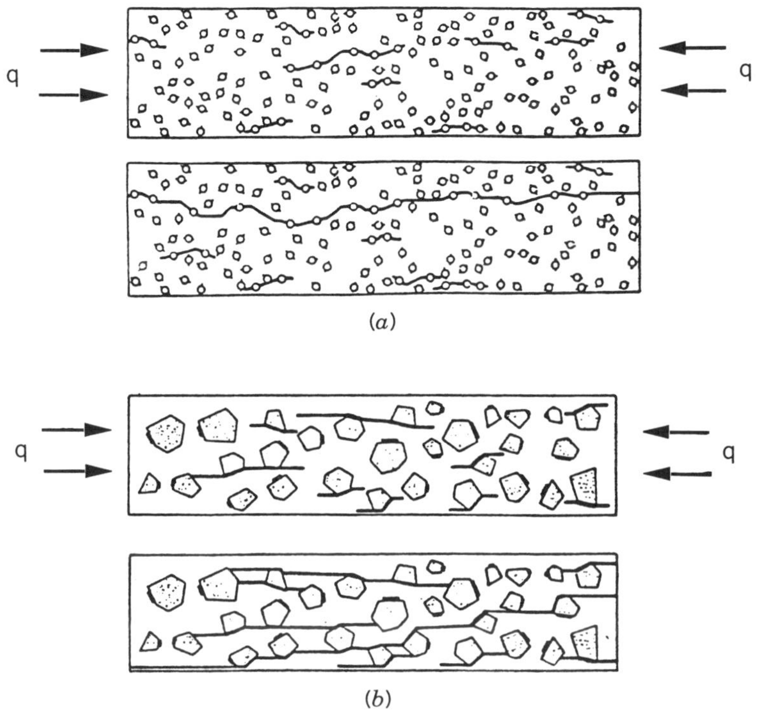

Figure 10-13 Crack patterns for different cementitious materials: (*a*) a hardened cement paste and (*b*) a normal strength concrete.[21]

10.4 MIXED-MODE FRACTURE IN COMPRESSION

The compressive failure of concrete can also be described by using principles of mixed-mode fracture presented in Chapter 8. This has been done by Taşdemir et al.[24,25] for a two-dimensional analysis. Their approach will be presented below.

Consider an infinite-size plate containing an inclined crack, as shown in Fig. 10-14. The length of the inclined crack is $2a$, and the plate is subjected to a uniaxial compressive stress q, where q is defined as negative for compression. The inclined crack may be a bond crack at the interface of an aggregate. After resolution of stresses, the interfacial crack may be considered as subjected to a normal stress σ_β and a shear stress τ_β. Values of σ_β and τ_β are given by

$$\sigma_\beta = -q \sin^2 \theta_k \qquad \tau_\beta = -q \sin \theta_k \cos \theta_k \tag{10.12}$$

where θ_k is defined in Fig. 10-14.

Propagation and stability of the interfacial crack could be evaluated using the concept of fracture mechanics. Under compressive loading, the interfacial

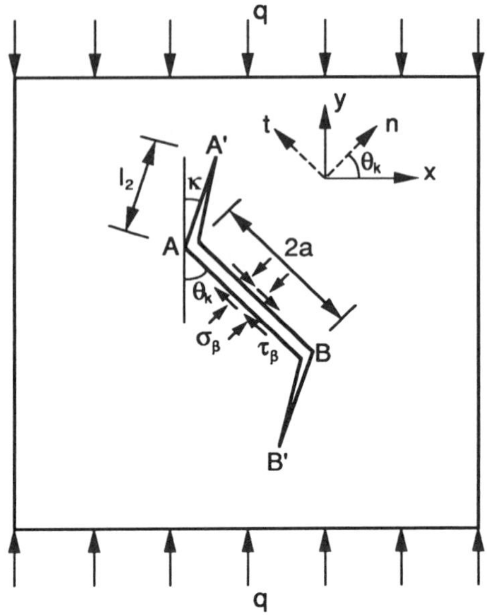

Figure 10-14 Initial crack, stresses at onset of interface debonding and kinks.[24]

crack with the inclination shown in Fig. 10-14 cannot open because the normal stress σ_β is compressive. It cannot close either because the crack surfaces are already in contact. Therefore, the mode I (opening) stress intensity factor at the tip of the interfacial crack, K_{I}, is zero. This result was verified using a FE analysis. Since the actual crack faces were already in contact, they would only be capable of sliding. The mode II (sliding) stress intensity factor at the tip of the initial interfacial crack is given by[26]

$$K_{\mathrm{II}} = -q\sqrt{\pi a}(\sin\theta_k \cos\theta_k - f\sin^2\theta_k) \tag{10.13}$$

where f is the friction coefficient at the interface. It is noted that for sliding to occur, $|\tau_\beta| > f|\sigma_\beta|$. This indicates that there is a critical value of θ_k given by $\theta_{k0} = \operatorname{arccot} f$. Here, K_{II} is zero for $\theta_k > \theta_{k0}$, whereas sliding can occur and the corresponding K_{II} is given by Eq. (10.13) for $\theta_k \le \theta_{k0}$.

The crack will propagate when the applied compressive stress increases. The direction of propagation of the existing crack subjected to a mixed mode is denoted by κ. Generally, κ does not coincide with θ_k. Several criteria have been proposed to determine the value of κ. For example, one may assume that the crack will extend in the direction of the maximum mode I stress intensity factor. The stress intensity factors at the tip of an infinitesimal kink that extends from the initial crack tip at the angle κ can be written as follows[27,28]:

$$K_{\mathrm{I}}^{\mathrm{kink}}(\kappa) = -\tfrac{3}{4}K_{\mathrm{II}}[\sin(\kappa/2) + \sin(3\kappa/2)] \tag{10.14}$$

and

$$K_{\mathrm{II}}^{\mathrm{kink}}(\kappa) = \tfrac{1}{4}K_{\mathrm{II}}[\cos(\kappa/2) + 3\cos(3\kappa/2)] \tag{10.15}$$

where $K_{\mathrm{I}}^{\mathrm{kink}}(\kappa)$ and $K_{\mathrm{II}}^{\mathrm{kink}}(\kappa)$ are the mode I and mode II stress intensity factors at the tip of the kink, respectively, and K_{II} is given by Eq. (10.13). It is noted that after kinking the value of $K_{\mathrm{I}}^{\mathrm{kink}}(\kappa)$ is not zero. Since it is assumed that the crack will extend in the direction of the maximum mode I stress intensity factor, the maximum value of $K_{\mathrm{I}}^{\mathrm{kink}}(\kappa)$ occurs at $\kappa = 71°$ for all values of θ_k and can be obtained from $\partial K_{\mathrm{I}}^{\mathrm{kink}}(\kappa)/\partial\kappa = 0$. Maji et al.[25] also discussed the use of different fracture criteria for extension of the kink.

An approximate solution has also been proposed by Steiff.[29] His expressions were based on the assumption that crack faces slide parallel to themselves, and the stress intensity factors at the tip of the kink for far-field uniaxial compression are given by

$$\begin{aligned} K_{\mathrm{I}}^{\mathrm{kink}}(\kappa) = {} & -q\sin^2(\theta_k - \kappa)\sqrt{\frac{\pi l_2}{2}} \\ & + \frac{3}{4}q\sqrt{\frac{\pi}{2}}\left(\sin\frac{\kappa}{2} + \sin\frac{3\kappa}{2}\right) \\ & \cdot (\sin\theta_k\cos\theta_k - f\sin^2\theta_k)(\sqrt{2a + l_2} - \sqrt{l_2}) \end{aligned} \tag{10.16}$$

and

$$\begin{aligned} K_{\mathrm{II}}^{\mathrm{kink}}(\kappa) = {} & -\frac{q}{2}\sin^2(\theta_k - \kappa)\sqrt{\frac{\pi l_2}{2}} \\ & - \frac{q}{4}\sqrt{\frac{\pi}{2}}\left(\cos\frac{\kappa}{2} + 3\cos\frac{3\kappa}{2}\right) \\ & \cdot (\sin\theta_k\cos\theta_k - f\sin^2\theta_k)(\sqrt{2a + l_2} - \sqrt{l_2}) \end{aligned} \tag{10.17}$$

where l_2 is the length of the kink, as shown in Fig. 10-14. These equations describe the variation of $K_{\mathrm{I}}^{\mathrm{kink}}(\kappa)$ and $K_{\mathrm{II}}^{\mathrm{kink}}(\kappa)$ at the tip of the kink. Equations (10.14) and (10.15) can be obtained by substituting $l_2 = 0$ into Eqs. (10.16) and (10.17). Typical variations of $K_{\mathrm{I}}^{\mathrm{kink}}(\kappa)$ and $K_{\mathrm{II}}^{\mathrm{kink}}(\kappa)$ for $\theta_k = 36°$ and $f = 0.25$ are shown in Fig. 10-15. Note that these equations are based on solutions of an infinite-size plate containing an inclined crack. It is seen that the value of κ for which $K_{\mathrm{I}}^{\mathrm{kink}}(\kappa)$ is maximum decreases with increasing value of l_2/a, and this value of κ corresponds to $K_{\mathrm{II}}^{\mathrm{kink}}(\kappa) = 0$.

Taşdemir et al.[24,25] also used a FE method to analyze finite-size plates sub-

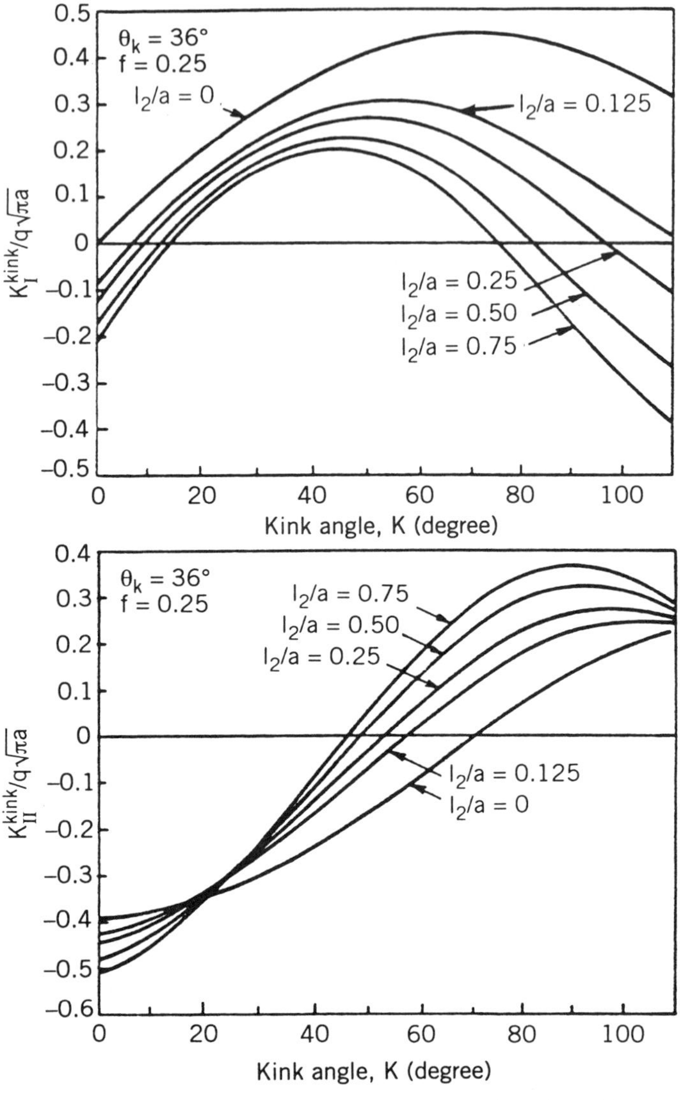

Figure 10-15 Stress intensity factors at tip of kink as function of kink angle.[29]

jected to mixed-mode fracture using the software FRANC, which has been briefly presented in Chapter 5. They also used laser holographic techniques to measure experimentally mixed-mode crack propagation. A two-dimensional elastic analysis was performed using six-noded triangular and eight-noded quadrilateral elements. The dimensions of the specimen and the initial element mesh are shown in Fig. 10-16*a*. The nodes on the top and bottom faces of the crack were fixed in the vertical *Y* direction to simulate absence of closure (Fig.

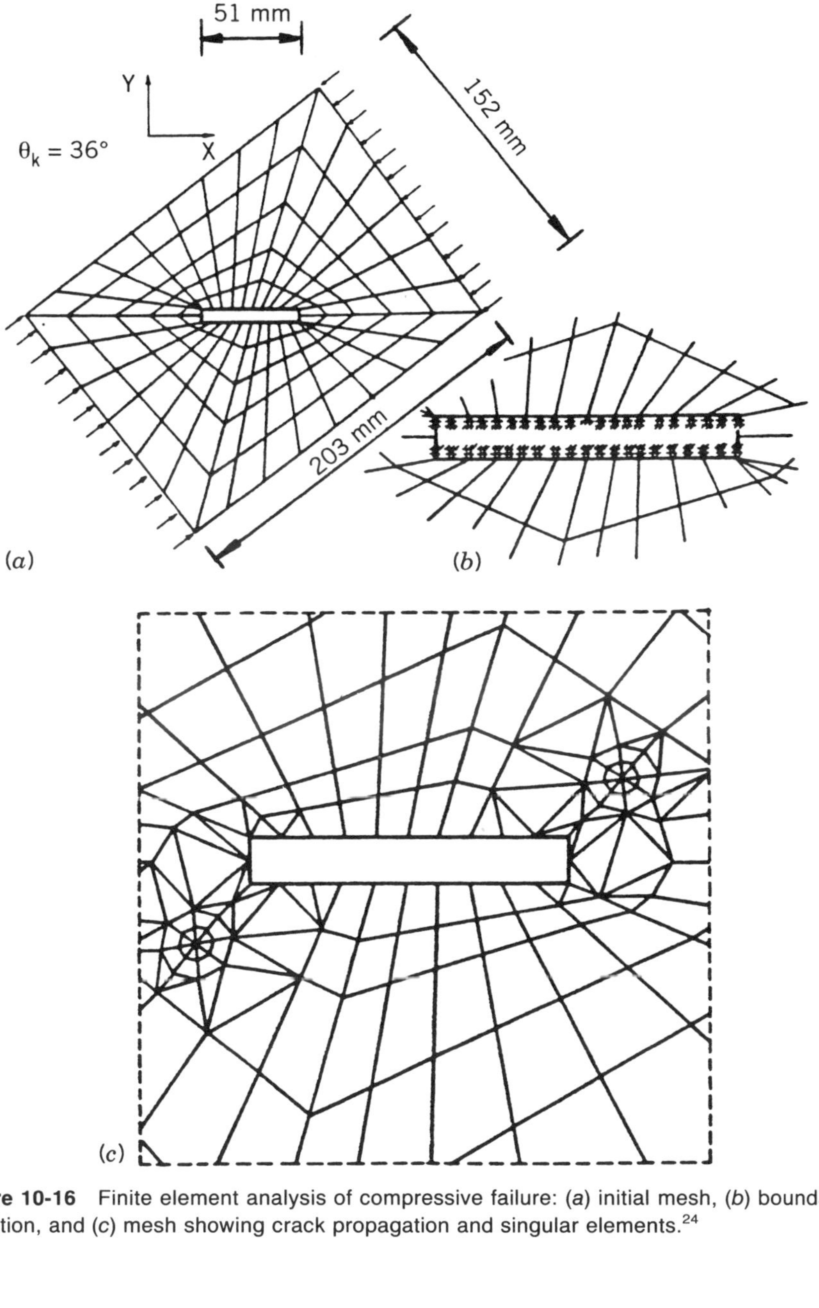

Figure 10-16 Finite element analysis of compressive failure: (*a*) initial mesh, (*b*) boundary condition, and (*c*) mesh showing crack propagation and singular elements.[24]

10-16*b*). One middle node at the upper side of the crack was fixed in the *X* direction to prevent rigid body translation. Hence, the crack faces were allowed to slide, but not allowed to close and open. The FE analysis was done based on the experimentally observed crack path. The loads were applied at specimen ends to simulate a uniaxial compression. A second set of runs was made where the top and bottom crack faces were loaded with shear tractions to simulate the aggregate matrix friction. The applied shear stress used was of the magnitude of $f\,q\,\sin^2\theta_k$, where $f = 0.25$ was determined experimentally and q was the applied uniaxial compressive stress. These stresses were lumped at the interfacial nodes. Superposition of these two runs resulted in the solutions. A typical mesh with kink extension of about 12.2 mm (0.5 in.) and $\theta_k = 36°$ is shown in Fig. 10-16*c*. Results of the FE analysis were compared with Eq. (10.13) and shown in Fig. 10-17. Note that for this comparison the crack was simulated by a single line representing a sharp crack in the FE mesh. As shown in Fig. 10-17, the results obtained by the FE analysis for the given dimensions are reasonably close to the infinite-size plate solution given by Eq. (10.13). As in previous discussions, it was assumed that the kink will extend

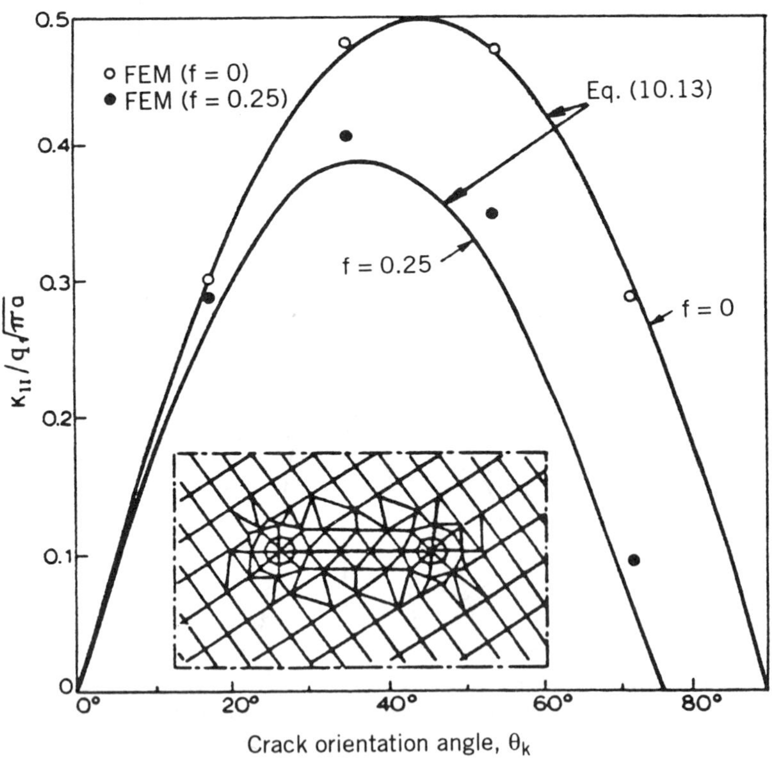

Figure 10-17 Normalized mode II stress intensity factor as function of crack orientation angle.[24]

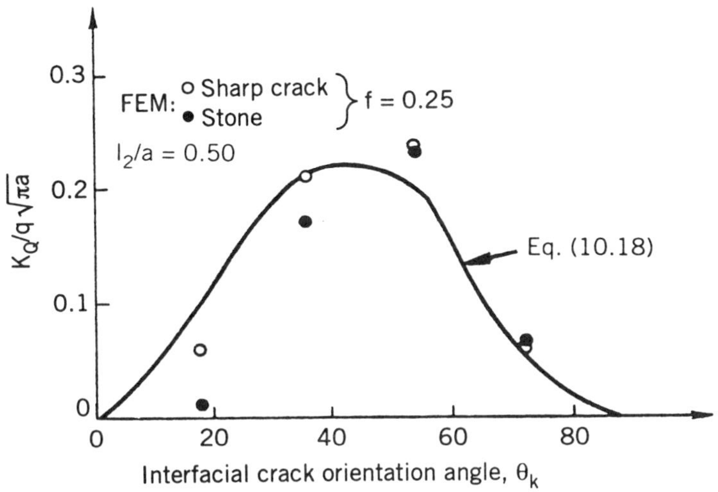

Figure 10-18 Comparison of values of K_Q between Eq. (10.18) and FE analysis.[24]

in the direction of the maximum value of $K_{\mathrm{I}}^{\mathrm{kink}}(\kappa)$. This maximum value of $K_{\mathrm{I}}^{\mathrm{kink}}(\kappa)$ is denoted by K_Q, which is given by

$$K_Q = \max K_{\mathrm{I}}^{\mathrm{kink}}(\kappa) = -1.1547q\sqrt{\pi a}(\sin\theta_k\cos\theta_k - f\sin^2\theta_k) \qquad (10.18)$$

based on Eq. (10.16) using the condition $\partial K_{\mathrm{I}}^{\mathrm{kink}}(\kappa)/\partial\kappa = 0$. Values of the apparent mode I stress intensity factor, K_Q, obtained by the FE analysis for the plate with given dimension are compared to Eq. (10.18) in Fig. 10-18. For the FE analysis, values of both the sharp crack and the stone–matrix interface are given. The analytical solution correctly predicts the trend for the finite-size plate for the given dimension, but the values of K_Q do not always match. For a given value of the applied stress q, the corresponding value of K_Q can be calculated either by Eq. (10.18) or by a FE method. Based on LEFM, if the value of K_Q is greater than the critical stress intensity factor $K_{\mathrm{I}c}$, which is a material fracture parameter, the crack will propagate. It is noted that the crack will propagate in the direction of the maximum value of the mode I stress intensity factor, which is determined from the condition $\partial K_{\mathrm{I}}^{\mathrm{kink}}(\kappa)/\partial\kappa = 0$. By repeating this procedure for different values of q, a relationship between the applied load and the crack extension can be obtained. Taşdemir et al.[24] also modified an LEFM FE method to take both normal and shear traction forces on cracked faces into account.

10.5 SUMMARY

Mechanisms for compressive failure of concrete specimens have been discussed based on various experimental observations. The compressive failure is characterized by strain localization. Cracks usually initiate from the interface between the aggregate and cementitious matrix and then coalesce into one or several major cracks that are in the same direction as the applied compressive load. Since the major crack(s) accompanied by the strain localization is in the same direction as the applied compressive load, modeling of the compressive failure using fracture mechanics is more difficult than when compared to the tensile failure of concrete. Several models based on fracture mechanics have been presented to account for this fracture process. It is noted that because of the difficulty in modeling the compressive failure, these models are primarily based on LEFM. Sophisticated models in this area will depend on more understanding of compressive failure mechanisms and on development of new analytical methods.

REFERENCES

1. Hsu, T. T. C., Slate, F. O., Sturman, M., and Winter, G., "Microcracking of Plain Concrete and the Shape of the Stress–Strain Curve," *ACI Journal*, Vol. 60, No. 2, 1963, pp. 209–2224.
2. Shah, S. P., and Chandra, S., "Critical Stress, Volume Change and Microcracking of Concrete," *ACI Journal*, Vol. 65, No. 9, 1968, pp. 770–781.
3. Shah, S. P., and Sankar, R., "Internal Cracking and Strain-Softening Response of Concrete under Uniaxial Compression," *ACI Materials Journal*, Vol. 84, No. 3, 1987, pp. 200–212.
4. Torrenti, J. M., Desrues, J., Benaija, E. H., and Boulay, C., "Stereophotogrammetry and Localization in Concrete under Compression," *Journal of Engineering Mechanics, ASCE*, Vol. 117, No. 7, 1991, pp. 1455–1465.
5. Bascoul, A., Benaija, E. H., Berthaud, Y., Torrenti, J. M., and Zizi, Z., "Analysis of Localization in Concrete through Stereophotogrammetry, Speckle Laser and Replica," *Cement and Concrete Research*, Vol. 23, 1993, pp. 1340–1350.
6. Maji, A. K., and Shah, S. P., "Application of Acoustic Emission and Laser Holography to Study Microfracture in Concrete," in *Nondestructive Testing*, ACI SP-112, edited by Lew, H. S., American Concrete Institute, Detroit, MI, 1988, pp. 83–109.
7. Kupfer, H., Hilsdorf, H. K., and Rusch, H., "Behavior of Concrete under Biaxial Stresses," *ACI Journal*, Vol. 66, No. 8, 1969, pp. 656–666.
8. Kotsovos, M. D., "Effect of Testing Techniques on Post-Ultimate Behavior of Concrete in Compression," *Materials and Structures*, Vol. 16, No. 91, 1983, pp. 3–12.
9. van Mier, J. G. M., "Multiaxial Strain-Softening of Concrete, Part I: Fracture," *Materials and Structures*, Vol. 19, No. 111, 1986, pp. 179–190.
10. Kupfer, H. B., and Gerstle, K. H., "Behavior of Concrete under Biaxial Stresses," *Journal of the Engineering Mechanics Division, ASCE*, Vol. 99, No. EM4, 1973, pp. 853–866.
11. Kotsovos, M. D., "A Mathematical Model of Deformational Behavior of Con-

crete under Generalized Stress Based on Fundamental Material Properties,'' *Materials and Structures*, Vol. 13, No. 76, 1980, pp. 289–298.

12. Vonk, R. A., ''A Micromechanical Investigation of Softening of Concrete Loaded in Compression,'' *Heron*, Vol. 38, No. 3, 1993.
13. Mehta, P. K., and Monteiro, P. J. M., *Concrete: Structure, Properties, and Materials*, Prentice-Hall, Englewood Cliffs, NJ, 1993.
14. Nilsen, A. U., and Monteiro, P. J. M., ''Concrete: A Three Phase Material,'' *Cement and Concrete Research*, Vol. 23, 1993, pp. 147–151.
15. Hsu, T. T. C., and Slate, F. O., ''Tensile Bond Strength between Aggregate and Cement Paste or Mortar,'' *ACI Journal*, Vol. 60. No. 4, 1963, pp. 465–486.
16. Shah, S. P., and Winter, G., ''Inelastic Behavior and Fracture of Concrete,'' *ACI Journal*, Vol. 63, No. 9, 1966, pp. 925–930.
17. Hsu, T. C., ''Mathematical Analysis of Shrinkage Stresses in a Model of Hardened Concrete,'' *ACI Journal*, Vol. 60, No. 3, 1963, pp. 371–390.
18. Buyukozturk, O., Nilson, A. H., and Slate, F. O., ''Stress–Strain Response and Fracture of a Concrete Model in Biaxial Loading,'' *ACI Journal*, Vol. 68, No. 8, 1971, pp. 590–599.
19. Palaniswamy, R. G., and Shah, S. P., ''A Model for Concrete Subjected to Triaxial Stresses,'' *Cement and Concrete Research*, Vol. 5, 1975, pp. 273–284.
20. Zaitsev, Y. B., and Wittmann, F. H., ''Simulation of Crack Propagation and Failure of Concrete,'' *Materials and Structures*, Vol. 14, No. 83, 1981, pp. 357–365.
21. Zaitsev, Y. B., ''Crack Propagation in a Composite Material,'' in *Fracture Mechanics of Concrete*, edited by Wittmann, F. H., 1983, Elsevier Science, Amsterdam, 1983, pp. 251–299.
22. Diaz, S. S., and Hilsdorf, H. K., ''Fracture Mechanisms of Concrete under Compressive Loads,'' *Cement and Concrete Research*, Vol. 3, 1973, pp. 363–388.
23. Kotsovos, M. D., and Newman, J. B., ''Fracture Mechanics and Concrete Behavior,'' *Magazine of Concrete Research*, Vol. 33, No. 115, 1981, pp. 103–112.
24. Taşdemir, M. A., Maji, A. K., and Shah, S. P., ''Crack Propagation in Concrete under Compression,'' *Journal of Engineering Mechanics, ASCE*, Vol. 116, No. 5, 1990, pp. 1058–1076.
25. Maji, A. K., Taşdemir, M. A., and Shah, S. P., ''Mixed Mode Crack Propagation in Quasi-Brittle Materials,'' *Engineering Fracture Mechanics*, Vol. 38, No. 2/3, 1991, pp. 129–145.
26. Melville, P. H., ''Fracture Mechanics of Brittle Materials in Compression,'' *International Journal of Fracture*, Vol. 13, 1977, pp. 532–534.
27. Cottrell, B., and Rice, J. R., ''Slightly Curved or Kinked Cracks,'' *International Journal of Fracture*, Vol. 16, No. 2, 1980, pp. 155–169.
28. Horii, H., and Nemat-Nasser, S., ''Brittle Failure in Compression: Splitting, Faulting, and Brittle-Ductile Transition,'' *Philosophical Transactions of the Royal Society of London*, Series A 319, 1986, pp. 337–374.
29. Steiff, P. S., ''Crack Extension under Compressive Loading,'' *Engineering Fracture Mechanics*, Vol. 20, No. 3, 1984, pp. 463–473.

11

DAMAGE MECHANICS AND FRACTURE MECHANICS

11.1 INTRODUCTION

Although the mechanisms of deformation for failure of concrete are not clearly understood, experimental observations have indicated that stress–deformation relationships for a concrete test specimen subjected to either uniaxial tension or compression may be divided into several stages based on initiation and propagation of internal flaws, as shown in Fig. 11-1 and discussed in Chapters 4 and 10.[1,2] The first stage is before point A, which is at about 30% of the peak load. Propagation of internal flaws, which primarily consist of interfacial voids between cementitious matrix and aggregate, is negligible during the first stage. The second stage is from point A to point B, which is at about 80% of the peak load. The internal cracks initiate and propagate during this stage. In this stage, these cracks are isolated and randomly distributed over the specimen volume, and distribution of strain in the loading direction is approximately uniform over the specimen length. The third stage is between point B and point C. During this stage, the internal cracks start to localize into a major crack when subjected to tension or several major cracks when subjected to compression. The major crack(s) propagate with increasing load. When strain localization occurs, distribution of strain is no longer uniform over the specimen length. Furthermore, the tensile strain of the material within the localized damage band (the major crack) continuously increases, whereas unloading may occur for the material outside the damage band. Propagation of the major crack is stable up to peak load (point C). Here the term *stable crack growth* means that the major crack proagates only when the load increases. The fourth stage is after the peak load. The major crack continuously propagates even though the load decreases. It can be understood from the above discussion that failure of concrete is char-

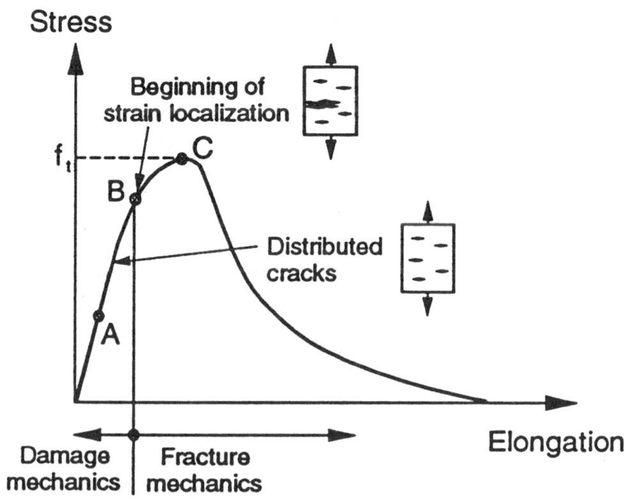

Figure 11-1 Stages of crack propagation in concrete under uniaxial tension.

acterized by a strain localization. Previous chapters have shown how fracture mechanics can be successfully used to simulate mechanical behavior of concrete after strain localization. However, continuous damage mechanics may be more suitable for modeling stress–strain response of concrete before the strain localization because the cracks basically are randomly and isolatedly distributed and the strain distribution is approximately uniform up to point *B*. Different areas for applications of damage mechanics and fracture mechanics are indicated in Fig. 11-1. It is noted that damage mechanics may be also used to model failure of concrete after the strain localization if some additional assumptions are introduced. This will be discussed later.

Continuous damage mechanics describes the influence of accumulation of material damage such as microcracks, dislocations, and slip on the mechanical response of a solid. Since the concept of damage mechanics was proposed by Kachanov[3] in 1958, various damage mechanics models have been developed, and a comprehensive overview of these models has been given by Krajcinovic.[4] Damage mechanics uses a damage variable to describe the extent of material damage. The damage variable can be related to stress or strain through a damage evolution law. By incorporating the damage variable into stress analysis, relationships among stresses (or strains), the damage variable, and applied load can be derived.

There are many ways to phenomenologically define or micromechanically derive damage variables. Generally, the damage variable can be a scalar,[5] a vector,[6] a second-order tensor,[7] or a fourth-order tensor[8] based on the nature of the problem. The scalar damage variable is related to volume change of microcracks in a material and has been widely used to describe isotropic damage due to its simplicity. The vectorial damage variable can account for both volume and orientation of microcracks and can be used to describe anisotropic damage. However, the vectorial representation cannot include the influence of shape of a microcrack, and mathematical treatments of the vectorial damage variable are not as simple as those using the scalar or the second-order tensorial representation. A second-order damage tensor is attractive since both area and orientation of microcracks can be simulated. However, it is incapable of describing general anisotropy. The fourth-order (or higher) tensor may be appropriate for simulating general anisotropic damage but may be too complicated to use. As previously discussed, experimental observations have shown that under uniaxially applied loading the internal cracks in concrete are primarily subjected to mixed-mode deformations and are randomly and isolatedly distributed before the beginning of the strain localization.[9, 10] This may imply that the damage process up to the strain localization in concrete subjected to uniaxial loading can be regarded as "isotropic." Micromechanical analysis[11] has shown that a scalar damage variable is sufficient to describe such an isotropic damage. However, it should be indicated that the scalar damage representation is intrinsically correct only if defects are randomly distributed spheroidal cavities (isotropic damage). For other types of defects, the scalar damage variable is only an approximate representation of material damage. Furthermore, the scalar representation of damage inherently implies a self-similar growth of cavities. Concrete becomes anisotropic when it is subjected to multiaxial loads. This anisotropy further depends on loading history. Since the presentation of general anisotropic damage mechanics, which may be based on high-order damage tensors, is beyond the scope of this book, only some damage mechanics models based on the scalar damage variable will be presented in this chapter.

Phenomenological scalar damage models, which may be suitably used to describe damage of concrete under uniaxial loading, will be first discussed. The discussion will be extended to include ways to define a scalar damage variable based on micromechanics. Then a micromechanical scalar damage model for damage of concrete subjected to uniaxial and biaxial tension will be presented. Some damage mechanics approaches to model tension-softening behavior of concrete will be mentioned. The influence of microcracks on the effective fracture toughness will also be discussed.

11.2 PHENOMENOLOGICAL MODELS BASED ON SCALAR DAMAGE VARIABLE

11.2.1 Scalar Representation of Material Damage

The physical meaning of a scalar damage variable may be demonstrated by considering a representative volume element from a material (or point), as shown in Fig. 11-2. The element has a gross cross-sectional area A. Due to the existence of some defects, which may result from initial material flaws and

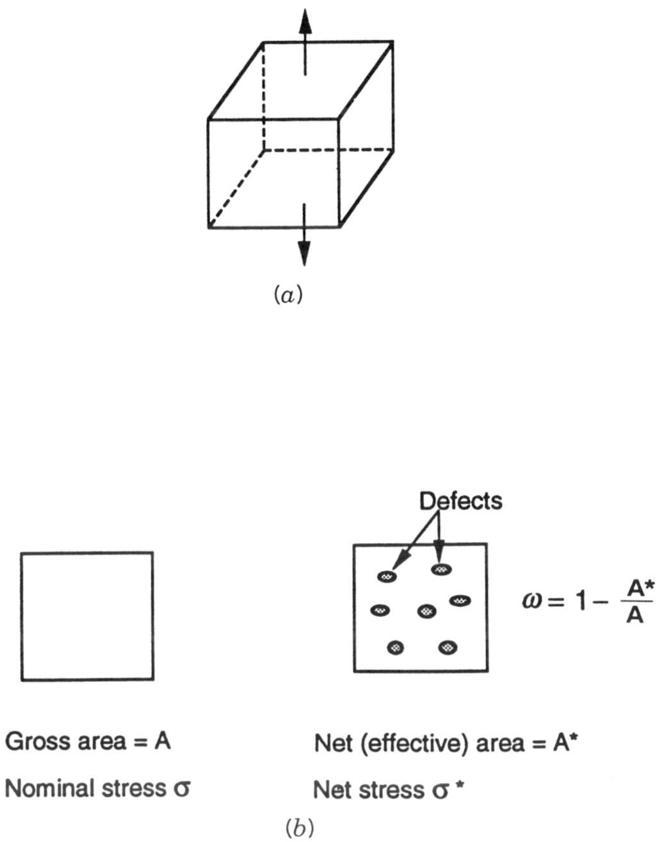

Figure 11-2 Illustration of scalar damage variable: (*a*) representative volume element of a material, and (*b*) the concept of effective stress.

load-induced cracks, the net cross-sectional area of the element is A^* instead of A. The value of A^* equals the cross-sectional area minus the area of the defects. It is noted that since the net cross section A^* excludes the area of the defects, it can be regarded as the quantity corresponding to the undamaged state of the material. If only the volume of the defects is considered, the damage of a material can be described by a ratio shown below[12]:

$$\Omega = \frac{A - A^*}{A} = 1 - \frac{A^*}{A} \tag{11.1}$$

where Ω is a scalar damage variable that accounts for the extent of material damage. The value of Ω changes with loading history and varies from zero to unity. The value of $\Omega = 0$ corresponds to an undamaged material, whereas the value of $\Omega = 1$ indicates the complete separation of material that corresponds to zero stress in the postpeak region of a stress–deformation curve. Based on such a definition of the damage variable, a net stress σ^* is defined as the total force on the material divided by the net cross-sectional area, whereas a nominal-stress σ is equal to the total force divided by the gross cross-sectional area. Noting the condition $\sigma A = \sigma^* A^*$, a relationship between the net stress σ^* and the nominal stress σ can be obtained as

$$\sigma = \sigma^* \frac{A^*}{A} = \sigma^*(1 - \Omega) \tag{11.2}$$

A linear relationship between the net stress and the strain is usually adopted in continuous damage mechanics. It is noted again that the net stress can be regarded to correspond to the undamaged state of the material.

Although the scalar damage variable given by Eq. (11.1) is clear for explaining the concept of net stress, it may not be convenient for practically measuring the value of Ω. To overcome this, Lemaitre and Chaboche[5] further phenomenologically assumed

$$\Omega = 1 - \frac{E}{E^*} \tag{11.3}$$

where E is effective modulus of elasticity for the material with damage and E^* is modulus of elasticity for the corresponding undamaged material.

Use of the scalar damage variable to describe failure of concrete will be discussed below.

11.2.2 Scalar Damage Model by Løland

Løland[13] used a scalar damage variable to describe failure of concrete under uniaxial tension, as shown in Fig. 11-3*a*. It is noted that continuous damage mechanics considers that the relationship between the net stress and the strain is linear elastic before the peak stress. For the one-dimensional specimen shown in Fig. 11-3*a*, this leads to

$$\sigma^* = E^* \epsilon \tag{11.4}$$

This linear elastic net stress–strain curve is shown in Fig. 11-3*b*. Substituting Eq. (11.2) into Eq. (11.4) results in

$$\sigma = E^*(1 - \Omega)\epsilon \tag{11.5}$$

Equation (11.5) describes a relationship among σ, ϵ, and Ω for a uniaxial stress condition before the peak load. If the value of Ω corresponding to a given strain level is determined, the value of stress can be calculated using Eq. (11.5).

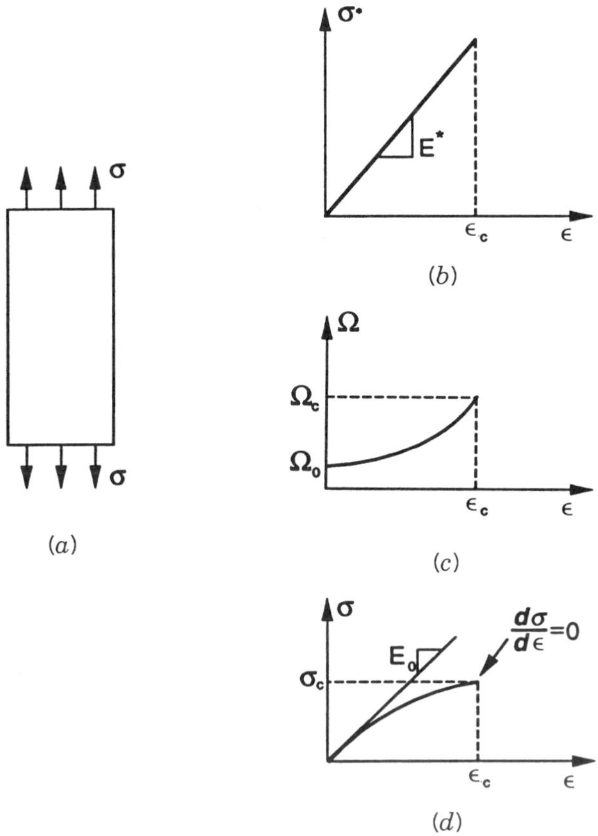

Figure 11-3 Damage model proposed by Løland.[13]

To calculate the uniaxial stress–strain curve, the following relationship between the damage variable and the strain was further assumed:

$$\Omega(\epsilon) = \Omega_0 + A_0 c^{\Lambda} \tag{11.6}$$

where Ω_0 is the initial damage immediately before load is applied and A_0 and Λ are constants to be determined. The value of Ω_0 can be related to the microstructure of the material. This relationship is illustrated in Fig. 11-3*c*.

By substituting Eq. (11.6) into Eq. (11.5), the prepeak stress–strain relationship can be obtained as

$$\sigma = E^*(1 - \Omega_0 - A_0 \epsilon^{\Lambda})\epsilon \tag{11.7}$$

Differentiating Eq. (11.7) and using the condition $d\sigma/d\epsilon = 0$ at $\epsilon = \epsilon_c$ as shown in Fig. 11-3*d* results in

$$\epsilon_c = \left[\frac{1 - \Omega_0}{A_0(\Lambda + 1)}\right]^{1/\Lambda} \tag{11.8}$$

where ϵ_c is the strain corresponding to the peak stress. Substituting the condition $\sigma = \sigma_c$ at $\epsilon = \epsilon_c$ into Eq. (11.7) and using Eq. (11.8) result in

$$\sigma_c = A_0 \Lambda E^* \left[\frac{1 - \Omega_0}{A_0(\Lambda + 1)}\right]^{(\Lambda+1)/\Lambda} = \frac{\Lambda E^*(1 - \Lambda_0)\epsilon_c}{\Lambda + 1}. \tag{11.9}$$

where σ_c is the peak stress. According to Eq. (11.3), the initial modulus of elasticity E_0, which corresponds to $\Omega = \Omega_0$ at initial loading, is

$$E_0 = E^*(1 - \Omega_0) \tag{11.10}$$

The value of Λ can then be determined by substituting Eq. (11.10) into Eq. (11.9):

$$\Lambda = \frac{\sigma_c/\epsilon_c}{E_0 - \sigma_c/\epsilon_c} \tag{11.11}$$

Based on Eq. (11.11), the value of Λ can be determined from the initial modulus of elasticity, the peak stress, and the corresponding strain. After obtaining the value of Λ, the constant A_0, which is coupled with Ω_0, can be obtained from Eq. (11.8) if the value of Ω_0 is known. After values of A_0 and Λ are determined, the prepeak stress–strain relationship can be predicted by Eq. (11.7). If only the shape of stress–strain relationship is of interest, the value of Ω_0 may be assumed to be zero. Løland has shown that this model can reasonably predict the prepeak stress–strain curve of concrete under uniaxial tension.

Løland[13] also proposed a stress–strain relationship for the postpeak region. His model for the postpeak region will not be presented here because it does not consider the strain localization that is dominant within the postpeak region.

When values of E_0, Ω_0, σ_c, and ϵ_c for a given material are determined, this simple model can explicitly describe failure of a concrete specimen under uniaxial loading without using any numerical methods for stress analysis. Another advantage of this model is that it includes the initial damage Ω_0. Therefore, it is possible to relate the value of Ω_0 to the microstructure of a material and to damage resulting from chemical attack or weathering before loading is applied. Ouyang et al.[14,15] have applied this model to describe the failure of cementitious materials under chemical attack.

The damage model proposed by Løland only involves a single stress component σ, which implies that the model assumes that all microcracks are mode I. It is also assumed by the model that these microcracks always open during loading. Since many mixed-mode and mode II cracks exist even though concrete is subjected to uniaxial tension, this model only phenomenologically fits the experimental stress–strain curve, rather than represents the actual failure mechanism. A scalar damage model proposed by Mazars[16] partially overcomes these problems. The Mazars model is presented below.

11.2.3 Scalar Damage Model by Mazars

In elasticity, nominal stresses at a material point may be represented by a stress tensor σ_{ij}, where subscripts $i, j = 1, 2, 3$. The nominal stress tensor σ_{ij} is symmetric for an isotropic material, and its six independent components are σ_{11}, σ_{22}, σ_{33}, σ_{12}, σ_{13}, and σ_{23}, as shown in Fig. 11-4. Mazars[16] proposed that for microcracks of mode I or mixed mode (modes I and II), which may occur during uniaxial loading of a concrete specimen, as shown in Fig. 11-5, evolution of damage may be simply described by a single damage scalar. By analogy with Eq. (11.2), the net stress tensor is

$$\sigma_{ij}^* = \frac{\sigma_{ij}}{1 - \Omega} \tag{11.12}$$

where σ_{ij}^* is the net stress tensor. The superscript asterisk indicates quantities for an undamaged material at initial loading from hereafter.

For an undamaged material, the relationship between the net stress and strain is linear elastic and can be expressed as

$$\sigma_{ij}^* = C_{ijkl}^* \epsilon_{kl} \tag{11.13}$$

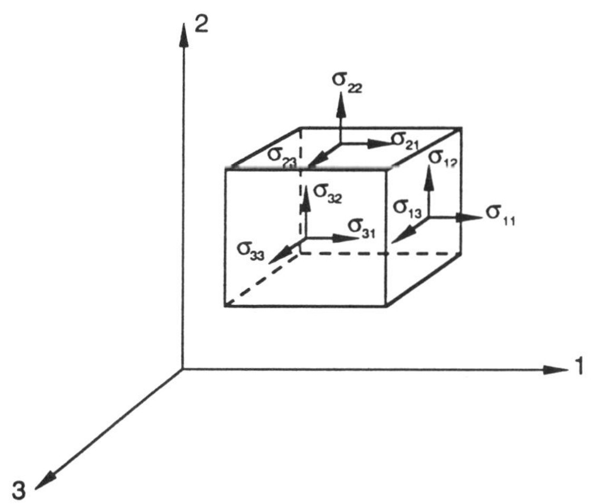

Figure 11-4 Components of stress tensor in cartesian coordinate system.

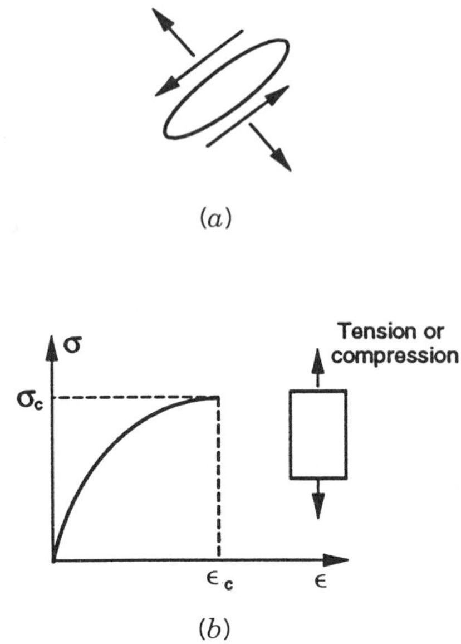

Figure 11-5 Local damage model and global mechanical response for damage model proposed by Mazars: (*a*) mode I or mixed mode (I & II) and (*b*) mechanical response.[16]

where ϵ_{kl} is the strain tensor and C^*_{ijkl} is the compliance tensor for the undamaged material at initial loading, given by[11]

$$C^*_{ijkl} = \lambda^*\delta_{ij}\delta_{kl} + \nu^*(\delta_{ik}\delta_{jl} + \delta_{il}\delta_{jk}) \tag{11.14}$$

where $\lambda^* = \mu^*E^*/[(1 - 2\mu^*)(1 + \mu^*)]$ and $\nu^* = E^*/[2(1 + \mu^*)]$ are Lamé constants and μ^* is Poisson's ratio for the undamaged material at initial loading. The notation δ_{ij} is the Kronecker delta, whose value equals unity for $i = j$ and zero for $i \neq j$.

As a result, the strain energy density of the damaged material is

$$U = \tfrac{1}{2}\sigma_{ij}\epsilon_{ij} = \tfrac{1}{2}(1 - \Omega)\sigma^*_{ij}\epsilon_{ij} = \tfrac{1}{2}(1 - \Omega)\epsilon_{ij}C^*_{ijkl}\epsilon_{kl} \tag{11.15}$$

It can be seen from Eq. (11.15) that the effective compliance of the damaged material is represented by $(1 - \Omega)C^*_{ijkl}$ in this model.

In the framework of thermodynamics, variables associated with ϵ_{ij} and Ω are σ_{ij} and Y, respectively, which can be obtained in terms of the strain energy density:

$$\sigma_{ij} = \frac{\partial U}{\partial \epsilon_{ij}} = (1 - \Omega)C^*_{ijkl}\epsilon_{kl} \tag{11.16}$$

and

$$Y = \frac{\partial U}{\partial \Omega} = -\tfrac{1}{2}\epsilon_{ij} C^*_{ijkl} \epsilon_{kl} \tag{11.17}$$

where $-Y$ is the damage strain energy release rate.

Mazars[16] proposed a loading surface as a condition for damage evolution based on the criterion of maximum principal strain. This loading surface is given by

$$F(\epsilon_{ij}, \Omega) = \epsilon^e - H(\Omega) \tag{11.18}$$

where ϵ^e is the equivalent strain and $H(\Omega)$ is the hardening–softening parameter. The value of ϵ^e is defined as

$$\epsilon^e = \sqrt{\langle \epsilon_{ii}\epsilon_{ii} \rangle_+} \tag{11.19}$$

where ϵ_{ii} with $i = 1, 2, 3$ are the principal strains and $\langle \epsilon_{ii} \rangle_+ = \epsilon_{ii}$ when $\epsilon_{ii} > 0$ and $\langle \epsilon_{ii} \rangle_+ = 0$ when $\epsilon_{ii} < 0$. It is noted that a summation convention is implied in Eq. (11.19) by the repeated subscript ii. The parameter $H(\Omega)$ takes the largest value of the equivalent strain ϵ^* ever reached by the material at the considered point to retain the previous loading history. Initially, $H(\Omega)$ is the threshold H_0. The loading surface given by Eq. (11.18) restrains the validity of this model to the uniaxial loading condition shown in Fig. 11-5.

In order to describe damage evolution with stress or strain, a relationship between the damage variable and the stress or the strain should be specified. This relationship is termed as the damage evolution law, which depends on internal structure, fracture mechanisms, and fracture parameters of a material. Since current knowledge of the fracture process of materials is very limited, development of the damage evolution law is the weakest link in damage mechanics. As a result, most current development of damage evolution laws has to rely on intuition, observation, and interpretation of experimentally observed trends. An empirical relationship between the damage variable and the equivalent strain is assumed here:

$$\Omega = 1 - \frac{(1 - B_1)H_0}{\epsilon^e} - \frac{B_1}{\exp\,[B_2(\epsilon^e - H_0)]} \tag{11.20}$$

where B_1 and B_2 are two empirical constants dependent on concrete material properties. These two values may be determined by fitting the theoretical prediction with some experimental data.

The above model may be used in conjunction with a FE method. For a given load level, the FE method is first used to compute distributions of stresses and strains for a given structure. Values of the equivalent strain ϵ^e are calculated for various material points using Eq. (11.19). At each material point (element), the obtained value of ϵ^e is substituted into the loading surface given by Eq. (11.18). If $F \leq 0$, damage does not grow in the element and the element compliance matrix remains unchanged. On the other hand, if $F > 0$, damage evolves in the element. A new value of Ω can be determined based on ϵ^e using Eq. (11.20); then the element compliance matrix corresponding to the new damage level can be determined based on the effective compliance $(1 - \Omega)C^*_{ijkl}$. The effective compliance matrices obtained in this manner for all elements can then be assembled to obtain a new global compliance matrix, which can be used for another FE analysis to obtain new distributions of stresses and strains. The above procedure can be repeated for different loading levels until the peak load of the structure is reached. Mazars has reported that the proposed model can reasonably predict the damage process in concrete under uniaxial loading.

Since a multiple stress state, which will occur even under uniaxial loading, has been included, this model is capable of accounting for evolution of both mode I and mode II microcracks (see Fig. 11-5*a*). However, the use of the loading surface given by Eq. (11.18) restrains the validity of this model to the uniaxial loading condition shown in Fig. 11-5*b*. It is noted that although the effective compliance $(1 - \Omega)C^*_{ijkl}$, which is obtained based on the concept of the net stress, is symmetric, its inverse may not be symmetric. This problem will be discussed again later.

Mazars et al.[17–19] have extended this model to different loading conditions and included the probabilistic description of damage initiation.

In both the Løland[13] and Mazars[16] models, the extent of material damage is described by a phenomenological scalar damage variable based on the concept of the net stress. The micromechanical analysis given in the next section will show that if microcracks are randomly and isolatedly distributed, a scalar damage variable, which may be a function of the accumulated volume of microcracks, is sufficient to describe the damage process of a material.

11.3 MICROMECHANICAL EVALUATION OF ISOTROPIC DAMAGE

Micromechanical approaches have been widely used to evaluate the effect of microcracks on compliance of a material. Some micromechanical approaches are presented here in order that one has a better understanding of how to use the scalar damage variable for description of isotropic damage based on micromechanics. Only a three-dimensional solid containing penny-shaped microcracks will be discussed in this section.[11,20]

Consider an isotropic and homogeneous solid containing an isolated penny-shaped microcrack subjected to a far-field stress σ, as shown in Fig. 11-6. The

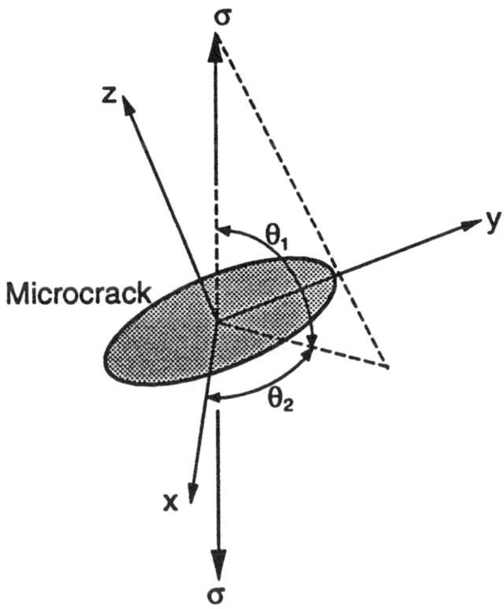

Figure 11-6 Three-dimensional solid containing penney-shaped crack subjected to uniaxial loading.

penny-shaped crack has a radius r_c that is much smaller than the size of the solid. The normal to the plane of the penny-shaped crack is assumed to be in the z direction. The far-field applied stress σ is oriented at angles θ_1 and θ_2, where θ_1 is the angle between the x-y plane and the radius vector of the applied stress and θ_2 is the azimuthal angle, as shown in Fig. 11-6. Stress intensity factors at the crack tip for this case have been given by Rice[21]:

$$
\begin{aligned}
K_{\mathrm{I}} &= \frac{2}{\pi} \sigma\sqrt{\pi r_c} \sin^2\theta_1 \\
K_{\mathrm{II}} &= \frac{4}{\pi(2 - \mu_n)} \sigma\sqrt{\pi r_c} \sin\theta_1 \cos\theta_1 \cos\theta_2 \\
K_{\mathrm{III}} &= \frac{4(1 - \mu_n)}{\pi(2 - \mu_n)} \sigma\sqrt{\pi r_c} \sin\theta_1 \cos\theta_1 \sin\theta_2
\end{aligned}
\tag{11.21}
$$

where K_{I}, K_{II}, and K_{III} are the stress intensity factors for opening (mode I), shearing (mode II), and tearing (mode III) modes of deformation, respectively, and μ_n is the intrinsic Poisson ratio for the material around the crack tip. Correspondingly, loss of the strain energy due to creating the crack in the plane-strain condition is[22]

$$
\Delta U_1 = \frac{2(1 - \mu_n^2)}{E_n} \int_0^{r_c} \left(K_{\mathrm{I}}^2 + K_{\mathrm{II}}^2 + \frac{K_{\mathrm{III}}^2}{1 - \mu_n} \right) \pi r_c \, dr_c \tag{11.22}
$$

where E_n is the intrinsic modulus of elasticity for the material around the crack tip. Substituting Eq. (11.21) into Eq. (11.22) results in

$$\Delta U_1 = \frac{8(1-\mu_n^2)}{3E_n}\sigma^2 r_c^3 \left[\sin^4\theta_1 + \frac{4}{(2-\mu_n)^2}\sin^2\theta_1\cos^2\theta_1\cos^2\theta_2 + \frac{4(1-\mu_n)}{(2-\mu_n)^2}\sin^2\theta_1\cos^2\theta_1\cos^2\theta_2\right] \tag{11.23}$$

It is noted that ΔU_1 is a function of angles θ_1 and θ_2 here.

If the solid contains N penny-shaped cracks in a unit volume with a mean crack length of r_c and the angles θ_1 and θ_2 for different cracks randomly distribute from 0 to 2π, respectively, the total loss of the strain energy due to creating N cracks, ΔU_N, can be estimated by taking spherical averages over θ_1 and θ_2 based on Eq. (11.23):

$$\Delta U_N = \frac{N}{2\pi}\int_0^{2\pi} d\theta_2 \frac{1}{2\pi}\int_0^{2\pi} \Delta U_1(\theta_1, \theta_2)\, d\theta_1 = \frac{8N\sigma^2 r_c^3}{45E_n}\frac{(1-\mu_n^2)(10-3\mu_n)}{(2-\mu_n)} \tag{11.24}$$

It is noted here that although the orientation effect of microcracks is initially considered for computing the loss of the strain energy, this effect is eliminated by the spherical average approach if the microcracks are randomly distributed.

On the other hand, strain energy densities for the same solid without any cracks and with N cracks can be evaluated respectively from elasticity:

$$U* = \frac{\sigma^2}{2E*} \quad \text{and} \quad U = \frac{\sigma^2}{2E} \tag{11.25}$$

where $U*$ and $E*$ are the strain energy density and the modulus of elasticity for the material without cracks and U and E are the strain energy density and the effective modulus of elasticity for the material with N cracks. The condition of energy equilibrium during cracking requires $U* - U = -\Delta U_N$. Substituting Eqs. (11.24) and (11.25) into this condition results in

$$\frac{\sigma^2}{2E*} - \frac{\sigma^2}{2E} = -\frac{8N\sigma^2 r_c^3}{45E_n}\frac{(1-\mu_n^2)(10-3\mu_n)}{(2-\mu_n)} \tag{11.26}$$

Equation (11.26) can be used to evaluate values of the effective modulus of elasticity E and the effective Poisson ratio μ for the solid with N microcracks.

If interaction among microcracks is neglected (Taylor's model), it may be physically regarded that microcracks are simultaneously formed in the solid with initial parameters E^* and μ^*. Therefore, substituting conditions of $E_n = E^*$ and $\mu_n = \mu^*$ into Eq. (11.26) results in

$$\frac{E^*}{E} = 1 + \frac{16}{45} \frac{(1 - \mu^{*2})(10 - 3\mu^*)}{2 - \mu^*} \omega \tag{11.27}$$

where the crack density $\omega = Nr_c^3$ has been used. For a given value of ω, the effective modulus of elasticity E can be solved based on the initial material parameters E^* and μ^* from Eq. (11.27).

The following relationship between μ and μ^* can be further obtained using Castigliano's theorem:

$$\frac{\mu}{\mu^*} = \frac{E}{E^*} \left[1 + \frac{16}{45} \frac{(1 - \mu^{*2})}{(2 - \mu^*)} \omega \right] \tag{11.28}$$

On the other hand, the interaction among microcracks may be taken into account by a self-consistent approach. This approach assumes that when cracks are formed, the intrinsic material parameters are equal to the effective material parameters. As a result, substituting conditions $E_n = E$ and $\mu_n = \mu$ into Eq. (11.26) leads to

$$\frac{E}{E^*} = 1 - \frac{16}{45} \frac{(1 - \mu^2)(10 - 3\mu)}{(2 - \mu)} \omega \tag{11.29}$$

It is noted that for a given value of ω, Eq. (11.29) includes two unknowns, E and μ. To obtain both values of E and μ, Eq. (11.29) should be solved simultaneously with another equation given in the following:

$$\omega = \frac{45}{16} \frac{(\mu^* - \mu)(2 - \mu)}{(1 - \mu^2)[10\mu^* - \mu(1 + 3\mu^*)]} \tag{11.30}$$

The derivation of Eq. (11.30) can be found in Budiansky and O'Connell.[11]

If values of the initial modulus of elasticity E^* and the initial Poisson ratio μ^* are known, for a given value of the crack density ω the values of the effective modulus of elasticity E and the effective Poisson ratio μ can be solved either from Eqs. (11.27) and (11.28) or from Eqs. (11.29) and (11.30) based on the self-consistent approach. Both values of E and μ decrease with increasing value of ω.

From the view of damage mechanics, the crack density ω can be regarded as the damage variable. Therefore, Eqs. (11.27) and (11.28) indicate that a scalar damage variable is sufficient to describe damage resulting from randomly and isolatedly distributed microcracks.

Evaluations of the effective modulus of elasticity using Taylor's model (noninteracting microcracks) and the self-consistent model are compared in Fig. 11-7, where the value of μ^* is assumed to equal 0.2. It is seen that Taylor's model predicts a much slower decrease in the effective modulus of elasticity than the self-consistent model does.

A similar analysis can be performed for a unit thickness plate (the plane-stress condition) containing a line crack (slit) of length $2a$ subjected to a far-field stress σ. In this case, the effective modulus of elasticity for the plate with N slits per unit area is expressed as

$$\frac{E}{E^*} = 1 - \omega \tag{11.31}$$

where $\omega = N\pi a^2$ is the crack density.

It is interesting that for the plane-stress condition, Eq. (11.31) obtained from the self-consistent evaluation is identical to Eq. (11.3) in terms of the phenomenological assumption, although the damage variable now is ω, which is based on the micromechanics, instead of Ω. Further analyses based on micromechanics can result in micromechanical-based scalar damage models.[4,23] One such micromechanical damage model will be presented in the next section.

It is noted that only an isotropic and homogeneous material weakened by randomly distributed cracks is discussed above. More general analyses for an anisotropic and homogeneous solid have been given by Horii and Nemat-Nasser[24] and Sumarac and Krajcinovic.[25]

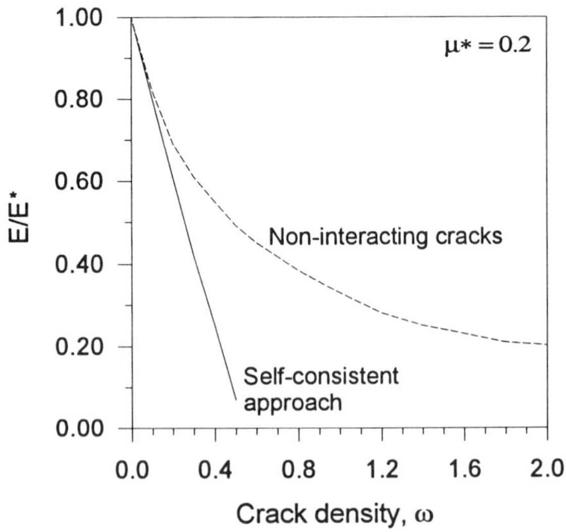

Figure 11-7 Evaluation of effective modulus of elasticity based on different approaches.[20]

11.4 MICROMECHANICAL SCALAR DAMAGE MODEL BY KARIHALOO ET AL.

Karihaloo et al.[26,27] have proposed a micromechanical scalar damage model for tensile response of concrete. The prepeak tensile response is divided into two stages in their model, which is similar to Fig. 11-1. Concrete behaves elastically before point A, whereas internal cracks in concrete initiate and propagate after point A. They assumed that the cracks are isolatedly and randomly distributed up to point C. This implies that strain localization appears at point C instead of point B, as shown in Fig. 11-1. After point C, a major crack forms. Karihaloo et al.[27] first proposed a regression formula for the strain at point A. A scalar damage approach was developed for behavior between points A and C. Their model will be presented here.

11.4.1 Threshold Strain for Initiation of Microcracks

Concrete contains pores and microcracks even in its unstressed state. Under low tensile stresses, however, their influence on mechanical response of the material is not noticeable. As previously discussed, these microcracks start to open and propagate at point A (Fig. 11-1). Therefore, mechanical behavior of concrete up to point A is approximately linearly elastic. The stress–strain relationship becomes nonlinear after point A. To predict the stress–strain relationship of concrete, the threshold strain at point A should be known.

The threshold strain corresponding to initiation of microcracks, which are primarily interfacial cracks between the cementitious matrix and coarse aggregates, was measured by Kaplan[28] under several tensile loading combinations. He found that this strain value depends only on the volume fraction of the coarse aggregate in the mix but does not vary with other microstructural parameters, such as the type and texture of coarse aggregate or the water-to-cement ratio. Likewise, in the case of biaxial tension the major principal threshold strain $\epsilon_{11}^{\text{th}}$ was found not to depend on the biaxial tension ratio ($r_b = \sigma_{22}/\sigma_{11}$). On the other hand, the minor principal threshold strain $\epsilon_{22}^{\text{th}}$ increased with increasing r_b and reached the value of $\epsilon_{11}^{\text{th}}$ at $r_b = 1$. Based on Kaplan's experimental observation, Karihaloo and Fu[29] gave a regression formula for values of $\epsilon_{11}^{\text{th}}$ in uniaxial tension as

$$\epsilon_{11}^{\text{th}} = 79.5 + 43.4V_a - 406.1V_a^2 + 707.2V_a^3 - 1150V_a^4 + 763V_a^5 \quad (11.32)$$

where V_a is the volume fraction of the coarse aggregate in the mix and $\epsilon_{11}^{\text{th}}$ is expressed in microstrain. It follows from Eq. (11.32) that the value of $\epsilon_{11}^{\text{th}}$ decreases with increasing value of V_a.

The response of a concrete specimen subjected to uniaxial or biaxial tension can be elastically analyzed using initial elastic constants up to the value of $\epsilon_{11}^{\text{th}}$.

11.4.2 Nonlinear Response Prior to Peak Load

The interfacial microcracks usually form first between the cementitious matrix and the coarse aggregates. With a further increase in stress, not only do the existing microcracks propagate stably but also more microcracks are progressively nucleated at interfaces between the cementitious matrix and small-size aggregates. It was assumed by Karihaloo and Huang[27] that this stable propagation of microcracks continues until the peak load is reached. Since microcracks are isolatedly and randomly distributed during this stage, this prepeak nonlinear stress and strain response may be described by using a scalar damage mechanics approach.

At a given stress level $\sigma_{ij} \geq \sigma_{ij}^{\text{th}}$, the strain ϵ_{ij} can be expressed by

$$\sigma_{ij} = C_{ijkl}\epsilon_{kl} \tag{11.33}$$

where C_{ijkl} is the effective-elastic tensor, which is a function of elastic constants and the damage state. Similar to the elastic tensor for the undamaged material ($\sigma_{ij} < \sigma_{ij}^{\text{th}}$) given in Eq. (11.14), the effective-elastic tensor corresponding to the given stress and damage state is

$$C_{ijkl} = \lambda\delta_{ij}\delta_{kl} + \nu(\delta_{ik}\delta_{jl} + \delta_{ij}\delta_{jk}) \tag{11.34}$$

where $\lambda = \mu E/[(1 - 2\mu)(1 + \mu)]$ and $\nu = E/[2(1 + \mu)]$ are the effective Lamé constants and μ is the effective Poisson's ratio. Based upon the micromechanical analysis previously discussed, both E and μ can be determined using Eqs. (11.27) and (11.28). These obtained values of E and μ result in

$$\frac{3\lambda^* + 2\nu^*}{3\lambda + 2\nu} = 1 + \frac{16(1 - \mu^{*2})}{9(1 - 2\mu^*)}\,\omega \tag{11.35}$$

and

$$\frac{\nu^*}{\nu} = 1 + \frac{32(1 - \mu^*)(5 - \mu^*)}{45(2 - \mu^*)}\,\omega \tag{11.36}$$

where ω is the microcrack density, which can be regarded as a damage variable.

With a further increase in the stress σ_{ij} from the current level, additional interfacial microcracks are nucleated. For concrete, it may be reasonable to assume that the density of nucleated flaws is some scalar function of the current accumulated effective strain ϵ^a, which is defined as

$$\epsilon^a = \sqrt{\epsilon_{ij}\epsilon_{ij}} \tag{11.37}$$

From an extensive analysis of the tension test data, Karihaloo and Fu[26] proposed the following isotropic scalar function for the evolution of microcracks:

$$\omega = \begin{cases} \omega_0 & \epsilon^a \leq \epsilon^{a\text{th}} \\ \omega_0 + C_1 \left(1 - \dfrac{\epsilon^{a\text{th}}}{\epsilon^a}\right)^{C_2} & \epsilon^a > \epsilon^{a\text{th}} \end{cases} \tag{11.38}$$

where $\epsilon^{a\text{th}} = (\epsilon_{ij}^a \epsilon_{ij}^a)^{1/2}$ is the effective strain threshold below which no nucleation takes place, and C_1 and C_2 are material constants that depend on the specific shape and distribution of flaws.[30] For randomly distributed penny-shaped cracks $C_1 = 0.16$ and $C_2 = 2.30$, whereas for randomly distributed spherical microvoids $C_1 = 0.24$ and $C_2 = 2.60$. It is noted that the constants C_1 and C_2 were determined for concrete whose elastic modulus E^* varies between 29 and 39 GPa and Poisson's ratio μ^* varies between 0.175 and 0.21. The value of $\epsilon^{a\text{th}}$ could reasonably be calculated using Eq. (11.32).

Nucleation will only occur if

$$\frac{\partial \omega}{\partial \epsilon^a} \Delta \epsilon^a \geq 0 \tag{11.39}$$

otherwise ω will remain unchanged. When the nucleation condition Eq. (11.39) is satisfied, the incremental stress–strain relation may be written as

$$\Delta \sigma_{ij} = C_{ijkl}^n \Delta \epsilon_{kl} \tag{11.40}$$

where $\Delta\sigma_{ij}$ and $\Delta\epsilon_{kl}$ are incremental stress and strain tensors, respectively, and C_{ijkl}^n is the effective compliance tensor for the nth strain increment. This can be calculated by

$$C_{ijkl}^n = C_{ijkl}^{n-1} + \Delta C_{ijkl}^n \tag{11.41}$$

where C_{ijkl}^{n-1} is the effective compliance tensor corresponding to the damage accumulated thus far in the deformation history given by Eq. (11.34) and ΔC_{ijkl}^n is the nth incremental effective-elastic tensor due to crack nucleation:

$$\Delta C_{ijkl}^n = \frac{dC_{ijmn}^{n-1}}{d\omega} \frac{d\omega}{d\epsilon_{kl}} \epsilon_{mn} \tag{11.42}$$

One consequence of the choice of the scalar function [Eq. (11.38)] for ω is that the incremental effective-elastic tensor preserves diagonal symmetry and normality during nucleation. Substituting Eq. (11.42) into Eq. (11.41) leads to

$$C_{ijkl}^{n} = C_{ijkl}^{n-1} + \frac{dC_{ijmn}^{n-1}}{d\omega}\frac{d\omega}{d\epsilon^{a}}\frac{\epsilon_{kl}\epsilon_{mn}}{\epsilon^{a}} \tag{11.43}$$

where the following relationship has been used based on the definition of ϵ^{a} [Eq. (11.40)]:

$$\frac{d\omega}{d\epsilon_{kl}} = \frac{d\omega}{d\epsilon^{a}}\frac{d\epsilon^{a}}{d\epsilon_{kl}} = \frac{d\omega}{d\epsilon^{a}}\frac{\epsilon_{kl}}{\epsilon^{a}} \tag{11.44}$$

Thus, when Eq. (11.39) is satisfied, the nucleation of microcracks occurs. Based on Eq. (11.34) the elastic constants λ, ν (or E, μ) are functions of the crack density ω. By using Eqs. (11.34)–(11.36), the derivative of the effective-elastic tensor with respect to ω is obtained:

$$\frac{dC_{ijmn}^{n-1}}{d\omega} = \left[\frac{E*}{3(\omega + \mu*)^{2}} - \frac{E*}{3(\omega - 2\mu*)^{2}}\right]\delta_{ij}\delta_{mn} - \frac{E*}{2(\omega + \mu*)^{2}}(\delta_{im}\delta_{jn} + \delta_{in}\delta_{jm}) \tag{11.45}$$

The incremental stress–strain relationship including the nucleation of interfacial microcracks is therefore given by Eqs. (11.40), (11.43), and (11.45). It is noted that the incremental stress–strain relationship depends on a single damage parameter ω.

For the strain level $\epsilon_{kl} = \epsilon_{kl}^{\text{th}}$, since the damage variable $\omega = \omega_0$, where ω_0 is the initial damage level before loading; then $C_{ijkl}^{n} = C_{ijkl}^{*}$, which can be evaluated from Eq. (11.34) using Lamé constants λ^* and ν^* for the material with the initial damage. The corresponding stresses then are calculated using Eq. (11.40) with $\Delta\epsilon_{kl} = \epsilon_{kl}^{\text{th}}$. For any given consecutive strain increment $\Delta\epsilon_{kl}$ associated with a damage density ω, the current effective-elastic tensor can be calculated using Eqs. (11.43) and (11.45). The corresponding stress increment $\Delta\sigma_{ij}$ can then be obtained from Eq. (11.40). It is noted that although C_{ijkl}^{n} is symmetric, its inverse may be no longer symmetric. However, to assure that the inverse of C_{ijkl}^{n} is symmetric, one may assume that the complementary strain energy of the damaged material is equal to that of an equivalent undamaged material.[31]

The prepeak nonlinear response of a concrete mix under biaxial tension predicted using this approach is shown in Fig. 11-8. The agreement between the predicted and test results is very good, thereby lending support to the accuracy of the proposed scalar function for the evolution of damage in the material. It is noted that this model has regarded a material at initial loading as the undamaged material. This means the damage state before loading is neglected.

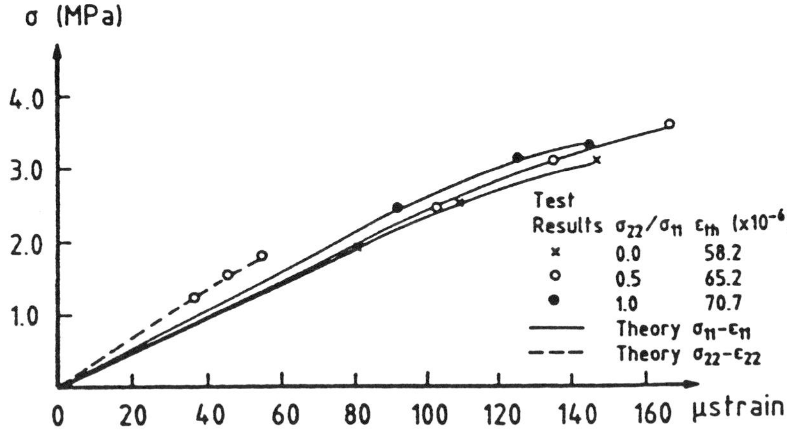

Figure 11-8 Predicted stress–strain response of concrete mix in biaxial tension up to peak stress.[27]

11.5 TENSION-SOFTENING RESPONSE AFTER PEAK LOAD

A macroflaw (major crack) will form from coalescence of microcracks as the load increases. The major crack will propagate unstably after the peak stress. As a result, a softening type of stress–strain curve will be observed. Several models have been proposed to describe the intrinsic tension softening of quasi-brittle materials. Most of the models are based on the assumption that unbroken ligaments bridge across a discontinuous macroflaw. Models proposed by Horii et al.,[32] Ortiz,[33] and Karihaloo et al.[34] will be briefly described in the following.

11.5.1 Model by Horii et al. and Ortiz

In this model[32,33] the tension-softening behavior is modeled by an infinite row of collinear cracks along the eventual fracture plane, which is subjected to a remote uniaxial tension σ, as shown in Fig. 11-9. It is assumed that at the onset of strain localization each segment of the discontinuous macroflaw is size $2l = 2l_0$ and is separated from the neighboring segment by an unbroken ligament of length $d_H - 2l_0$. Thus $2l_0/d_H$ may be regarded as equal to the accumulated damage per unit area (volume) at the onset of strain localization, which is assumed to correspond to point C in Fig. 11-1 in this model. This leads to $2l_0/d_H = \omega_p$, where ω_p is the value of the damage density at the peak stress.

Based on LEFM, the stress intensity factor K_{I} at each crack tip in the row is given by

$$K_{\text{I}} = \sigma \sqrt{d_H \tan \frac{\pi l}{d_H}} \tag{11.46}$$

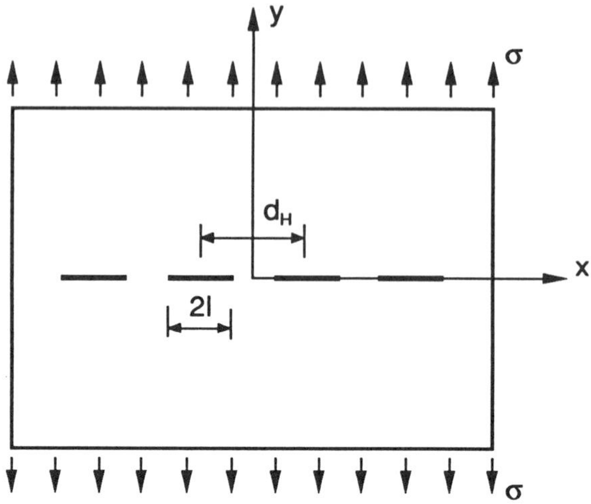

Figure 11-9 Tension-softening model proposed by Horii et al.[32] and Ortiz.[33]

and the crack opening displacement w by

$$w = \frac{4\sigma d_H}{\pi E} \log \left[\frac{\cos (\pi x/d_H) + \sqrt{\cos^2 (\pi x/d_H) - \cos^2 (\pi l/d_H)}}{\cos (\pi l/d_H)} \right] \tag{11.47}$$

where $2l$ is the current length of each segment. Since the discontinuous macroflaw propagates unstably in the tension-softening stage, the stress intensity factor $K_{\rm I}$ is always equal to the critical stress intensity factor K_m of the matrix material. Thus a relationship between the applied stress and the flaw size during the tension softening stage can be obtained from $K_{\rm I} = K_m$,

$$\frac{\sigma}{f_t} = \left[\frac{\tan (\pi l_0/d_H)}{\tan (\pi l/d_H)} \right]^{1/2} \qquad l \geq l_0 \tag{11.48}$$

where the ultimate tensile strength f_t is given by

$$f_t = \frac{K_m}{\sqrt{d_H \tan (\pi l_0/d_H)}} \tag{11.49}$$

The measured deformation is less than the opening displacement of each of the segments because of unbroken bridging ligaments. As an approximation, it may be calculated by averaging the crack opening displacement over the period d_H to account for the bridging ligaments:

$$w_a = \frac{1}{d_H}\int_{-l}^{l} w\,dx = \frac{4\sigma d_H}{\pi E}\log\left[\sec\left(\frac{\pi l}{d_H}\right)\right] \tag{11.50}$$

where w_a is the average crack opening displacement. The net deformation due to propagation of the macroflaw in the tension-softening stage is given by $w = w_a - w_p$,

$$\frac{w}{w_p} = \frac{\sigma}{f_t}\frac{\log\,[\sec\,(\pi l/d_H)]}{\log\,[\sec\,(\pi l_0/d_H)]} - 1 \tag{11.51}$$

where w_p is the deformation at the peak stress given by

$$w_p = \frac{4d_H f_t}{\pi E}\log\,[\sec\,(\pi l_0/d_H)] \tag{11.52}$$

Equations (11.48) and (11.51) provide the tension-softening relation between the stress σ and the deformation w in terms of the material constants at the peak stress, namely the accumulated damage $\omega_p = 2l_0/d_H$ and the matrix fracture toughness K_m. For a given value of the crack length l, the value of σ is calculated using Eq. (11.48). Then the corresponding value of w is determined from Eq. (11.51). Repeating this procedure for different values of l, the softening stress–separation curve is obtained. The predicted tension-softening curves for several values of ω_p are shown in Fig. 11-10. It is noted that at low stresses the deformation attains the maximum value of ω_p at which complete rupture must take place in order to maintain the relationship $K_{\text{I}} = K_m$ in the

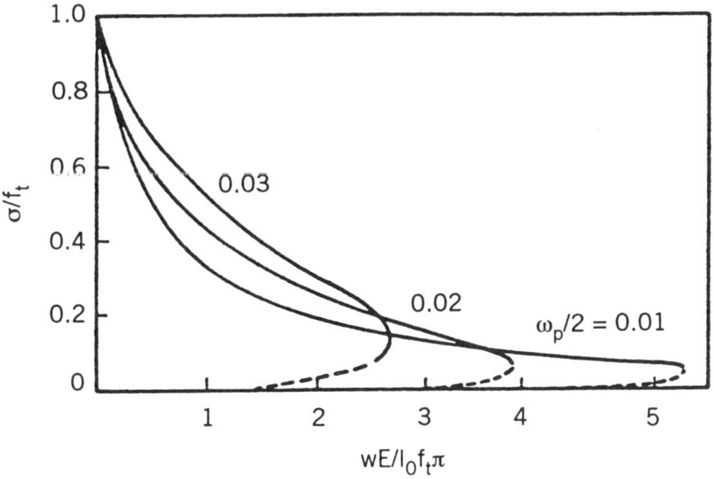

Figure 11-10 Tension-softening curves predicted by model proposed by Horii et al.[32] and Ortiz.[33]

tension-softening stage. This should not be regarded as a material characteristic but is a consequence of the chosen model.

11.5.2 Model by Karihaloo et al.

In this approach,[34] the tension-softening behavior of quasi-brittle materials is modeled by a discontinuous macroflaw assumed to consist of an infinite row of collinear circular micropores of radius r_0 with edge crack of length l_0, as shown in Fig. 11-11, thereby including the propensity of cracks to be attracted to voids in the material. Based on LEFM, the stress intensity factor K_I at each crack tip in the array is given by

$$K_{\mathrm{I}} = \sigma\sqrt{\pi(l + r)}\,f_1(\lambda_1, \omega_1) \tag{11.53}$$

where $2(l + r)$ is the current flaw size corresponding to the stress σ and

$$f_1(\lambda_1, \omega_1) = \left(1 + \sum_{A=1}^{3}\sum_{B=0}^{3} Q_{AB}\lambda_1^A\omega_1^B\right)\left[\frac{\tan(\pi\omega_1/2)}{(\pi\omega_1/2)}\right]^{1/2} \tag{11.54}$$

with $\lambda_1 = r/(r + l)$ and $\omega_1 = 2(r + l)/d_H$. The coefficients Q_{AB} are given in Table 11-1. The parameter λ_1 is a measure of the relative sizes of microvoids and microcracks. Thus, when λ_1 approaches unity, this corresponds to microvoids with infinitesimally small edge cracks, whereas $\lambda_1 = 0$ corresponds to microcracks without voids.

Likewise, the average crack opening displacement is given by

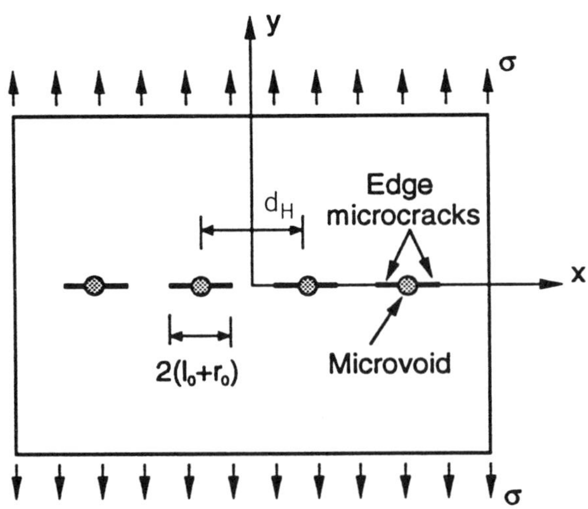

Figure 11-11 Tension-softening model proposed by Karihaloo et al.[34]

Table 11-1 Coefficients Q_{AB} and S_{AB}

A	B	Q_{AB}	S_{AB}
1	0	−0.0131246	1.0012000
1	1	−0.1801658	−0.8720340
1	2	0.3349874	0.5115431
1	3	1.3436450	0.8064520
2	0	0.6175492	−1.1061990
2	1	0.8620229	1.1268440
2	2	−2.0971480	−2.4623200
2	3	−3.1807730	−1.3658990
3	0	−0.8855224	0.4455536
3	1	−0.5714183	−0.7251084
3	2	0.8950143	2.4867950
3	3	3.0786150	0.5254524

$$w_a = \frac{1}{d_H}\int_{-(l+r)}^{(l+r)} w\,dx = \frac{\sigma d_H}{2E} f_2(\lambda_1, \omega_1) \tag{11.55}$$

where

$$f_2(\lambda_1, \omega_1) = \left(1 + \frac{1}{\sqrt{1-\omega_1}} \sum_{A=1}^{3} \sum_{B=0}^{3} S_{AB}\lambda_1^A \omega_1^B\right)\left[\frac{8}{\pi}\log\left(\sec\frac{\pi\omega_1}{2}\right)\right] \tag{11.56}$$

The coefficients S_{AB} are also given in Table 11-1. The functions $f_1(\lambda_1, \omega_1)$ and $f_2(\lambda_1, \omega_1)$ are obtained by solving the boundary value problem corresponding to the model (Fig. 11-11).

By using a procedure similar to that of the Horii et al. and Ortiz model,[32,33] Karihaloo et al.[34] obtained the following expressions for the tension-softening response:

$$\frac{\sigma}{f_t} = \sqrt{\frac{l_0 + r_0}{l + r}\,\frac{f_1(\lambda_{10}, \omega_{10})}{f_1(\lambda_1, \omega_1)}} \tag{11.57}$$

and

$$\frac{w}{w_p} = \frac{\sigma}{f_t}\,\frac{f_2(\lambda_1, \omega_1)}{f_2(\lambda_{10}, \omega_{10})} \tag{11.58}$$

where $\omega_{10} = 2(l_0 + r_0)/d_H$, $\lambda_{10} = r_0/(l_0 + r_0)$, and

$$f_t = \frac{K_m}{\sqrt{\pi(l_0 + r_0)} f_1(\lambda_{10}, \omega_{10})} \tag{11.59}$$

The results for an infinite array of collinear cracks (Fig. 11-9) are immediately recovered by setting $\lambda_1 = 0$ and $\omega_1 = 2l/d_H$. It is noted that in this model the neighboring segments are separated by unbroken ligaments of length $d_H - 2(l_0 + r_0)$, so that $\omega_{10} = 2(l_0 + r_0)/d_H$ may be regarded as equal to the accumulated damage per unit area (volume) at the peak stress. The value of $\omega_1 = 2(r + l)/d_H$ then increases with increasing applied load. Equations (11.57) and (11.58) can be easily rewritten as

$$\frac{Ew}{f_t d_H} = \frac{f_1(\lambda_{10}, \omega_{10})}{2} \left[\left(\frac{\omega_{10}}{\omega_1} \right)^{1/2} \frac{f_2(\lambda_1, \omega_1)}{f_1(\lambda_1, \omega_1)} - \frac{f_2(\lambda_{10}, \omega_{10})}{f_1(\lambda_{10}, \omega_{10})} \right] \tag{11.60}$$

The tension-softening curves predicted by this model for several values of the accumulated damage ω_{10} at the peak load are shown in Fig. 11-12. A comparison of these curves with those shown in Fig. 11-10 reveals that the model given in Fig. 11-11 predicts a higher fracture deformation without any instability.

It is noted from Eq. (11.60) in conjunction with Eqs. (11.54) and (11.56) that for $\lambda_1 > 0$ the ratio $f_2(\lambda_1, \omega_1)/f_1(\lambda_1, \omega_1)$ approaches infinity as ω_1 tends to unity, whereas for $\lambda_1 = 0$ the ratio $f_2(\lambda_1, \omega_1)/f_1(\lambda_1, \omega_1)$ approaches zero as ω_1 tends to unity. This result verifies the instability inherent to the collinear mi-

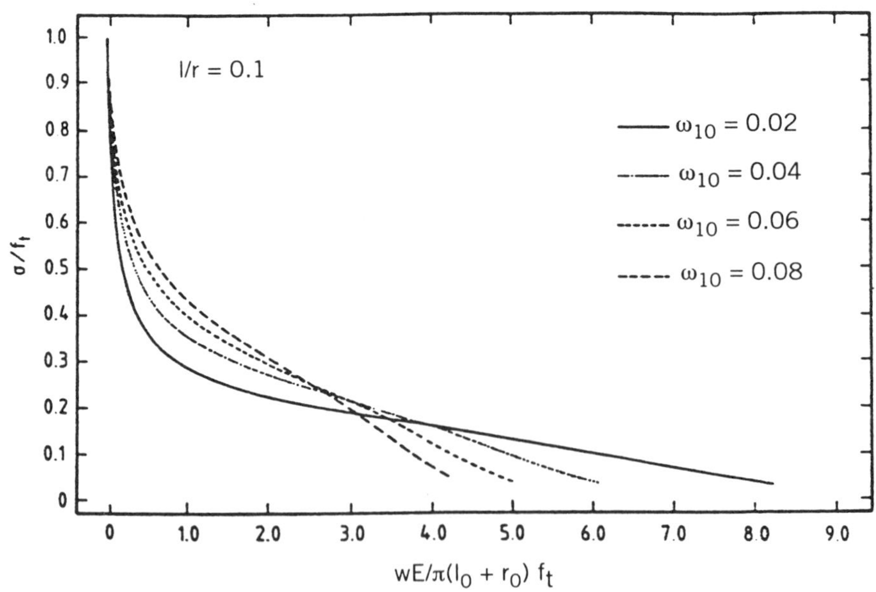

Figure 11-12 Tension-softening curves predicted by model proposed by Karihaloo et al.[34]

crocrack model. Therefore, it is important to realize that the tension-softening models (based on Figs. 11-9 and 11-11) cannot predict the exact value of the critical crack opening displacement w_c at complete rupture ($\omega_1 = 1$). A notional value of w_c may be estimated by limiting ω_1 to $1 - \delta$, where δ is a very small constant (say $\delta = 0.001$).

11.6 INFLUENCE OF MICROCRACKS ON FRACTURE TOUGHNESS

Based on the previously presented damage model for tension softening, Karihaloo et al.[34] calculated the fracture energy per unit area, G_d. Since G_d is the energy rate required for complete fracture of a material, it may be evaluated from the area under a tension-softening stress–elongation curve such as the one given in Fig. 11-12. As a result, the fracture energy per unit area G_d is given as

$$G_d = \int_{f_t}^{0} w(\sigma)\, d\sigma = \int_{0}^{w_c} \sigma(w)\, dw \tag{11.61}$$

where $\sigma(w)$ is the tension-softening stress–elongation curve and is a function of the accumulated damage at the peak load. By using Eq. (11.60), G_d can be expressed as

$$G_d = \int_{f_t}^{0} w(\sigma)\, d\sigma = \int_{\omega_{10}}^{1} w(\omega_1) \frac{\partial \sigma}{\partial \omega_1}\, d\omega_1 \tag{11.62}$$

where for a fixed value of λ_1,

$$\frac{\partial \sigma}{\partial \omega_1} = -\frac{f_t}{2} \frac{f_{10}}{f_1} \sqrt{\frac{\omega_{10}}{\omega_2^3}} \left(1 + \frac{2\omega_1}{f_1} \frac{\partial f_1}{\partial \omega_1}\right) \tag{11.63}$$

with

$$\frac{1}{f_1} \frac{\partial f_1}{\partial \omega_1} = \frac{\pi}{2} \left[\frac{1}{\sin(\pi\omega_1)} - \frac{1}{\pi\omega_1}\right] + \frac{\sum_{A=1}^{3} \sum_{B=0}^{3} B Q_{AB} \lambda_1^A \omega_1^{B-1}}{1 + \sum_{A=1}^{3} \sum_{B=0}^{3} Q_{AB} \lambda_1^A \omega_1^B} \tag{11.64}$$

where $f_1 = f_1(\lambda_1, \omega_1)$ and $f_{10} = f_1(\lambda_{10}, \omega_{10})$ have been used.

Since the accumulated damage at the peak stress is ω_{10}, the value of G_d

obtained in such a way represents the energy per unit cracked area on the creation of a traction-free crack of relative length $(1 - \omega_{10})$.

Based on the value of G_d obtained, Karihaloo et al.[34] further calculated the effective fracture toughness G_{Ic}. It is recalled the G_{Ic} is the critical energy release rate (per unit cracked length for a constant-thickness specimen) at the peak load, and G_d corresponds to the relative crack length $1 - \omega_{10}$. By simply equating the total energies (per unit thickness), the following relationship can be obtained:

$$G_d = G_{Ic}(1 - \omega_{10}) = \frac{(K_{Ic})^2(1 - \omega_{10})}{E} \tag{11.65}$$

where K_{Ic} is the effective critical stress intensity factor. Equation (11.65) is for the plane-stress condition. For the plane-strain condition, E should be replaced by $E/(1 - \mu^2)$.

Using Eqs. (11.58), (11.63), and (11.64), Eq. (11.62) can be rewritten as

$$G_d = \frac{(K_m)^2}{\pi E f_{10}\omega_{10}} \left(W_d^* - \frac{f_{20}}{f_{10}} \right) \tag{11.66}$$

where

$$W_d^* = \frac{f_{10}\omega_{10}}{2} \int_{\omega_{10}}^{1} \frac{f_2}{(f_1\omega_1)^2} \left[\frac{\pi\omega_1}{\sin(\pi\omega_1)} + \frac{2 \sum_{A=1}^{3} \sum_{B=0}^{3} Q_{AB}\lambda_1^A\omega_1^B}{1 + \sum_{A=1}^{3} \sum_{B=0}^{3} Q_{AB}\lambda_1^A\omega_1^B} \right] d\omega_1 \tag{11.67}$$

and for brevity, $f_2 = f_2(\lambda_1, \omega_1)$ and $f_{20} = f_2(\lambda_{10}, \omega_{10})$.

By substituting Eqs. (11.66) and (11.67) into Eq. (11.65), a relationship between the effective fracture toughness K_{Ic} and the intrinsic matrix fracture toughness K_m is obtained:

$$\frac{K_{Ic}}{K_m} = \sqrt{\frac{W_d^* - f_{20}/f_{10}}{\pi f_{10}\omega_{10}(1 - \omega_{10})}} \tag{11.68}$$

Karihaloo et al.[34] numerically calculated this relationship in ranges of $0.02 \leq \omega_{10} \leq 0.25$ and $0 \leq \lambda_1 \leq 1$. They found that in these ranges the obtained numerical values of the toughening ratio K_{Ic}/K_m may be fitted with a polynomial approximation (error $< 1\%$)

Table 11-2 Coefficient Y_{AB}

A	B	Y_{AB}
0	0	1.003944
0	1	1.408276
0	2	−3.878354
0	3	3.201900
1	0	0.326490
1	1	0.928382
1	2	−2.502774
1	3	1.921036

$$\frac{K_{Ic}}{K_m} = \sum_{A=0}^{1} Y_{A0}\omega_{10}^{A} + \sum_{A=0}^{1}\sum_{B=1}^{3} Y_{AB}\omega_{10}^{A}\lambda_1^{B} \tag{11.69}$$

where the coefficients Y_{AB} are given in Table 11-2.

The ratios K_{Ic}/K_m for several levels of the accumulated damage at peak load ω_{10} as a function of the parameter λ_1 are shown in Fig. 11-13. It is recalled that λ_1 is a measure of the relative fractions of microvoids and microcracks. Thus, $\lambda_1 = 0$ denotes the absence of any microvoids, whereas $\lambda_1 \approx 1$ implies microvoids with negligibly small edge microcracks. The curves in Fig. 11-13

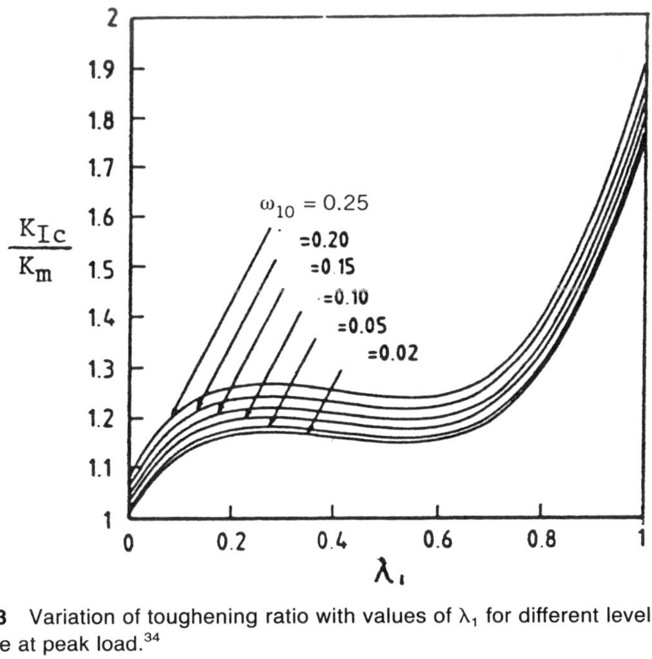

Figure 11-13 Variation of toughening ratio with values of λ_1 for different levels of accumulated damage at peak load.[34]

have an initial primary region of increasing K_{Ic}/K_m with increasing $\lambda_1 < 0.15$, an intermediate secondary or steady-state region for $0.15 \leq \lambda_1 \leq 0.6$ in which K_{Ic}/K_m remains practically constant, and a tertiary region in which this ratio increases sharply with increasing $\lambda_1 > 0.6$.

The variation of K_{Ic}/K_m with ω_{10} for a fixed value of λ_1 is fairly moderate, as shown in Fig. 11-14. Thus, it appears that microvoids with small edge microcracks have a far larger toughening than microcracks alone. Using the coefficients given in Table 11-2, the ratio K_{Ic}/K_m for $\lambda_1 = 0$, which represents no microvoids such as in high-strength concrete, can be explicitly expressed as

$$\frac{K_{Ic}}{K_m} = 1.0039 + 0.3265\omega_{10} \tag{11.70}$$

It is recalled that K_{Ic} is the effective fracture toughness of concrete, whereas K_m is the intrinsic fracture toughness of the cementitious matrix. Thus, the ratio of K_{Ic}/K_m primarily represents the bridging mechanism of aggregates. For a material such as a high-strength concrete, even when the accumulated damage ω_{10} at peak load is up to 0.25, the toughening provided by this bridging mechanism would be as large as 8.5% according to Eq. (11.70). On the other hand, for normal-strength concrete, which may be expected to have a λ_1 value in the range of $0.15 \leq \lambda_1 \leq 0.6$, the toughening provided by the bridging mechanism would range between 15 and 20%, as indicated in Fig. 11-14.

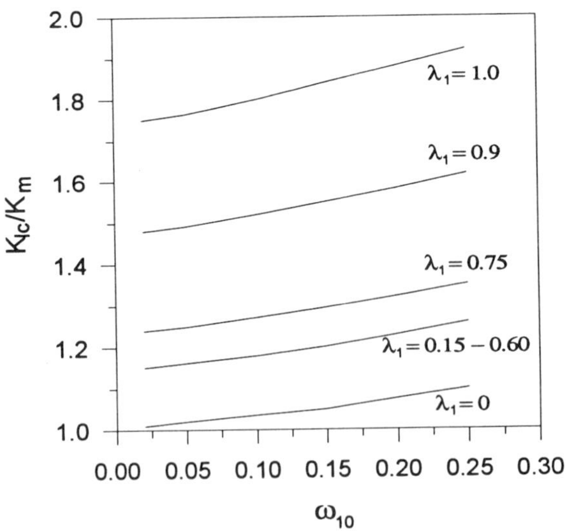

Figure 11-14 Variation of toughening ratio with values of ω_{10} for different relative fractions of microvoids and microcracks.[34]

11.7 SUMMARY

Since microcracks in concrete up to the onset of the strain localization are randomly and isolatedly distributed, damage mechanics can be used to describe the mechanical behavior of concrete in this stage. Micromechanical analyses indicate that a scalar crack density is sufficient to describe the damage process for such randomly distributed cracks. Applications of some damage mechanics models using a scalar damage variable to concrete have been briefly presented in this chapter. It has been shown that by specially arranging the microcracks damage mechanics can also be used to predict the tension-softening response of concrete. The influence of microcracks on the fracture toughness is discussed based on the presented model.

REFERENCES

1. Gopalaratnam, V. S., and Shah, S. P., "Softening Response of Plain Concrete in Direct Tension," *ACI Journal*, Vol. 82, No. 3, 1985, pp. 310–323.
2. Shah, S. P., and Chandra, S., "Critical Stress, Volume Change and Microcracking of Concrete," *ACI Journal*, Vol. 65, No. 9, 1968, pp. 770–781.
3. Kachanov, L. M., "Time of Rupture Process under Creep Conditions," *Izvestia Akademii Nauk, USSR*, Vol. 8, 1958, pp. 26–31 (in Russian).
4. Krajcinovic, D., "Damage Mechanics," *Mechanics of Materials*, Vol. 8, 1989, pp. 117–197.
5. Lemaitre, J., and Chaboche, J. L., "Aspect phénoménologique de la rupture par endommangement," *Journal de Mécanique Appliqueé*, Vol. 2, No. 3, 1978, pp. 317-365.
6. Krajcinovic, D., "Constitutive Theory of Damaging Materials," *Journal of Applied Mechanics*, Vol. 50, 1983, pp. 355–360.
7. Kachanov, M., "A Continuum Model for Medium with Cracks," *Journal of Engineering Mechanics Division, ASCE*, Vol. 106, No. 5, 1980, pp. 1039–1051.
8. Ju, J. W., "On Energy-Based Coupled Elastoplastic Damage Theories: Constitutive Modeling and Computational Aspects," *International Journal of Solids and Structures*, Vol. 25, No. 7, 1989, pp. 803–833.
9. Landis, E., and Shah, S. P., "Recovery of Microcrack Parameter in Mortar Using Quantitative Acoustic Emission," *Journal of Nondestructive Evaluation*, Vol. 12, No. 4, 1993, pp. 219–232.
10. Li, Z., and Shah, S. P., "Localization of Microcracking in Concrete under Uniaxial Tension," *ACI Materials Journal*, in press.
11. Budiansky, B., and O'Connell, R. J., "Elastic Moduli of a Cracked Solid," *International Journal of Solids and Structures*, Vol. 12, 1976, pp. 81–97.
12. Rabotnov, Y. N., "On the Equations of State for Creep," in *Progress in Applied Mechanics*, Prager Anniversary Volume, MacMillan, New York, 1963, p. 307.
13. Løland, K. E., "Continuous Damage Model for Load-Response Estimation of Concrete," *Cement and Concrete Research*, Vol. 10, 1980, pp. 395–402.

14. Ouyang, C., Suaris, W., and Chang, W. F., ''Effects of Sulfate Attack on Compressive Properties of Cement Mortar Containing Phosphogypsum,'' *ACI Materials Journal*, Vol. 85, No. 2, 1988, pp. 82–89.
15. Ouyang, C., ''A Damage Model for Sulfate Attack of Cement Mortars,'' *Cement, Concrete and Aggregates*, Vol. 11, No. 2, 1989, pp. 92–99.
16. Mazars, J., ''Application de la Mécanique de L'endommangement au Comportement non Linéaire et à la Repture du Béton de Structure,'' Thése de Doctorate d'Etat, L.M.T., Université Paris, France, 1984.
17. Mazars, J., ''A Description of Micro- and Macroscale Damage of Concrete Structures,'' *Engineering Fracture Mechanics*, Vol. 25, Nos. 5/6, 1986, pp. 729–737.
18. Mazars, J., and Pijaudier-Cabot, G., ''Continuum Damage Theory—Application to Concrete,'' *Journal of Engineering Mechanics, ASCE*, Vol. 115, No. 2, 1989, pp. 345–365.
19. Mazars, J., Pijaudier-Cabot, G., and Saouridis, C., ''Size Effect and Continuous Damage in Cementitious Materials,'' *International Journal of Fracture*, Vol. 51, 1991, pp. 159–173.
20. Kemeny, J., and Cook, N. G. W., ''Effective Moduli, Non-Linear Deformation and Strength of a Cracked Elastic Solid,'' *International Journal of Rock Mechanics, Mining Science & Geometrical Abstract*, Vol. 23, No. 2, 1986, pp. 107–118.
21. Rice, J. R., ''Mathematical Analysis in the Mechanics of Fracture,'' in *Fracture: A Advanced Treatise*, Vol. II, edited by Liebowits, H., Academic, New York, 1968, pp. 191–311.
22. Irwin, G. R., ''Analysis of Stresses and Strains Near the Ends of a Crack Traversing a Plate,'' *Journal of Applied Mechanics*, Vol. 24, 1957, pp. 361–364.
23. Ju, J. W., ''Isotropic and Anisotropic Damage Variables in Continuum Damage Mechanics,'' *Journal of Engineering Mechanics, ASCE*, Vol. 116, No. 12, 1991, pp. 2764–2770.
24. Horii, H., and Nemat-Nasser, S., ''Overall Moduli of Solids with Microcracks: Load-Induced Anisotropy,'' *Journal of Mechanics and Physics of Solids*, Vol. 31, No. 2, 1983, pp. 155–171.
25. Sumarac, D., and Krajcinovic, D., ''A Self-Consistent Model for Microcrack-Weakened Solids,'' *Mechanics of Materials*, Vol. 6, 1987, pp. 39–52.
26. Karihaloo, B. L., and Fu, D., ''A Damage-Based Constitutive Law for Plain Concrete in Tension,'' *European Journal of Mechanics, A/Solids*, Vol. 8, No. 5, 1989, pp. 373–384.
27. Karihaloo, B. L., and Huang, X., ''Tensile Response of Quasi-Brittle Materials,'' *PAGEOPH*, Vol. 137, No. 4, 1991, pp. 461–487.
28. Kaplan, M. F., ''Strains and Stresses of Concrete at Initiation of Cracking and Near Failure,'' *ACI Journal*, Vol. 60, 1963, pp. 853–880.
29. Karihaloo, B. L., and Fu, D., ''An Anisotropic Damage Model for Plain Concrete,'' *Engineering Fracture Mechanics*, Vol. 35, 1990, pp. 205–209.
30. Karihaloo, B. L., and Fu, D., ''Orthotopic Damage Model for Plain Concrete in Tension,'' *ACI Materials Journal*, Vol. 87, 1990, pp. 62–67.
31. Chow, C. L., and Wang, J., ''An Anisotropic Theory of Elasticity for Continuum Damage Mechanics,'' *International Journal of Fracture*, Vol. 33, 1987, pp. 3–16.

32. Horii, H., Hasegawa, A., and Nishino, F., "Fracture Process and Bridging Zone Model and Influencing Factors in Fracture of Concrete," in *Fracture of Concrete and Rock*, edited by Shah, S. P., and Swartz, S. E., Springer-Verlag, New York, 1989, pp. 205–219.
33. Ortiz, M., "Microcrack Coalescence and Macroscopic Crack Growth Initiation in Brittle Solids," *International Journal of Solids and Structures*, Vol. 24, 1988, pp. 231–250.
34. Karihaloo, B. L., Fu, D., and Huang, X., "Modeling of Tension Softening in Quasi-Brittle Materials by an Array of Circular Holes with Edge Cracks," *Mechanics of Materials*, Vol. 11, 1991, pp. 123–134.

12

OTHER AREAS IN APPLICATION OF FRACTURE MECHANICS

12.1 CHARACTERIZATION OF TOUGHENING IN FIBER-REINFORCED CONCRETE

12.1.1 Toughening Mechanisms in Fiber-Reinforced Concrete

Concrete has low tensile strength and corresponding strain. The tensile strain capacity of concrete can be improved by the addition of fibers. Fiber-reinforced concrete has been used for a variety of constructed facilities, such as highway pavements, blast-resistant slabs, hydraulic structures, tunnel lining, and wall panels in building structures. Fibers used in concrete are primarily made of steel, glass, or polymer.

Short fibers (about 25 mm in length) with a volume fraction of less than 2% are normally used in concrete for engineering applications. For such composites, the influence of fibers is primarily evident in the postpeak response. The tensile strength of the composite is not appreciably different from that of matrix. Recently, with a variety of processing techniques, it has become possible to incorporate up to 15% fibers by volume into concrete. Fibers in such large quantities fundamentally alter the nature of the concrete, and the inherent tensile strength and strain capacity of the matrix itself are actually enhanced. The stress–strain curves of cement-based matrices reinforced with 8.7 and 13.4% volume fractions of polypropylene fiber are shown in Fig. 12-1.[1] The contribution of the matrix (shown in the inset) was approximately derived by subtracting the contribution of the fibers from the composite's response using the law of mixtures. It is seen from Fig. 12-1 that the tensile strength of the cement-based matrix reinforced with 13.4% fiber is approximately 15 MPa, whereas the tensile strength of the plain matrix is only approximately 4 MPa. Similar results have also been observed in cement-based matrixes reinforced

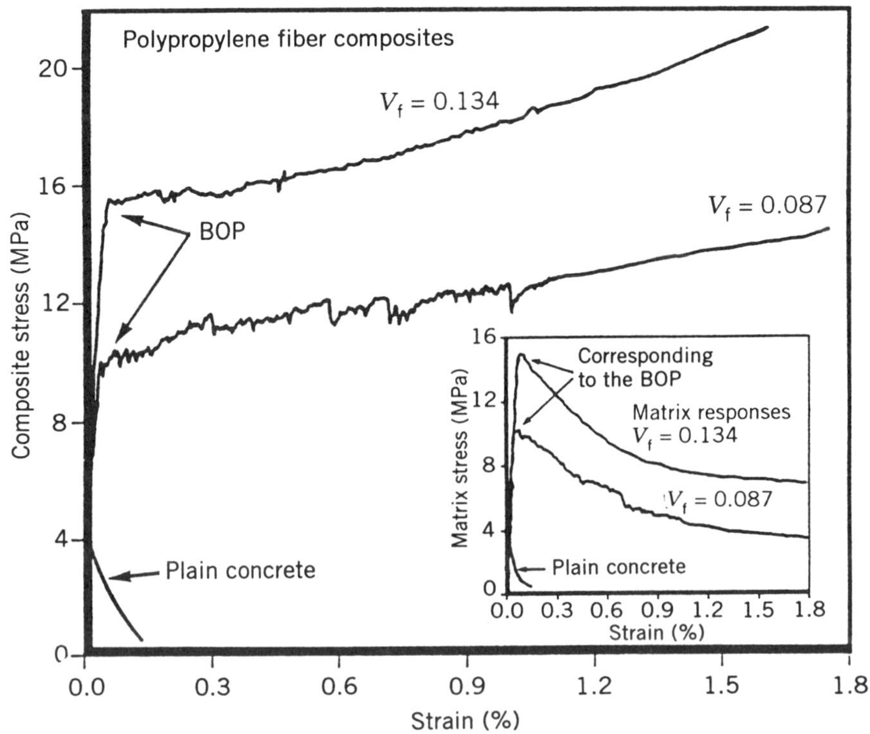

Figure 12-1 Tensile response of polypropylene fiber-reinforced mortar.[1]

by steel fibers and glass fibers.[2,3] With such high-volume concentration of fibers, microcracking is stabilized and a homogeneous distribution of microcracks can be found even at very high strain levels (1% strain, which is about 100 times the strain at localization in plain concrete). Thus the mechanical behavior of the composite is fundamentally different from the behavior of cement-based matrices. Such beneficial interactions can lead to a new class of fiber-reinforced concrete structure.

Many experimental studies, which include quantitative optical microscopy[4] and laser holographic interferometry,[5] have been conducted to understand toughening mechanisms of cement-based matrices reinforced by high-volume fraction of fibers. Based on the crack development observed using these two procedures, it appears that cracks initiate at a very low load level and are few and widely dispersed, no more than 1 mm in length, as schematically shown in Fig. 12-2*a*. Cracks begin to localize at about the point when the matrix contribution is maximum; this point is referred to as the bend-over point (BOP) (see Fig. 12-2*b*). Note that the debonding and sliding at the fiber–matrix interface that may occur are omitted in Fig. 12-2. With further straining, the localized bands seem to move closer (Fig. 12-2*c*). This process eventually

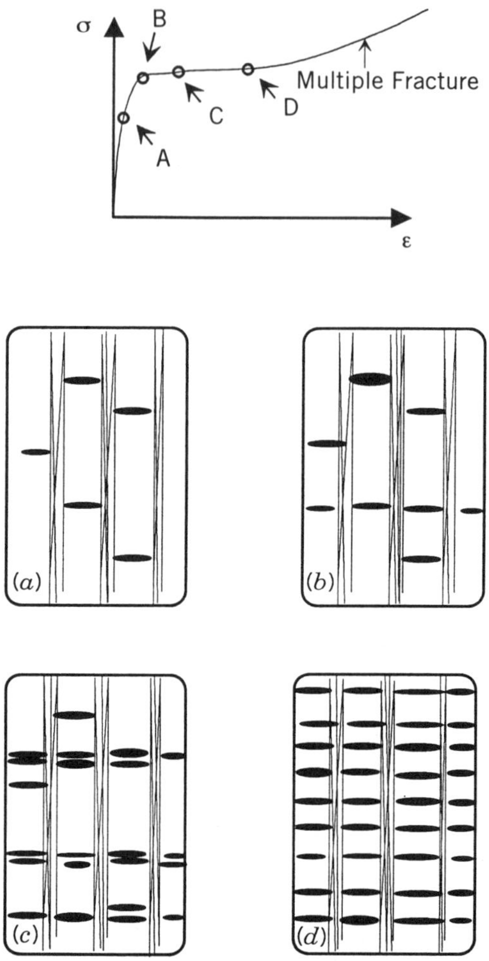

Figure 12-2 Loading stages for fiber-reinforced composite.[5]

leads to homogenization of microcracking (Fig. 12-2*d*). As a result, the toughening mechanisms of fibers may include (i) the suppression of the localization phenomenon, which may lead to the observed enhanced tensile strength of the matrix (BOP, *B* in Fig. 12-2), and (ii) homogeneous distribution of cracking observed at very high strain levels (point *D* in Fig. 12-2), which may explain the observed relatively high level of matrix strength apparent at this level.

Since only a single band of microcracks, which can be regarded as a major crack, has been observed at bend-over point *B*, fracture mechanics can be used to characterize toughening behavior of fiber-reinforced concrete up to point *B*. On the other hand, the toughening response due to the homogenization of microcracks at point *D* may be appropriately described by a damage mechanics approach. Aveston et al.[6] may have been the first to apply the Griffith-type fracture criterion to failure of continuous fiber-reinforced composites. A general overview of mechanics behavior of fiber-reinforced concrete has recently been given by Shah and Ouyang.[7] Only fracture mechanics approaches used to characterize the toughening behavior of fiber-reinforced concrete at point *B* are discussed in this section.

As shown in Fig. 12-3, to model the behavior of fiber-reinforced concrete at the BOP, one needs to know fracture properties of the plain concrete matrix and the influence of fibers on the composite response. The fracture properties of the plain concrete matrix (Fig. 12-3*b*) have been discussed in Chapters 5 and 7. The effect of fibers may be simulated by a closing pressure on cracked surfaces (Fig. 12-3*c*). Since both debonding and sliding at the fiber–matrix interface usually accompany the composite's cracking, as shown in Fig. 12-3*d*, the distribution and magnitude of the closing pressure are related to the pullout

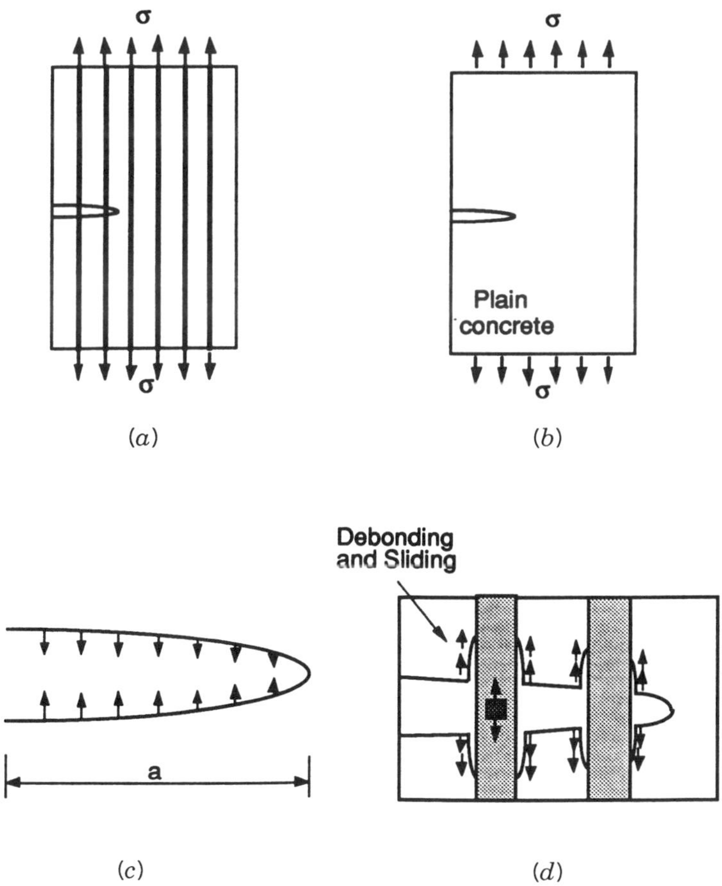

Figure 12-3 Modeling of composite response at BOP.[7]

behavior of the fibers from the concrete. As a result, one can combine the fracture properties of a plain concrete matrix and the pullout behavior of fibers to simulate the composite response of fiber-reinforced concrete. These will be discussed in the following.

12.1.2 Mechanical Behavior of Fiber–Matrix Interface

12.1.2.1 Pullout of Aligned Fibers Pullout of a fiber from a cement-based matrix has been used to characterize the interface bond between the matrix and the fiber, as shown in Fig. 12-4*a*. Typical pullout force and slip curves for a single fiber, an average of 16 fibers being simultaneously pulled out at different inclination angles, are shown in Fig. 12-4*b*. The pullout load initially almost linearly increases with the slip. Nonlinearity in the pullout load and slip curves, which is often regarded as an indication of propagation of an interfacial crack, is usually observed before the peak load. After the pullout load reaches the maximum value, it decreases slowly with increasing slip. This result may indicate that when fibers are pulled out from a cementitious matrix, an interfacial crack initiates at some point before the peak. This interfacial crack stably propagates up to peak load; that is, the crack propagates only when the pullout load increases. The interfacial crack length corresponding to the peak load is usually termed as the critical crack length. After the peak load, unstable crack growth occurs, which means that the crack automatically grows even though the pullout load decreases. As a result, softening-type load–slip curves are observed from the pullout test.

To quantify results of pullout test of fibers, it was initially assumed that shear stresses are uniformly distributed. The average shear strength, which is equal to the maximum pullout load divided by the debonded surface area of the fiber, was initially used to characterize the interface response.[8] However, experimental results[9] have indicated that this average shear strength depends on embedded length and diameter of fibers.

To better characterize fiber–matrix interface behavior, a shear–lag approach[10,11] has been used to analyze stress and strain distributions in the interface. Stang et al.[12] proposed a shear–lag model that takes the effect of the weak zone between the fiber and the cement-based matrix. Based on the stress distributions obtained from a shear–lag analysis, they evaluated the strain energy release rate for pullout of fibers G_{int} as

$$G_{\text{int}} = \frac{(P - \tau_F a_d)^2 \coth^2[\omega_{\text{int}}(L_f - a_d)]}{4E_f \pi^2 r_f^3} - \frac{\tau_F (P - \tau_F a_d)}{2E_f \pi^2 r_f^3 \omega_{\text{int}}} \coth[\omega_{\text{int}}(L_f - a_d)] + \frac{\tau_F}{4\pi r_f k_{\text{int}}} \qquad (12.1)$$

where P is the pullout load of the fiber, τ_F is the frictional stress between the fiber and the matrix, a_d is the debonded length at the fiber–matrix interface, L_f

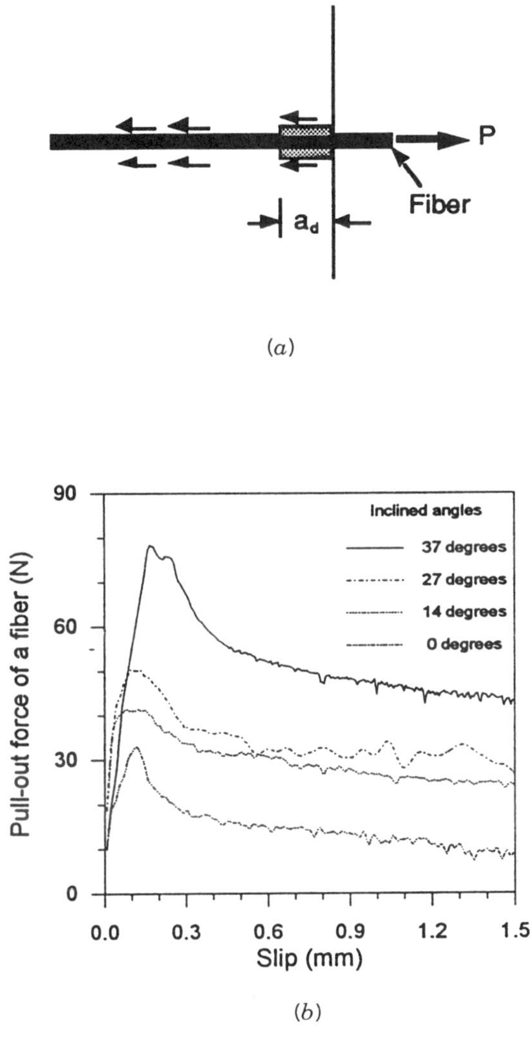

Figure 12-4 Pullout of fibers from cementitious matrix and typical behavior.[14]

is the fiber-embedded length, r_f is the fiber radius, E_f is the modulus of elasticity of the fiber, k_{int} is the initial slope between the shear stress and the slip, and ω_{int} is a parameter to quantify the stiffness of the weak interface layer between the fiber and the matrix and is defined as

$$\omega_{\text{int}} = \sqrt{\frac{k_{\text{int}}}{E_f \pi r_f^2}} \tag{12.2}$$

The subscript int represents the interface from hereafter.

By introducing a Griffith-type debonding criterion (LEFM), Li et al.[13] have used Eq. (12.1) to evaluate interfacial parameters based on their experimental results. They reported that the value of ω_{int} is basically independent of the fiber-embedded length L_f. However, the value of the critical debonding energy obtained depends on L_f. Then a question raised is whether or not the LEFM criterion is suitable for describing debonding between the fiber and the matrix. Actually, the substantial prepeak nonlinearity as seen in Fig. 12-4*b* indicates that some stable propagation of the debonding crack must exist before the peak load. Ouyang et al.[14] have used the *R*-curve discussed in Chapter 7 to account for this stable process of debonding. Their approach will be presented here.

When a fiber is pulled out from a matrix, the interface between the fiber and the matrix first debonds. Then some slip occurs due to a relative movement between the fiber and the matrix. When the debonded length at the peak load is smaller than L_f (the fiber–matrix is not fully debonded), the slip prior to the peak load may be relatively small, and the effect of frictional stress on the pullout behavior may be neglected during this stage. As a result, Eq. (12.1) can be simplified as ($\tau_F = 0$)

$$G_{int} = \frac{P^2 \coth^2[\omega_{int}(L_f - a_d)]}{4E_f \pi^2 r_f^3} \tag{12.3}$$

The *R*-curve given in Eqs. (7.31) and (7.32) is interpreted as an envelope of a series of G_q-curves with different specimen sizes but the same initial crack length. For pullout of fibers from a matrix (mode II fracture), the *R*-curve can be interpreted as an envelope of a series of G_{int}-curves with different fiber-embedded lengths but the same initial debonded length, as shown in Fig. 12-5. If only the length effect of fibers is concerned, for the *G*-curve given by Eq. (12.3) the *R*-curve given by Eqs. (7.31) and (7.32) can be further simplified by letting $\beta_1 = 0$ (because the R-curve should be concave downward, $\beta_2 = 0$ is not allowable). This leads to

$$\begin{aligned} R_{int} &= \beta_2 (a_d - a_{d0})^{d_2} \\ d_2 &= \frac{1}{2} + \frac{\alpha - 1}{\alpha} - \left[\frac{1}{4} + \frac{\alpha - 1}{\alpha} - \left(\frac{\alpha - 1}{\alpha}\right)^2\right]^{1/2} \end{aligned} \tag{12.4}$$

where $R = R_{int}$ and $\alpha = a_{dc}/a_{d0}$ are used, a_{d0} is the initial debonded crack, and a_{dc} is the critical debonded crack at the peak load. Two parameters α and β_2 should be known before one can use Eq. (12.4). These two constants can be determined from instability conditions $G_{int} = R_{int}$ and $\partial G_{int}/\partial a_d = \partial R_{int}/\partial a_d$ in conjunction with Eqs. (12.3) and (12.4). It will be proven that for a pullout problem with the strain energy release rate given by Eq. (12.3), the use of Eq. (12.4) for the R_{int}-curve can exactly satisfy the two instability conditions.

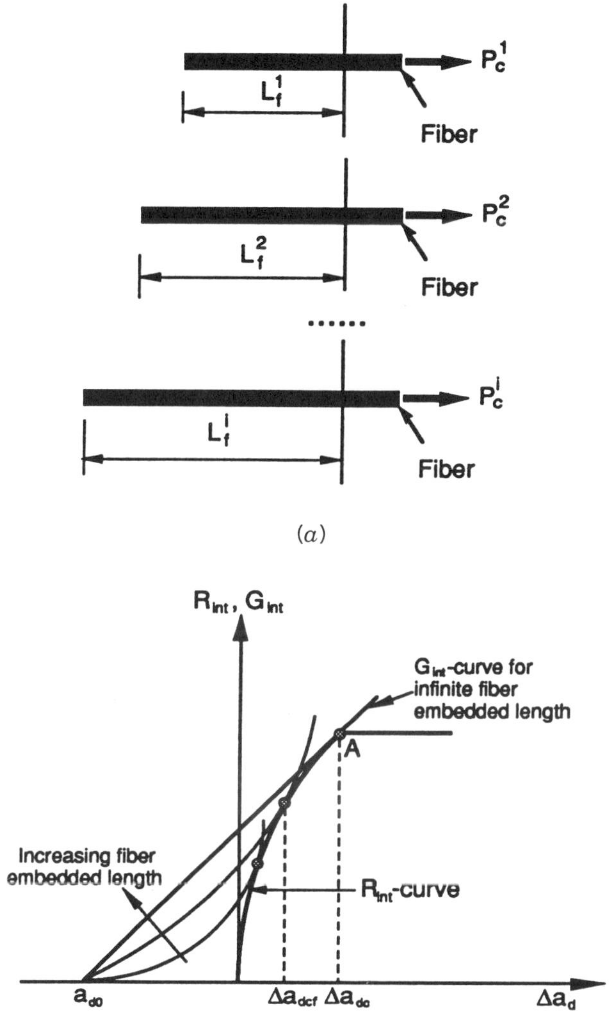

Figure 12-5 The R_{int}-curve as envelope of G_{int}-curves for series of specimens with same initial debonding length but different embedded lengths.[14]

The values of α and β_2 can be determined from Eqs. (12.3) and (12.4). The two conditions of $G_{int} = R_{int}$ and $\partial G_{int}/\partial a_d = \partial R_{int}/\partial a_d$ lead to two simultaneous equations to determine the values of α and β_2. By dividing these two equations into each other, one obtains

$$\frac{a_{dc}^* - a_{d0}}{d_2} - \frac{1}{2\omega_{int}} \coth[\omega_{int}(L_f^* - a_{dc}^*)]\sinh^2[\omega_{int}(L_f^* - a_{dc}^*)] = 0 \quad (12.5)$$

where L_f^* and a_{dc}^* are the fiber-embedded length and the corresponding critical debonded length at the maximum load for a reference specimen. It is noted that Eq. (12.5) is independent of the maximum load.

To determine the parameters α and β_2, one needs to test a reference specimen with the embedded fiber length of L_f^* to obtain its maximum pullout load P_c^*. The value of a_{dc}^* can be obtained by substituting the value of L_f^* into Eq. (12.5) and solving this nonlinear equation. It is seen from Eq. (12.4) that the term d_2 is a function of a_{dc}^* in Eq. (12.5). The value of α ($= a_{dc}^*/a_{d0}$) can thus be calculated. The value of β_2 can then be determined from the condition of $G_{\text{int}} = R_{\text{int}}$ and the measured value of P_c^*:

$$\beta_2(a_{dc}^* - a_{d0})^{d_2} = \frac{(P_c^*)^2 \coth^2[\omega_{\text{int}}(L_f^* - a_{dc}^*)]}{4E_f\pi^2 r_f^3} \tag{12.6}$$

The R_{int}-curve obtained here is interpreted as the envelope of G-curves, at points corresponding to critical debonded lengths, for fibers with the same radius but different embedded lengths for a given matrix. This means that a series of specimens with the same fiber diameter but different embedded lengths follow the same R_{int}-curve but have different critical values corresponding to the peak loads. This will be demonstrated later.

After knowing both G- and R-curves, the pullout behavior of aligned fibers from cement-based matrices can be predicted. From the condition $R_{\text{int}} = G_{\text{int}}$, the relationship between the pullout load and the debonded length can be obtained:

$$P = \left\{ \frac{4E_f\pi^2 r_f^3 \beta_2(a_d - a_{d0})^{d_2}}{\coth^2[\omega_{\text{int}}(L_f - a_d)]} \right\}^{1/2} \tag{12.7}$$

Equation (12.7) predicts that the pullout load increases with increasing debonded length up to the peak load for an aligned fiber. The peak pullout load is automatically obtained from maximizing the P values based on Eq. (12.7). For a given matrix, the peak pullout loads for a series of fibers, which have the same diameters but different embedded lengths, can be predicted by substituting different L_f-values into Eq. (12.7).

To calculate the slip corresponding to the maximum pullout load, one needs to evaluate the compliance governing the pullout load–slip relationship. Based on LEFM, the strain energy release rate can also be expressed as[15]

$$G_{\text{int}} = \frac{P^2}{4\pi r_f} \frac{\partial C_{\text{int}}}{\partial a_d} \tag{12.8}$$

where C_{int} is the compliance for the pullout of a fiber from a matrix. Equation (12.8) can be rewritten as the following due to $G_{\text{int}} = R_{\text{int}}$:

$$\frac{\partial C_{\text{int}}}{\partial a_d} = \frac{4\pi r_f G_{\text{int}}}{P^2} = \frac{4\pi r_f \beta_2 (a_d - a_{d0})^{d_2}}{P^2} \tag{12.9}$$

The value of the compliance can be obtained by integrating Eq. (12.9):

$$C_{\text{int}} = \int_0^{a_d} \frac{4\pi r_f \beta_2 (a_d - a_{d0})^{d_2}}{P^2}\, da_d \tag{12.10}$$

After obtaining the compliance, the slip is equal to the pullout load multiplied by the corresponding compliance.

Somayaji and Shah[16] have reported a series of results on pullout of aligned steel fibers from mortars. The diameter of the fiber used was 1.59 mm, and the modulus of elasticity for the fiber was 210 GPa. The embedded lengths of the fibers varied from 38 to 83 mm. Note that the initial debonded length a_{d0} depends on the microstructure of the fiber–matrix interface. A previous study[17] has proven that the given R_{int}-curve is not sensitive to the selection of a_{d0}. Therefore the value of $a_{d0} = 0.02$ mm was used in this study for all predictions. Based on Stang et al.,[12] the value of ω_{int} for pullout of the steel from the mortar at the age of 28 days is 0.075 mm^{-1}. To predict peak pullout loads for the fibers with different embedded lengths, the R_{int}-curve should first be determined based on a reference pullout specimen. A specimen with 83 mm fiber embedded length was chosen as the reference sample, and the corresponding peak pullout load was 729 N ($L_f^* = 83$ mm and $P_c^* = 729$ N). By substituting the values of L_f^*, ω_{int}, and a_{d0} into Eq. (12.5), the critical debonded length for the reference specimen, a_{dc}^*, can be solved. As a result, the value of $\alpha = a_{dc}^*/a_{d0}$ was obtaincd. Thc obtained value of α was used in the second equation of Eq. (12.4) to calculate the value of d_2. These quantities of L_f^*, a_{d0}, a_{dc}^*, d_2, and P_c^* were substituted into Eq. (12.6) to obtain the value of β_2. The R_{int}-curve for this series of specimens with different embedded lengths was then determined from Eq. (12.4), and relationships between the R_{int}-curve and the G_{int}-curves are shown in Fig. 12-6*a*. The G_{int}-curves for L_f values of 38, 63, and 83 mm tangent to the R_{int}-curve at points *A*, *B*, and *C*, respectively, correspond to the peak loads for different specimens. These critical values of G_{int} (at points *A*, *B*, and *C*) vary with the fiber-embedded lengths. However, the Griffith fracture criterion assumes that these critical values of G_{int} are a constant that is independent of the fiber-embedded length. Using the obtained R_{int}-curve in conjunction with Eq. (12.7), the peak pullout loads for different fiber-embedded lengths can be predicted. The predicted peak pullout loads are compared with those experimentally obtained in Fig. 12-6*b*. It is seen that although the frictional stress is neglected in this study, the proposed single *R*-curve is capable of predicting the peak pullout loads for different fiber-embedded lengths. It has been determined that the proposed R_{int}-curve is not sensitive for the selection of the reference specimen.

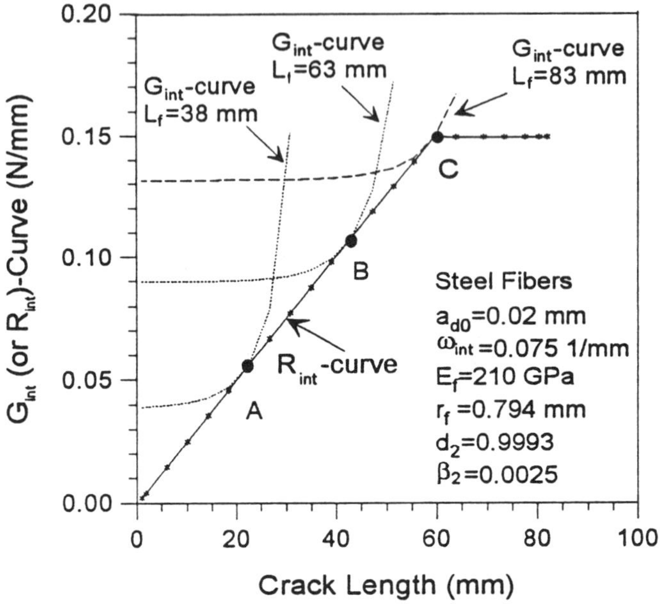

(*a*)

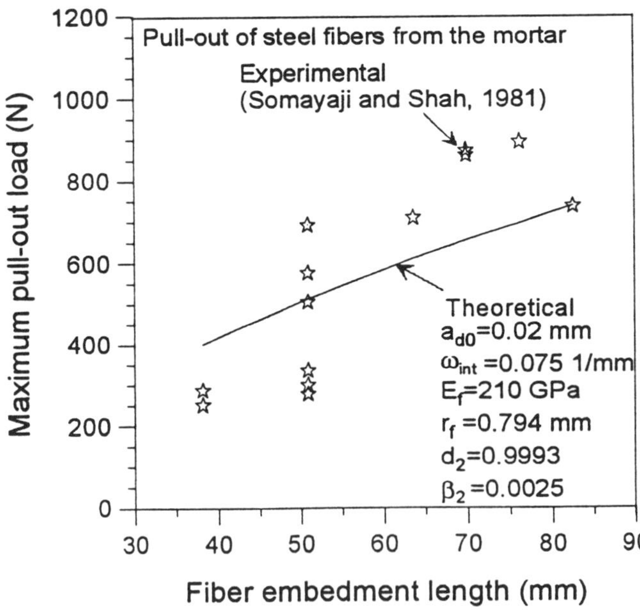

(*b*)

Figure 12-6 An R_{int}-curve for pullout of steel fibers from mortar: (*a*) relationship between G_{int}-curves and the R_{int}-curve, and (*b*) relationship between pull-out load and debonded length.[14]

12.1.2.2 Pullout of Inclined Fibers When a fiber is pulled out from a matrix at an inclined angle of θ_i as shown in Fig. 12-7, the pullout load P_θ can be divided into two components, P_x and P_y. When the value of θ_i is smaller than 45°, pullout resistance of inclined fibers is usually higher than that of aligned fibers. This is first due to the fact that only a component in the fiber axis, P_x, is dedicated to pullout of the inclined fiber. This may be referred to as the pulley effect. There are several models proposed to explain the role of the other component, P_y. For pullout of inclined metal fibers from a matrix, Morton and Groves[18] proposed that bending of the fiber at its exit from the matrix and spalling of the matrix are major mechanisms to account for this increase. This is termed the bending mechanism. On the other hand, Brandt[19] considered that fiber yielding and matrix spalling at the fiber exit point from the matrix should be responsible for this increase of pullout resistance. This is

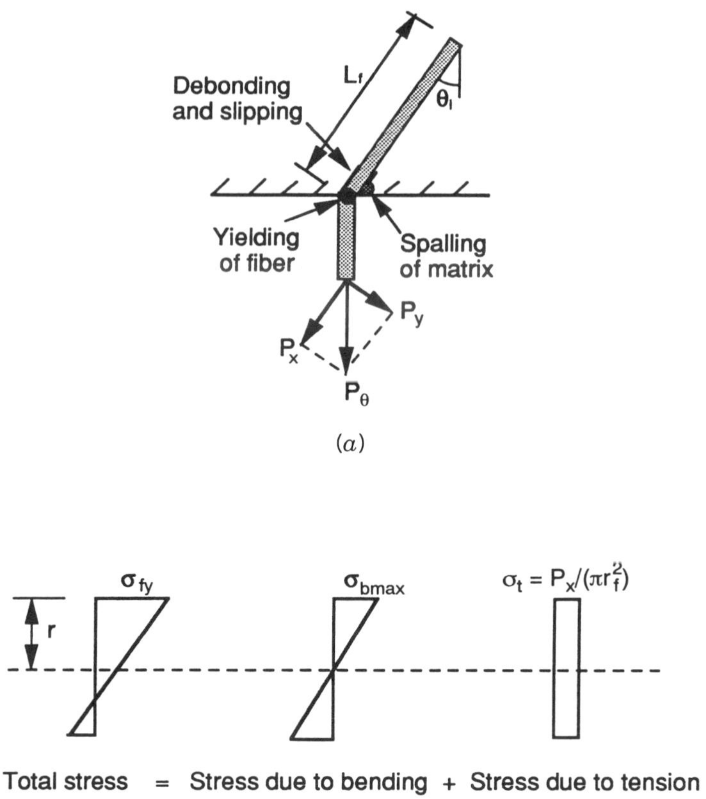

Figure 12-7 Pullout of inclined fiber from matrix: (*a*) failure mechanisms with pull-out of an inclined fiber, and (*b*) stress distributions at yielding section of a metal fiber.[14]

referred to as the yielding mechanism. For pullout of inclined synthetic fibers, Li et al.[20] reported that the increase of pullout resistance is due to an additional friction stress generated.

The pulley effect always exists for pullout of all types of inclined fibers. The question needed to be answered is what kind of mechanism is most suitable to explain the role of the component P_y in pullout of metal fibers from a cementitious matrix. Both the bending mechanism and the increase of frictional stress may exist during fiber pullout. Since the fiber will almost immediately be bent when a pullout force starts to apply, the bending mechanism is present even though the pullout load is small. On the other hand, frictional stress at the fiber–matrix interface will be generated only when the fiber and the matrix start to slip (the interface first debonds, then the fiber and the matrix slip). Since the slip value prior to peak load is relatively small, the effect of frictional stress on pullout of inclined metal fibers may not be dominant before the peak load. Therefore, both the pulley effect and the bending mechanism may primarily account for increase of peak pullout load for inclined fibers. However, the influence of frictional stress becomes more and more important after the peak load because more slipping occurs. This is particularly evident by the pullout load and slip curves of the specimen shown in Fig. 12-4*b*, where increments in pullout loads due to fiber inclination increase after the peak load as more slip occurs. Based on this discussion, Ouyang et al.[14] have extended their approach to predict the peak pullout load of inclined fibers. In their approach only the bending mechanism is taken into account. This approach is briefly presented below.

In order to predict change of load capacity due to inclination, additional energy due to fiber bending should be taken into account. A simple approach based on the yield stress of the fiber is used to calculate the additional bending energy. It has been realized that bending of the fiber is a localized phenomenon.[17,18] The fiber is considered as an elastic material. It is assumed that the maximum tensile stress in the section of the fiber reaches the yield stress of material under the pullout load. Therefore, the maximum tensile stress due to bending, $\sigma_{b,\max}$, can be calculated as (Fig. 12-7*b*)

$$\sigma_{b,\max} = \sigma_{fy} - \sigma_t = \sigma_{fy} - \frac{P_x}{\pi r_f^2} \tag{12.11}$$

where σ_{fy} is the yield stress of the fiber and σ_t is the stress resulting from tension. The bending stress at any point, y, along the bent section of the fiber, σ_{by}, can be obtained as

$$\sigma_{by} = \frac{\sigma_{b,\max} y}{r_f} \tag{12.12}$$

Therefore, the total strain energy ϕ_b due to bending is

$$\phi_b = \int_{V_b} \frac{(\sigma_{by})^2}{2E_f} dV_b$$
$$= \int_0^{L_b} \int_0^{\pi} \int_0^{r_f} \frac{\sigma_{b,\max}^2 r^2 \sin^2 \zeta}{r_f^2 E_f} r\, dr\, d\zeta\, dx = \frac{\sigma_{b,\max}^2 \pi L_b r_f^2}{8E_f} \tag{12.13}$$

where V_b and L_b $(0 \leq x \leq L_b)$ are the volume and the length of the bent zone of the fiber, respectively, and r and ζ are polar coordinates for a circular section. Note that the relationship $y = r \sin \zeta$ has been used in Eq. (12.13). The value of L_b is assumed to be proportional to the pullout slip at the load point. Based on Stang et al.'s solution,[12] the following expression for the value of L_b was obtained if the frictional stress is neglected:

$$L_b = m_b u_{\text{int}} = \frac{m_b P_x \coth[\omega_{\text{int}}(L_f - a_d)]}{\omega_{\text{int}} \pi r_f^2 E_f} - \frac{m_b P_x a_d}{\pi r_f^2 E_f} \tag{12.14}$$

where u_{int} is the load point slip and m_b is a proportional coefficient.

Substituting Eq. (12.14) into Eq. (12.13) and noting $\sigma_i << \sigma_{fy}$ result in

$$\phi_b = \frac{m_b P_x}{8} \left(\frac{\sigma_{fy}}{E_f}\right)^2 \left\{\frac{\coth[\omega_{\text{int}}(L_f - a_d)]}{\omega_{\text{int}}} + a_d\right\} \tag{12.15}$$

As a result, the additional strain energy release rate, $G_{\text{int},b}$, due to bending of the fiber is given by

$$G_{\text{int},b} = \frac{1}{2\pi r_f} \frac{\partial \phi_b}{\partial a_d} = \frac{m_b P_x}{16\pi r_f} \left(\frac{\sigma_{fy}}{E_f}\right)^2 \left\{1 + \frac{1}{\sinh^2[\omega_{\text{int}}(L_f - a_d)]}\right\} \tag{12.16}$$

It is noted that for the inclined fiber the load P in Eq. (12.3) should be replaced by the load component $P_x = P_\theta \cos \theta_i$. Therefore, the strain energy release rate for pullout of the inclined fiber, $G_{\text{int},\theta}$, is calculated as

$$G_{\text{int},\theta} = \frac{P_x^2 \coth^2[\omega_{\text{int}}(L_f - a_d)]}{4E_f \pi^2 r_f^3} = \frac{(P_\theta \cos \theta_i)^2 \coth^2[\omega_{\text{int}}(L_f - a_d)]}{4E_f \pi^2 r_f^3} \tag{12.17}$$

The energy balance, $G_{\text{int},\theta} - G_{\text{int},b} = R_{\text{int}}$, for the pullout of the inclined fiber then results in

$$\frac{P_\theta^2 \cos^2\theta_i \coth^2[\omega_{\text{int}}(L_f - a_d)]}{4\pi^2 r_f^3 E_f} - \frac{m_b P_\theta \cos\theta_i}{16\pi r_f}\left(\frac{\sigma_{fy}}{E_f}\right)^2 \cdot \left\{1 + \frac{1}{\sinh^2[\omega_{\text{int}}(L_f - a_d)]}\right\} - \beta(a_d - a_{d0})^{d_2} = 0 \quad (12.18)$$

It is assumed that the R_{int}-curve does not change with fiber inclination in Eq. (12.18). For each value of debonded length a_d, the corresponding value of P_θ can be solved from Eq. (12.18). Equation (12.18) includes both the pulley effect and the bending mechanism. The former is considered by the coefficient of $\cos\theta_i$, and the latter is accounted for by σ_{fy}. Since Eq. (12.18) includes the bending mechanism, it is valid only when the fiber is yielded. It can be expected that the minimum inclination angle for fiber yielding should be very small for metals based on the small deformation assumption used in mechanics of materials.

Since the effect of fiber bending is often negligible due to their low bending stiffness, the value of $G_{\text{int},b}$ is approximately equal to zero for synthetic fibers. The following relationship between P_θ for the inclined fiber and P for the aligned fiber can be derived for the synthetic fibers from the conditions $G_{\text{int},\theta} = R_{\text{int}}$ and $G_{\text{int}} = R_{\text{int}}$:

$$\frac{P_\theta}{P} = \frac{1}{\cos\theta_i} \quad (12.19)$$

It is seen from Eq. (12.19) that the pulley effect of inclined fiber is accounted for by the factor $1/\cos\theta_i$, which is different from the exponential function proposed by Li et al.[20]

It is noted that the above analysis may not apply to the case when a matrix wedge failure also occurs as the fiber is pulled out. This usually occurs when $\theta_i > 45°$.

It was assumed in the above analysis that the R_{int}-curve only depends on fiber–matrix interface properties (ω_{int} value) and fiber geometry and does not change due to fiber inclination. This means that one can determine the R_{int}-curve from an aligned fiber and then use this R_{int}-curve to predict maximum loads of inclined fibers with the same geometry. After the R_{int}-curve was determined based on a reference-aligned fiber, Eq. (12.18), instead of Eq. (12.7), was used to calculate the component of pullout load P_x. The total pullout load was obtained using $P_x = P_\theta \cos\theta_i$. The value of yield stress of the fiber, σ_{fy}, should be known before solving Eq. (12.18).

The theoretical predictions are also compared to the experimental results of inclined steel fibers pulled out from a cement paste.[14] The steel fiber had a diameter of 0.408 mm, and the embedment length of the fiber was 12.7 mm. The value of $m_b \approx 20$ was used. The values of $a_0 = 0.02$ mm and $\omega_{\text{int}} = 0.048\ \text{mm}^{-1}$ were used. The value of $P_c = 33.5$ N for the aligned fiber was used as the reference value for the theoretical prediction. Both the values of $\sigma_{fy} = 1700$ MPa and $\sigma_{fy} = 0$ were used for theoretical prediction. The value of $\sigma_{fy} = 0$ corresponds to the case where the bending effect on fiber pullout load is negligible. As a result, the bending effect can be separated as shown in Fig. 12-8. It is noted that this bending effect depends on the diameter and the yield strength of the fibers used.

The proposed model can also be applied to synthetic fiber-reinforced concrete by setting $\sigma_{fy} = 0$.

Pacios et al.[21] have extended the above model to predict the influence of loading rate on pullout behavior.

12.1.3 Tensile Response of Fiber-Reinforced Concrete

The presence of fibers provides an additional toughening on concrete. This is shown in Fig. 12-9. As previously mentioned, the influence of fibers can be modeled by a closing pressure on cracked surfaces (see Fig. 12-3) that tends to close the crack. Based on the two-parameter fracture model, Mobasher et al.[22] have applied the R-curve given by Eq. (12.4) to predict the tensile stress at the BOP for a concrete matrix reinforced by fibers (now R_{int} and a_d should be replaced by R and a, respectively). By using the principle of superposition, two failure criteria for concrete with fiber reinforcement become

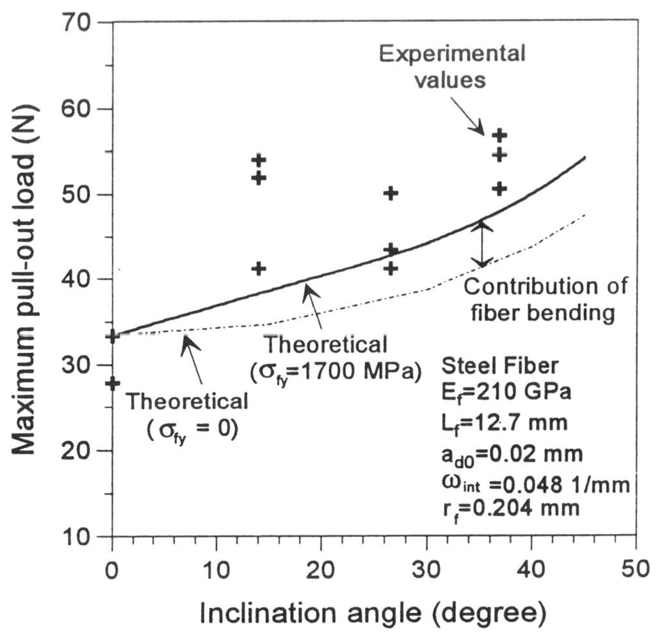

Figure 12-8 Experimental and theoretical comparison of maximum pullout load for steel fibers.[14]

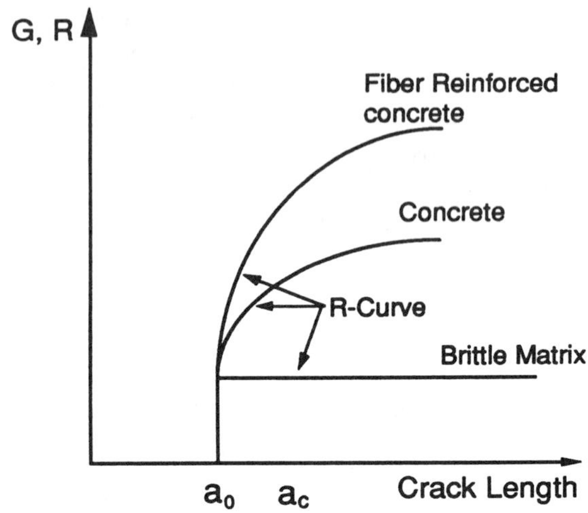

Figure 12-9 An *R*-curve for fiber-reinforced concrete.

$$K_{\mathrm{I}c}^{s} = K_{\mathrm{I}m}(\sigma_{mc}, a_c) - \int_0^{a_c} p_f(x,a_c)K_F(x, a_c)\, dx \tag{12.20}$$

and

$$\mathrm{CTOD}_c = \mathrm{CTOD}_m(\sigma_{mc}, a_c, a_0) - \int_{a_0}^{a_c} p_f(x, a_c)Q(\sigma_{mc}, a_c, a_0)\, dx \tag{12.21}$$

where σ_{mc} is the stress at the BOP resisted by the matrix with fiber reinforcement, Q is Green's function for the crack closure at point a_0 (initial crack mouth) due to a unit force applied at point x along the crack length, p_f is the closure force at point x along the crack face, and K_F is the stress intensity due to a unit load applied at point x along the crack surface. The parameters $K_{\mathrm{I}m}$, CTOD_m, K_F, and Q can be obtained based on LEFM. Two unknowns, σ_{mc} and a_c, can be solved from Eqs. (12.20) and (12.21). Then, the parameters $\alpha = a_c/a_0$ can be determined. The value of β_2 can be obtained from the condition of $R_c = (K_{\mathrm{I}c}^{s})^2/E = G_c$, where G for the single-edge notch tensile member should be used for G_c. After both α and β_2 are determined, the *R*-curve is known. The mechanical response of matrices reinforced by fibers can be predicted using $R = G$. It is noted that in the above analysis, one can use either the experimental pullout slip curve as shown in Fig. 12-4*b* or the theoretical pullout slip curve as previously presented as the closing pressure.

The results obtained for a concrete reinforced by continuous glass fibers are shown in Fig. 12-10. The matrix material properties, $E = 28.9$ GPa, $K_{\mathrm{I}c}^{s} = 26.3\ \mathrm{N/mm^{3/2}}$ [i.e., $G_c = (K_{\mathrm{I}c}^{s})^2/E = 0.024$ N/mm], and $\mathrm{CTOD}_c = 0.0094$

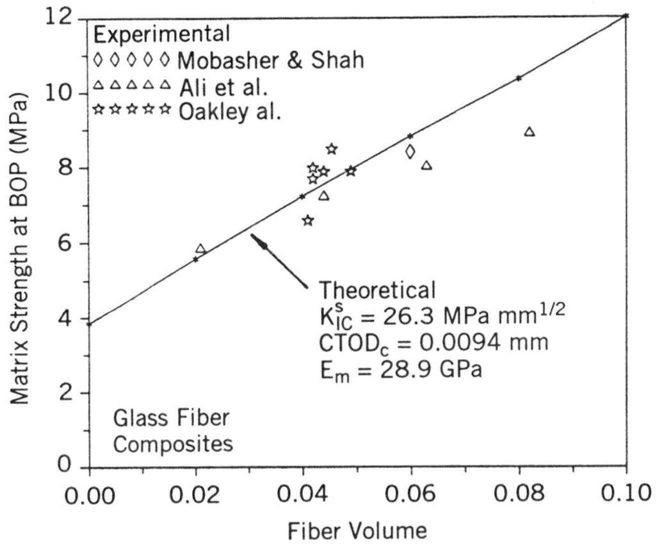

Figure 12-10 Experimental and theoretical comparison of stress at BOP for continuous glass fiber-reinforced cementitious materials.[22]

mm were used here. The value of a_0 = 0.5 mm was assumed. The experimental pullout slip curve for glass fiber reported by Stang et al.[12] was used as the closing pressure. It is seen that the model predicts the matrix stress at the BOP quite well.

Ouyang and Shah[23] have extended this model to predict tensile behavior of concrete reinforced by discontinuous, short fibers.

12.1.4 Summary

Applications of fracture mechanics to fiber-reinforced concrete have been discussed in this chapter. The mechanical response of fiber-reinforced concrete is greatly influenced by the fiber–matrix interface, which is usually evaluated by a test of fibers being pulled out from a matrix. Modeling of interfacial debonding and slipping based on fracture mechanics has been presented. Debonding of the interface is regarded as mode II fracture. By using a simple fracture mechanics *R*-curve approach, the fiber length effect on the pullout behavior can be simulated. Response of inclined fibers pulled out from the matrix can be predicted by introducing a term accounting for bending energy into the fracture criteria.

When a crack propagates in fiber-reinforced concrete, fibers can provide bridging to prevent opening of the crack. This effect of fibers can be modeled by a closing pressure on the crack surfaces. By incorporating the effects of the closing pressure on both the stress intensity factor and the crack opening displacement, mechanical response of fiber-reinforced concrete can be predicted.

12.2 FATIGUE AND FRACTURE OF CONCRETE

Fatigue of plain and reinforced concrete structural members subjected to uniaxial stress states has been studied for many years, see references 24 and 25 for comprehensive summaries. The process of fatigue, which is due to progressive and irreversible deterioration in the material when subjected to repeated loads, may lead to excessive deformations, excessive crack widths, debonding of reinforcement, and rupture of the reinforcement or matrix leading to structural collapse.

As an example of this, consider the fatigue behavior of a small, model, reinforced concrete beam, as shown in Fig. 12-11. Two such beams were tested by Swartz and Noory.[26] The reinforcement consisted of smooth bars anchored at the beam ends to steel plates. Cracks were monitored using a photoelastic coating (Photoelastic PS-2A) bonded to the beam surface at midspan as shown. The coating had a thickness of 3 mm, a k factor of 0.13, and a sensitivity of 750 μm/m/fringe. Prior to starting fatigue testing the beam was loaded to its computed service level (see Chapter 1) and cracks were induced in the central portion. Following this the beam was cyclically loaded at 4 Hz with $P_{max} = 3.56$ kN and $P_{min} = 2.67$ kN, giving a load ratio of 0.75. Crack growth versus load cycles is shown in Table 12-1. These data were obtained from the photoelastic coating photographs, two of which are shown in Fig.

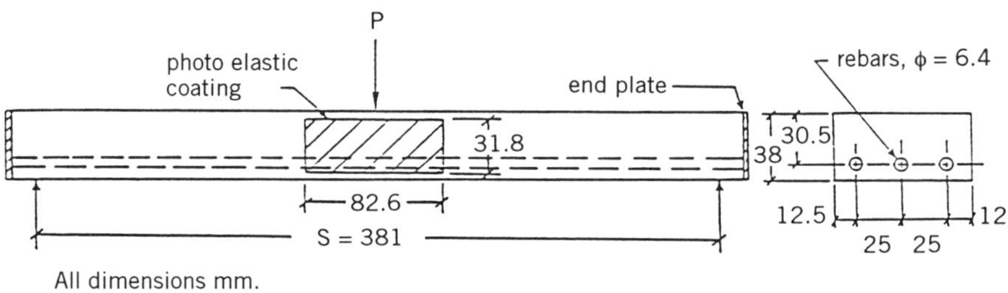

Figure 12-11 Model beam with end-anchored reinforced tested in fatigue.[26]

Table 12-1 Crack Size versus Number of Cycles[a]

Number of Cycles, $N \times 10^3$	Ratio of Crack Depth to Beam Depth	Percent Change in Crack Width
0	0.232	—
25	0.344	5
850	0.376	11
1174	0.400	74
4300	0.600	137

[a] From ref. 26.

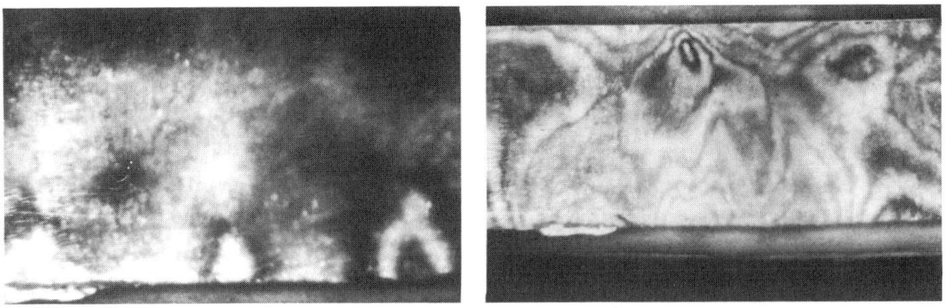

(a) (b)

Figure 12-12 Crack growth in model reinforced concrete beam[26]: (a) prior to start fatigue loading and (b) after 4.5×10^6 cycles.

12-12. The failed beam is shown in Fig. 12-13. It is seen that the primary mode of failure was by penetration of a flexural crack completely through the compressed region after which the load was carried solely by reinforcement. This induced a yield failure in the steel bars. The fatigue life for the beam was 4.83×10^6 cycles. Crack growth versus cycles is shown in Fig. 12-14. The second beam failed at 17.68×10^6 cycles.

Recent work on characterizing the fatigue behavior of beams is categorized

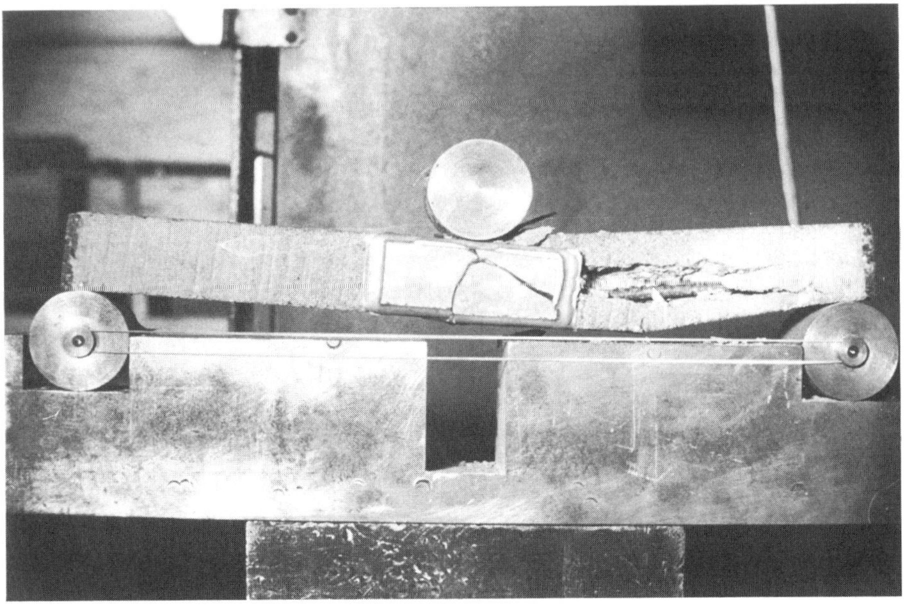

Figure 12-13 Fatigue failure of reinforced concrete beam.[26]

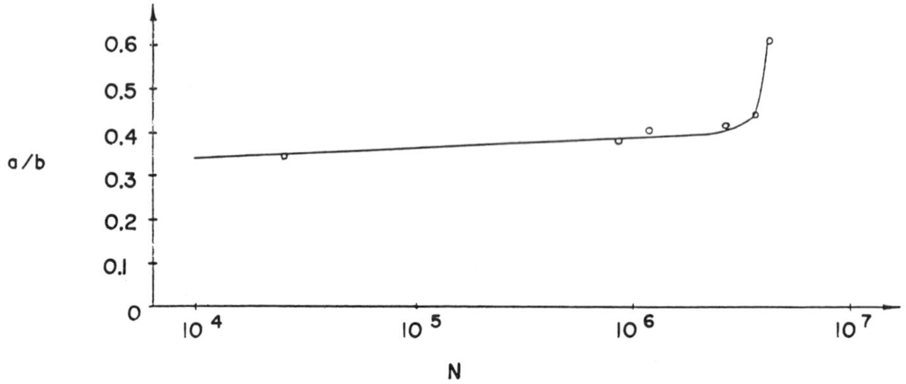

Figure 12-14 Crack growth with load cycles for reinforced concrete model beam.[26]

here as statistical methods for life prediction, damage methodologies, and fracture mechanics methods.[27] These approaches are discussed below.

12.2.1 Statistical Methods

Fatigue life estimates for concrete have traditionally been based on test data expressed in the form of Wöhler, or S_P–N, curves in which each curve represents the locus of points of average values of life N obtained at different maximum stress levels S_P. Different curves are obtained for different stress ranges, types of load cycle/frequency, and so on. A recent study in which the frequency of occurrence, or data dispersion, is obtained in terms of stress level has been presented by Oh.[28] The general situation is shown in Fig. 12-15. The determination of confidence limits of stress level is based on a Weibull probability density function $f_N(n_F)$ and the related cumulative distribution function F_N, which is given by

$$F_N(n_F) = 1 - \exp\left[-\left(\frac{n_F}{u_F}\right)^{\alpha_F}\right] \qquad n_F \geq 0 \tag{12.22}$$

By plotting $\ln\{\ln[1/(1 - F_N)]\}$ versus $\ln(n_F)$, values of α_F and u_F (the characteristic extreme value) may be obtained provided the plotted data fall on a straight line. The term α_F is the shape parameter and is related to variance or standard deviation s_s. When $\alpha_F \to \infty$, $s_s \to 0$.

The variation of α_F with stress level, $f_r^{\max}/f_r'$ (the maximum stress in cycle/static strength), as given by Oh[28] for three stress ranges is shown in Fig. 12-16. This indicates that the dispersion of data increases with increasing fatigue life and therefore $F_N(n_F)$ itself is a function of stress ratio $f_r^{\max}/f_r'$.

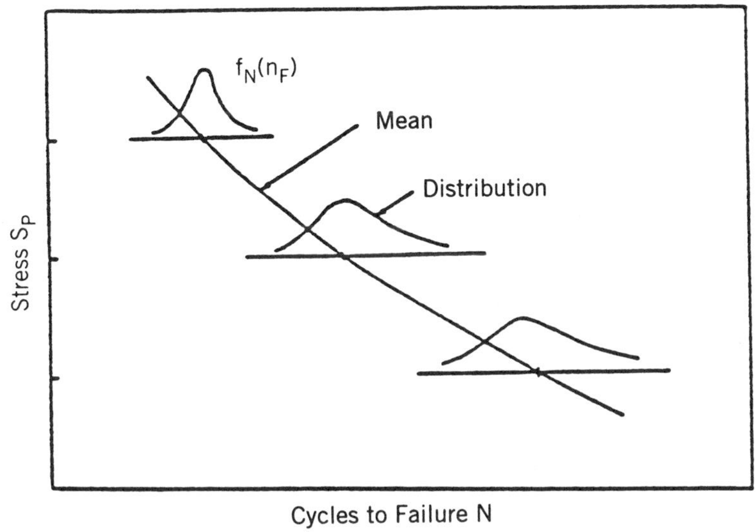

Figure 12-15 Fatigue life distributions at different stress levels.[28]

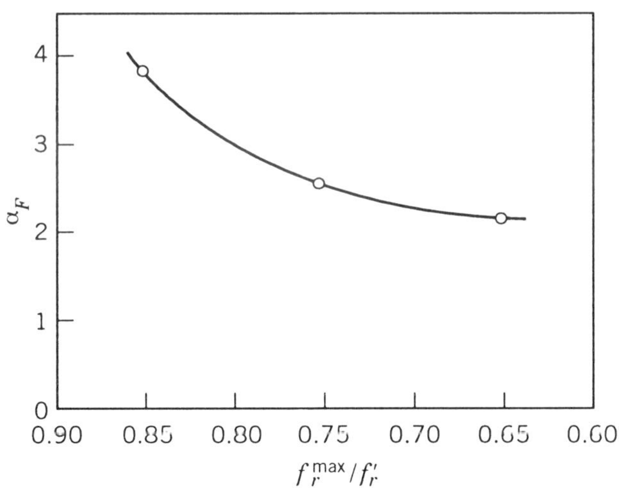

Figure 12-16 Variation of shape parameter according to applied fatigue stress levels.[28]

12.2.2 Damage Methodologies

The use of the method of damage mechanics is directed to obtaining damage indices to describe the reduced function of the structural component in terms of load cycles. Typically, the damage function is of the form

$$\varphi_F = \varphi_{F0}(1 - \Omega_F) \tag{12.23}$$

where φ_F is the current stiffness parameter, φ_{F0} is the original (considered to be undamaged) reference state, and Ω_F is a fatigue damage parameter. As noted by Alliche and Francois,[29] this approach to describing damage, as opposed to considering stress, yields less scatter, and therefore strain ϵ_t, or equivalently deflection, may be better correlated with damage. They describe three distinct stages as shown in Fig. 12-17. Using stages II and III, they assume $d\Omega_F/dN$ to be a scalar function of strain and stress ratio ($P_{\min}/P_{\max}$). The life estimates obtained from this approach showed a reasonable correlation with their experimental data.

Recent work presented by Hordijk and Reinhardt[30] uses a damage model obtained from observed cyclic behavior of a cracked element with uniaxial stresses as shown in Fig. 12-18. This was obtained from deformation-controlled tension tests. Empirical expressions were given to characterize the softening σ–w relation in terms of f_t, w_c, w_{eu}, σ_{eu}, w_L, σ_L, w_{inc}, and $\Delta\sigma$. This was used in Hillerborg's fictitious crack approach. The basic idea is that damage accumulates in each load cycle, where $P_{\max} < P_c$ (critical static load), until some cycle when $P_{\max}$ reaches the same value on the softening branch of the static load–deflection curve. At this point, the crack propagation is unstable and failure occurs.

This model was implemented in the DIANA finite element program[31] and applied to a beam in four-point bending with a central notch of half the beam height (Fig. 12-19). It is seen that the deflection response with load cycle N is very similar to that observed experimentally (cf. Fig. 12-17). Although this approach is very time consuming (computing intensive), it nevertheless appears to be realistic.

Expressions for the damage parameter Ω_F have been proposed by Zhang et al.[32] for concrete beams loaded cyclically. Fracture mechanics concepts were used and a threshold parameter based on opening mode stress intensity ΔK_{th}

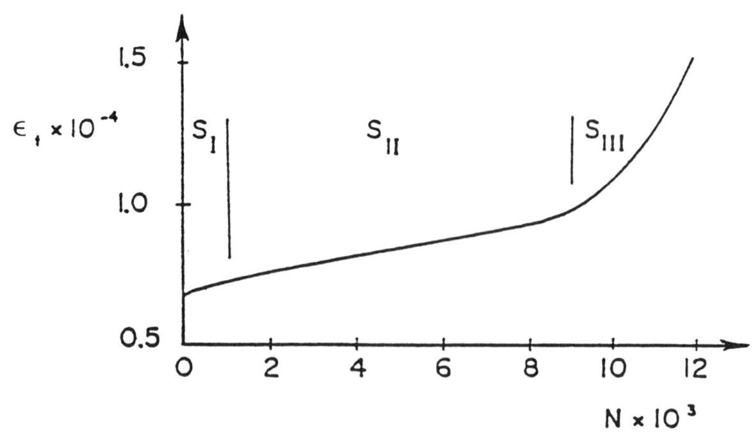

Figure 12-17 Evolution of ϵ_t with number of cycles.[29]

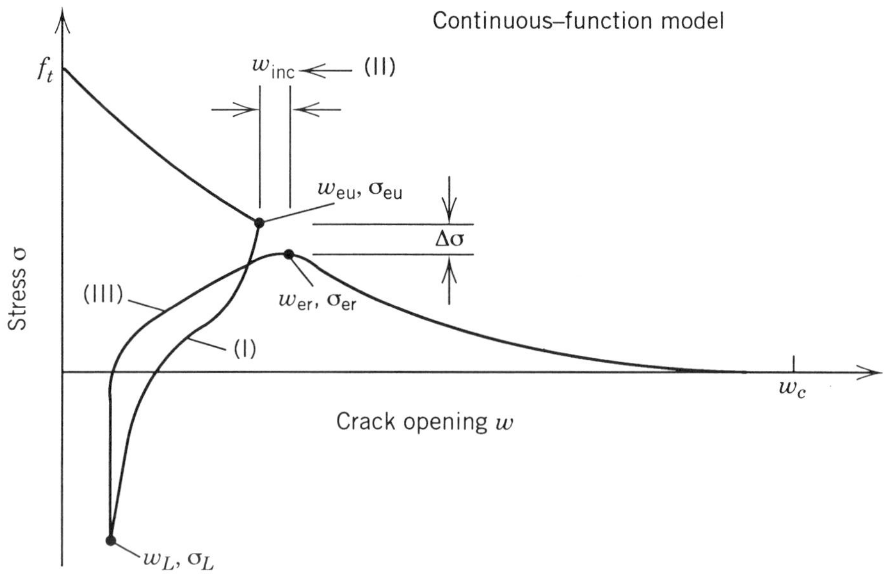

Figure 12-18 Crack cyclic behavior model[30]: eu, envelope unloading; er, envelope reloading; inc, increase.

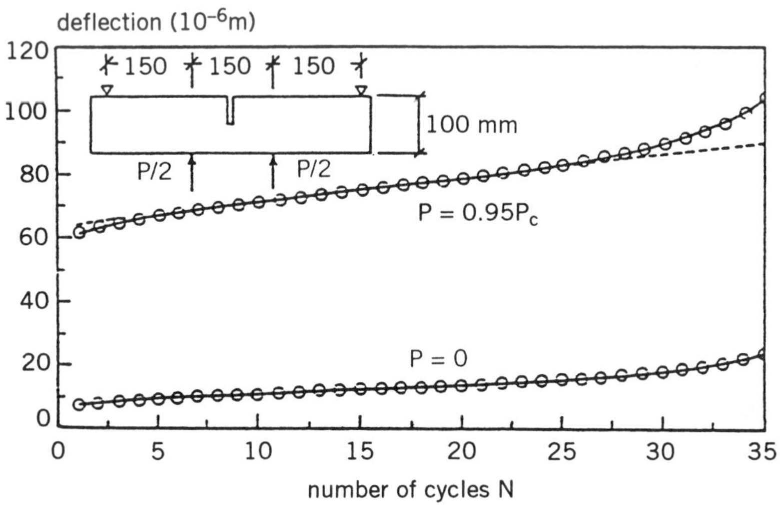

Figure 12-19 Deflection as function of number of cycles.[30]

below which the elastic modulus E_d remains constant. For $\Delta K > \Delta K_{th}$, E_d decreases linearly with load cycle, according to their data, until a value of $N = N_L$. After this, E_d decreases nonlinearly, as shown in Fig. 12-20, where $S_d = E_d/E_{d0}$ and E_{d0} is a reference value.

The limit of linear response is denoted E_{dL} (or S_{dL}). It is seen that the value of N_L is very sensitive to the variation of ΔK. After introducing $S_d = 1 - \Omega_F$, the following relationships are presented[32]:

$$
\begin{aligned}
&\Omega_F = 0 \quad \text{for } \Delta K < \Delta K_{th} \quad \text{(no damage)} \\
&\Omega_F = \alpha_d N \quad \text{for } \Delta K_{th} < \Delta K < K_{Icd}(1 - k_f) \quad \text{and} \quad N < N_L \quad \text{(linear damage)} \\
&\Omega_F = f(N) \quad \text{for } \Delta K_{th} < \Delta K < K_{Icd}(1 - k_f) \quad \text{and} \\
&\qquad\qquad N_L < N < N_f \quad \text{(nonlinear damage)} \\
&\Omega_F = 1 \quad \text{for } \Delta K = K_{Icd}(1 - k_f) \quad \text{(failure)}
\end{aligned}
\tag{12.24}
$$

where K_{Icd} is the dynamic, mode I fracture toughness of concrete and $k_f = K_{min}/K_{max}$ (= 0.2 in this study). The variation of S_d with N is given as

$$\frac{dS_d}{dN} = \frac{\alpha_d(\Delta K - \Delta K_{th})^{n_d}}{K_{Icd}(1 - k_f) - \Delta K} = f(\Delta K) \tag{12.25}$$

and

$$N_f = \frac{S_{dL} - 1}{f(\Delta K)} \tag{12.26}$$

is the fatigue life. The parameters α_d and n_d are empirical constants.

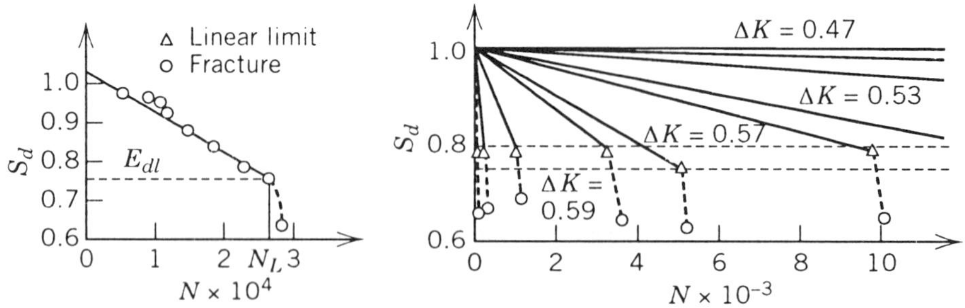

Figure 12-20 Relationship between S_{dL}, ΔK, and N.[32]

12.2.3 Fracture Mechanics Methods

Some of the methods of the previous section incorporated linear and nonlinear fracture mechanics into the damage models. In this section, the explicit use of fracture mechanics concepts to evaluate fatigue data will be discussed. An approach commonly used for metals relates crack growth with number of cycles according to the so-called Paris law[33,34]:

$$\frac{da}{dN} = C_F(\Delta K)^{m_F} \tag{12.27}$$

where C_F and m_F are experimental constants and K is the stress intensity factor. In the logarithmic form, Eq. (12.27) becomes

$$\log \frac{da}{dN} = \log C_F + m_F \log(\Delta K) \tag{12.28}$$

and the resulting data should plot as a straight line. The difficulties implementing this approach are measurement of crack length a, calculation of $K = K(P, a, \text{geometry})$, and calculation of da/dN and ΔK. Inherent to this are the questions of the influence of process zone size and use of LEFM.

The research presented by Baluch et al.[35] describes a compliance–calibration technique using cracked specimens with the crack length measurement based on surface observations. Work by Swartz and Refai[36] indicates this approach will be approximately valid for determining Δa if the process zone is fully developed and if a is less than about 70% of the beam depth. Since the maximum size of the process zone appears to be approximately constant for a given aggregate size, then clearly a size effect will be present with this approach.

Nevertheless, the concept is useful and probably can be modified (e.g., use of the two-parameter fracture model[37] to determine an effective crack length).

Using this approach to determine da/dN and ΔK, results from reference[35] are presented in Fig. 12-21*a* for three different stress ratios R_F. The straight lines were fitted to data values and all have about the same slope m_F. Similar work by Perdikaris and Calomino[38] is presented in Fig. 12-21*b* along with the data values that show the scatter. Their values of a and K were based on notch beam compliance calibration and LEFM. Although the data in Fig. 12-21*b* appear to be fitted fairly well by the straight line, the authors point out that the error in crack growth rate or life expectancy could be as high as 100%.[38]

In related work by Perdikaris et al.,[39] it was shown that the measurement of a and calculation of K are influenced by the nonplanar crack front as revealed by dye (described also in ref.[36]). They found that surface measurements of crack length a were 25–30% higher than obtained by compliance or by the average cracked depth of the dyed surface. They also found that values of fracture toughness K_{Ic} obtained from specimens loaded in fatigue until $a \approx$

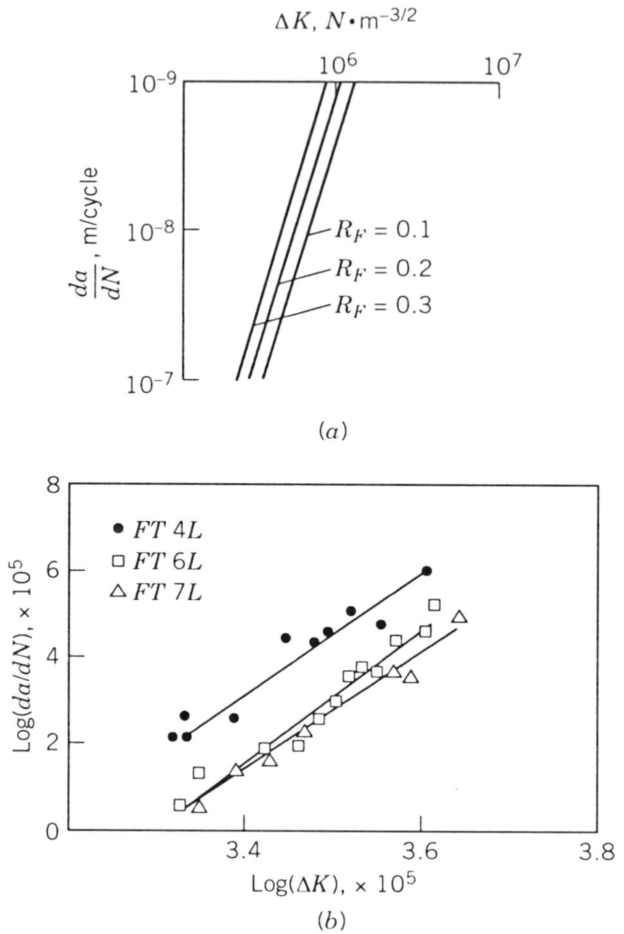

Figure 12-21 Crack growth *da/dN* versus ΔK: results from (*a*) ref. 35 and (*b*) ref. 38.

0.5 depth (N ranged from 268 to 112,740) were generally higher than for companion beams tested statically (with a short-time load).

Work by Bazant and Xu[40] that exploited the principle of the size effect model resulted in a modified, or generalized, size-adjusted Paris law,

$$\frac{\Delta a}{\Delta N} = C_F \left(\frac{\Delta K_i}{K_{\mathrm{I}c}} \right)^{m_F} \tag{12.29}$$

where $K_{\mathrm{I}c}$ is a size-dependent fracture toughness given by

$$K_{\mathrm{I}c} = K_f \left(\frac{\beta_b}{1 + \beta_b} \right)^{1/2} \tag{12.30}$$

In this K_f is the fracture toughness of an infinitely large beam and the brittleness number $\beta_b = D/D_0$, where D is the beam depth and D_0 is an empirical constant considered to be the transition size between LEFM and strength behavior (see Chapter 5). Results are plotted in Fig. 12-22*a* for a value of $D_0 = 2.86$ in. (72.6 mm) obtained from static tests as described in Chapter 5. It is seen in Fig. 12-12*a* that a size effect still is present. It is postulated[40] that D_0 may vary

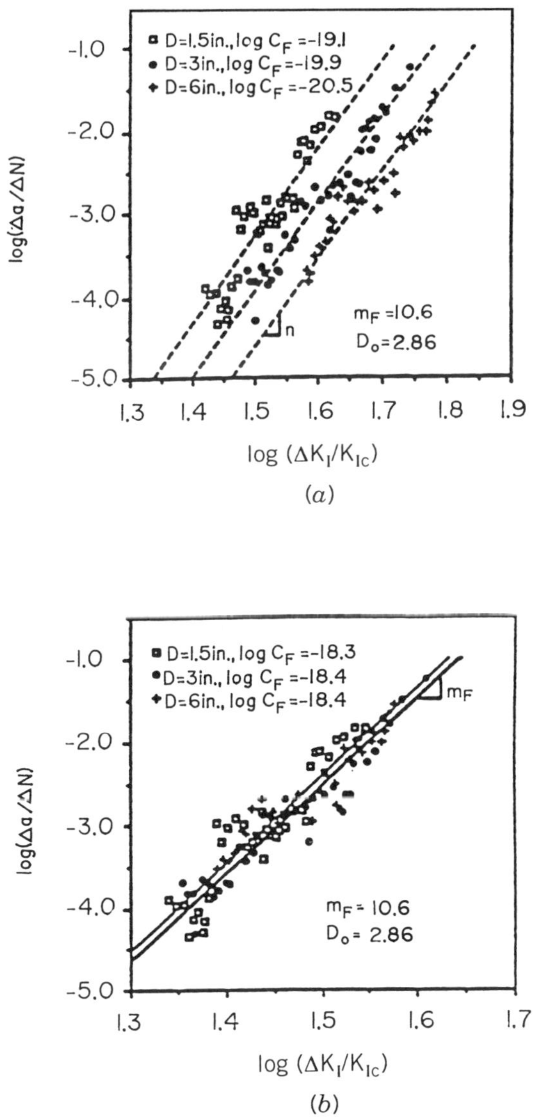

Figure 12-22 Crack growth in three different sizes of beams.[40]

according to the ratio of $\Delta K_I/K_{Ic}$. In their tests this was 0.8 and the minimum load was zero. By arbitrarily adopting a value of $D_0 = 28.6$ in. (726 mm), or 10 times the value for static loading, improved results are shown in Fig. 12-22*b*, in which it is seen that the size effect virtually disappears. Thus, they conclude that the size of the fracture process zone is greatly enlarged by load cycling.

Swartz et al.[41] tested beams in three- and four-point bending using two sizes and two mix designs. The crack lengths were determined using compliance calibration.[41,42] The fatigue loading was done using CMOD strain control and was stopped at $N = 1 \times 10^6$ cycles. A typical group of data plots is shown in Fig. 12-23. The objective was to grow true cracks of different lengths but with about the same elapsed time ($1 \times 10^6/4$ Hz $= 0.25 \times 10^6$ s). After cracking, the beams were loaded to failure and P–CMOD plotted. In each case a companion beam was cracked such that the slope of the ascending P–CMOD curve matched that of the fatigued beam. Then this beam was also loaded to failure. In both cases, K_{Ic} was computed using P_{max} and LEFM with the crack length corresponding to the slope of the ascending curve.

The results of these tests are presented in Table 12-2. In this K_{Ic} was also determined using the method of Jenq and Shah.[37] It is seen that both methods indicate that fatigue precracking yields K_{Ic} values somewhat smaller than those obtained from static precracking. The results are the opposite of those reported in reference.[39]

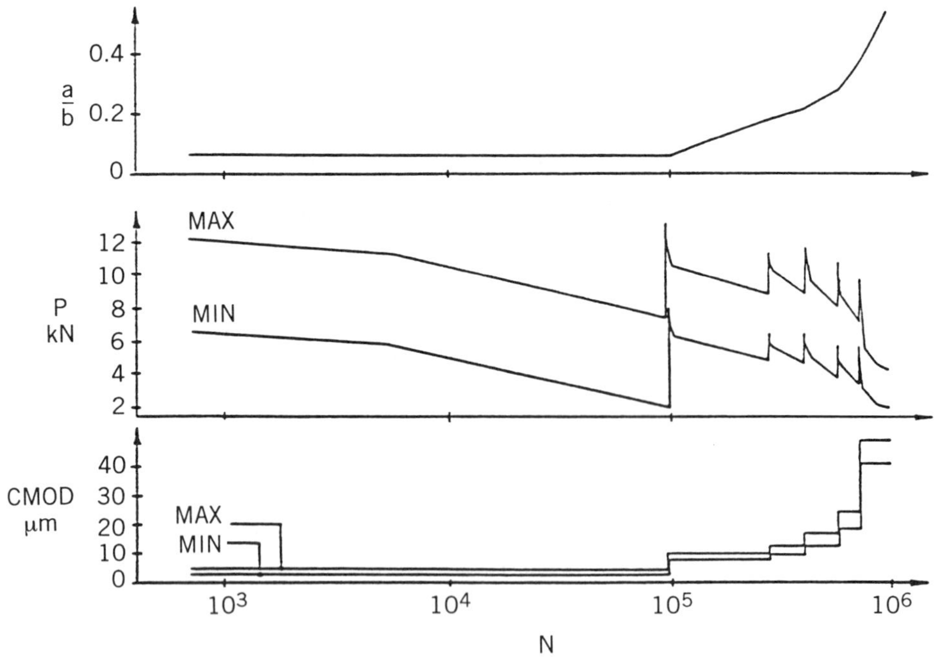

Figure 12-23 Fatigue crack growth under displacement control.[41]

Table 12-2 Stress Intensity Factors at Unstable Crack Growth, Fatigue, and Static Precracking[a]

b (mm)[b]	Load Type[c]	f'_c (MPa)	P_m (kN)	a_c/b^d	K^a_{Ic} $(kN/m^{3/2})^d$	F/S^d	a/b^e	K_{Ic} $(kN/m^{3/2})^e$	F/S^e
102	F3	48.9	4.54	0.352	1192	1.07	0.195	769	0.84
102	S3	48.9	4.94	0.297	1116	1.07	0.245	916	0.84
102	F3	54.2	1.92	0.525	740	0.95	0.510	727	0.93
102	S3	54.2	2.02	0.527	780	0.95	0.520	782	0.93
102	F3	51.7	2.06	0.605	1117	0.89	0.612	1102	0.86
102	S3	51.7	2.31	0.605	1252	0.89	0.618	1276	0.86
203	F3	22.5	12.37	0.072	595	0.93	0.160	842	0.83
203	S3	23.4	13.71	0.065	642	0.93	0.180	1016	0.83
203	F3	24.8	5.70	0.350	631	0.98	0.490	880	0.91
203	S3	24.8	5.83	0.350	646	0.98	0.515	972	0.91
203	F3	52.7	11.35	0.230	902	0.94	0.374	1263	0.95
203	S3	52.7	10.24	0.290	963	0.94	0.438	1329	0.95
203	F3	53.1	3.90	0.590	857	0.58	0.740	1670	0.70
203	S3	53.1	4.05	0.760	1710	0.58	0.742	1752	0.70
203	F3	53.2	1.07	0.770	1110	0.58	0.830	779	0.70

[a] From ref. 41.
[b] For b = 102 mm, t = 76 mm, S = 381 mm; for b = 203 mm, t = 102 mm, S = 610 mm.
[c] F = fatigue precracking, S = static precracking, 3 = three-point bend.
[d] Method of Jenq and Shah (ref. 37).
[e] Compliance calibration (ref. 42).

The results also contradict those by Bazant and Xu.[40] If the size of the process zone is associated with the load ratio used in the fatigue test, then the effective crack length a_c should be substantially different for the statically loaded beams as compared to the fatigue-loaded beams. Yet a review of the results obtained from the Jenq–Shah model in Table 12-2 shows this is not the case.

12.2.4 Summary

The following can be summarized from the above discussion:

1. Recent advances in the understanding of the static cracking behavior of concrete subjected to mode I deformation (e.g., the size effect model[40] and the two-parameter fracture model[37]) have shown promise of more realistic modeling of fracture behavior in fatigue using fracture mechanics.
2. Nevertheless, problems remain in understanding the influence of the fatigue mechanism on crack formation, coalescence and bridging, that is, the proces zone effect. The size effect model evidently predicts a large increase in the process zone due to fatigue,[40] whereas the equivalent elastic crack model does not. Clearly, more detailed experimental evi-

dence of the actual crack growth process, especially due to low-level, repeated loads, is needed.

3. When these problems are resolved, it appears that the Paris law, in some modified form, may reasonably model crack growth versus load cycles for plain concrete beams.

12.3 CREEP OF CONCRETE AND ITS EFFECT ON CRACK GROWTH AND FRACTURE

Creep may be defined as a time-dependent response of solids subjected to stress. It is a particularly important effect in concrete structures and is insidious in that it can lead to failure after a very long period of time. If these failures were in fact initiated through design or construction errors, then the length of time to collapse could have major consequences with respect to liability of the parties involved and recovery to the owner and victims. It is thus highly desirable to understand the mechanisms of creep and how to incorporate these into a design philosophy. Because of this importance, many references appear in the literature and this effect is considered explicitly, but in a limited way, in codes of practice. A relatively recent book[43] that deals with modeling this behavior contains 5 committee reports and 50 papers covering most aspects of structural behavior. However, very little is presented on the subject of fracture mechanics behavior of concrete as influenced by the effect of creep.

As an example of this effect on structural performance, it is mentioned in reference[44] that the influence of creep on the strength of concrete is important for dams. The Zhexi Dam in the Hunan province in China was observed to contain many vertical cracks shortly after construction in 1961 but before the storage water was introduced. The dam appeared to be safe; however, a large fracture area developed in 1969 followed by failure of the dam.[45] Results of many studies confirmed that a small crack on the upstream surface propagated with time and finally produced the failure. Notice that the time scale of this failure is considerably greater than that of the Kölnbrein Dam described in Chapter 9,[46] although it may be expected that for that dam also some creep effects were present. These effects were, however, not modeled in the failure analysis.

In the following sections, results of studies dealing with sustained load–deflection response, time to failure, and size effects will be presented.

12.3.1 Behavior of Concrete Beams Subjected to Sustained Loads

12.3.1.1 Results by Liu et al. Liu et al.[44] have performed creep fracture tests to observe the time-dependent relations for displacement and failure and other effects. Prior to creep testing, the specimens—beams in three-point bending—were precracked and dye inserted. The fracture surface revealed by dye was then used to develop a strain-softening model. This was used in conjunction with a creep model in a FE analysis to predict time-dependent crack growth and fracture based on viscoelasticity and fracture mechanics.

Plain concrete beams were tested in three-point bending (TPB) for instantaneous fracture behavior as well as for long-term creep response and failure. The beam dimensions were width $t = 76$ mm, depth $b = 102$ mm, span $S = 384$ mm, and length $L = 406$ mm. The LEFM concepts modified to include time-dependent behavior and, to a limited extent, process zone effects were used in the analysis.

All beams were cast in steel molds, removed one day after casting, and placed in a 100% humidity curing room for 90 days until the start of the various tests. Companion 76 × 152 mm cylinders were made and the following material properties obtained: compressive strength $f_c' = 48.7$ MPa, split cylinder strength $f_t' = 4.41$ MPa, and secant modulus $E = 32.7$ GPa.

The beam tests were in seven groups according to the following load ratios: $P_k/P_f = 0.95$, 0.90, 0.85, 0.80, 0.75, 0.70, 0.65, where P_k and P_f are the sustained load level and the failure strength, respectively. For each group, two replicate series of four tests were conducted: three beams in each series were precracked and dye inserted. They were then loaded to failure using the same load frame and dead weights as for the creep tests. The remaining beam was precracked and loaded to P_k equal to the desired fraction of P_f, the average maximum load of the three tested instantaneously. The creep test of this beam was done using dead weights in the creep frame arrangement shown in Fig. 12-24. The crack mouth opening displacement (CMOD) and load point displacement (LPL) were measured using dial gages with 1×10^{-4} in. (25.4×10^{-4} mm) and 1×10^{-3} in. (25.4×10^{-3} mm) direct least readings, respectively, as shown in Fig. 12-25.

Specimen preparation included the following steps:

1. A notch of depth 12.7 mm and width 3 mm was cut at the center of the beam.
2. The cracking procedure was designed to produce precracked beams with a peak load P_f of approximately 2.2 kN.
3. As mentioned, all beams had dye inserted to reveal the fracture surface after failure.

The time to failure of the various beams is plotted in Fig. 12-26. The creep modeling predicted that below a load ratio of 0.78, failure by creep will not occur. The LPD and CMOD curves are plotted against time in Fig. 12-27. Three stages of deformation behavior are apparent: (1) initial, nonlinear transient response; (2) a lengthy, minimum-time response; and (3) a final, nonlinear response to failure. It is seen that this type of behavior mimics the fatigue-damage behavior, as shown in Figs. 12-14 and 12-17. The second-stage re-

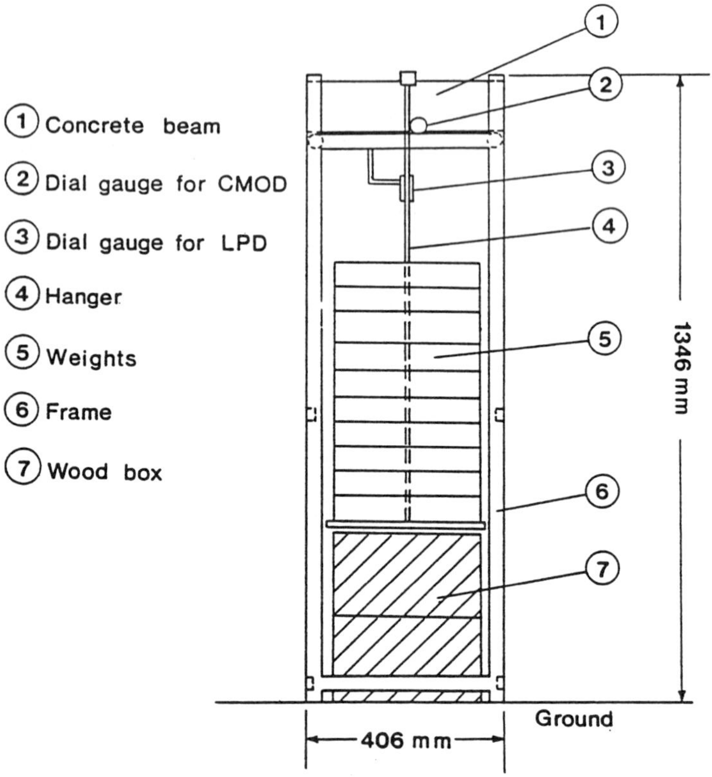

Figure 12-24 Load frame for three-point bend beams.[44]

sponses shown for load ratios of 0.85 and 0.80 indicate "jumps" in the deformations at relatively constant time intervals.

An LEFM approach to modeling the above observed responses was attempted. In this, the opening-mode stress intensity factor as a function of time was assumed to be

$$K_{\text{I}}(t) = K_{\text{I}}^{\text{ref}} f_\sigma(t) \tag{12.31}$$

where $K_{\text{I}}^{\text{ref}}$ is a reference, time-independent opening-mode stress intensity factor and

$$\begin{aligned} f_\sigma(t) &\neq 0 \quad \text{for } t \geq 0 \\ f_\sigma(t) &= 0 \quad \text{for } t = 0 \end{aligned} \tag{12.32}$$

is a time-varying function. By use of LEFM equations (see Chapter 2), the stresses and strains at the crack tip will be functions of $K_{\text{I}}(t)$ and the crack tip energy release rate is given as

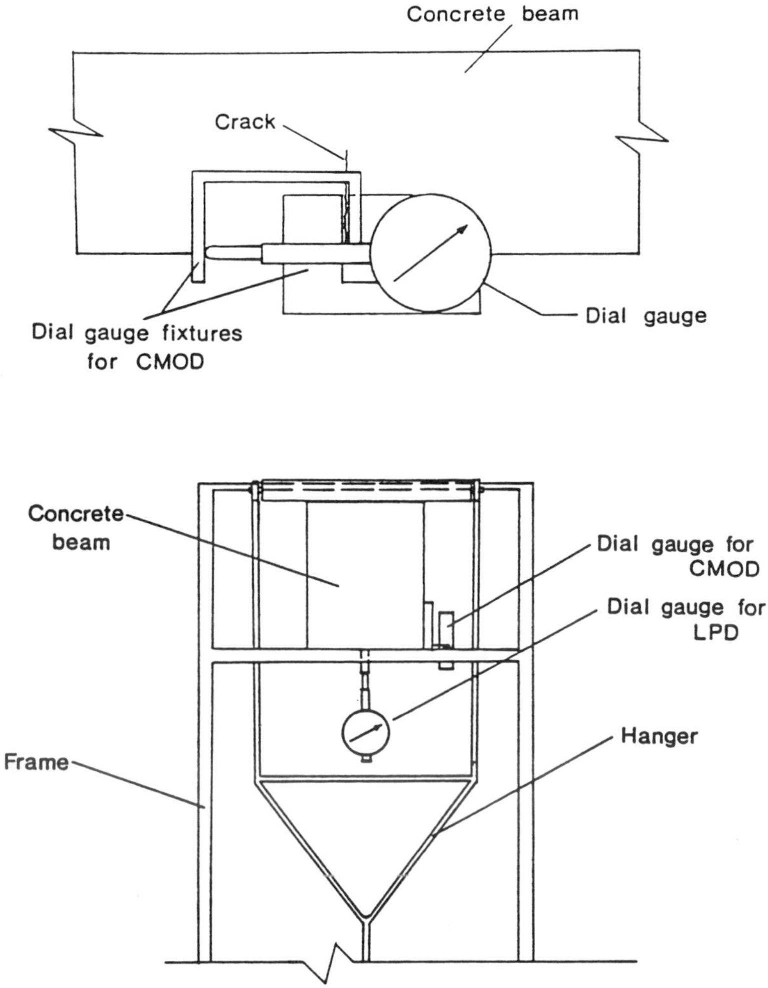

Figure 12-25 Displacement measurement for three-point bend beams.[44]

$$G_{\mathrm{I}}(t) = \frac{(K_{\mathrm{I}}^{\mathrm{ref}})^2}{E} f_{\sigma}^2(t) = \frac{(K_{\mathrm{I}}^{\mathrm{ref}})^2}{E} f_{st}(t) \tag{12.33}$$

The term $f_{st}(t)$ is a time decay function. This was taken to be

$$f_{st}(t) = 1 + Es_1(1 - e^{-s_2 t}) \tag{12.34}$$

where s_1 and s_2 are creep constants. When $G_{\mathrm{I}} \geq G_{\mathrm{I}c}$, the critical energy release ratio at $t = 0$, crack propagation will occur. Possible changes in $G_{\mathrm{I}c}$ with time were not considered.

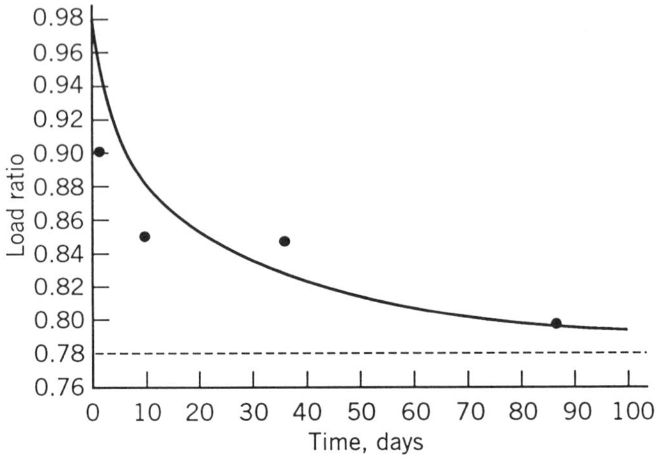

Figure 12-26 Time to failure for various load ratios.[44]

A simple FE model with constant-strain, triangular elements was used in conjunction with the time relationship described above. In addition, the stress distribution ahead of the crack was modeled using information obtained from the dyed surface area to simplify the complicated, nonuniform stress variation through the width of the cross section. The critical energy release rate values given in Table 12-3 were calculated using the peak load and average depth of the crack front, a_i, as shown by dye (see Chapter 6 for use of the dye technique). Values of the initial energy release rate G_{I0} $[=(K_I^{ref})^2/E]$ associated with each load P_k and crack length a_i are also given in Table 12-3. Calculated stress distributions ahead of the crack at different times for a beam with $P_k/P_f = 0.85$ are shown in Fig. 12-28. Finally, failure time estimates for beams with $P_k/P_f = 0.95$ and 0.90 are given in Table 12-4.

12.3.1.2 Results by Hansen Hansen[47] performed creep experiments on notched, cantilever beams in which deformations in the region of the fracture were monitored. The overall test setup is shown in Fig. 12-29*a*. Three transducers shown in Fig. 12-29*b* are used to measure the (CMOD), the crack tip opening displacement (CTOD) and the crack path opening displacement (CPOD) positioned 25 mm ahead of the notch tip across the expected crack path. In addition to these transducers, the region of crack propagation and process zone development was monitored using a video camera and holographic techniques.

The concrete mix had a water–cement ratio of 0.48 and ratios of cement–sand–gravel of 1:2.25:2.4. Beams were cast separately and also cut from slabs. The dimensions were 100 × 100 × 1200 mm and all beams were cut in half to make test specimens 600 mm long. Two test series were conducted: ages 28 days and 90 days. The compressive strengths (100 mm cubes) were

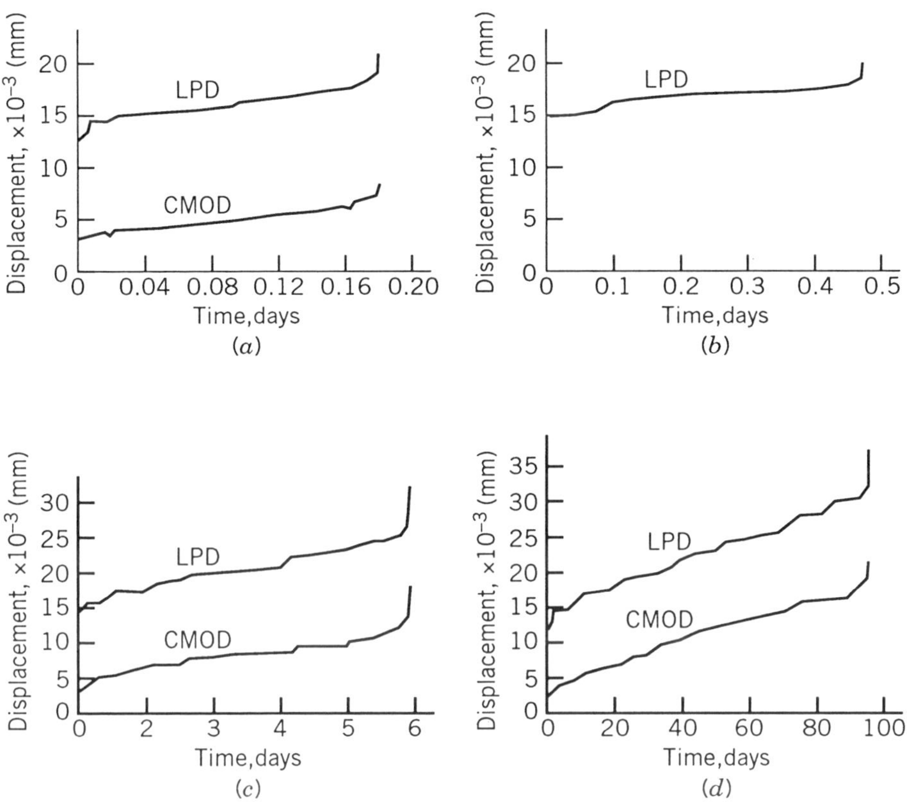

Figure 12-27 Displacement–time response for three-point bend beams subjected to sustained loads.[44] Load ratios: (*a*) 0.95, (*b*) 0.90, (*c*) 0.85, and (*d*) 0.80.

65.3 and 74.2 MPa for 28- and 90-day specimens, respectively. Tensile strengths at the different ages were the same and were 4.0 MPa. The secant modulus at 50% peak stress was 33.0 GPa and the dynamic modulus was 43.5 Gpa, both at 90 days.

As shown in Fig. 12-29*a*, half the beam was clamped and a notch of depth

Table 12-3 Calculated Values of Initial and Critical Energy Release Rates for Three-Point Bend Beams

P_k/P_f	P_f (kN)	a_i/b	G_{Ic} (N/m)	G_{I0} (N/m)
0.95	1.019	0.56	21.6	19.5
0.90	1.088	0.65	77.9	62.5
0.85	1.154	0.56	27.0	19.3
0.80	1.084	0.62	42.4	27.2

[a]From ref. 44.

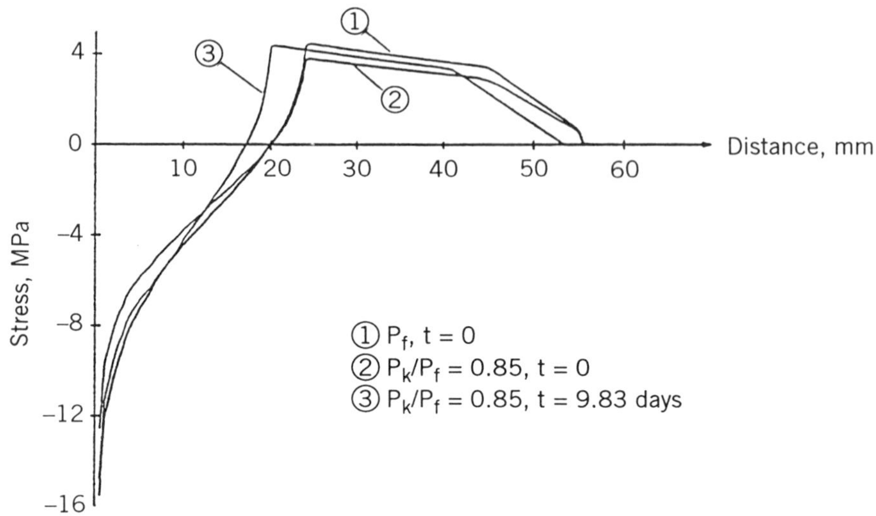

Figure 12-28 Calculated stress distributions along crack line.[44]

Table 12-4 Time to Creep Failure[a] (days)

	$P_k/P_f = 0.95$	$P_k/P_f = 0.90$	$P_k/P_f = 0.85$	$P_k/P_f = 0.80$
Experimental	0.180	0.468	9.83	86.1
Theoretical	0.191	0.515		

[a]From ref. 44.

50 mm and width 2 mm was introduced by sawing. The load application for ‘‘zero’’ time, or normal loading, was controlled to reach peak load in 2 or 3 min. For beams with sustained loads the same deformation rate was used up to a level of about 90% of the peak load. At that time the deformation was halted and a constant load applied using deformation control. When the microcracking reached a critical level such that the load could no longer be maintained, the deformation rate was increased to the initial rate and the beam was failed. Fracture energy results using this procedure, but for different sustained times, are presented in Table 12-5. In this, it is seen that the values of G_F for the 28-day tests increased about 24% when the sustained time was in the range of 20–120 min. The increase for the 90-day tests was more modest, about 6%. The creep deformation curves shown in Fig. 12-29*c* are quite similar for the two test ages and also similar to the curves in Fig. 12-27.[44]

Based on these studies the following conclusions were made:[47]

1. The fracture energy was influenced by a short period of sustained load, which implies that this is time dependent. This effect, however, decreased with an increase in the test age.

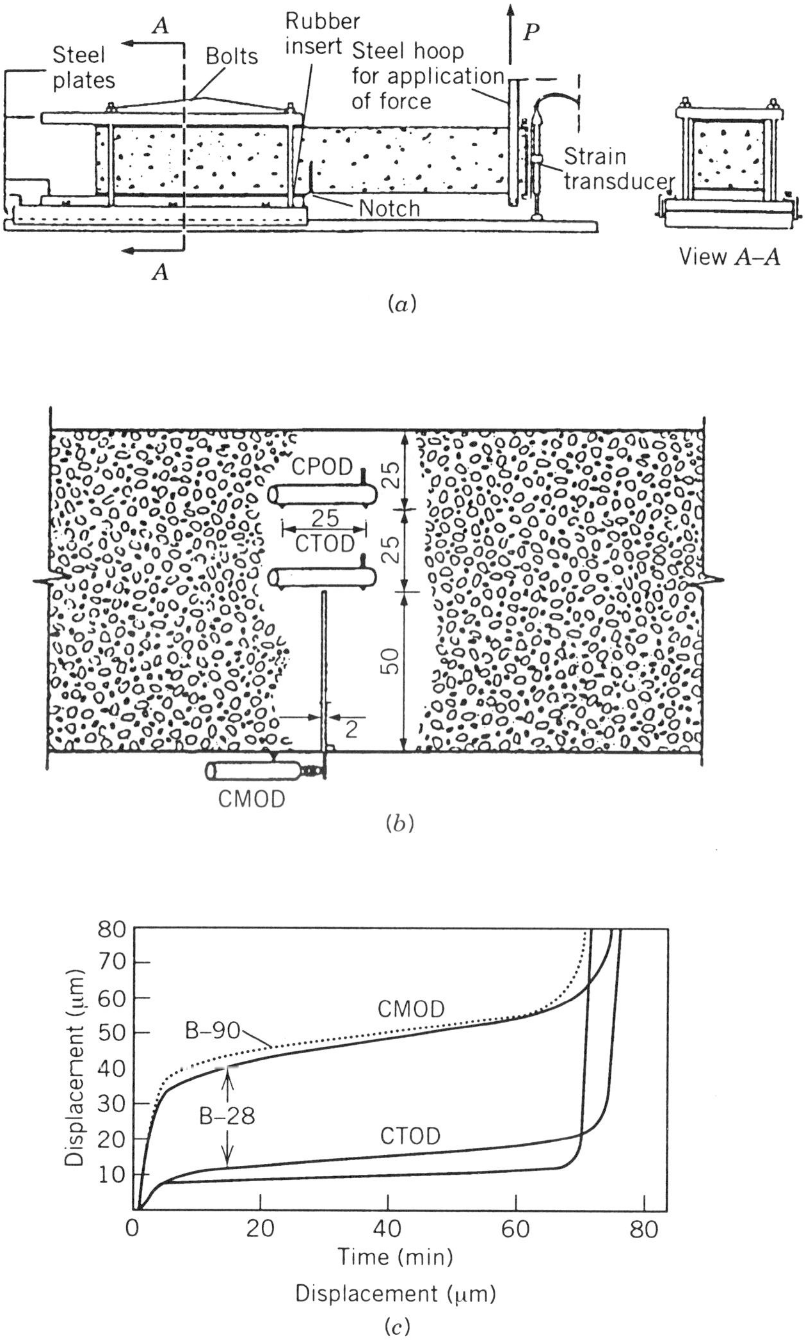

Figure 12-29 Creep tests of Hansen[47]: (*a*) sketch of fracture energy determination unit, (*b*) position of transducers, and (*c*) CMOD and CTOD versus time for beams B-28 and B-90.

Table 12-5 Average Fracture Energy (G_F) with Standard Deviation[a]

Test Age (days)	Constant Load (min)	Number of Beams	G_F (N/m)	Standard Deviation
28	0	11	150	0.14
	2–6	7	156	0.07
	20–120	6	185	0.04
90	0	10	147	0.17
	1–3	6	133	0.06
	20–40	7	153	0.07
	70–240	4	157	0.11

[a]From ref. 47.

2. For the 28-day test beams, the width of the fracture zone increased significantly during the period of sustained load. This effect appeared not to be present for the 90-day beams.

12.3.1.3 Results by Zhou Zhou[48] conducted an extensive study investigating the creep behavior of concrete. This included experiments using tension, compression, and bending specimens as sketched in Fig. 12-30. In terms of time to peak load, fracture energy, measured as the area under load–displacement curves shown in Fig. 12-31, varied directly with load rate, as shown in Fig. 12-32. The time to reach peak load varied from about 5 s for the fastest deflection rate of 50 μm/s to 5000 s for a rate of 0.05 μm/s. It is seen that for this range G_F changes from about 67 to about 87 N/m (Fig. 12-32). On the other hand, Zhou's analysis indicates that for sustained loads, fracture energy decreases to just 95.4% as compared to that for static (start time) loads when the sustained load is as low as 76% of the static peak load. This is in contradistinction to the experimental results of Hansen,[47] which show an increase in G_F with sustained loads (Table 12-5). The measured basic material properties were $f_c' = 38$ MPa, $f_t' = 2.8$ MPa, and $E = 36$ GPa. Results for beams subjected to constant loads (see Fig. 12-30 for dimensions) that were kept at constant load until failure are presented in Table 12-6. A trend of failure with time is similar to that prescribed by Liu et al.[44] in Fig. 12-26. However, a comparison of failure times given in Table 12-4 and those in Table 12-6 shows a difference of three to four orders of magnitude. The time-dependent displacement curves were also similar to those presented by Liu et al.[44] and Hansen.[47] A typical curve of CMOD versus time for a notched beam with a load ratio of 0.80 is shown in Fig. 12-33.

12.3.2 Load Rate Effects Determined Using Size Effect Model

Bazant and Gettu[49] have presented results of tests on TPB beams subjected to static loads applied at different rates. An example of the rate effect on beam

Geometry of specimen (mm)	No of test	Objective
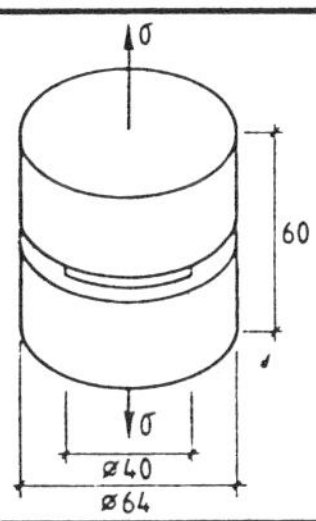	5	Tensile strength (f_t)
	6	Relaxation
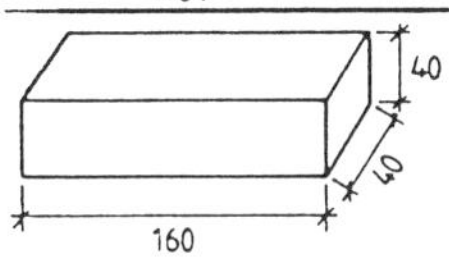	3	Modulus of elasticity (E)
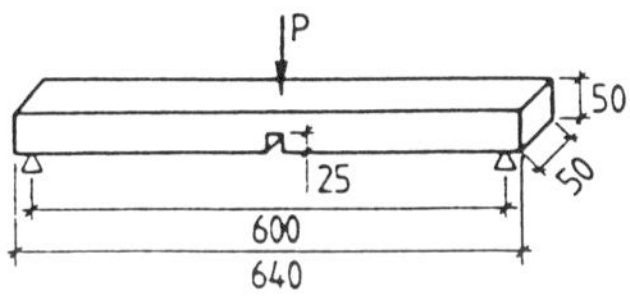	14	Rate effect on fracture energy (G_F)
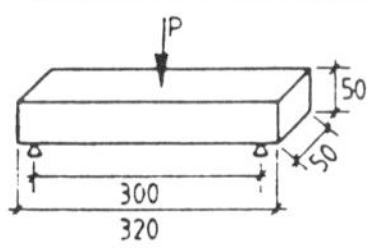	12	Flexural creep rupture
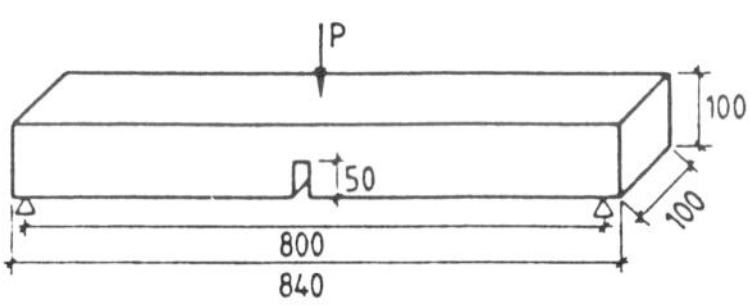	6	Fracture energy (G_F)
	14	Flexural creep rupture
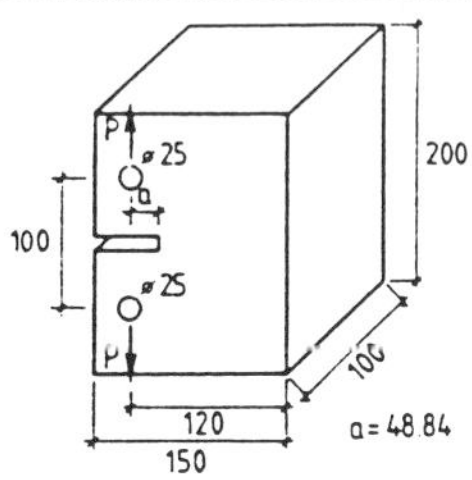	16	Compact tensile creep rupture
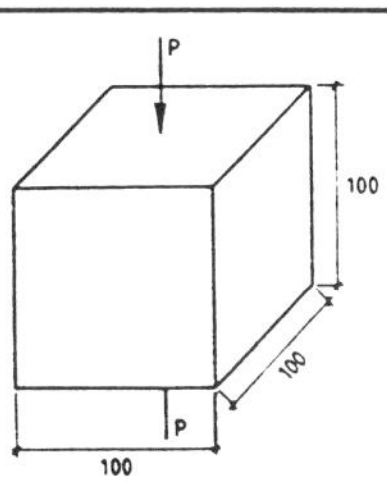	3	Compressive strength

Figure 12-30 Creep investigated by Zhou.[48]

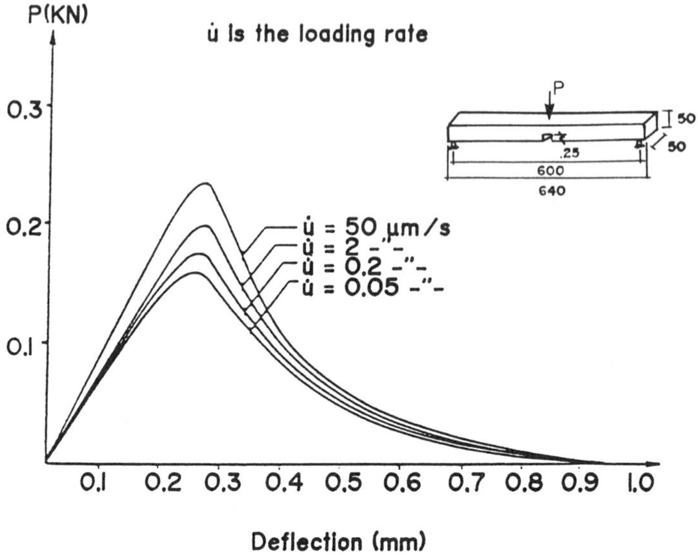

Figure 12-31 Load–deflection curves at various deflection rates.[48]

response obtained by them is shown in Fig. 12-34. The load–CMOD responses shown in part (*a*) for two beams loaded at much different times to peak load, t_p, are quite similar. However, the load–LPD plots are much different. It is stated that "this difference can be attributed to creep in the bulk of the specimen, since the load-line displacement reflects the cumulative response of the entire specimen, whereas CMOD is affected primarily by the deformations of the crack and fracture process zone."[49]

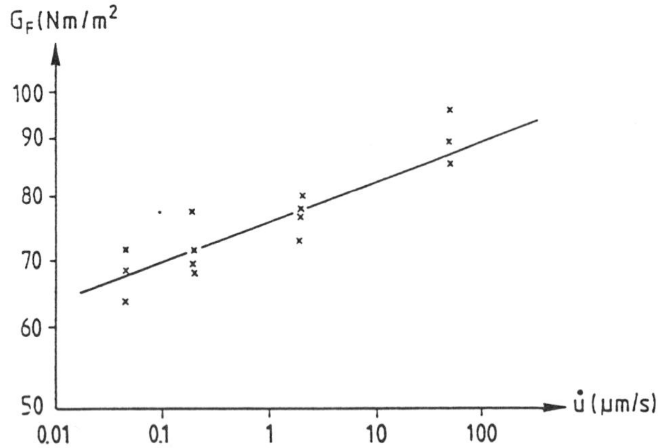

Figure 12-32 Rate effect on value of G_F.[48]

Table 12-6 Flexural Rupture Tests of Beams[a]

Beam type	Test Number	Load Level P_k/P_f	Failure Time, t_p Seconds	Days
Unnotched beams	BULC8	0.95	15	0.00017
	BULC4	0.90	105	0.00122
	BULC7	0.90	210	0.00243
	BULD4	0.85	540	0.00625
	BULD5	0.85	600	0.00694
	BULD7	0.80	960	0.01111
	BULB4	0.80	1200	0.01389
	BULD6	0.76	7200	0.08333
Notched beams	92A	0.92	20	0.00023
	92B	0.92	70	0.00081
	85A	0.85	190	0.00220
	85B	0.85	530	0.00613
	80A	0.80	550	0.00637
	80B	0.80	580	0.00671
	80C	0.80	360	0.00417
	80D	0.80	980	0.01134
	76A	0.76	930	0.01076
	76B	0.76	3200	0.03704
	76C	0.76	2400	0.02778

[a] From ref. 48.

To investigate this effect an extensive testing program was conducted using three sizes of beams: width $D = 38, 76, 152$ mm; thickness $t = 38$ mm, notch $a_0 = D/6$, and span $S = 2.5D$. The mix design used cement–sand–gravel–water ratios of 1:2:2:0.6. Type I cement, crushed limestone with a maximum size of 13 mm and standard No. 2 sand with maximum size of 5 mm, was used. All specimens were cured under water until testing to prevent loss of moisture. During testing, the times to peak load for a given series were kept approximately constant for all specimen sizes.

The size effect model, as described in detail in Chapter 5, was utilized for this application. The simplest version of this is Eq. (5.30) and is repeated here:

$$\sigma_{Nc} = \frac{Bf_t}{\sqrt{1 + \beta_b}} \qquad \beta_b = \frac{D}{D_0} \tag{12.35}$$

where σ_{Nc} is the nominal strength, D is the depth of the beam, f_t is the tensile strength of the material, and B and D_0 are parameters of the model. These parameters are determined from the failure data for the beams using a nonlinear regression analysis. Four series of tests corresponding to nominal loading rates t_p of 1.4, 500, 13,650, and 253,000 s, which represent time to peak load, were conducted. In the first three series, three replicate tests were performed for

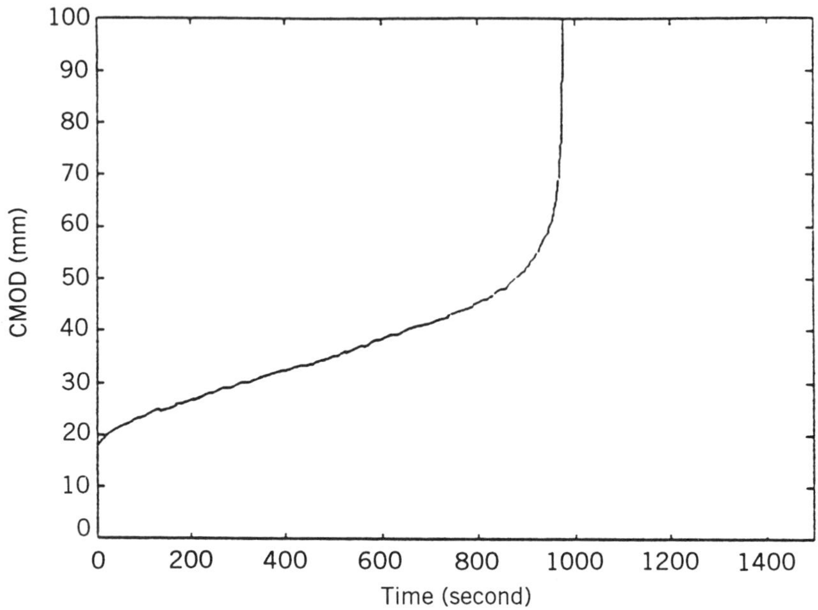

Figure 12-33 CMOD–time curve under sustained load level, $P_k/P_f = 0.80$.[48]

each beam size, or nine total tests for each loading rate. For the last series, only one beam of each size was tested. The average values of test results are presented in Table 12-7. All the results are normalized and plotted in Fig. 12-35. It is seen that the data are represented rather well by Eq. (12.35). Further, it is seen that the beams tested at very low load rates behave in accordance with LEFM; that is, the fracture is brittle. The implication then is that the fracture process zone is reduced in size as the loading rate decreases.

The size effect model was also used to determine other fracture parameters (see Chapter 5):

$$G_f = \frac{B^2 f_t^2 D_0}{E_{\text{eff}}}\, g\left(\frac{a_0}{D}\right)$$

$$K_f = B f_t \sqrt{D_0 g(a_0/D)} \tag{12.36}$$

$$c_f = \frac{D_0 g(a_0/D)}{g'(a_0/D)}$$

and

$$\delta_f = \frac{8K_f}{E_{\text{eff}}}\sqrt{\frac{c_f}{2\pi}} \tag{12.37}$$

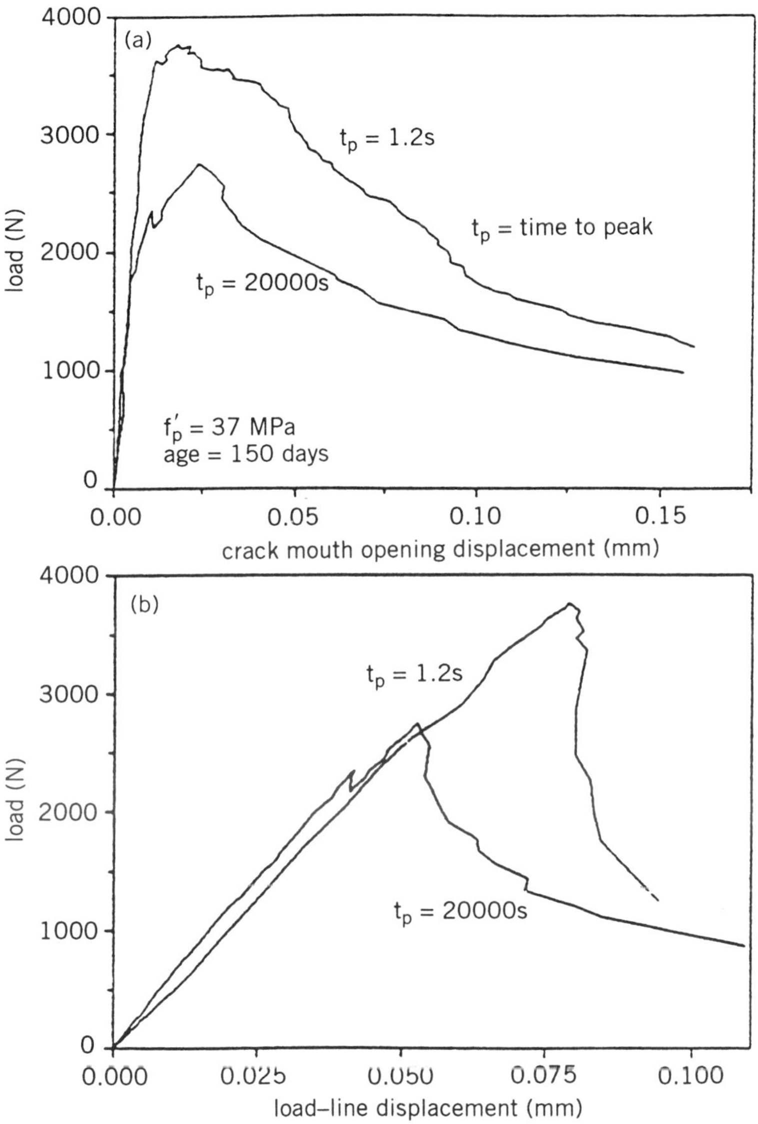

Figure 12-34 Rate effects on (*a*) load–CMOD response and (*b*) load–LPD response for three-point bend beams.[49]

where G_f is the critical energy release rate, K_f is the critical stress intensity factor, $g(a_0/D)$ is the LEFM geometry function, c_f is the effective length of the fracture process zone, and δ_f is the effective CTOD. The terms G_f, K_f, c_f, and δ_f are defined for an infinitely large size specimen. It is noted that an effective modulus of elasticity E_{eff} is used in place of E to account for creep

Table 12-7 Material Fracture Parameters from Size Effect Model[a]

Series	Average t_p (s)	Average Age (days)	f'_c (MPa)	Bf_t (MPa)	D_0 (mm)	K_f, (MPa mm$^{1/2}$)	c_f (mm)	E_{eff} (GPa)	G_f (N/m)	δ_f (mm)
Fast	1.4	28	36.6	1.60	102.5	39.5	17.2	36.0	43.4	0.0146
Usual	500	28	36.5	1.68	4.13	26.3	6.9	28.6	24.1	0.0077
Slow	13,650	38	37.2	2.94	13.3	26.1	2.2	24.1	28.4	0.0052
Very slow	253,000	106	36.9	3.47	8.5	24.6	1.4	22.4	26.9	0.0042

[a]From ref. 49.

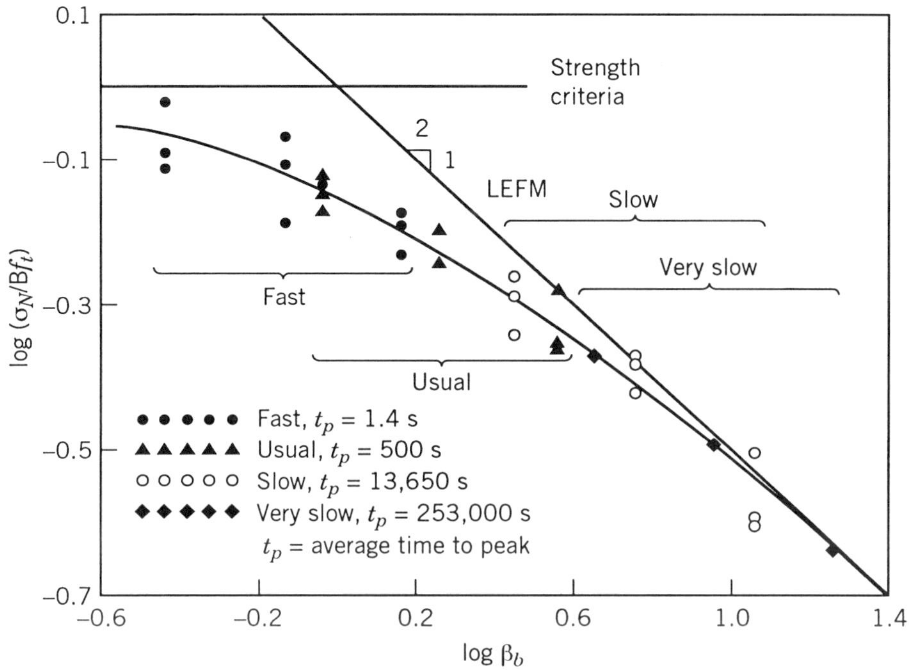

Figure 12-35 Change of failure mode with loading rate.[49]

effects in a simple way. The value of E_{eff} uses a creep function along with the initial value of E and time to peak load t_p.

All the results are presented in Table 12-7. From this it is clear that both G_f and K_f decrease from values at a high load rate to approximately constant values at usual to very long load rates. This is also true even if the effect of age of loading on values of E_{eff} and f'_c are considered. Of significant interest is that the effective length of the fracture process zone, c_f, decreases considerably as the loading rate proceeds to very slow. It is speculated that "the decrease in c_f is probably due to the relaxation of the high stresses in the material ahead of the crack tip, causing the stress drop to be more concentrated."[49]

Their conclusions include the following:

1. The size effect model agrees with the beam fracture test results for loading rates with time to peak loads varying from 1 to 250,000 s.
2. These test results show that a decrease in loading rate results in a higher brittleness in fracture behavior and closer correspondence to LEFM.
3. The fracture toughness, effective length of the fracture process zone, and effective critical CTOD all decrease with an increase in the time to peak load.
4. There is a strong interaction between fracture and creep in concrete. This

is very important for both failure and serviceability analysis of structures.

12.3.3 Summary

The studies presented in this section all dealt with time-dependent (creep) effects on fracture of concrete. They may be broadly classified in two ways. In one approach, the load is applied to the beam in a relatively short time up to some fraction of an estimated short-term failure load. The load is then held constant until failure occurs. Measured parameters included displacements, time to failure, and fracture energy (G_F). The length of time to failure varied from 15 s to 86 days. Under the conditions of constant load to failure, Hansen[47] found an increase in G_F whereas Zhou[48] found a decrease.

The other approach was to apply the load slowly to failure and measure fracture energy. This was done by Zhou[48] using a measured area under the load–displacement curve, whereas Bazant and Gettu[49] tested beams of different sizes and used the size effect model to obtain G_f with a modification for time dependency of the modulus of elasticity. The time of loading varied from 5 to 5000 s for Zhou[48] and 1 to 250,000 s for Bazant and Gettu.[49] Both investigations showed the fracture energy decreased with time. For long enough times this appears to become constant.[49]

Probably the most interesting result is the apparent increase in brittleness and agreement with LEFM with an increase in time.[49]

REFERENCES

1. Mobasher, B., Stang, H., and Shah, S. P., "Microcrack in Fiber Reinforced Concrete," *Cement and Concrete Research*, Vol. 20, 1990, pp. 665–676.
2. Somayaji, S., and Shah, S. P., "Bond Stress Versus Slip Relationship and Cracking Response of Tension Members," *Journal of American Concrete Institute*, Vol. 78, 1981, pp. 279–313.
3. Mobasher, B., and Shah, S. P., "Interaction between Fibers and Cement Matrix in Glass Fiber Reinforced Concrete," in *Thin-Section Fiber Reinforced Concrete and Ferrocement*, SP-124, American Concrete Institute, Detroit, MI, 1990, pp. 136–150.
4. Stang, H., Mobasher, B., and Shah, S. P., "Quantitative Damage Characterization in Polypropylene Fiber Reinforced Concrete," *Cement and Concrete Research*, Vol. 20, 1990, pp. 540–558.
5. Mobasher, B., Castro-Montero, A., and Shah, S. P., "A Study of Fracture in Fiber Reinforced Cement-Based Composites Using Laser Holographic Interferometry," *Experimental Mechanics*, Vol. 30, 1990, pp. 286–294.
6. Aveston, J., Cooper, G. A., and Kelly, A., "Single and Multiple Fracture in the Properties of Fiber Composites," in *Conference Proceedings of the National*

Physical Laboratory, IPC Science and Technology, Surrey, England, 1971, pp. 15–26.

7. Shah, S. P., and Ouyang, C., "Mechanical Behavior of Fiber-Reinforced Cement-Based Composites," *Journal of the American Ceramic Society*, Vol. 74, No. 11, 1991, pp. 2727–2738, 2947–2953.
8. Bartos, P., "Review Paper: Bond in Fiber Reinforced Cements and Concretes," *International Journal of Cement Composites and Lightweight Concrete*, Vol. 3, 1981, pp. 159–177.
9. Gray, R. J., "Analysis of the Effect of Embedded Fibre Length on Fibre Debonding and Pull-Out from an Elastic Matrix, Part I, Review of Theory," *Journal of Materials Science*, Vol. 19, 1984, pp. 861–870.
10. Lawrence, P., "Some Theoretical Considerations of Fibre Pull-Out from an Elastic Matrix," *Journal of Material Science*, Vol. 7, 1972, pp. 1–6.
11. Gopalaratnam, V. S., and Shah, S. P., "Tensile Failure of Steel-Fiber-Reinforced Mortar," *Journal of Engineering Mechanics Division, ASCE*, Vol. 113, No. 5, 1987, pp. 635–652.
12. Stang, H., Li, Z., and Shah, S. P., "Pullout Problem: Stress versus Fracture Mechanical Approach," *Journal of Engineering Mechanics*, Vol. 116, No. 10, 1990, pp. 2136–2150.
13. Li, Z., Mobasher, B., and Shah, S. P., "Characterization of Interfacial Properties in Fiber Reinforced Cementitious Composites," *Journal of the American Ceramic Society*, Vol. 74, No. 9, 1991, pp. 2156–2164.
14. Ouyang, C., Alvarez, A. P., and Shah, S. P., "Pull-out of Aligned and Inclined Fibers from Cement-Based Matrices," *Journal of Engineering Mechanics*, ASCE, Vol. 120, No. 12, 2641–2659, Dec. 1994.
15. Shah, S. P., and Ouyang, C., "Mechanical Behavior of Fiber Reinforced Cement-Based Composite," *Journal of the American Ceramic Society*, Vol. 74, No. 11, 1991, pp. 2727–2738, 2947–2953.
16. Somayaji, S., and Shah, S. P., "Bond Stress Versus Slip Relationship and Cracking Response of Tension Members," *ACI Journal*, Proceedings, Vol. 78, No. 3, 1981, pp. 217–225.
17. Ouyang, C., and Shah, S. P., "Geometry Dependent *R*-Curve for Quasi-Brittle Materials," *Journal of the American Ceramic Society*, Vol. 74, No. 11, 1991, pp. 2831–2836.
18. Morton, J., and Groves, G. W., "The Cracking of Composites Consisting of Discontinuous Ductile Fibers in a Brittle Matrix—Effect of Fiber Orientation," *Journal of Materials Science*, Vol. 9, 1974, pp. 1439–1445.
19. Brandt, A. M., "On the Optimal Direction of Short Metal Fibers in Brittle Matrix Composites," *Journal of Materials Science*, Vol. 20, 1985, 3835–3841.
20. Li, V. C., Wang, Y., and Backer, S., "Effect of Inclining Angle, Bundling, and Surface Treatment on Synthetic Fiber Pull-Out from a Cement Matrix," *Composites*, Vol. 21, No. 2, 1990, pp. 132–140.
21. Pacios, A., Ouyang, C., and Shah, S. P., "Rate Effect on Interfacial Response between Fibers and Matrix," *Materials and Structures, RILEM*, Vol. 28, 1995, pp. 83–91.
22. Mobasher, B., Ouyang, C., and Shah, S. P., "Modeling of Fiber Toughening in

Cementitious Materials Using an *R*-Curve Approach,'' *International Journal of Fracture*, Vol. 50, 1991, pp. 199–219.

23. Ouyang, C., and Shah, S. P., ''Toughening of High Strength Cementitious Matrix Reinforced by Discontinuous Short Fibers,'' *Cement and Concrete Research*, Vol. 22, 1992, pp. 1201–1215.
24. Shah, S. P., Ed., *Fatigue of Concrete Structures*, ACI SP-75, American Concrete Institute, Detroit, MI, 1982.
25. ACI Committee 215, ''Consideration for Design of Concrete Structures Subjected to Fatigue Loading,'' *ACI Journal*, Vol. 71, No. 3, 1974, pp. 97–121.
26. Swartz, S. E., and Noory, A. K., ''Photoelastic Coatings to Monitor Crack Growth in Concrete,'' *Experimental Techniques*, Vol. 5, No. 3, 1981, pp. 2-3.
27. Swartz, S. E., ''Concrete Fatigue and Fracture: Plain and Reinforced Beams,'' in *Fatigue and Fracture in Steel and Concrete Structures*, edited by Madhava Rao, A. G., and Appa Rao, T. V. S. R., Madras, India, 1991, pp. 407–421.
28. Oh, B. H., ''Fatigue-Life Distributions of Concrete for Various Stress Levels,'' *ACI Materials Journal*, Vol. 88, No. 2, 1991, pp. 122–128.
29. Alliche, A., and François, D., ''Fatigue Damage of Concrete,'' in *Fracture of Concrete and Rock*, edited by Shah, S. P., and Swartz, S. E., Springer-Verlag, New York, 1989, pp. 88–95.
30. Hordijk, D. K., and Reinhardt, H. W., ''Growth of Discrete Cracks in Concrete under Fatigue Loading,'' in *Toughening Mechanisms in Quasi-Brittle Materials*, edited by Shah, S. P., Kluwer Academic, Dordrecht, 1991, pp. 541–554.
31. Deborst, R., Kusters, G. M. A., Nauta, P., and de Witte, F. C., ''DIANA—A Comprehensive But Flexible Finite Element System,'' in *Finite Element Systems Handbook*, 3rd ed., Springer-Verlag, Berlin, 1983.
32. Zhang, B., Zhu, Z., and Wu, K., ''Fatigue Rupture of Plain Concrete Analyzed by Fracture Mechanics,'' in *Fracture of Concrete and Rock*, edited by Shah, S. P., and Swartz, S. E., Springer-Verlag, New York, 1989, pp. 58–63.
33. Paris, P. C., ''The Growth of Fatigue Cracks Due to Variation in Load,'' Ph.D. Dissertation, Lehigh University, Bethlehem, PA, 1962.
34. Paris, P. C., and Erdogan, F., ''A Critical Analysis of Crack Propagation Laws,'' *Journal of Basic Engineering, ASME*, Vol. 85, 1963, pp. 528–534.
35. Baluch, M. A., Qureshy, A. B., and Azad, A. K., ''Fatigue Crack Propagation in Plain Concrete,'' in *Fracture of Concrete and Rock*, edited by Shah, S. P., and Swartz, S. E., Springer-Verlag, New York, 1989, pp. 80–87.
36. Swartz, S. E., and Refai, T. M. E., ''Cracked Surface Revealed by Dye and Its Utility in Determining Fracture Parameters,'' in *Fracture Toughness and Fracture Energy*—Test Methods for Concrete and Rock, edited by Mihashi, H., Takahashi, H., and Wittmann, F. H., Balkema, Rotterdam, 1989, pp. 509–520.
37. Jenq, Y. S., and Shah, S. P., ''Two-Parameter Fracture Model for Concrete,'' *Journal of Engineering Mechanics, ASCE*, Vol. 111, No. 10, 1985, pp. 1227–1241.
38. Perdikaris, P. C., and Calomino, A. M., ''Kinetics of Crack Growth in Plain Concrete,'' in *Fracture of Concrete and Rock*, edited by Shah, S. P., and Swartz, S. E., Springer-Verlag, New York, 1989, pp. 64–69.
39. Perdikaris, P. C., Calomino, A. M., and Chudnovsky, A., ''Effect of Fatigue on

Fracture Toughness of Concrete,'' *Journal of Engineering Mechanics, ASCE*, Vol. 112, No. 8, 1986, pp. 776–790.

40. Bazant, Z. P., and Xu, K., ''Size Effect in Fatigue Fracture of Concrete,'' *ACI Materials Journal*, Vol. 88, No. 4, 1991, pp. 390–399.
41. Swartz, S. E., Huang, C.-M. J. and Hu, K.-K., ''Crack Growth and Fracture in Plain Concrete—Static versus Fatigue Loading,'' in *Fatigue of Concrete Structures*, ACI SP-75, edited by Shah, S. P., American Concrete Institute, Detroit, MI, 1982, pp. 47–69.
42. Swartz, S. E., Hu, K.-K., and Jones, G. L., ''Compliance Monitoring of Crack Growth in Concrete,'' *Journal of the Engineering Mechanics Division, ASCE*, Vol. 104, No. EM4, 1978, pp. 789–800.
43. Bazant, Z. P., Ed., *Proceedings of Fourth RILEM International Symposium on Creep and Shrinkage of Concrete: Mathematical Modeling*, Northwestern University, Evanston, IL, August 26–29, 1986.
44. Liu, Z.-G., Swartz, S. E., Hu, K.-K., and Kan, Y.-C., ''Time-Dependent Response and Fracture of Plain Concrete Beams,'' in *Fracture of Concrete and Rock: Recent Developments*, edited by Shah, S. P., Swartz, S. E., and Barr, B., Elsevier Applied Science, London, 1989, pp. 577–586.
45. Yuan, L.-W., Liu, Z.-G., Zhou, Z.-B., and Li, Z.-D., ''Rheological Fracture Approach to the Concrete Dam on Zhexi Hydropower Station in Hunan,'' in *Proceedings SEM-RILEM International Conference on Fracture of Concrete and Rock*, edited by Shah, S. P., and Swartz, S. E., Houston, TX, June 17–19, 1987, p. 388.
46. Linsbauer, H. N., Ingraffea, A. R., Rossmanith, H. P., and Wawrzynek, P. A., ''Simulation of Cracking in Large Arch Dam: Part I, Part II,'' *Journal of Structural Engineering, ASCE*, Vol. 115, No. 7, 1989, pp. 1599–1630.
47. Hansen, E. A., ''Influence of Sustained Load on the Fracture Energy and the Fracture Zone of Concrete,'' in *Fracture Processes in Concrete, Rock and Ceramics*, edited by van Mier, J. G. M., Rots, J. G., and Bakker, A., E & F Spon, London, 1991, pp. 829–838.
48. Zhou, F. P., ''Time Dependent Crack Growth and Fracture in Concrete,'' Ph.D. Dissertation, Report TVBM-1011, Lund Institute of Technology, Lund, Sweden, 1992.
49. Bazant, Z. P., and Gettu, R., ''Rate Effects and Load Relaxation in Static Fracture of Concrete,'' *ACI Materials Journal*, Vol. 89, No. 5, 1992, pp. 456–468.

PROBLEMS

Chapter 1

1 Describe the classical questions addressed by fracture mechanics.

2 What is the meaning of the term *notch sensitivity*?

3 Why is it that linear elastic fracture mechanics cannot be directly used for fracture of quasi-brittle materials?

4 Using the Euler–Bernoulli principle, derive Eqs. (1.8)–(1.11).

5 Verify the results of the cracked beam shown in Fig. 1-9. How do you suppose Eqs. (1.8)–(1.11) could be modified to approximate cracking conditions more accurately?

6 Derive Eqs. (1.14), (1.15), and (1.17).

7 What are problems associated with the currently used service state design methods for reinforced concrete structures?

8 What are major problems associated with the currently used strength state design methods for reinforced concrete structures?

Chapter 2

1 Derive Eq. (2.16) by separating the real and imaginary parts of Eq. (2.15).

2 An infinite-size plate as shown in Fig. P2-2 is subjected to a far-field stress $\sigma = 50$ MPa. The plate contains a line crack (slit) of length $2a = 8$ mm. Calculate and plot the distribution of the stress component σ_y along the x axis for $r < a$.

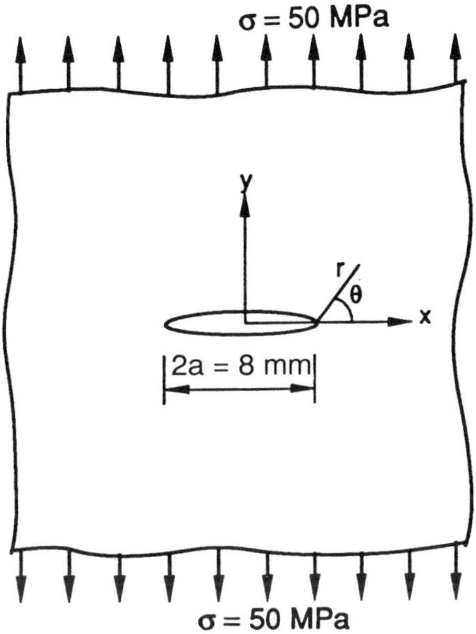

Figure P2-2 Infinite-size plate with crack subjected to tension.

3 A unit-thickness finite-size plate is subjected to a far-field stress $\sigma = 20$ MPa. (i) If the plate has a width $2b = 40$ mm and contains a center crack of length $2a = 6$ mm as shown in Fig. P2-3a, calculate the stress intensity factor at the crack tip. (ii) If the plate has a width $b = 20$ mm and contains a single-edge crack of length $a = 3$ mm as shown in Fig. P2-3b, calculate

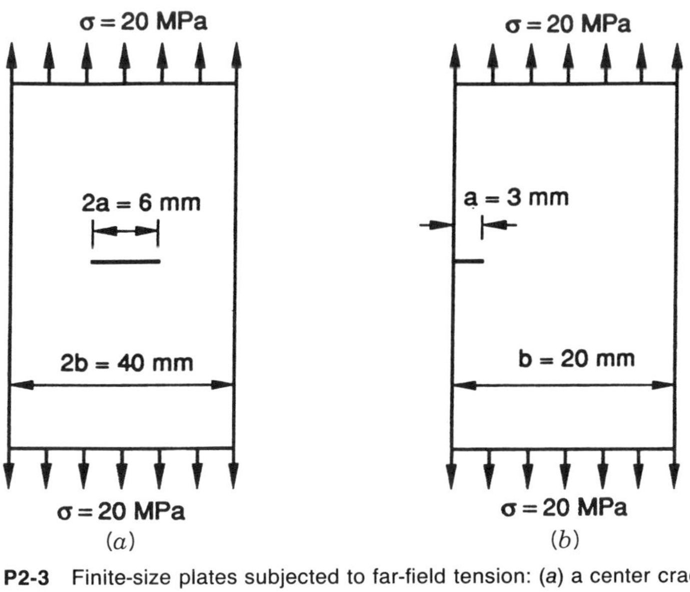

Figure P2-3 Finite-size plates subjected to far-field tension: (a) a center cracked-plate and (b) a single edge cracked-plate.

the stress intensity factor and compare both values of the stress intensity factor. (iii) The material has a yield stress $\sigma_{sy} = 420$ MPa. Calculate the size of the plastic zone for the single-edge crack plate using Irwin's second-order estimation and the Dugdale estimation.

4 For the infinite-size specimen given in Fig. P2-4, derive the stress intensity factor at the crack tip using the superposition principle.

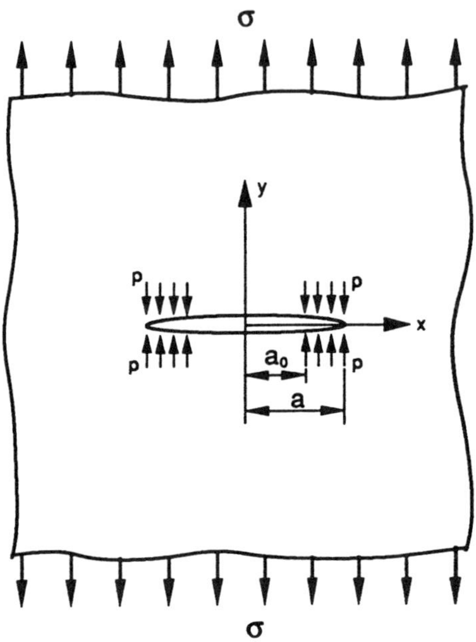

Figure P2-4 Infinite-size plate with center crack.

5 A double-cantilever beam specimen as shown in Fig. P2-5 is often used to measure fracture toughness of a material. Derive the strain energy release rate of the specimen using the compliance approach [Eq. (2.65)]. (*Hint:* The deflection for a cantilever beam is given by $u = 4Pa^3/Eb^3t$).

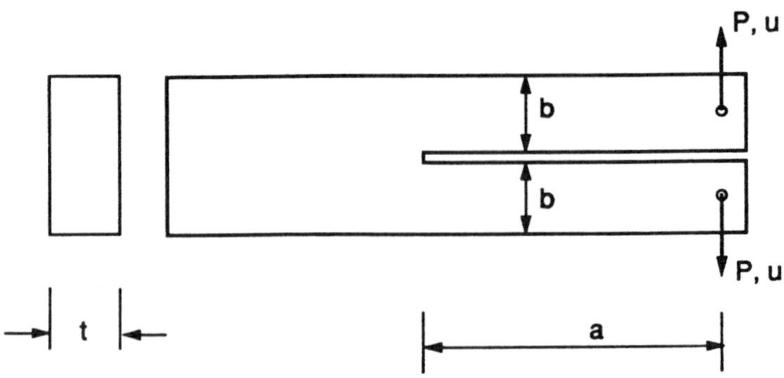

Figure P2-5 Double-cantilever beam specimen.

6 A three-point bend beam is made of a steel with a yield stress $\sigma_{sy} = 600$ MPa and a critical stress intensity factor $K_{Ic} = 80$ MPa·mm$^{1/2}$. The rectangular cross-sectional beam has a span $S = 800$ mm, a depth $b = 200$ mm, and a thickness $t = 50$ mm. The safety factor F is assumed to be 2.0. (i) By assuming that the critical section is the midspan, determine the maximum loading capacity (the concentrated load P) of the beam based on the stress design criterion. (ii) If the material contains a 10-mm-long through-thickness initial flaw, determine the maximum loading capacity based on the fracture mechanics design criterion. For the most unsafe case, it may be assumed that the initial flaw is located in the middle bottom of the beam.

Chapter 3

1 A crack of length $2a$ is located in a thin elastic plate as shown in Fig. P3-1, where $L \gg a$. The boundary conditions of the plate are (i) $u_y = \Delta u$ and $u_x = 0$ for $y = +h$ and (ii) $u_y = -\Delta u$ and $u_x = 0$ for $y = -h$. Explain, if possible, how to use the J-integral for computation of K_I.

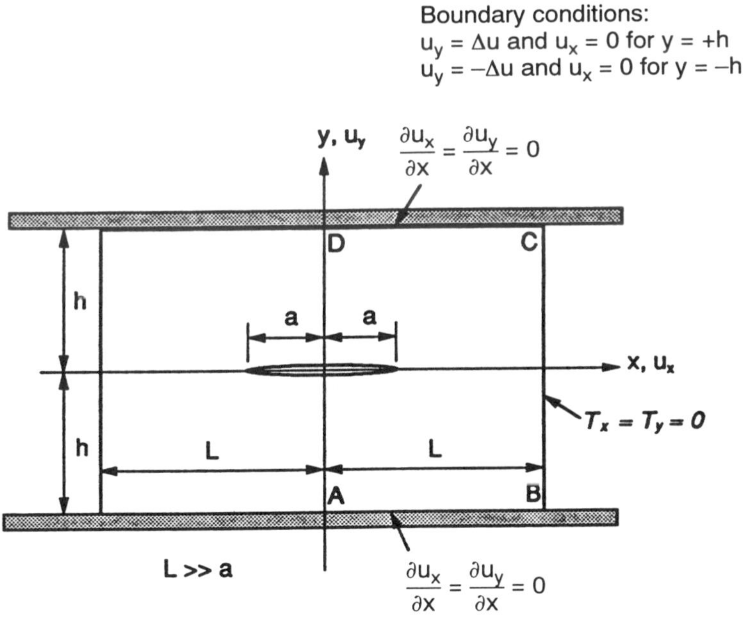

Figure P3-1 Cracked plate with boundary movement.

2 An elastic plate of unit thickness (plane-stress condition) is fixed along its upper and lower boundaries as shown in Fig. P3-2. A horizontal crack extends from one side through the thickness of the plate due to a constant movement u_0 at the upper boundary. Both $a >> H$ and $b >> H$ are as-

sumed. Evaluate the J-integral of the plate. [*Hint:* The only far-field, nonzero stress and strain components in the uncracked portion of the plate are $\epsilon_y = u_0/h$ and $\sigma_y = E\epsilon_y/(1 - \mu^2)$.]

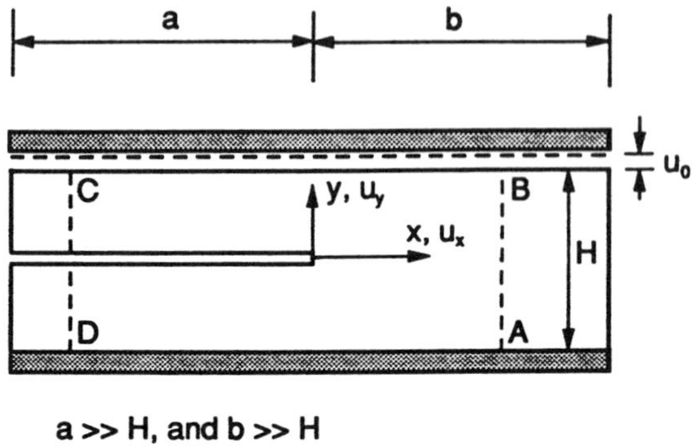

Figure P3-2 Calculation of J-integral for plate with boundary movement.

3 An infinite-size plate with a center crack of size a_0 is subjected to a far-field tensile stress σ. The plate is made of a material with an R-curve of the form $R = A(a - a_0)^B$, where A and B are two constants. Determine the critical crack length and the corresponding load.

4 A three-point bend beam is made of a steel with $\sigma_{sy} = 2500$ MPa, $\kappa_p = 0.3$, E = 210,000 MPa, and $\text{CTOD}_c = 0.0002$ mm. The rectangular cross-sectional beam has a span $S = 800$ mm, a depth $b = 200$ mm, a thickness $t = 50$ mm, and $a_0 = 10$ mm. If it is assumed that $\text{CMOD}_p = 0.15$ CMOD, determine the load for initiation of the crack using Eq. (3.15).

Chapter 4

1 Briefly describe internal porosities and cracks in cementitious materials prior to loading.

2 Briefly describe relationships between crack propagation and mechanical response for concrete under uniaxial tension.

3 What is strain localization and when does it occur? Which approach can be used to describe strain localization.

4 What are the major toughening mechanisms in the fracture process zone?

5 In some studies, researchers used an optical device for measuring the length of the surface crack. Then, this surface crack length was used for calculating fracture toughness. May this method be applied to concrete?

6 What are three distinctive phenomena of the mechanical behavior of quasi-brittle materials?

7 A series of three-point bend beams have been tested for measuring fracture toughness of a concrete. The beams have different dimension but are made of the same concrete. Dimensions and measured peak loads of each beam are listed in Table P4-7. From information given in the table, use linear elastic fracture mechanics to calculate the critical stress intensity factor of the concrete. The initial notch lengths are used as the critical crack lengths for calculation. What may you conclude from these calculations?

Table P4-7 Dimensions and Measured Peak Loads of Various Concrete Beams

S (mm)	b (mm)	t (mm)	a_0 (mm)	P_c (N)
320	80	30	24	820
640	160	60	48	2540
960	240	90	72	5750

8 Plain, concrete beams of dimensions $t = 76.2$ mm, $b = 305$ mm, and $S = 1143$ mm are tested in three-point bending with critical crack lengths (midspan) of 101.6, 152.4, and 203.2 mm. If $K_{Ic} = 1320$ $\text{kN} \cdot \text{m}^{3/2}$ for this concrete, determine the following if self-weight is neglected.

a. expected peak load for each beam

b. corresponding tensile strength σ_c at the net section

9 For plain concrete beams of width $t = 76.2$ mm and the following depths, spans are tested in three-point bending. Each beam is cracked at midspan to a critical length equal to one-half the depth. If $K_{Ic} = 1320$ $\text{kN} \cdot \text{m}^{3/2}$ determine P_c and σ_c (net section) for each beam. Neglect self-weight.

b, mm	S, mm
203	762
305	1143
406	1524
508	1905

Chapter 5

1 What are some major difficulties of applying linear elastic fracture mechanics to concrete? Why does one need to use an effective crack length to model fracture of concrete?

2 Briefly describe types of energy dissipation mechanisms included in Eqs. (5.1) and (5.3) and the physical meaning of each energy dissipation mechanism.

3 A concrete double-cantilever beam with a fracture process zone is shown in Fig. P5-3. Assuming that the crack tip opening displacement w_t exceeds the value of w_c during the softening stage, (i) determine the softening load–crack length relationship by the Hillerborg model using a linear stress–separation curve and (ii) determine the softening load–crack length relationship by the crack band model. (*Hint:* It is known from Problem 5 of Chapter 2 that $G_q = 12P^2a^2/Eb^3t^2$ for the double-cantilever beam.)

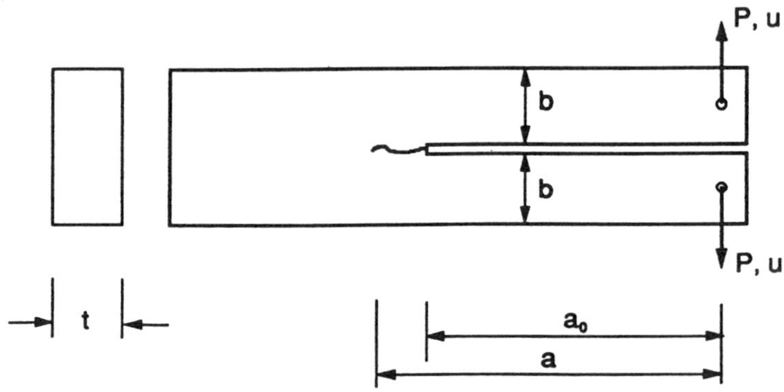

Figure P5-3 Double-cantilever beam specimen with crack extension.

4 Determine the peak load and corresponding crack length for the double-cantilever beam shown in Fig. P5-3, where a_0 = 800 mm, b = 80 mm, t = 40 mm, K_{Ic}^s = 31.62 MPa·mm$^{1/2}$, and CTOD$_c$ = 0.015 mm. [*Hint:* For the double-cantilever beam CTOD = $4P(a_0^3 - 3a^2a_0 + 2a^3)/Eb^3t$.]

5 Briefly describe advantages and disadvantages of the discrete crack approach and the smear crack approach.

Chapter 6

1 A three-point bend beam was tested to measure the value of G_F. The beam was loaded downward as shown in Fig. 6-1. The dimensions of the beam are S = 800 mm, b = 100 mm, and t = 100 mm. The initial notch of the beam, a_0, was equal to 50 mm. The values of W_0 = 650 N·mm and δ_0 = 1.5 mm were measured. The weight of concrete is 2300 kg/m^3. Calculate the value of G_F based on the above given results.

2 A three-point bend beam was tested to measure values of E, K_{Ic}^s, and CTOD$_c$. The dimensions of the beam were S = 600 mm, b = 150 mm, and t = 80 mm. An initial notch of a_0 = 50 mm was used. The values of C_i = 0.000007 mm/N, C_u = 0.000009 mm/N, and P_c = 3500 N were

measured. The weight of the concrete is 2300 kg/m^3. Assuming $HO = 0$, determine the values of E, K_{Ic}^s, and CTOD$_c$ based on the two-parameter fracture model.

3 Three different sizes of geometrically similar beams were tested to measure the value of G_f for a concrete. Three identical specimens were tested for each beam size. All beams had the same thickness of 40 mm and a span–depth ratio of 4. The value of $a_0/b = 0.25$ was used for all beams. The beams were loaded under three-point bending. The obtained maximum loads and size of the beams are listed in Table P6-3. The value of $E =$ 30,000 MPa for the concrete was separately measured. Determine the values of G_f and c_f based on the above given data.

Table P6-3 Maximum Loads Measured

Depth, b (mm)	Maximum Loads, P^0 (N) 1	2	3
50	2410	2450	2480
100	3630	3540	3500
200	5250	5310	5230

4 A three-point bend beam was tested to measure the values of K_{Ic}^e. The dimensions of the beam are $S = 600$ mm, $b = 150$ mm, and $t = 80$ mm. The maximum size of aggregate is 3 mm. An initial notch of $a_0 = 50$ mm was used. The values of $P_c = 3500$ N and $E = 27{,}300$ MPa were measured. The weight of the concrete is 2300 kg/m^3. Determine the value of K_{Ic}^e using the effective crack model of Karihaloo and Nallathambi.

Chapter 7

1 What is a disadvantage if one measures an R-curve based on the concept of stable crack propagation?

2 A material has properties of $m = 1$, $d_m = 0.1$ mm, $E = 30{,}000$ MPa, $K_{Ic} = 0.015$ MPa mm$^{1/2}$, $w_c = 0.0015$ mm, and $f_t = 4$ MPa. Determine the K_R-curve using the Mai et al. simplified model.

3 Prove that Eq. (7.30) can be solved from Eq. (7.26) for $n = 1$.

4 A three-point bend beam has dimensions of $S = 600$ mm, $b = 150$ mm, and $t = 80$ mm. The initial notch a_0 is 50 mm. Values of $E = 26{,}280$ MPa, $K_{Ic}^s = 27.55$ MPa$\cdot$mm$^{1/2}$, and CTOD$_c$ = 0.0095 mm have been found. Determine R-curves for the beam using the Ouyang and Shah model.

5 A series of geometrically similar three-point bend beams were made of a concrete with $G_f = 0.869$ N/mm and $c_f = 2.670$ mm. Each beam has

dimensions of $S = 600$ mm and $D = 150$ mm. The value of a_0 is 50 mm. The value of $E = 30{,}000$ MPa was separately measured. Determine the R-curve using the Bazant and Kazemi model.

Chapter 8

1 Derive Eq. (8.2).

2 Derive Eqs. (8.3)–(8.5).

3 Explain the difference in philosophy between the maximum circumferential stress criterion and the maximum circumferential strain criterion. Will these two criterion yield the same results?

4 Using the Mohr–Coulomb criterion [Eq. (8.25)] determine C_1, C_2 in terms of $K_{\mathrm{I}c}$ or $K_{\mathrm{II}c}$, or a combination.

5 The Iosipescu test geometry shown schematically in Figs. 8-20 and 8-21 is purported to give true mode II crack propagation failures for certain materials. What are important differences between this test geometry and those used by various investigators for concrete (Figs. 8-22, 8-29, 8-30, 8-35, 8-36, and 8-37)?

6 Describe mode II crack propagation behavior. How should $K_{\mathrm{II}c}$ be defined? How can it be measured?

Chapter 9

1 What are major factors that influence the failure mode of reinforced concrete structural members?

2 What are major problems associated with current conventional design methods for reinforced concrete structures?

3 Three different widths (D) of uniaxial tensile specimens made of the same concrete mix were tested to determine the strength of the concrete. The obtained nominal strengths were $\sigma_{Nc} = 3.7$ MPa for $D = 100$ mm, $\sigma_{Nc} = 3.4$ MPa for $D = 200$ mm, and $\sigma_{Nc} = 3.1$ MPa for $D = 300$ mm. The value of f_t is 4.1 MPa. (i) Determine constants B and D_0 for the size effect equation given by Eq. (9.6). (ii) Calculate the value of σ_{Nc} for $D = 150$ mm based on the obtained size effect equation. [*Hint:* Equation (9.6) can be rewritten as $Y = A_B D + C_B$, where $Y = (f_t/\sigma_{Nc})^2$, $A_B = 1/B^2 D_0$, and $C_B = 1/B^2$. Two coefficients A_B and C_B can be obtained from the slope and Y intercept of the linear regression curve based on the given data.]

4 A reinforced concrete tensile member has dimensions of $L = 763$ mm, $b = 305$ mm, and $t = 178$ mm. The member is reinforced by eight deformed steel bars with a diameter of $r_s = 5.65$ mm ($\rho = 0.015$). Values of $E_s = 210{,}000$ MPa, $E = 33{,}300$ MPa, $k = 0.25$, and $w_c = 0.12$ mm are used.

Values of $R_{\mathrm{I}cf} = 0.018$ N/mm, $a_{cf} = 17.2$ mm, and $f_{tf} = 2.76$ MPa have been determined from an R-curve analysis. Values of $\gamma_d = 0.004$ N/mm and $\tau_s = 2.5$ MPa are assumed. Determine the values of the concrete stress σ_c, the steel stress σ_s, the applied stress σ, and corresponding crack width w_0 for the number of cracks $N = 4$.

5 A reinforced concrete tensile member has dimensions of $L = 763$ mm, $b = 305$ mm, and $t = 178$ mm. Grade 40 steel bars used have the values of $E_s = 210{,}000$ MPa and $f_y = 276$ MPa. The concrete used has the value of $E = 33{,}300$ MPa. The values of $R_{\mathrm{I}cf} = 0.018$ N/mm, $a_{cf} = 17.2$ mm, and $f_{tf} = 2.76$ MPa have been determined from an R-curve analysis. Determine the minimum reinforcement ratio for the given tensile member.

6 A concrete beam has a cross section of $b = 200$ mm and $t = 120$ mm. The beam is made of a concrete with $G_F = (K_{\mathrm{I}c})^2/E = 0.16$ N/mm, $f_t = 4$ MPa, and $E = 25{,}000$ MPa. A relationship of $f_t = 0.556(f_c)^{1/2}$ (f_c in megapascals) is used. A linear softening stress–separation curve is assumed. A grade 40 steel with $f_y = 276$ MPa is used. Determine the minimum reinforcement ratio of the beam using Eqs. (9.53) and (9.54), respectively.

7 A concrete beam reinforced for bending but not for shear is tested in the laboratory with a central, concentrated load. The following results were obtained: span = 1829 mm, $b = 305$ mm, $b_1 = 254$ mm, $t = 152$ mm, $f'_c = 41.3$ MPa, $K_{\mathrm{I}c} = 1320$ kN/mm$^{3/2}$, $E = 31$ GPa, $G_F = 8.75$ N/m, and failure load = 82.8 KN (shear failure). What would be the expected failure load for a beam with $t = 457$ mm, $b_1 = 1143$ mm, and $b = 1219$ mm?

(i) Method of Hillerborg and Gustafson using G_F and $G_{\mathrm{I}c}$.

(ii) Method based on the size effect model.

(iii) Method based on Jenq and Shah model.

(iv) ACI Code using $v_c = 2bt(f'_c)^{1/2}$.

Chapter 10

1 Briefly describe the failure process of a uniaxial compressive concrete specimen.

2 Describe differences in failure mechanisms for both uniaxial tensile and compressive failures of a square concrete plate.

3 A concrete plate with two aggregates is shown in Fig. 10-12. An initial crack AB has a length of $2L_1 = 20$ mm and an inclination angle of $\theta_k = 45°$. Values of $K^M_{\mathrm{I}c} = 20$ MPa mm$^{1/2}$, $K^{\mathrm{IF}}_{\mathrm{II}c} = 20$ MPa mm$^{1/2}$, $f = 0.2$ and, $\theta_a = 50°$ are given. (i) Determine the relationship between the applied stress q and the crack length l_2 and (ii) if assuming the crack propagates as mode II when it meets with the second aggregate, determine the relationship between the applied stress q and the crack length l_2.

4 A concrete plate with an inclined crack with kinks is shown in Fig. 10-14. Values of $q = -10$ MPa, $\theta_k = 40°$, $\kappa = 30°$, $f = 0.15$, $a = 20$ mm, and $l_2 = 3$ mm are known. Determine values of $K_{\mathrm{I}}^{\mathrm{kink}}$ and $K_{\mathrm{II}}^{\mathrm{kink}}$ using (i) Eqs. (10.14) and (10.15) and (ii) Eqs. (10.16) and (10.17), respectively.

Chapter 11

1 What are the different roles of damage mechanics and fracture mechanics in describing the mechanical behavior of concrete.

2 Values of $\sigma_c = 4$ MPa, $\epsilon_c = 0.0008$, and $E_0 = 25{,}000$ MPa were measured from a concrete uniaxial tensile specimen. The value of $\Omega_0 = 0.2$ was measured. Determine the prepeak stress–strain relationship based on Løland's model.

3 An isotropic and homogeneous plate under the plane-stress condition contains N linear cracks (slits) with an average length of $2a$ per unit area. The cracks are randomly and isolatedly distributed. Stress intensity factors for the plate are

$$K_{\mathrm{I}} = \sigma\sqrt{\pi a}\,\sin^2\theta \qquad K_{\mathrm{II}} = \sigma\sqrt{\pi a}\,\sin\theta\,\cos\theta \tag{P11.3a}$$

where σ is the far-field stress and θ is the angle between the crack plane and the direction of the far-field stress. The loss of the strain energy due to creating a crack in the plate is given by

$$\Delta U_1 = \frac{2}{E_n}\int_0^a (K_{\mathrm{I}} + K_{\mathrm{II}})\,da \tag{P11.3b}$$

(i) Derive the relationship between the effective Young's modulus E and the crack density ω $(= N\pi a^2)$ using the self-consistent approximation [see Eq. (11.31)]. (ii) Derive the same relationship using Taylor's approach. (iii) Comment on these two approaches by comparison.

4 If values of $E = 23{,}000$ MPa, $K_m = 22$ MPa mm$^{1/2}$, $l_0 = 3$ mm, and $d_H = 7$ mm are assumed for a concrete (see Fig. 11-9), plot the softening stress–elongation curve using the model by Horii and Ortiz.

Chapter 12

1 Describe the influence that the volume fraction of fibers has on the matrix apparent toughness. Can this concept be transferred to large, heavily reinforced concrete structural members, e.g., columns and pipe columns?

2 By using the conditions $G_{int} = R_{int}$ and $\partial G_{int}/\partial a_d = \partial R_{in}t/\partial a_d$ and Eqs. (12.3) and (12.4), derive Eq. (12.5).

3 Assuming the Paris equation as modified by the size effect model for fatigue crack growth is valid [Eqs. (11.29) and (11.30)], propose a testing program to obtain the experimental parameters.

4 If the constants associated with Eqs. (11.29) and (11.30) are known, how can these equations be used in the design of concrete structures?

5 In general terms, describe the influence of strain rate on the development of the fracture process zone. What are the implications with respect to the suitability of LEFM? Will the structural size for which LEFM is valid increase or decrease with increasing strain rate?

INDEX